Benchmark Papers in Geology

Series Editor: Rhodes W. Fairbridge
Columbia University

Volume

1 ENVIRONMENTAL GEOMORPHOLOGY AND LANDSCAPE CONSERVATION, Volume I: Prior to 1900 / Donald R. Coates
2 RIVER MORPHOLOGY / Stanley A. Schumm
3 SPITS AND BARS / Maurice L. Schwartz
4 TEKTITES / Virgil E. Barnes and Mildred A. Barnes
5 GEOCHRONOLOGY: Radiometric Dating of Rocks and Minerals / C. T. Harper
6 SLOPE MORPHOLOGY / Stanley A. Schumm and M. Paul Mosley
7 MARINE EVAPORITES: Origin, Diagenesis, and Geochemistry / Douglas W. Kirkland and Robert Evans
8 ENVIRONMENTAL GEOMORPHOLOGY AND LANDSCAPE CONSERVATION, Volume III: Non-Urban / Donald R. Coates
9 BARRIER ISLANDS / Maurice L. Schwartz
10 GLACIAL ISOSTASY / John T. Andrews
11 GEOCHEMISTRY OF GERMANIUM / Jon N. Weber
12 ENVIRONMENTAL GEOMORPHOLOGY AND LANDSCAPE CONSERVATION, Volume II: Urban Areas / Donald R. Coates
13 PHILOSOPHY OF GEOHISTORY: 1785–1970 / Claude C. Albritton, Jr.
14 GEOCHEMISTRY AND THE ORIGIN OF LIFE / Keith A. Kvenvolder
15 SEDIMENTARY ROCKS: Concepts and History / Albert V. Carozzi
16 GEOCHEMISTRY OF WATER / Yasushi Kitano
17 METAMORPHISM AND PLATE TECTONIC REGIMES / W. G. Ernst
18 GEOCHEMISTRY OF IRON / Henry Lepp
19 SUBDUCTION ZONE METAMORPHISM / W. G. Ernst
20 PLAYAS AND DRIED LAKES: Occurrence and Development / James T. Neal
21 GLACIAL DEPOSITS / Richard P. Goldthwait
22 PLANATION SURFACES: Peneplains, Pediplains, and Etchplains / George F. Adams
23 GEOCHEMISTRY OF BORON / C. T. Walker
24 SUBMARINE CANYONS AND DEEP-SEA FANS: Modern and Ancient / J. H. McD. Whitaker
25 ENVIRONMENTAL GEOLOGY / Frederick Betz, Jr.
26 LOESS: Lithology and Genesis / Ian J. Smalley
27 PERIGLACIAL PROCESSES / Cuchlaine A. M. King
28 LANDFORMS AND GEOMORPHOLOGY: Concepts and History / Cuchlaine A. M. King

29 METALLOGENY AND GLOBAL TECTONICS / *Wilfred Walker*

30 HOLOCENE TIDAL SEDIMENTATION / *George deVries Klein*

31 PALEOBIOGEOGRAPHY / *Charles A. Ross*

32 MECHANICS OF THRUST FAULTS AND DÉCOLLEMENT / *Barry Voight*

33 WEST INDIES ISLAND ARCS / *Peter H. Mattson*

34 CRYSTAL FORM AND STRUCTURE / *Cecil J. Schneer*

35 OCEANOGRAPHY: Concepts and History / *Margaret B. Deacon*

36 METEORITE CRATERS / *G. J. H. McCall*

37 STATISTICAL ANALYSIS IN GEOLOGY / *John M. Cubitt and Stephen Henley*

38 AIR PHOTOGRAPHY AND COASTAL PROBLEMS / *Mohamed T. El-Ashry*

39 BEACH PROCESSES AND COASTAL HYDRODYNAMICS / *John S. Fisher and Robert Dolan*

40 DIAGENESIS OF DEEP-SEA BIOGENIC SEDIMENTS / *Gerrit J. van der Lingen*

41 DRAINAGE BASIN MORPHOLOGY / *Stanley A. Schumm*

42 COASTAL SEDIMENTATION / *Donald J. P. Swift and Harold D. Palmer*

43 ANCIENT CONTINENTAL DEPOSITS / *Franklyn B. Van Houten*

44 MINERAL DEPOSITS, CONTINENTAL DRIFT AND PLATE TECTONICS / *J. B. Wright*

45 SEA WATER: Cycles of the Major Elements / *James I. Drever*

46 PALYNOLOGY, PART I: Spores and Pollen / *Marjorie D. Muir and William A. S. Sarjeant*

47 PALYNOLOGY, PART II: Dinoflagellates, Acritarchs, and Other Microfossils / *Marjorie D. Muir and William A. S. Sarjeant*

48 GEOLOGY OF THE PLANET MARS / *Vivien Gornitz*

49 GEOCHEMISTRY OF BISMUTH / *Ernest E. Angino and David T. Long*

50 ASTROBLEMES—CRYPTOEXPLOSION STRUCTURES / *G. J. H. McCall*

51 NORTH AMERICAN GEOLOGY: Early Writings / *Robert Hazen*

52 GEOCHEMISTRY OF ORGANIC MOLECULES / *Keith A. Kvenvolden*

53 TETHYS: The Ancestral Mediterranean / *Peter Sonnenfeld*

54 MAGNETIC STRATIGRAPHY OF SEDIMENTS / *James P. Kennett*

55 CATASTROPHIC FLOODING: The Origin of the Channeled Scabland / *Victor R. Baker*

56 SEAFLOOR SPREADING CENTERS: Hydrothermal Systems / *Peter A. Rona and Robert P. Lowell*

57 MEGACYCLES: Long-Term Episodicity in Earth and Planetary History / *G. E. Williams*

58 OVERWASH PROCESSES / *Stephen P. Leatherman*

59 KARST GEOMORPHOLOGY / *M. M. Sweeting*

60 RIFT VALLEYS: Afro-Arabian / *A. M. Quennell*

61 MODERN CONCEPTS OF OCEANOGRAPHY / *G. E. R. Deacon and Margaret B. Deacon*

62 OROGENY / *John G. Dennis*

63 EROSION AND SEDIMENT YIELD / *J. B. Laronne and M. P. Mosley*

64 GEOSYNCLINES AND PLATE TECTONICS / *F. L. Schwab*

**Benchmark Papers
in Geology/63**

A BENCHMARK® Books Series

EROSION AND
SEDIMENT YIELD

Edited by

JONATHAN B. LARONNE

**Ben-Gurion University of the Negev,
Israel**

and

M. PAUL MOSLEY

**Ministry of Works and Development,
New Zealand**

Hutchinson Ross Publishing Company

Stroudsburg, Pennsylvania

84 83 82   1   2   3   4   5
Manufactured in the United States of America.

LIBRARY OF CONGRESS CATALOGING IN PUBLICATION DATA
Main entry under title:
Erosion and sediment yield.

   (Benchmark papers in geology; 63)
   Includes bibliographical references and index. 1. Erosion—Addresses,
   essays, lectures.   2. Sediment transport—Addresses, essays, lectures.   I.
Laronne, Jonathan B.   II. Mosley, M. Paul.   III. Series.
QE571.E77   551.3   81-6456
ISBN 0-87933-409-6   AACR2

Distributed world wide by Academic Press,
a subsidiary of Harcourt Brace Jovanovich,
Publishers.

009670

CONTENTS

Series Editor's Foreword   xi
Preface   xiii
Contents by Author   xv

Introduction   1

PART I: GENERAL PRINCIPLES

Editors' Comments on Papers 1 Through 6   4

1   **WOLMAN, M. G., and J. P. MILLER:**   Magnitude and Frequency of
Forces in Geomorphic Processes   13
Jour. Geology **68**:54–74 (1960)

2   **TRICART, J.:**   Discontinuities in Erosion Processes   34
Translated from *Internat. Assoc. Sci. Hydrology Bull.* **59**:
33–43 (1962)

3   **RAPP, A.:**   Recent Development of Mountain Slopes in Kärkevagge
and Surroundings, Northern Scandinavia   44
Geog. Annaler **42**:184–187 (1960)

4   **YOUNG, A.:**   Present Rate of Land Erosion   49
Nature **224**:851–852 (1969)

5   **SCHUMM, S. A.:**   The Disparity Between Present Rates of
Denudation and Orogeny   51
U.S. Geol. Survey Prof. Paper 454-H, 1963, pp. H1–H13

6   **GILLULY, H.:**   Atlantic Sediments, Erosion Rates, and the Evolution
of the Continental Shelf: Some Speculations   64
Geol. Soc. America Bull. **75**:483–492 (1964)

PART II: PROCESSES OF EROSION AND SEDIMENT TRANSPORT

Editors' Comments on Papers 7 Through 11   76

7   **WOLLAST, R.:**   Kinetics of the Alteration of K-Feldspar in
Buffered Solutions at Low Temperature   86
Geochim. et Cosmochim. Acta **31**:635–648 (1967)

8   **ELLISON, W. D.:**   Studies of Raindrop Erosion   100
Agricultural Engineering **25**:131–136, 181–182 (1944)

Contents

9 CHEPIL, W. S.: Dynamics of Wind Erosion: I. Nature of Movement of Soil by Wind   108
Soil Sci. **60**:305–320 (1945)

10 EINSTEIN, H. A.: Formulas for the Transportation of Bed Load   124
Am. Soc. Civil Engineers Trans. **107**:561–573 (1942)

11 GRISSINGER, E. H.: Resistance of Selected Clay Systems to Erosion by Water   137
Water Resources Research **2**:131–138 (1966)

PART III: CONTROLS UPON EROSION RATES

Editors' Comments on Papers 12 Through 19   146

12 ZINGG, A. W.: Degree and Length of Land Slope as It Affects Soil Loss in Runoff   155
Agricultural Engineering **21**:59–64 (1940)

13 ANDERSON, H. W.: Suspended Sediment Discharge as Related to Streamflow, Topography, Soil, and Land Use   161
Am. Geophys. Union Trans. **35**:268–281 (1954)

14 FOURNIER, F.: Climatic Factors in Soil Erosion   175
Translated from *Assoc. Géographes Français Bull.* **203**:97–103 (1949)

15 LANGBEIN, W. B., and S. A. SCHUMM: Yield of Sediment in Relation to Mean Annual Precipitation   181
Am. Geophys. Union Trans. **39**:1076–1084 (1958)

16 MILLER, J. P.: Solutes in Small Streams Draining Single Rock Types, Sangre de Cristo Range, New Mexico   190
U.S. Geol. Survey Water-Supply Paper 1535F, 1961, pp. F-1–F-9, F-16, F-17–F-23

17 GIBBS, R. J.: Amazon River: Environmental Factors That Control Its Dissolved and Suspended Load   205
Science **156**:1734–1737 (1967)

18 CLEAVES, E. T., A. E. GODFREY, and O. P. BRICKER: Geochemical Balance of a Small Watershed and Its Geomorphic Implications   209
Geol. Soc. America Bull. **81**:3015–3032 (1970)

19 SCHUMM, S. A.: Speculations Concerning Paleohydrologic Controls of Terrestrial Sedimentation   227
Geol. Soc. America Bull. **79**:1573–1588 (1968)

PART IV: MAN'S INFLUENCE ON EROSION

Editors' Comments on Papers 20, 21, and 22   248

20 WOLMAN, M. G.: A Cycle of Sedimentation and Erosion in Urban River Channels   255
Geog. Annaler **49A**:385–395 (1967)

21 DUNNE, T.: Sediment Yield and Land Use in Tropical Catchments   266
Jour. Hydrology **42**:281–300 (1979)

22 **LIKENS, G. E., F. H. BORMANN, N. M. JOHNSON, D. W. FISHER, and
R. S. PIERCE:** Effects of Forest Cutting and Herbicide
Treatment on Nutrient Budgets in the Hubbard Brook
Watershed-Ecosystem   286
Ecol. Monogr. **40**:23–25, 41–47 (1970)

PART V: RATES OF EROSION

Editors' Comments on Papers 23 Through 27   298

23 **EVEREST, R.:** Some Observations on the Quantity of Earthy
Matter Brought Down by the Ganges River   303
Asiatic Soc. of Bengal Jour. **1**:238–242 (1832)

24 **DOLE, R. B., and H. STABLER:** Denudation   308
U.S. Geol. Survey Water-Supply Paper 234, 1909, pp. 78–93

25 **HOLEMAN, J. N.:** The Sediment Yield of Major Rivers of the World   324
Water Resources Research **4**:737–747 (1968)

26 **LI, Y.-H.:** Denudation of Taiwan Island Since the Pliocene Epoch   335
Geology **4**:105–107 (1976)

27 **MEYBECK, M.:** Concentrations des eaux fluviales en éléments
majeurs et apports en solution aux océans   338
Rev. Geol. Syn. Geog. Phys. **21**:215–237, 243–246 (1979)

Author Citation Index   365
Subject Index   373
About the Editors   377

SERIES EDITOR'S FOREWORD

The philosophy behind the Benchmark Papers in Geology is one of collection, sifting, and rediffusion. Scientific literature today is so vast, so dispersed, and, in the case of old papers, so inaccessible for readers not in the immediate neighborhood of major libraries that much valuable information has been ignored by default. It has become just so difficult, or so time consuming, to search out the key papers in any basic area of research that one can hardly blame a busy person for skimping on some of his or her "homework."

This series of volumes has been devised, therefore, as a practical solution to this critical problem. The geologist, perhaps even more than any other scientist, often suffers from twin difficulties—isolation from central library resources and immensely diffused sources of material. New colleges and industrial libraries simply cannot afford to purchase complete runs of all the world's earth science literature. Specialists simply cannot locate reprints or copies of all their principal reference materials. So it is that we are now making a concerted effort to gather into single volumes the critical materials needed to reconstruct the background of any and every major topic of our discipline.

We are interpreting "geology" in its broadest sense: the fundamental science of the planet Earth, its materials, its history, and its dynamics. Because of training in "earthy" materials, we also take in astrogeology, the corresponding aspect of the planetary sciences. Besides the classical core disciplines such as mineralogy, petrology, structure, geomorphology, paleontology, and stratigraphy, we embrace the newer fields of geophysics and geochemistry, applied also to oceanography, geochronology, and paleoecology. We recognize the work of the mining geologists, the petroleum geologists, the hydrologists, and the engineering and environmental geologists. Each specialist needs a working library. We are endeavoring to make the task of compiling such a library a little easier.

Each volume in the series contains an introduction prepared by a specialist (the volume editor)—a "state of the art" opening or a summary of the object and content of the volume. The articles, usually some twenty to fifty reproduced either in their entirety or in significant extracts, are selected in an attempt to cover the field, from the key papers of the last century to fairly recent work. Where the original works are in foreign languages, we have endeavored to locate or commission translations. Geologists, because of their global subject, are often acutely aware of the oneness of our world. The selections cannot therefore be restricted to any one country, and whenever possible an attempt is made to scan the world literature.

To each article, or group of kindred articles, some sort of "highlight commentary" is usually supplied by the volume editor. This commentary should serve to bring that article into historical perspective and to emphasize its particular role in the growth of the field. References, or citations, wherever possible, will be reproduced in their entirety—for by this means the observant reader can assess the background material available to that particular author, or, if desired, he or she too can double check the earlier sources.

A "benchmark," in surveyor's terminology, is an established point on the ground that is recorded on our maps. It is usually anything that is a vantage point, from a modest hill to a mountain peak. From the historical viewpoint, these benchmarks are the bricks of our scientific edifice.

RHODES W. FAIRBRIDGE

PREFACE

The study of erosion, which encompasses the weathering, detachment, entrainment, and transportation of particles of rock and soil, is but one branch of geology. The processes of erosion, however, are an essential step in the conversion of rocks into sediments, and it is these same processes of erosion that are responsible for the physiographic features over a large part of the surface of the continents. Benchmark Papers in Geology cover the whole span of the science of geology, from metamorphism and plate tectonics to aspects of sedimentation, sedimentary rocks, and geomorphology. Thus we regard the present volume as an essential link joining the broad areas of "hard rock geology," "soft rock geology," and geomorphology.

As well as having a central role within geology, the study of erosion is an important meeting place for a wide range of disciplines, both academic and applied. Much of the early work in the field of erosion and sediment transport was undertaken by hydraulic engineers concerned with the practical problems of canal lining, navigation, and flood mitigation. Our knowledge of erosion processes on hillsides owes a great deal to agricultural engineers, soil conservationists, and other land managers. Scientists, both pure and applied, in fields such as ecology, forestry, and aqueous chemistry have made important contributions by their efforts to understand the functioning of ecosystems and the biosphere, while geologists and, particularly, geomorphologists as well as physical geographers have intensively studied rates and processes of erosion as part of their efforts to understand the lithosphere. Only recently have these various groups of workers become generally aware of the contributions of the others and have utilized each other's methods and results. This series is well suited to further this increase in awareness of broader interrelationships. The selection of articles for this book on erosion and sediment yield attempts to reflect the range of contributions from many disciplines. In particular we hope that it indicates the importance of biological processes, which have been hitherto rather neglected in the study of the lithosphere.

Geology has become an increasingly quantitative science, and students of erosion have been in the forefront in applying quantitative methods. Perhaps this is because many processes of erosion function at measurable rates on a human time scale, because they are accessible and relatively obvious, and because they are of great significance to many practical problems such as reservoir sedimentation or navigation in rivers and estuaries. This book concentrates upon the identification of specific processes of erosion and upon quantitative determination of the spatial and temporal variability of their rates of operation. Companion volumes (*Denudation*, edited by E. H. Muller, and *Landforms and Geomorphology*, edited by C. A. M. King) have dealt with the history and philosophy of research into erosion, denudation, and landforms.

In presenting this volume on erosion and sediment yield, we wish to acknowledge the influence of Stanley Schumm, under whom we studied at Colorado State University. His work, some of which we have included, exemplifies the points made in this preface.

JONATHAN B. LARONNE
M. PAUL MOSLEY

CONTENTS BY AUTHOR

Anderson, H. W., 161
Bormann, F. H., 286
Bricker, O. P., 209
Chepil, W. S., 108
Cleaves, E. T., 209
Dole, R. B., 308
Dunne, T., 266
Einstein, H. A., 124
Ellison, W. D., 100
Everest, R., 303
Fisher, D. W., 286
Fournier, F., 175
Gibbs, R. J., 205
Gilluly, H., 64
Godfrey, A. E., 209
Grissinger, E. H., 137

Holeman, J. N., 324
Johnson, N. M., 286
Langbein, W. B., 181
Li, Y.-H., 335
Likens, G. E., 286
Meybeck, M., 338
Miller, J. P., 13, 190
Pierce, R. S., 286
Rapp, A., 44
Schumm, S. A., 51, 181, 227
Stabler, H., 308
Tricart, J., 34
Wollast, R., 86
Wolman, M. G., 13, 255
Young, A., 49
Zingg, A. W., 155

EROSION AND SEDIMENT YIELD

INTRODUCTION

Perhaps more has been written on erosion and its consequences, sediment yield, than on any other geomorphological topic. The relevant literature is scattered through a multitude of books, memoirs, manuals, conference proceedings, government reports, and journals dealing with general geology, sedimentology, geochemistry, geography, agricultural engineering, forestry, soil science, hydraulics, and hydrology. The selection of benchmark contributions to the study of erosion and sediment yield has presented difficulties in proportion to this volume of material.

We have not necessarily selected for this book contributions that are difficult to find in a modern, well-equipped library; indeed only two papers were published before 1940. Rather the papers were chosen because they incorporate new ideas, succinct syntheses, exceptional analysis of information (particularly quantitative information), or data unavailable until their publication. Many have initiated fruitful scientific debate or have led to completely new lines of inquiry. Our seeming bias toward work published by North American authors has arisen because until recently much of the progress made in this field had been achieved in North America. The combination of an immensely important and productive agricultural land resource base, great national wealth, and severe human problems caused by erosion and sedimentation led to levels of funding of erosion research that have not been matched anywhere else in the world.

The book is divided into five parts. The first includes six papers that deal with the basic principles of erosion, clarify its temporal, spatial, and dynamic aspects, and establish a broad frame of reference. The papers in Part II may be viewed as the essence of the book in that they deal with the fundamental processes of erosion. However, despite the progress that has been made toward understanding these processes, their interrelationships and complexities are such that a more empirical approach to erosion has frequently

1

been necessary. This type of approach has been particularly char-acteristic of work intended to define the factors that control rates of erosion (Part III). A distinction is frequently made between "normal" or geologic erosion on the one hand and accelerated man induced, or anthropogenic erosion on the other. The effects of human activity have been briefly dealt with in Part IV; they have been considered separately from the other controls upon erosion because they may be regarded as an abnormal disruption of the natural functioning and evolution of the landscape. Human activ-ity may have been relatively unimportant on a geologic time scale, particularly in comparison with the effects of climatic shifts, changes in sea level, and so on. Nevertheless, there is increasing evidence that currently it is the dominant control upon erosion, sediment or solute yields, and landscape change in many parts of the world. A result of much erosion research and an interface with other parts of the earth and engineering sciences is the quan-titative determination of rates of erosion and sediment or solute yields. This forms the subject matter of Part V.

Many of the twenty-seven papers included in this book are not only benchmarks in the progress of erosion research but also provide valuable introductions to specific subfields of this branch of science. In addition to placing these papers in a broader con-text and evaluating their contribution, our own commentaries to each part provide an introduction to and review of other signifi-cant work and point out the limitations of results achieved to date.

The entire field of erosion and sediment yield cannot be covered by a single book. Fortunately, several other volumes in the Benchmark Papers in Geology series are complementary and include material of interest to students of erosion. Other material is referred to in our introductions to the individual parts, but spe-cific mention must be made here of the following basic textbooks: M. A. Carson's *The Mechanics of Erosion* (Pion Ltd, London, 1971), the American Society of Civil Engineers' manual *Sedimentation Engineering* (edited by V. A. Vanoni, 1975), and *Manual on Erosion and Sediment Yield* (International Commission on Continental Erosion of IAHS) by R. F. Hadley and D. E. Walling (1982).

Part I

GENERAL PRINCIPLES

Editors' Comments
on Papers 1 Through 6

1   **WOLMAN and MILLER**
    *Magnitude and Frequency of Forces in Geomorphic
    Processes*

2   **TRICART**
    Discontinuities in Erosion Processes

3   **RAPP**
    *Excerpt from Recent Development of Mountain Slopes in
    Kärkevagge and Surroundings, Northern Scandinavia*

4   **YOUNG**
    Present Rate of Land Erosion

5   **SCHUMM**
    *The Disparity Between Present Rates of Denudation and
    Orogeny*

6   **GILLULY**
    *Atlantic Sediments, Erosion Rates, and the Evolution of the
    Continental Shelf: Some Speculations*

Uniformitarianism, a fundamental principle of geology, is fundamental also to the study of erosion and landscape formation. Without the assumption that current erosion processes and their rates of operation are representative of a longer span of time, their measurement is of rather ephemeral interest—useful perhaps in the design of reservoirs but of limited value in providing an understanding of the evolution of the earth's surface on a geologic time scale. Nevertheless, the "uniformitarian-catastrophist" controversy of the mid-nineteenth century (Chorley, Dunn, and Beckinsale, 1964) is by no means settled; students of erosion and denudation currently are involved in the debate and by providing quantitative data on the rates of operation of geologic and geomorphic processes are helping to resolve it. As with most other controversies, the truth is proving to lie somewhere in between the two extreme standpoints of uniformitarianism and catastrophism, and the first two papers in this part convey both moderation and compromise.

Tricart points out that the phenomena responsible for erosion and landscape formation range from those that are truly uniform or continuous in operation (such as removal of matter in solution by groundwater flow) to those that are truly catastrophic (such as a major flood caused by an extreme combination of meteorological and hydrologic conditions) (Paper 2). Wolman and Miller suggest that it is the periodic phenomena on Tricart's continuum that are responsible for transport of the greatest proportion of the sediment carried by fluvial, aeolian, and coastal processes; in other words, the frequent events of moderate magnitudes do the greatest proportion of the work expended in eroding the landscape (Paper 1). This view has been confirmed by several recent studies of fluvial sediment transport (Pickup and Warner, 1976; Andrews, 1980). Other modes of erosion and denudation may not, however, conform to Wolman and Miller's model. For example, where a large part of the material exported from a watershed is carried in solution, erosion may actually be accomplished primarily by low flows—that is, by "continuous" phenomena on Tricart's continuum. Thus, Gerson (1974) has shown that although the landforms in the eastern Upper Galilee, Israel, are sculptured by periodic fluvial action, most of the sediment is contributed by continuously active karstic processes. On the other hand, infrequent mass movement processes on hill slopes may dominate erosion and landscape formation in some environments (Selby, 1974). Starkel (1972) has shown how a catastrophic storm in the Darjeeling Hills caused widespread mass movement on hillsides that was responsible for removal of the equivalent of about 100,000 m^3 km^{-2} from the slopes, while stream channel incision and lateral erosion were responsible for the removal of a similar quantity of sediment from the valley bottoms. By comparison, sediment yields under normal conditions are probably only about 600 to 700 m^3 km^{-2} y^{-1}. The recurrence interval of the storm was estimated to be about twenty to twenty-five years, a sporadic event in Tricart's terminology. In the White Mountains of California, Beaty (1974) stated that "most gradation in the region, in short, is spectacular, cataclysmic, catastrophic!" This may also be the case in extremely arid environments (Schick, 1974). Beaty suggested that the definitions of "catastrophic" and "uniformitarian" may require careful examination because the repetition of the same erosion process, although of extreme magnitude and at very infrequent intervals, conforms to the notion of uniformitarianism, although individual events may be catastrophic in the normal sense of the word. Most suspended sediment transport by a river may be achieved by flows occurring once or twice a year on average, but it is clear that the periodicity and frequency of the dominant, and possibly the formative, erosional events in a given area may vary widely, depending upon the processes involved and the variables that control thresholds of activity and recovery rates (Wolman and Gerson, 1978; Brunsden and Thornes, 1979).

That processes and rates of erosion and deposition change with time is clear from the evidence afforded by vertical changes in stratigraphic columns. An abrupt change or discontinuity (that is, a catastrophe) is a relative concept that depends upon the time scale considered. For example, the Quaternary glaciations were abrupt on the geologic time scale of the Cainozoic era, but would be considered slow changes on a human time scale. Erosion and sediment yields resulting from recent geologic events like the Lake Missoula and Bonneville floods (Baker, 1973) were on a different scale from those associated with the processes and rates of erosion characteristic of these locations at the present day, but they were typical of conditions in the Columbia Plateau during the Pleistocene. Thus, we cannot accept present-day measurements of sediment yields as representative of geologic rates of erosion. Measurements provide information on the rates of erosion by given processes under known conditions, whether these processes be uniform annual rilling (Schumm, 1956) or catastrophic flood surges (Scott and Gravlee, 1968). With this information and with knowledge or assumptions regarding the conditions operating in the past, postdiction of past rates is possible.

Just as erosion is discontinuous through time, it also displays a high degree of spatial variability. For example, mass movement processes on hillsides can deliver huge quantities of sediment to the drainage network, but because the stream channel is unable to carry the material, it is stored in the valley bottom (Mosley, 1978). Tricart places great emphasis upon the influence of sediment storage within a geomorphic system on net erosion rates (Paper 2); the notion that erosion on the watershed may greatly exceed the export of sediment by the drainage system has led to the development of the concept of the sediment delivery ratio (Maner, 1958; Roehl, 1962). Renfro (1975) listed several factors that may control the proportion of eroded sediment that is actually exported from a watershed; they include the type, magnitude, and proximity to the drainage network of the sediment source, the characteristics of the stream system, the grain size of the sediment, the availability of sediment storage areas, and watershed characteristics such as mean slope. The sediment delivery ratio has been shown to be strongly dependent on the well-known relation between sediment yield and watershed area A, $Y \propto A^b$, where the exponent b is negative, approximately -0.12 (Glymph, 1951), but approaches zero for large drainage basins. Sediment yield from a watershed appears to be limited by the process with the lowest capacity for erosion. In some places processes such as rapid mass movement or sheet erosion operating on hillsides may supply more sediment than the stream system is able to transport, whereas in others hillsides may be affected only by slow processes such as soil creep, and the stream's potential capacity for sediment transport is underutilized. Carson and Kirkby (1972) called these two cases transport limited and supply limited, respectively.

Although overall sediment yield from a watershed upon which a variety of processes is acting is determined by measuring fluvial sediment and solute loads, estimation of the relative contributions of each process requires that they be presented in identical units and placed on a comparable basis so that the effects of different frequencies, intensities, and durations of activity may be identified. Rapp solved this problem in his classic study of the geomorphic processes active in Karkevagge by measuring for each the net mass transfer in units of tonne-meters per year (Paper 3). The distance used was in the vertical because he was primarily considering slope processes. Tricart, on the other hand, discussed the relative importance of fluvial transport of fine and coarse sediment in terms of mass transfer along the course of the channel (in the horizontal). Mass transfer as measured by Rapp may include merely redistribution of material within a watershed and not necessarily export entirely from the watershed. Rapp's data indicate that transport in solution produces a mass transfer within Kärkevagge roughly equal to that of all the other processes combined, but because the other processes considered only redistributed material within the watershed, transport in solution completely dominates net ground lowering in the valley because the solutes are carried from the watershed.

Caine (1976) proposed that Rapp's method be extended by defining erosion and sediment transport as geomorphic work and presenting measurements in units of work, or joules. The

7

extension is achieved by introducing the gravitational constant into the computations so that a rate of erosion (or of ground lowering, hence change in the potential energy of the landscape) is defined as power and measured in watts (joules per second). A similar approach has already been used extensively in studies of energy, resistance, and sediment transport in rivers (Bagnold, 1966) and more generally in fluvial geomorphology (Leopold, Wolman, and Miller, 1964).

Erosion, sediment yield, and landscape formation are closely interrelated, and Wolman and Miller (Paper 1) extended their discussion to a consideration of the processes and events that are dominant in controlling the form of the landscape. For example, they suggested that river channel form appears to be primarily controlled by periodic events of moderate magnitude, implying that the channel can achieve a time-independent form that rapidly adjusts to changing environmental conditions. There is now much evidence that river channels, at least in temperate areas, quickly recover from the effects of major flood events (Moss and Kochel, 1978; Wolman and Gerson, 1978), although Baker (1977) and Wolman and Gerson (1978) indicated that streams draining small watersheds or in semiarid areas may show the effects of sporadic events for much longer periods. The concepts of time-independence and dynamic equilibrium have been very influential, but many authors have questioned Hack's (1960) notion that the present-day landscape is truly in equilibrium with currently active processes. Young has considered the relationship of erosion rates, the time required for formation of specific landscape features, and the time available for their formation under constant climatic conditions, concluding that "the revised ideas about rates of landform evolution put forward since 1940 are basically correct" but that "landforms have taken somewhat longer to evolve than is sometimes thought to have been the case" (Paper 4). One must be careful not to use data on erosion rates collected at the present day to make conclusions about the evolution of landscape features that may have formed under earlier climatic conditions (or about past rates of erosion). This is especially the case because, as Young and Tricart pointed out, present-day fluvial sediment loads have been severely modified by agricultural and industrial development (see Part IV). Conversely Graustein (1979) has suggested that past climatic changes may produce errors in estimates of present-day erosion rates. Commenting upon a paper examining landscape reduction by chemical erosion (Owens and Watson, 1979), he pointed out that soils may not be in equilibrium with present-day weathering and erosion processes and may reflect environmental conditions that existed many thousands of years ago. Estimates of chemical erosion that use a comparison of the chemical composition of the bedrock and of stream runoff, assuming that the soil is in a steady state, may be in error because some solutes in the runoff may be coming from further weathering of the soil. Similarly valleys of underfit streams underwent erosion and were formed by events having different magnitudes, frequencies, and durations from those of the present day (Dury, 1973).

The concepts of time independence and dynamic equilibrium of the landscape are closely tied up with the view of the landscape as an open system through which energy and matter are cycled. The landscape achieves a time-independent state of dynamic equilibrium in which the landscape-forming processes convert the inputs (precipitation, dustfall, solar energy) into the system into outputs (stream flow, sediment, solutes), the landscape itself changing only very slowly. However, with only a slight change in viewpoint the landscape may be viewed as a closed system. With an intitial input of potential energy provided by rapid uplift, the system undergoes progressive and irreversible change toward a maximum entropy state, the final result being a peneplain. This time-dependent view of the landscape is, of course, enshrined in the Davisian cycle of erosion. Schumm has compared the two viewpoints using quantitative data on erosion rates, concluding that because rates of orogeny may greatly exceed denudation rates and because rates of fluvial and hill-slope processes are widely different, time-independent landforms are unlikely (Paper 5). More recent work in the Soviet Union has provided evidence for wide differences in rates of orogeny and denudation over large areas (Lisytsina, 1977). The balance between rates of erosion and uplift in the Soviet Union is controlled by both tectonic setting and the climate-vegetation regime, but Lisytsina's data indicated that in few areas is there a balance between the processes of uplift and downwasting. On the other hand, Adams (1979) has presented data for a mountain range in a tectonically active part of the earth's surface, the southern Alps of New Zealand, which indicate both that there is a balance between rates of uplift and of erosion and that the mountain landscape has been in a steady state since 1.5 million years after uplift began. Adams noted that such a balance exists in other tectonically active areas on plate margins, such as Taiwan (Li, 1976, Paper 27) and Japan (Tanaka, 1976), and hence that there need not

9

be a disparity between rates of erosion and uplift. Schumm's work predates the recent widespread adoption of the plate tectonic model, and although he saw no reason in a later work (Schumm, 1977) to modify his views, the implications for rates of erosion and landscape formation of plate tectonics and of periodicity in tectonic, eustatic and climatic activity need further examination along the lines followed by Melhorn and Edgar (1975; to be reprinted in *Denundation,* Benchmark Papers in Geology, edited by E. H. Muller).

The link joining continental erosion rates, sedimentation, and the processes acting in the interior of the earth was also explored by James Gilluly, a colleague of Schumm in the U.S. Geological Survey. In 1963, at a time when the theory of continental drift was receiving increasing support, Gilluly (Paper 6) used information on rates of denudation on a geological time scale (provided by sedimentary evidence) and at the present day (provided by data on river sediment loads; see Paper 25) to suggest that subcrustal flow must have reinforced the isostatic sinking of the crust under the continental shelf of the Atlantic Coast, North America. He noted that his evidence was in accord with (though did not require) the postulated westward drift of America away from the mid-Atlantic ridge. In a later contribution, Gilluly and his coworkers (1970) used more extensive data of the type used in Paper 6 to consider such topics as the chemical composition of the oceans and the evidence for sea floor spreading and subduction. They concluded that sediments must have been recycled up to six times by repeated deposition, uplift, and erosion during the Phanerozoic; subaerial erosion of the continents hence is intimately related to the fundamental processes operating in the earth's interior.

REFERENCES

Adams, J., 1979, Contemporary Uplift and Erosion of the Southern Alps, New Zealand: Summary, *Geol. Soc. America Bull.* **91**:2–4.

Andrews, E. D., 1980, Effective and Bankfull Discharges of Streams in the Yampa River Basin, Colorado and Wyoming, *Jour. Hydrology* **46**: 311–330.

Bagnold, R. A., 1966, An Approach to the Sediment Transport Problem from General Physics, *U.S. Geol. Survey Prof. Paper 422-I,* 37p.

Baker, V. R., 1973, Erosional Forms and Processes for the Catastrophic Pleistocene Missoula Floods in Eastern Washington, in *Fluvial Geo-*

morphology, ed. M. Morisawa, State University of New York, Publications in Geomorphology, Binghamton, New York, pp. 123–148.

Baker, V. R., 1977, Stream-Channel Response to Floods, with Examples from Central Texas, *Geol. Soc. America Bull.* **88**:1057–1071.

Beaty, C. B., 1974, Debris Flows, Alluvial Fans, and a Revitalised Catastrophism, *Zeitschr. Geomorphologie* **22**:39–51.

Brunsden, D., and J. B. Thornes, 1979, Landscape Sensitivity and Change, *Inst. British Geographers Trans.* n.s., **4**:463–484.

Caine, N., 1976, A Uniform Measure of Subaerial Erosion, *Geol. Soc. America Bull.* **87**:137–140.

Carson, M. A., and M. J. Kirkby, 1972, *Hillslope Form and Process,* Cambridge University Press, London.

Chorley, R. J., A. J. Dunn, and R. P. Beckinsale, 1964, *The History of the Study of Landforms,* vol. 1, Methuen, London.

Dury, G. H., 1973, Magnitude-Frequency Analysis and Channel Morphology, in *Fluvial Geomorphology,* ed. M. Morisawa, State University of New York, Publications in Geomorphology, Binghamton, New York, pp. 91–122.

Gerson, R., 1974, Karst Processes of the Eastern Upper Galilee, Northern Israel, *Jour. Hydrology* **21**:131–152.

Gilluly, J., J. C. Reed, and W. M. Cady, 1970, Sedimentary Volumes and Their Significance, *Geol. Soc. America Bull.* **81**:353–376.

Glymph, L. M., 1951, Relation of Sedimentation to Accelerated Erosion in the Missouri River Basin, *U.S. Dept. Agriculture Soil Conservation Service Tech. Paper 103,* 23p.

Graustein, W. C., 1979, Comment on "Landscape Reduction by Weathering in Small Rhodesian Watersheds," by L. B. Owens and J. P. Watson (1979), *Geology* **7**:515–516.

Hack, J. T., 1960, Interpretation of Erosional Topography in Humid Temperate Regions, *Am. Jour. Sci.* **258-A**:80–97.

Leopold, L. B., M. G. Wolman, and J. P. Miller, 1964, *Fluvial Processes in Geomorphology,* W. H. Freeman, San Francisco.

Lisytsina, K. N., 1977, The Effects of Exogenous and Endogenous Factors on Water Erosion Development in the USSR, *Internat. Assoc. Sci. Hydrology Pub.* **122**:67–74.

Maner, S. B., 1958, Factors Affecting Sediment Delivery Rates in the Red Hills Physiographic Area, *Am. Geophys. Union Trans.* **39**:669–675.

Melhorn, W. N., and D. E. Edgar, 1975, The Case for Episodic, Continental-Scale Erosion Surfaces: A Tentative Geodynamic Model, in *Theories of Landform Development,* ed. W. N. Melhorn and R. C. Flemal, State University of New York, Publications in Geomorphology, Binghamton, New York, pp. 243–276.

Mosley, M. P., 1978, Erosion in the Ruahine Range and Its Implications for Downstream River Control, *New Zealand Jour. Forestry* **23**:21–48.

Moss, J. H., and R. C. Kochel, 1978, Unexpected Geomorphic Effects of the Hurricane Agnes Storm and Flood, Conestoga Drainage Basin, Southeastern Pennsylvania, *Jour. Geology* **86**:1–11.

Owens, L. B., and J. P. Watson, 1979, Landscape Reduction by Weathering in Small Rhodesian Watersheds, *Geology* **7**:281–284.

Pickup, G., and R. F. Warner, 1976, Effects of Hydrologic Regime on Magnitude and Frequency of Dominant Discharge, *Jour. Hydrology* **29**:51–75.

Renfro, G. W., 1975, Use of Erosion Equations and Sediment-Delivery Ratios for Predicting Sediment Yields, *U.S. Dept. Agriculture, Agricultural Res. Service ARS-S-40*, pp. 35–45.

Roehl, J. W., 1962, Sediment Source Areas, Delivery Ratios, and Influencing Morphological Factors, *Internat. Assoc. Sci. Hydrology Pub.* **59**:202–213.

Schick, A. P., 1974, Formation and Obliteration of Desert Stream Terraces—Conceptual Analysis, *Zeitschr. Geomorphologie* **21**:88–103.

Schumm, S. A., 1956, Evolution of Drainage Systems and Slopes in Badlands at Perth Amboy, New Jersey, *Geol. Soc. America Bull.* **67**:597–646.

Schumm, S. A., 1977, *The Fluvial System,* Wiley-Interscience, New York.

Scott, K. M., and G. C. Gravlee, 1968, Flood Surge on th Rubicon River, California—Hydrology, Hydraulics and Boulder Transport, *U.S. Geol. Survey Prof. Paper, 422-M,* 40p.

Selby, M. J., 1974, Dominant Geomorphic Events in Landform Evolution, *Internat. Assoc. Engineers Geol. Bull.* **9**:85–89.

Starkel, L., 1972, The Role of Catastrophic Rainfall in the Shaping of the Relief of the Lower Himalaya (Darjeeling Hills), *Geographica Polonica* **21**:103–147.

Tanaka, M., 1976, Rate of Erosion in the Tanzawa Mountains, Central Japan, *Geog. Annaler* **58A**:155–163.

Wolman, M. G., and R. Gerson, 1978, Relative Scales of Time and Effectiveness of Climate in Watershed Geomorphology, *Earth Surf. Proc.* **3**:189–208.

1

MAGNITUDE AND FREQUENCY OF FORCES IN GEOMORPHIC PROCESSES[1]

M. GORDON WOLMAN AND JOHN P. MILLER

Johns Hopkins University and Harvard University

ABSTRACT

The relative importance in geomorphic processes of extreme or catastrophic events and more frequent events of smaller magnitude can be measured in terms of (1) the relative amounts of "work" done on the landscape and (2) in terms of the formation of specific features of the landscape.

For many processes, above the level of competence, the rate of movement of material can be expressed as a power function of some stress, as for example, shear stress. Because the frequency distributions of the magnitudes of many natural events, such as floods, rainfall, and wind speeds, approximate log-normal distributions, the product of frequency and rate, a measure of the work performed by events having different frequencies and magnitudes will attain a maximum. The frequency at which this maximum occurs provides a measure of the level at which the largest portion of the total work is accomplished. Analysis of records of sediment transported by rivers indicates that the largest portion of the total load is carried by flows which occur on the average once or twice each year. As the variability of the flow increases and hence as the size of the drainage basin decreases, a larger percentage of the total load is carried by less frequent flows. In many basins 90 per cent of the sediment is removed by storm discharges which recur at least once every five years.

Transport of sand and dust by wind in general follows the same laws. The extreme velocities associated with infrequent events are compensated for by their rarity, and it is found that the greatest bulk of sediment is transported by more moderate events.

Many rivers are competent to erode both bed and banks during moderate flows. Observations of natural channels suggest that the channel shape as well as the dimensions of meandering rivers appear to be associated with flows at or near the bankfull stage. The fact that the bankfull stage recurs on the average once every year or two years indicates that these features of many alluvial rivers are controlled by these more frequent flows rather than by the rarer events of catastrophic magnitude. Because the equilibrium form of wind-blown dunes and of wave-formed beaches is quite unstable, the frequency of the events responsible for their form is less clearly definable. However, dune form and orientation are determined by both wind velocity and frequency. Similarly, a hypothetical example suggests that beach slope oscillates about a mean value related in part to wave characteristics generated by winds of moderate speed.

Where stresses generated by frequent events are incompetent to transport available materials, less frequent ones of greater magnitude are obviously required. Closer observation of many geomorphic processes is required before the relative importance of different processes and of events of differing magnitude and frequency in the formation of given features of the landscape can be adequately evaluated.

INTRODUCTION

Denudation of the earth's surface and modification of existing land forms involve forces which are ultimately controlled by highly variable atmospheric influences coupled with the unvarying effects of gravity. Almost any specific mechanism requires that a certain threshold value of force be exceeded. However, above this threshold or critical limit there occurs a wide range in magnitude of forces which results from variations in intensity of precipitation, wind speed, etc. The problem to be examined in this paper is the relative importance of extremes or catastrophic events and more ordinary events with regard to their geomorphic

effectiveness expressed in terms of material moved and modification of surface form. Thus this is a re-examination of the concept of "effective force" in landscape development.

It is widely believed that the infrequent events of immense magnitude are most effective in the progressive denudation of the earth's surface. Although this belief might seem to be supported by observations of some individual events, such as large floods, tsunamis, and dust storms, the catastrophic event is not necessarily the critical factor responsible for the development of land forms. Available evidence indicates that evaluation of the effectiveness of a specific mechanism and of the relative importance of different geomorphic processes in mold-

[1] Manuscript received May 4, 1959.

ing specific forms involves the frequency of occurrence as well as the magnitude of individual events.

Evidence related to the influence of frequent events of small magnitude is far less spectacular than the exciting descriptions of the Johnstown flood or the Galveston disaster. It may also be true that in many instances the importance of the latter actually is directly proportional to their grandeur. The purpose of this paper is not to play down any valid significance of the awesome catastrophes but to demonstrate by means of several examples that a more accurate picture of the over-all effectiveness of various geomorphic processes should include not only the rare extreme events but also events of moderate intensity which recur much more frequently.

The relative amount of "work" done during different events is not necessarily synonymous with the relative importance of these events in forming a landscape or a particular feature of the landscape. The effectiveness of an event of a given frequency in terms of its performance of work is measurable both by its magnitude and by the frequency with which it recurs. Thus the relative amounts of work performed by events such as floods of different magnitude and frequency are measurable in part, at least, by comparisons of the relative quantities of sediment transported. On the other hand, although related to the form of the landscape, the ranking of events in terms of the relative amounts of work performed is not necessarily directly correlated with their relative importance in the determination of river pattern, drainage density, slope form, or other aspects of the landscape. This paper deals first with the significance of frequency and magnitude in terms of "work done" and second in terms of the formation of specific features of the landscape.

Any discussion of the frequency of events of geomorphic significance clearly raises some concern about the length of the available record. On the geologic time scale any record of water and sediment discharge is infinitesimally short. On the other hand, where something is known about mechanical aspects of the process, a record of twenty-five to fifty years, considerably longer than most river records, may be sufficient to provide an adequate sample of a river's regimen of flow for certain kinds of analyses. The significance of the likely omission of some extremely high as well as extremely low values will vary with the measure of effective force used. Thus for the case of effective force measured in terms of competence, a "rare" event not experienced in historic time may have recurred a significant number of times in the geologic record. However, because of their relative rarity, such events are of less significance in analyses concerned with percentages of material moved by events of varying frequency and magnitude.

EROSION AND SEDIMENT TRANSPORT

GENERAL CASE

The movement of sediment by water or air is essentially dependent upon shear stress and, according to Malina (1941), Bagnold (1941), Brown (1954), etc., can be described by the equation

$$q = k \, (\tau - \tau_c)^n , \qquad (1)$$

where q is the rate of transport, k is a constant related to the characteristics of the material transported, τ is the shear stress per unit area, and τ_c is a critical or threshold shear stress required to move the material. In its simplest form, equation (1) is essentially a power function

$$q = x^n , \qquad (2)$$

where q is the rate of movement, and x is a variable, some responsible stress such as shear, etc., which exceeds the required threshold value. This relation is shown diagrammatically in figure 1, a.

The distribution in time of many hydrologic and meteorologic events, such as wind speeds or flood peaks, has been shown to approximate a log-normal distribution (see Chow, 1954, and Krumbein, 1955, for

14

numerous examples). These events may be visualized as cumulative applied stresses acting upon particular segments of the landscape. If the stress is log-normally distributed and continuous (fig. 1, b) and if the quantity or rate of movement is related to some power of this stress, then the relation between stress and the product of frequency times rate of movement must attain a maximum. The recurrence interval or frequency at which this maximum occurs is controlled by the relative rates of change of q with the

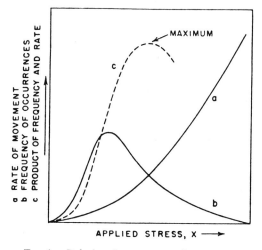

FIG. 1.—Relations between rate of transport, applied stress, and frequency of stress application.

stress x, and of x with time. This maximum, which can be derived mathematically, is shown diagrammatically in figure 1, c.

It should be repeated that this generalization holds only if the applied stress exceeds a threshold value. Below this value no work is done in moving material. This means, of course, that both the equation of transport and the frequency distribution must apply to all values of the applied stress above this threshold value. These conditions are met in several of the simple examples that follow. In these, the recurrence interval at which the product of frequency

and rate of movement is a maximum (fig. 1, c) and the cumulative percentage of the total work performed by successively larger events provide an indirect measure of the relative amounts of work performed by events of different magnitude and frequency.

TRANSPORT OF SEDIMENT BY RIVERS

Most rivers in flood carry large amounts of sediment. The relative importance of different flows can be evaluated by comparing the quantity of sediment transported by great, rare floods with the quantity carried by more frequent, lesser flows. If the quantities of material carried by flows of various magnitudes are known and the frequency of occurrence of each flow is also known, the per cent of the total transported by each flow can be computed from the equation shown at the bottom of this page. Cumulating these percentages gives the proportion of the total carried by successively higher flows.

The computation has been made for the Rio Puerco in the western United States and for Brandywine Creek in the eastern United States. The results of the analysis are given in table 1. Because the results vary depending upon the type of distribution used, two analyses are given for each stream. The use of volumes of runoff in individual peaks eliminates a large number of days of steady base flow in each of which varying amounts of sediment are transported. The results in this case exaggerate the effect of higher flows. Although the record of peaks includes higher flows, the results are of less general value because, unlike the duration curve which is a continuous record, the individual peaks are discrete and do not form a continuous sequence.

Despite climatic and physiographic differences, 50 per cent of the total suspended load of both streams was transported by

$$\frac{(\text{Sediment carried by given flow})(\text{Frequency of given flow})}{\text{Total sediment transported}} \cdot 100$$

$$= \text{Per cent of total sediment carried by flows of different magnitude .}$$

flows which occur on the average one day or more per year (table 1, col. 4). Although many fewer flows were required to transport the remaining 50 per cent of the sediment, the data indicate that at least half the suspended sediment is removed from these drainage basins by low and moderate flows.

Catastrophic floods are those which are truly rare. Arbitrarily we might say that they recur, on the average, once in fifty or a hundred years. Although less satisfactory than the data based on the continuous daily record of flow duration, the data in table 1

In each example the measured load probably represents a large percentage of the total sediment leaving the basin. On Brandywine Creek little of the very coarse sediment not included in the samples appears to leave the basin, whereas on the Rio Puerco most of the sediment is fine material, and the measured load is probably representative of the total load.

It must be emphasized that this analysis of the transport of suspended sediment does not give a picture of the effect of the removal of this material on the physiography of the basin. It does, however, provide some

TABLE 1

PERCENTAGE OF SUSPENDED SEDIMENT TRANSPORTED BY FLOWS OF DIFFERENT MAGNITUDES*

(1)	Distribution Measure of Flow Used in Analysis (2)	Magnitude of Flow below Which 50 Per Cent of Total Sediment Is Transported (3)	Frequency of Occurrence of Flow in Col. 3 (4)	Magnitude of Flow below Which 90 Per Cent of Total Sediment Is Transported (5)	Frequency of Occurrence of Flow in Col. 5 (Per Year) (6)	Remarks (7)
Rio Puerco at Rio Puerco, N.M......	Volume of flow, individual rises	1,800 cfs-days	2 times/yr	9,500 cfs-days	0.26 times	
Rio Puerco near Bernardo, N.M......	Daily discharge, duration curve	950 cfs	Equaled or exceeded 6 days/yr	3,400 cfs	0.7 days	Zero flow approxim^{a}tely 70 per cent of year
Brandywine Creek at Wilmington, Del...	Volume of flow, individual rises	9,000 cfs-days	0.3 per year	11,000 cfs-days	0.2 times	
Brandywine Creek at Wilmington, Del...	Daily discharge, duration curve	1,900 cfs	Equaled or exceeded 11 days/yr	8,200 cfs	0.2 days	

* Flow data are from *Water Supply Papers*, published annually by the U.S. Geological Survey, entitled "Surface Water Supply of the United States." Sediment data are from reports entitled "Quality of Surface Waters of the United States."

show that 90 per cent of the sediment in both these basins (col. 6) is transported by storm runoffs or discharges which recur at least once every five years. The relative proportions of load carried by flows of various magnitudes differ considerably in different rivers. However, for these examples, from both a humid and a semiarid region, by far the greatest part of the total sediment removed from the drainage basins during the period of record was carried by small to moderate flows and not by catastrophic floods. Although the extremely large floods carry greater quantities of sediment, they occur so rarely that from the standpoint of transport their over-all effectiveness is less than that of the smaller and more frequent floods.

measure of the relative amounts of work done by large and small flows. Their impact upon the form of the landscape is considered more specifically elsewhere in this paper.

Another way of investigating frequency relations of sediment transport is to consider load directly, apart from the water discharge. This is appropriate because maximum sediment loads generally do not coincide with peak flood discharges. Snow-fed rivers commonly carry their maximum discharges during the spring melt season and their largest sediment loads during heavy rains of summer and fall. Four streams, all of them in the West, were considered by this method. The Colorado River at Grand Canyon has a large drainage basin, derives the major part of its flow from snowmelt, and drains a variety of rock types which range widely in sediment-yielding potential. The Rio Puerco is an ephemeral stream which carries tremendous sediment loads, derived mostly from unconsolidated deposits and soft shale bedrock, during intense summer and fall rains. The Cheyenne River is a plains stream which drains poorly consolidated rocks and unconsolidated sediments. It shows wide variations in flow and experiences floods which result from summer rains. The Niobrara River is characterized by relatively uniform flow and load, which result from the regulating effects of groundwater storage in the Sand Hills.

recur more frequently than once in 10 years. A larger percentage of the total transport occurs during infrequent floods for the streams which have highly variable flow. Events which recur more than once per year account for 78–95 per cent of the total suspended load. Transport of half the average annual load takes only a few days for Rio Puerco and the Cheyenne River, about a month for the Colorado, and three months for the Niobrara.

It should be emphasized that the data referred to in figure 2 are for suspended load only. Except for the Niobrara, suspended load probably accounts for 90 per cent of the

TABLE 2

TIME REQUIRED TO TRANSPORT VARIOUS PERCENTAGES OF TOTAL SUSPENDED LOAD*

| RIVER AND STATION | DRAINAGE AREA (SQ. MILES) | PERCENTAGE OF TOTAL SUSPENDED LOAD CARRIED DURING | | | DAYS/YR REQUIRED TO TRANSPORT 50 PER CENT OF LOAD |
		Max. Day	10 Max. Days	Events Which Recur 1 Day/Yr	
Colorado River at Grand Canyon, Ariz..........	137,800	0.5	4	92	31
Rio Puerco at Rio Puerco, N.M.................	5,160	5	31	82	4
Cheyenne River near Hot Springs, S.D...........	8,710	5	28	78	4
Niobrara River near Cody, Neb.................	....	2	7	95	95

* Flow data are from *Water Supply Papers*, published annually by the U.S. Geological Survey, entitled "Surface Water Supply of the United States." Sediment data are from reports entitled "Quality of Surface Waters of the United States."

For each stream, daily suspended loads during the period of record were arranged in order of magnitude from largest to smallest, and the percentages of total load and total time were computed. For example, the largest daily load carried by the Colorado was 15.8×10^6 tons. This corresponds to 0.5 per cent of the total load ($3,062 \times 10^6$ tons) carried during 0.012 per cent of the total time (3,036 days). Both percentages of load and time were cumulated and the results, partially summarized in table 2, are plotted in figure 2. Because only the larger loads were considered, the curves in some cases do not account for much more than half the total time involved.

For all four streams, 98–99 per cent of the total load is carried during events which

total clastic load. Colby and Hembree (1957) have estimated that suspended load of the Niobrara near Cody amounts to roughly half the total sediment discharge.

The curves in figure 2 also suggest that the greater the variability of the runoff, the larger the percentage of the total load which is likely to be carried by infrequent flows. Because runoff becomes increasingly variable as drainage area is reduced, it is to be expected that the smaller the drainage area, the larger will be the percentage of sediment carried by the less frequent flows. Thus Culler (personal communication) has shown that for the Cheyenne River near Hot Springs, South Dakota (drainage area 8,700 square miles), 42 per cent of the average annual runoff is produced by storms having a frequency of once each year. In contrast, storms of similar frequency account for 78 per cent of the average annual runoff from drainage areas ranging in size from 0.1 to 3 square miles in the same region. This fact, combined with the higher discharge per square mile produced over small areas, would produce not only high-sediment discharges per square mile but would also increase the percentage of sediment carried by the less frequent events.

Southern California provides a similar example of the influence of size of drainage area on the frequency characteristics of sediment transport. The extreme quantities of sediment produced by small watersheds heading in the mountains near Los Angeles have been discussed by many authors. Because much of the sediment is coarse debris

and difficult to measure, the data on magnitude and frequency of transport are imperfect. A recent report by Ferrell *et al.* (1957) provides some new data and also presents a comprehensive review of efforts to apply all available information to planning control measures. All the 192 basins considered are less than 8 square miles in area. Using a 50-year synthetic rainfall record,[2] it is concluded that 87–91 per cent of the debris is moved by runoff from storms of recurrence interval > 5 years and half of the debris by floods of recurrence interval > 21–28 years. Considering only the basins for which there are records extending over 20 years yields somewhat different conclusions, however, as is shown by table 3.

[2] Based on precipitation data for stations with the area considered.

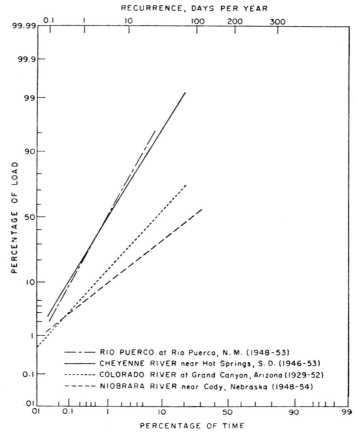

Fig. 2.—Plot of cumulative per cent of time against cumulative percentage of total suspended load

The maximum flood of one-day duration accounted for 29 per cent of the total debris moved during the period of twenty years, as compared with 4 per cent predicted from theoretical considerations. Thus even under these extreme conditions of slope and precipitation the available record indicates that more than 50 per cent of the load is carried by flows which recur once every two or three years.

TABLE 3

MOVEMENT OF SEDIMENT IN MOUNTAINOUS WATERSHEDS NEAR LOS ANGELES

RECURRENCE Interval (Yr.)	PERCENTAGE TOTAL DEBRIS MOVED BY FLOODS OF RECURRENCE INTERVAL EQUAL TO OR LESS THAN GIVEN VALUE	
	Actual*	Theoretical†
1................	39	5
2................	56	14
5................	59	33

* 13 of 20 years were drought years.
† 50-year synthetic record.

TABLE 4

SUSPENDED LOAD TRANSPORTED FROM SMALL DRAINAGE BASINS*

Location	Basin Area (Sq. Miles)	Maximum Daily Load (Per Cent Total for Period)
Kansas	15– 52	23–29
Missouri	95–200	7–22
North Carolina....	11–124	7–18
South Carolina....	106–351	4– 8
Oklahoma........	13–165	15–53
Texas............	30–166	14–29
Washington.......	7–132	10–22
Wisconsin........	77–119	15–36
Pennsylvania......	10– 15	2–39

* Data for Pennsylvania are from Culbertson (1957); all other data are from Love (1936).

It might be supposed that the regimen of sediment transport in regions of perennial stream flow is somewhat less sensitive to differences in drainage area. Data for the Yadkin River basin in North Carolina (drainage area 2,280 square miles) indicate that 90 per cent of the total sediment is transported by flows which occur, on the average, three or more days per year. Flows of similar frequency on the smaller Brandywine Creek in

eastern Pennsylvania (drainage area 312 square miles) transport only 54 per cent of the total load. For very small drainage basins, differences between humid and arid regions appear to be slight.

Data given by Love (1936) and Culbertson (1957) indicate that a larger fraction of the total load transported from small basins is carried by infrequent flows (see table 4). Suspended sediment was measured by Love during a fifteen-month period at stations in 36 basins which represent a wide range of climatic and physiographic conditions. A single day (0.2 per cent of time) accounted for 4–53 per cent of the total for the period, and in three-fourths of the basins more than 10 per cent of the total was carried in one day. Some of the streams in Missouri and Wisconsin transported 90 per cent of the total load in 10–12 days (2 per cent of time.) Culbertson's data refer to a single year of record.

In relating sediment transportation to frequency of runoff, it is interesting to note that Smith and Wischmeier (1957) have shown that the kinetic energy of rainfall decreases rapidly with increasing intensity (less frequent events). Because erosion was found to be a function of the product of precipitation intensity and kinetic energy (the time or frequency factor), as in the case of sediment transport, greater frequency compensates in large measure for the lower intensity.

In summary, these comparisons indicate that most of the work of moving sediment from the drainage basin is done by frequent flows of moderate magnitude. As used here, a "frequent" event recurs at least once each year or two and in many cases several or more times per year. The evidence also suggests that the more variable the regimen of flow of the stream, the larger the percentage of total sediment load which is likely to be carried by the infrequent flows. However, even for many small streams a large percentage of the sediment is carried by flows which recur at least once every five years.

The significance of frequent events of slight intensity is perhaps even better illustrated by a consideration of the dissolved load transported by rivers. Although the process of solution may be aided by floods and fast-flowing streams, it is more dependent on the presence of soluble, permeable rocks and abundant precipitation to percolate through them. Comparisons of the percentage of the total solids removed in solution and as suspended load from drainage basins underlain by diverse lithologies provide the general case for this discussion.

The percentage of the material carried in solution is not in itself a measure of the relative importance of frequent, moderate flows as opposed to larger and less frequent ones. However, this percentage should reflect the degree of influence of the lesser flows because (1) unlike the suspended load which increases both in concentration and volume with increasing flow, a large part of the dissolved load is contributed by ground-water flow (Durum, 1953), and thus concentration decreases with increasing flow thereby reducing in terms of volume the relative importance of the higher flows; (2) the frequency distribution of flows is skewed toward the smaller and more frequent flows. The high concentration experienced during low flows is supported by many observations (Durum, 1953). Figure 3 shows, for example, the rate of increase of dissolved load with increasing discharge at several stations on different rivers. As the slope of each curve on the graph is less than one, it is clear that in all cases concentration decreases with increasing discharge. Combining this information with data on the frequency of occurrence of flows of varying magnitude yields a relation similar to those described for the transport of suspended load. The example suggests that a very large part of the "work" done in the transport of dissolved load is by flows comparable to the mean or even the median flow of the stream.

It follows, then, that the higher the percentage of the total load which is carried in solution, the greater the relative importance of the frequent smaller flows.

It should be emphasized that acquisition and transport of dissolved load depend only slightly on stream discharge derived from surface runoff. Durum (1953) estimated that 74 per cent of the dissolved load carried annually by the Saline River at Russell, Kansas, is derived from ground water and the remainder from contact of stream water with the channel perimeter and particularly from solution of the clastic load. Ground water accounts for only 35 per cent of the annual water discharge at Russell. If the surface runoff contributes clastic sediment, the dissolved load may be increased and thereby affected by magnitude and frequency of flow. If the stream carries essentially no suspended load, then flow frequency has little bearing on transport of dissolved load. Most ground waters are unsaturated and hence discharges above base flow simply dilute the dissolved load. In any case, the concept of effective force is applicable to transport of dissolved load only indirectly through its bearing on transport of suspended load.

The percentage of the total load which is carried in solution varies with the geologic topographic, and climatic characteristics on the drainage basin. Thus for the Salief, River basin in Kansas (table 5), a relatively large river in the semiarid plains, 13 per cent of the total load is dissolved. The Bighorn River at Thermopolis, Wyoming, which derives part of its flow from the mountains, also carries 13 per cent of its total load in solution. The discharge at Thermoplis contains considerable salt derived from the return flow of irrigation water. Rates of solution in various tributary basins upstream from Thermopolis are highly variable. Although the streams in the headwaters of this basin are dilute, precipitation is great and hence the streams transport a large volume of material in solution—evidence that the basin is undergoing more

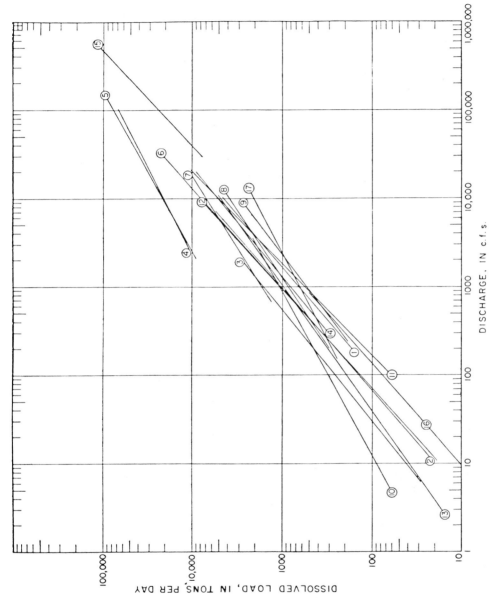

FIG. 3.—Relation of dissolved load to stream discharge at various measuring stations. Key to stations: *1.* Rio Grande at Otowi, N.M.; *2.* Rio Grande at San Acacia, N.M.; *3.* Rio Grande at San Marcial, N.M.; *4.* Colorado River at Lees Ferry, Ariz.; *5.* Colorado River at Grand Canyon, Ariz.; *6.* San Juan River at Bluff, Utah; *7.* Bighorn River at Thermopolis, Wyo.; *8.* Shoshone River at Byron, Wyo.; *9.* Wind River at Riverton, Wyo.; *10.* Saline River at Russell, Kans.; *11.* Iowa River at Iowa City, Iowa; *12.* White River near Kadoka, S.D.; *13.* Moreau River at Bixby, S.D.; *14.* Cedar River at Cedar Rapids, Iowa; *15.* Columbia River at International Boundary; *16.* Ute Creek near Bueyeros, N.M.; *17.* Allegheny River at Red House, N.Y.

rapid chemical denudation (Colby *et al.*, 1956, p. 139). However, from the lower semiarid parts of the basin large amounts of clastic load are derived. Thus, although the salt discharge at Thermopolis is high, it makes up a relatively small percentage of the total load (table 5). In general, the percentage of dissolved load increases with increasing precipitation and vegetative cover (table 5). However, geologic influences, such as erodibility and content of soluble constituents, may overshadow the effects of climate.

sheds in Pennsylvania more than 40 per cent of the total load is dissolved (table 5).

To the extent that these comparisons of dissolved and clastic load illustrate the relative apportionment of the "work done" by large and small flows in degrading drainage basins, the above comparisons indicate that much of the work is done by low flows of frequent occurrence. When combined with the observations of suspended sediment load alone, it is clear that very large amounts of material are transported from drainage basins in diverse climatic and physiographic

TABLE 5

PERCENTAGE OF TOTAL LOAD CARRIED IN SOLUTION FROM SELECTED DRAINAGE BASINS

River and Location	Drainage Area (Sq. Miles)	Approximate Mean Annual Discharge (cfs per (Sq. Miles)	Average Total Load (Tons per Year)	Per Cent Total Load Carried in Solution
Colorado River at Grand Canyon, Ariz.	137,800	0.1	177,000,000	6
Colorado River near Cisco, Utah..	24,100	0.3	22,000,000	20
Green River at Green River, Utah.	40,600	0.2	25,000,000	10
San Juan River near Bluff, Utah..	23,000	0.1	35,000,000	3
Gunnison River near Grand Junction, Colo.	8,020	0.3	341,000	45
Bighorn River at Thermopolis, Wyo.	8,080	0.2	5,750,000	13
Wind River at Riverton, Wyo.....	2,320	0.4	703,000	27
Pogo Agie River near Riverton, Wyo.	2,010	0.3	555,000	45
Wind River near Dubois, Wyo....	233	0.8	174,400	13
Saline River at Russell, Kan......	1,502	0.1	645,000	13
Iowa River at Iowa City, Iowa....	3,721	0.5	1,720,000	17
Bixler Run, central Pa.*.........	15	1.0	2,800	59
Corey Creek, central Pa.*........	12	0.6	1,260	44
Elk Run, central Pa.*...........	10	0.7	1,110	37

* Preliminary data, approximate only.

In humid regions perennial stream flow released from ground-water storage results in a large percentage of dissolved load because production of clastic sediment is inhibited by vegetation (Langbein and Schumm, 1958). Unfortunately, suitable data for comparing basins of different types and sizes are not available. Contamination of stream waters by wastes from industrial and domestic sources is the most serious obstacle to such comparisons. Also, there are very few stations in humid regions with adequate records of both dissolved and suspended load, and most of these are for rather large drainage areas. Preliminary estimates by Rainwater (personal communication) indicate that on three small water-

environments by frequent events of small magnitude. Insufficient information is presently available with which to compare the relative importance of infrequent and frequent events in different kinds of drainage basins.

TRANSPORT BY WIND

Thus far the discussion of relative amounts of work done by events of varying magnitude has been based upon measurements of the transport of material by rivers. This is due primarily to the simplicity of the illustration and also to the fact that a large volume of data is available for analysis. In emphasizing the importance of the more frequent but smaller events, it should be noted that, to the extent that the log-normal distribution describes the frequencies of wind velocities or other currents beyond the necessary threshold, the results of the analyses of river data presumably should apply also to other similar phenomena. The analyses of Zingg (1949), Thom (1954), and many others seem to indicate that a lognormal frequency distribution of wind velocities is a general rule.

Bagnold (1941) shows that movement of dune sand at the Kharga Oasis in Egypt is primarily effected by winds of less than sandstorm velocity (in the terms used here, non-catastrophic) but greater than those which occur a large part of the time (table 6). Similar analyses for the transport of dust by wind can also be made. A measure of the generality of this example is indicated by the data in table 7, which show that a critical wind speed of 20 miles per hour is equaled or exceeded from 5 to 25 per cent of the time at several diverse localities.

In summary, these examples of sediment transport by wind supplement the earlier examples already given from alluvial environments. Both suggest that much of "work" upon the landscape is performed by events of moderate magnitude and relatively frequent occurrence. The common assumption that the rare and infrequent events do the significant work seems to require some modification.

MAGNITUDE AND FREQUENCY OF EVENTS RELATED TO THE FORMATION OF SPECIFIC LANDSCAPE FEATURES

The discussion thus far has considered only process, specifically the effectiveness of various events expressed in terms of the relative amounts of material transported. The following sections describe relations of

TABLE 6

MOVEMENT OF DUNE SAND, KHARGA OASIS, EGYPT*

	A Sand Movement on Dunes Only	B Sand Driving across Country and onto Dunes	C Sand Storms
Velocity, meters/sec†..............	5.8–10	10–13.5	13.5–15.7
Estimated mean wind velocity, V_m, for categories A, B, and C............	7.9	11.3	14
Mean velocity less critical velocity required for movement, $V_m - V_c$.....	3.5	6.9	9.6
Number of hours of north wind less number of hours of south wind, T..	2,015	960	132
Sand movement, q, tons per year proportional to $T (V_m - V_c)^3$.........	8.7×10^4	32×10^4	12×10^4

* Data from Bagnold, 1941, p. 215–216, unidirectional wind assumed.

† 1 meter per second = 2.24 miles per hour.

TABLE 7

APPROXIMATE FREQUENCY OF WIND SPEEDS EXCEEDING 20 MILES PER HOUR AT SELECTED LOCALITIES

Location	Source of Information	Approximate Frequency with Which Wind Speed of 20 mph Is Exceeded (Per Cent)	Type of Data
Dodge City, Kan......	Zingg, 1949	10	April av. wind velocity, 73 yrs.
Beersheba, Israel......	Rosenan, 1954	5	Data in personal communication, 7 yrs.
Lake Hefner, Okla.....	Harbeck, 1952	5	Continuous, 16 mos.
Washington, D.C......	U. S. Weather Bureau, 1945	24	Days of max, 5 min.; speed, 40 yrs.

processes to specific features of the landscape. These processes are ones in which the applied stress is provided by climatically controlled events, such as river discharges, rainstorms, or wind velocities. The frequency distribution of these events normally includes a relatively large number of events of small magnitude and a small number of events of large magnitude. An understanding of the relative importance of these events in molding specific land forms is dependent upon a rather detailed understanding of the mechanisms and processes by which such land forms are developed. Unfortunately, the number of forms which are understood to this degree is exceedingly small. Nevertheless, the available information appears to warrant a preliminary analysis.

DISCHARGES CONTROLLING RIVER CHANNEL SHAPE AND PATTERN

The shape and pattern of a river channel are to some degree adjusted to the discharge of water and sediment load provided by the drainage basin and to the specific size and shape of the materials provided by the geology of the region (Mackin, 1948; Leopold and Maddock, 1953; Wolman, 1955). Because of the range of discharge to which most natural channels are subject, it is logical to assume that the channel shape is affected by a range of flows rather than by a single discharge. Blench (1951, sec. 6.16), for example, refers to a dominant discharge as the "steady discharge that would produce the same result as the actual varying discharge." Melton (1936, p. 601) suggests that in a stream with low-water channel and floodplain the "floods of greatest geological effect should be of two types: those that nearly fill the channel but do not overtop its banks and those that rise to still greater heights and cover the plain." Using several examples, an attempt is made here to illustrate the way in which events of varying magnitude and frequency actually influence alluvial land forms and to define thereby what might be termed a range of physically effective discharges.

In an analysis of factors controlling bank erosion, Wolman (1959) has shown that lateral cutting of the cohesive channel bank of a small stream in Maryland occurs mostly during the winter months, when flows of a size which occurs eight to ten times per year attack previously wetted banks. Although summer thunderstorms in this area usually provide the highest peak discharge, the dry, hard riverbanks of the summer season are less susceptible to erosion. For this reason the lesser and much more frequent flows of midwinter commonly are able to erode as much as one foot from the channel banks at stages well below bankfull. Thus events of moderate magnitude and relatively frequent occurrence control the erosional form of the channel, including its size and shape. Analysis of transport data for several other rivers in the eastern United States also indicates that the largest amounts of suspended sediment are carried during the winter months. This is true for Brandywine Creek in Pennsylvania, Perkiomen Creek in Pennsylvania, and the Yadkin River in North Carolina. It appears, then, that those factors which influence erosion of the cohesive riverbank also help to determine the seasonal pattern of erosion on the surface of the drainage basin, particularly in drainage basins of moderate to large size where the variability of stream flow is not at a maximum.

Although the cross-sectional shape of a meandering channel is maintained by erosion of the concave or outer bank and deposition on the convex or inner bank, deposition need not be precisely in phase with erosion. For the channel bank to be built to its maximum height requires discharges of adequate stage. If, for example, the floodplain is overtopped only once each year, then, in terms of the maximum elevation of the floodplain surface, the effective discharge occurs only once each year. In the example from Maryland cited above, this effective "constructional" discharge is less frequent than the effective erosional discharge.

It has been shown (Wolman and Leopold, 1956) that many rivers of all sizes flowing in diverse physiographic and climatic regions attain the bankfull stage once each year or once every two years. This uniform frequency of flooding suggests that, if the regimen of the stream remains constant, no change takes place in the relative elevation of the surface of the floodplain and the bed of the stream. The constancy of the relation indicates that progressive deposition by overbank flows is not responsible for the formation of the floodplain. Instead, the principal mechanism appears to be lateral movement of the channel and deposition of point bars or deposits of lateral accretion.

of the fact that both the width of the river and its pattern must be related to a discharge approximating the bankfull stage. These observations indicate, then, that not only the form but also the pattern of the river channel in alluvium is related to events of moderate frequency.

Development of the alluvial land forms described requires that the river be competent to move the materials comprising the bed or banks of the channel. Because few if any channels are composed of unigranular material, there is a range of discharges over which material of different sizes begins to move. Some observations on bank erosion were cited above. Although the complexity of

TABLE 8

COMPUTATION OF SIZE OF MATERIAL MOVED BY SENECA CREEK,
DAWSONVILLE, MARYLAND, AT BANKFULL STAGE

Bankfull Discharge (cfs)	Recurrence Interval (Yr.)	Tractive Force at Bankfull Stage (Lb/sq ft)	Critical Tractive Force (Lb/sq ft)	Size d_s Moved at Bankfull Stage (Ft.)	Per Cent of Bed Material Less than d_s
1,160	1.01	0.175	$\tau_c = 3.5\, d_s$	0.05	25

NOTE: τ = tractive force = γDs; τ_c = critical tractive force (O'Brien and Rindlaub, 1934); γ = 62.4 lbs/cu ft.; D = depth = 4.0 ft.; s = slope = .0007; d_s = grain size.

Because the bankfull stage is equal to the elevation of the floodplain surface, and overbank flows contribute only a small part of the floodplain sediment, the bankfull discharge appears to be the "effective" discharge controlling the development of the floodplain. This discharge is attained each year or every other year. Hence, to the extent that these relations hold true, the river floodplain as well as the channel shape is controlled by a force of moderate magnitude which recurs frequently rather than by rare discharges of unusual magnitude.

Within the channel itself, there is a close relation between wave length of bends in sinuous channels and the bankfull width of channel (Leopold and Wolman, 1956). Presumably, flows responsible for making the wave pattern must follow the path of the wave. Because flows well above the banks do not follow the sinuous pattern of the channel itself, the very good correlation between wave length and width is evidence

relations in natural channels makes it impossible to define precisely the critical tractive force at which specific sizes of material will begin to move, the following example does provide a rough approximation of the possible range of discharges competent to alter the form of the channel.

The longitudinal profile of the bed of a river channel is generally made up of a succession of pools and riffles. At low flow the slope of the water surface consists of alternating flat segments over the pools and steep segments over the riffles. With increasing stage these discontinuities of the water surface tend to be smoothed out. In many natural channels the surface becomes nearly uniform at two-thirds of bankfull to near-bankfull stage. This equalization of slope tends to equalize the tractive force in the pool and in the riffle. On Seneca Creek near Dawsonville, Maryland, for example, computations based on an equation for critical tractive force indicate that the tractive force at a discharge near bankfull stage (table 8) is sufficient to move particles which are equal to or greater in size than 25 per cent of the material on the bed. This size is considerably larger than material making up the floodplain. In this and similar cases, the bankfull discharge is fully competent to move the material required in the formation of the river channel.

The foregoing examples indicate that the floodplain and the shape and pattern of the river channel are related to discharges approximating the bankfull stage. This view is in close accord with the opinions of both Franjii (1946, p. 122) and Inglis (1941, p. 112). As the bankfull discharge recurs on the average once each year or two, it may reasonably be concluded that significant alluvial land forms are formed by frequently recurring events of moderate intensity and not by rare floods of unusual magnitude. Some confirmation of this view is provided by observations following the record-breaking floods of August, 1955, in Connecticut. The effects of the flood in many river valleys were extremely spotty (Wolman and Eiler, 1958). In those places where the flood altered the valley materially, it appears to have destroyed those orderly features produced in the alluvium by the steady working of more moderate flows. These latter flows, rather than the extreme flood, appear to be responsible for the major land forms in stream valleys in Connecticut.

The chronologic details of recent stream history are poorly known. Certain features of channels and valleys seem to bear the marks of inheritance, but, unfortunately, many of these are difficult to separate definitely from the effects of modern stream action. Thus some cases of anomalous valley width have been attributed to underfit or overfit streams. However, there is at present no objective standard for judging the degree of adjustment between sizes of valleys and the channels flowing through them. Another kind of example is that provided by glaciated mountain valleys, where stream channels contain bed material which apparently is too large to be

moved by recorded floods (Miller, 1958). Such floodplains as occur are composed mostly of coarse gravels overlain by thin layers of sand and silt. Banks are poorly defined and channels generally lack pools and riffles. These properties may be interpreted as the reaction of a smaller modern stream to materials inherited from a period of greater stream competence which possibly existed during glacial times. Alternatively, catastrophic floods which recur infrequently under modern climatic conditions may be a critical factor affecting such streams. Direct observation of the competence of catastrophic floods is not available, but computations based upon known depths of high water suggest that this explanation is less likely.

FORM AND ORIENTATION OF SAND DUNES

Studies of the relation between dune orientation and wind direction have long included both frequency and magnitude of the wind. Thus Cooper (1958, p. 62) complains that in some explanations of the orientation of longitudinal dunes "emphasis is put upon frequency to the exclusion of other characteristics: velocity and turbulence." According to Bagnold (1941, p. 204), the height of the slip face of a dune is principally determined by the grain size and the wind speed. Because of the relative ease with which a dune can be deformed, its form is a function of a resultant wind made up of those winds which exceed a critical velocity for movement. The form prevailing the largest fraction of the time might be considered the normal or equilibrium one. Because increasing wind speeds occur with decreasing frequency, other things being equal, the form of the hypothetical normal dune should be related to a range of winds somewhat in excess of the competent speed and not simply to the strongest wind.

The rate of movement of sand above the threshold velocity is proportional to the third power of the velocity. Thus for equal periods of time, higher velocities produce greater effects. Bagnold (1941, p. 69) developed a formula to describe the "weighted" resultant wind, which includes the time during which the wind blows from various directions as well as a weighting factor based on the relation between the rate of sand movement and the velocity. Recent studies have shown that seif dunes in southern Israel are aligned parallel to the "weighted resultant wind direction" (Rosenan, 1953, p. 94; 1954). It is interesting to note, in passing, that the direction of the weighted resultant wind may differ considerably from the direction of the so-called "prevailing" wind.

Considering a unidirectional wind, the data in table 6 provide a simple example of the magnitude and frequency of the winds controlling dune form and orientation. All winds shown in the table exceed the critical. The critical velocity required to move the size of particles commonly found in dunes (0.25 mm.) is about 10 miles per hour. A speed of 10 mph at the ground corresponds roughly to from 17 to 20 mph twenty feet above the ground, where wind speeds are more often measured. Winds in class A (table 6) occur far more often than do the greater wind speeds in classes B and C. Net effects of the strongest winds (class C), however, are diminished by their reduced frequency. Without weighting, the product of velocity and time (a crude measure of the sand transported by each class) shows that by far the largest transport occurs in class B. When weighted by the cube of the velocity, maximum transport is still performed by winds in the middle range, but the weighting markedly reduces the per cent of the total sand transported by winds in this class. Cooper (1958, p. 54) does not weight the vectors but suggests that the orientation and form of a set of oblique dunes on the coast of Oregon may be related simply to the resultant wind formed from a parallelogram in which the direction of the vectors is given by the principal wind directions and their magnitude by the wind speed. Weighting tends to increase the proportionate effect of higher wind speeds in both the examples cited. However, net transport of sand, and also dune form and

orientation which depend on wind speed and direction, are adjusted to winds which recur often during the year; they are not controlled simply by the isolated rare event of extreme magnitude.

PROFILES OF BEACHES

Many workers have observed that a beach maintains, on the average, what is called an "equilibrium profile." However, like the dune form, and to a much greater degree than the form of a river channel, the profile of a beach is subject to rapid adjustment with variation in conditions controlling beach form. The equilibrium profile, therefore, must be considered as an average form around which rapid fluctuations occur. Waves from storms may periodically destroy the equilibrium form, but over a period of years there is an average equilibrium profile by which the beach may be characterized. The processes which control this profile are our concern here. In the following example an attempt will be made to show that the effective force determining such an equilibrium profile is provided indirectly by frequent winds of moderately high speeds and not by rare storm winds of maximum wind speeds.

The profile of a beach is primarily a function of grain size and the ratio of wave height to wave length. This ratio is designated as the "wave steepness," H/L (fig. 4, a). In general, the larger the size of material on the beach, the steeper the slope; and the greater the ratio H/L, the lower the slope (Bascom, 1951; Rector, 1954). For a given grain size, then, the slope is controlled by the ratio H/L. Because wave steepness in part is related to wind speed, a hypothetical example may be used to illustrate the frequency and magnitude of the effective winds controlling the beach profile.

Let the grain size on a given beach be 0.4 mm., a value within the range of beach sands on the West Coast of the United States (Bascom, 1951). Assuming that the beach is perpendicular to the winds controlling the wave steepness, the relation between the wave steepness, H/L, and the general foreshore slope at a constant grain size can be determined from equation (7) of Rector (1954, p. 24). The curve relating the slope of the foreshore to the wave steepness for an assumed grain size of 0.4 is shown in figure 4, a. The equilibrium profile for this beach presumably lies somewhere within this range of slope (fig. 4, a). For a given range of wind speeds, another curve (fig. 4, b) can be drawn relating H/L to wind velocity (U.S. Navy, Hydrographic Office, 1951). The range of wave steepness, H/L, used in the example is similar to the range of values observed by O'Brien (1951) at the Columbia River Lightship off the coast of the state of Washington. These values are also shown in figure 4, b.

Observations have shown that during the winter the beach may steepen as a result of short, choppy waves, whereas in the summer it should be expected to flatten. For the material used in his experiments, Johnson (1949) noted that the transition from an "ordinary" to a "storm" profile took place at a wave steepness, H/L, of about 2.5 per cent. Using a value of 2.5, the equilibrium profile in the hypothetical example can be determined from figure 4, a. For a value $H/L = 2.5$ at the assumed grain size, the equilibrium beach profile will have a slope of about 0.1. This wave steepness is also associated with a wind speed of about 12 knots (fig. 4, b).

At Eureka, California, which is a coastal station not at the most exposed coast, a velocity of 12 knots, here associated with the

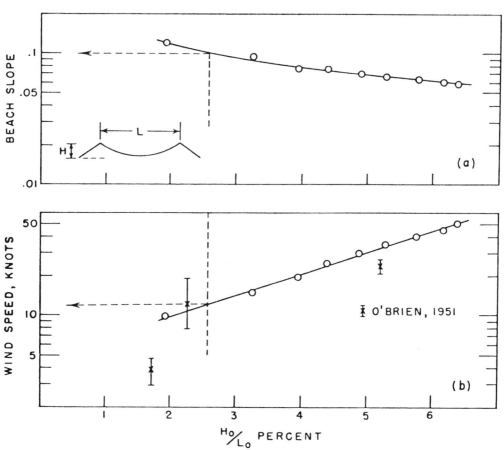

FIG. 4.—Interrelations of wind speed, wave characteristics, and beach slope

equilibrium beach profile, is exceeded by all maximum monthly wind velocities (table 9). This 20-year record indicates that a wind speed of 12 knots will be exceeded as a monthly maximum at least twelve times per year. At a more exposed station—North Head, Washington—the average hourly wind velocity is approximately 13 knots, roughly that associated with the beach profile described here. This example suggests that the effective stress to which the equilibrium profile of the beach is related is one produced by moderately strong winds which

comparisons have been made, have reached a rather mature stage and consequently changes within the rather brief period of historical record are slow and slight, and the level of wave attack during hurricanes has been higher than that most effective in changing the configuration of the shore." A similar observation was made by Nichols and Marston (1939) when they stated that "although the changes caused by the hurricanes of September 21, 1939, were greater than those resulting from many years of winter storms, it seems likely that in a few

TABLE 9

CUMULATIVE FREQUENCY DISTRIBUTION OF MAXIMUM MONTHLY WIND SPEEDS AT EUREKA, CALIFORNIA, DURING THE PERIOD 1930–1949*

Maximum Monthly Wind Speeds (Knots)	Cumulative Per Cent of Occurrence Less than Specified Speed
15	2
20	19
24	50
25	60
30	89
35	98

* Data from published records of the U.S. Weather Bureau.

TABLE 10

ENERGY TRANSMITTED BY LAKE AND OCEAN WAVES OF TWO FREQUENCIES, CONSIDERING WAVE SYSTEM TO BE UNIDIRECTIONAL*

	PERCENTAGE OF TOTAL ENERGY BY WAVES WITH RECURRENCE INTERVAL	
	1 Day per Month	1 Day per Year
Lake Michigan:		
Bailey's Harbor, Wis.	10	2
Milwaukee, Wis.	18	5
Chicago, Ill.	26	2
Muskegon, Mich.	26	7
Frankfort, Mich.	21	3
Lake Erie:		
Monroe, Mich.	32	5
Cleveland, Ohio	19	4
Erie, Pa.	23	1
Buffalo, N.Y.	26	5
Lake Ontario:		
Stony Point, N.Y.	25	3
Atlantic Ocean:		
Penobscot Bay, Me.	16	2
Nauset Beach, Mass.	19	1
New York Harbor Entrance	30	1
Chesapeake Bay Entrance	30	1

* Data are from Saville (Beach Erosion Board Tech. Mem. 36, 37, 38, and 55) for 3 years of record.

generate moderate storm waves rather than by winds which accompany infrequent catastrophic storms.

Several observations offer some support for this conclusion. Johnson (1952, p. 950) states that "contrary to popular belief, it is not the relatively steep storm waves that cause a relatively large littoral transport, but rather the intermediate or summer waves which are the major factor in shoreline processes. . . . The greatest transport of sediment along the beach occurs when the beach profile is in equilibrium with waves whose steepness is approximately 0.025." Explaining the fact that the shoreline and offshore depth changes are not as marked as might be expected following a New England hurricane, a report of the Chief of Engineers (U.S. Army, 1950, p. 17) suggests that "the evolutionary processes within the normal tidal range, and under water, where

years the beaches will be in essentially the same condition as they were prior to the hurricane." Data on wave energies given in table 10 likewise indicate that forces that affect beaches are of moderate magnitude most of the time.

SIGNIFICANCE OF CATASTROPHIC EVENTS

In the preceding discussion it has been argued that forces of moderate magnitude and frequency have greater net effect on land-form development than do intense, short-lived forces associated with catastrophic events. Clearly, such a general conclusion requires qualification to the extent that catastrophic events produce results that are (1) unique in some respect because of magnitude or (2) different in kind from effects of more ordinary occurrences. Several illustrative examples are discussed below.

Landslides and formation of new gullies are common occurrences during exceptional storms. Once formed, a gully continues to grow during more moderate storms, and thus the extreme event may have an enduring effect on drainage pattern and topography.

Changes in dimension and position of stream channels commonly occur during large floods. Many such cases were reported following the record Kansas-Missouri floods of July, 1951. Woolley (1946) mentions several cases where channels were downcut several tens of feet and widened a few hundred feet during cloudburst floods in Utah. These trenches, which are too large for the ordinary flows, persist for long periods after the extreme flood. Channel characteristics of many arroyos are clearly related to flows of different magnitude. Deposition occurs during flows smaller in volume than the losses by percolation into the channel; floods large enough to carry the full length of the arroyo cause scour of the channel bed and banks. Extensive migration of bends and meander cutoffs may also be associated with floods. Jahns (1947) cites a case in the Connecticut Valley where destruction of vegetation by exceptional floods in 1936 and 1938 resulted in accelerated bank erosion during later low-water stages. Channel islands of the Connecticut River were destroyed during the 1936 and 1938 floods, but new islands were deposited in approximately the same places as the floodwaters receded. Aggrading streams like the Rio Grande often undergo spectacular channel changes, called "avulsions." Deposition at stages below bankfull continues until the stream is literally flowing on a ridge. Then, during a flood the stream breaks out of its channel and is re-established in a lower part of its valley floodplain, where the building of a new channel ridge commences all over again.

According to Woolley (1946), channels of larger streams in Utah are sometimes dammed temporarily by coarse debris from tributaries. Rio Puerco and Rio Salado occasionally build fans at their mouths which divert the Rio Grande to the opposite side of its valley. Depending on the flow of the main stem, removal of these features may take years. Breaks in the profile of the main stem below tributaries that contribute flood debris apparently reflect downstream progression of sediment waves. Phenomena of this kind are not restricted to arid regions. Jahns (1947) mentions debris fans in the Connecticut River drainage, and they have been observed in other places.

Scouring of floodplains during exceptional floods is often localized. However, some of the features produced in this way are impressive because of their topographic relief. Jahns (1947), for example, describes scour channels and swirl pits 15–20 feet deep formed during the Connecticut River floods of 1936 and 1938. Like floodplain scour, overbank deposition is sporadic and localized. However, a floodwater bar composed of coarse materials, and several feet thick, may be a major topographic feature of the floodplain. A special kind of case requiring extreme flood conditions is deposition on high terrace surfaces. Such deposition has been reported in several Eastern stream valleys, among them the Connecticut (Jahns, 1947; Wolman and Eiler, 1958) and the Susquehanna.

Extreme floods accomplish transport of material that is impossible by more ordinary flows. For example, Woolley (1946) reports that huge boulders weighing more than 100 tons have been moved long distances on gentle slopes by mudflows in Utah. Countless comparable examples, from southern California, Arizona, and many other places, could be cited. With regard to the sizes of materials transported, the effects of floods appear to be directly proportional to their magnitude. This implies that alluvial fans owe many of their properties to extreme rather than moderate flows, although the frequency relations of stream flows and mudflows have not been adequately defined.

The previous discussion of equilibrium beaches emphasized seasonal variations. In many places during the summer, beaches build seaward, are composed of finer material, and have flatter slopes than during winter when they are eroded by higher waves. Where marine cliffs are fronted by broad beaches, erosion of the cliff may occur only during periods of extreme wave action. Barrier islands, which are more or less stable under ordinary conditions, may be eroded below mean tide level during extreme storms. Similarly, along shores of low coastal plains, building of beach ridges several feet above mean low water occurs only during infrequent events of great magnitude.

Because of their relative familiarity, an exhaustive enumeration of examples of the effect of infrequent catastrophic events on the landscape has not been given here. Placed in the context of the earlier discussion, these examples principally illustrate the known thesis that the rare or infrequent events become increasingly important as the threshold stress (competence) required to move the available masses of material increases.

CONCLUSIONS

The observations described in this paper suggest that the effectiveness of processes which control many land forms depends upon their distribution in time as well as their magnitude. It cannot be assumed that, simply because of their magnitude, the rare or infrequent events must be the most significant. Analyses of the transport of sediment by various media indicate that a large portion of the "work" is performed by events of moderate magnitude which recur relatively frequently rather than by rare events of unusual magnitude.

Examples which suggest the way in which some specific land forms are controlled by events of moderate magnitude and frequency were cited. The examples of land forms described in these terms are almost exclusively depositional. For the lone exception, the erosion of a cohesive river bank, a complex combination of conditions actually determines the frequency and magnitude of the principal effective stress. However, these two kinds of cases are related in that the depositional examples and those involving a combination of factors are indirectly determined through the control of the critical or threshold value.

There is a notable lack of examples demonstrating effectiveness of moderate events of frequent occurrence in molding erosional land forms. However, it seems apparent that in many valleys rivers scour to bedrock only during high and relatively infrequent flows. Similarly, from a dynamical standpoint, the movement of large boulders and the erosion of hard bedrock obviously require stresses which are attained during large floods occurring at relatively infrequent intervals.

As an example of the cohesive river banks shows, the threshold of erosion may also be modified by the complex interaction of several factors. The relative effectiveness of various climatic events is generally more complex than that measured in the examples given here. Pre-wetting of bare soil, by reducing the threshold, may prepare the soil for erosion. In most cases a given land form is related to several processes, each of which in turn is controlled by the interaction of precipitation, temperature, and vegetation. For example, the threshold of erosion of sand dunes may be markedly increased by the establishment of vegetation on the dunes. More often than not, however, as in the semiarid regions, the relation between climate, vegetation, and erosion is less direct. In a given region processes of diverse character control individual features of the landscape. The magnitude and frequency of the events responsible for one feature may be very different from the magnitude and frequency of events responsible for another.

Evaluation of the relative importance of various geomorphic processes in a given region, as well as the relative effectiveness of events of different frequency, will require more detailed observations of the land forms themselves and of the processes operative on them.

Perhaps the state of knowledge as well as the geomorphic effects of small and moderate versus extreme events may be best illustrated by the following analogy. A dwarf, a man, and a huge giant are having a wood-cutting contest. Because of metabolic peculiarities, individual chopping rates are roughly inverse to their size. The dwarf works steadily and is rarely seen to rest. However, his progress is slow, for even little

trees take a long time, and there are many big ones which he cannot dent with his axe. The man is a strong fellow and a hard worker, but he takes a day off now and then. His vigorous and persistent labors are highly effective, but there are some trees that defy his best efforts. The giant is tremendously strong, but he spends most of his time sleeping. Whenever he is on the job, his actions are frequently capricious. Sometimes he throws away his axe and dashes wildly into the woods, where he breaks the trees or pulls them up by the roots. On the rare occasions when he encounters a tree too big for him, he ominously mentions his family of brothers—all bigger, and stronger, and sleepier.

REFERENCES CITED

BAGNOLD, R. A., 1941, The physics of blown sand and desert dunes: London, Methuen and Co.

BASCOM, W. N., 1951, The relationship between sand size and beach-face slope: Am. Geophys. Union Trans., v. 32, p. 866–874.

BLENCH, THOMAS, 1951, Hydraulics of sediment-bearing canals and rivers: Vancouver, Evans Industries, Ltd.

BROWN, C. B., 1954, Sediment transportation, chap. xii; *in:* ROUSE, H. (ed.), Engineering hydraulics, p. 769–857: New York, John Wiley and Sons.

CHOW, VEN TE, 1954, The log-probability law and its engineering applications: Proc. Am. Soc. Civ. Eng., v. 80, Sept. No. 536.

COLBY, B. R., and HEMBREE, C. H., 1955, Computations of total sediment discharge Niobrara River near Cody, Nebraska: U.S. Geol. Survey Water Supply Paper 1357.

———, ———, and RAINWATER, F. H., 1956, Sedimentation and chemical quality of surface waters in the Wind River basin, Wyoming: U.S. Geol. Survey Water Supply Paper 1373.

COOPER, W. S., 1958, Coastal sand dunes of Oregon and Washington: Geol. Soc. America Mem. 72.

CULBERTSON, J. K., 1957, Hydrology and sedimentation in Bixler Run, Corey Creek, and Elk Run Watersheds, Pennsylvania: U.S. Geol. Survey Open File Rept.

DURUM, W. H., 1953, Relationship of the mineral constituents in solution to stream flow, Saline River near Russell, Kansas: Am. Geophys. Union Trans., v. 34, p. 435–442.

FERRELL, W. R., *et al.*, 1957, Debris potential appraisal of some small watersheds of the Los Angeles County coastal perimeter: Los Angeles County Flood Control Dist., 2d Prog. Rept. (mimeographed).

FRANJII, K. K., 1946, Dominant formation concept in non-boulder rivers: Central Board of Irrigation (India) Ann. Rept. (Tech.), pt. 1, p. 122.

HARBECK, G. E., 1952, Water loss investigations: Lake Hefner studies, technical report: U.S. Geol. Survey Prof. Paper 269.

INGLIS, C. C., 1941, Meandering of rivers: Central Board of Irrigation (India) Pub. No. 24, p. 98–117.

JAHNS, R. H., 1947, Geologic features of the Connecticut Valley, Mass. as related to recent floods: U.S. Geol. Survey Water Supply Paper 996.

JOHNSON, J. W., 1949, Scale of effects in hydraulic models involving wave motion: Am. Geophys. Union Trans., v. 30, p. 517–525.

——— 1952, Sand transport by littoral currents: Proc. 5th Hydraulics Conf., State Univ. of Iowa, Studies in Engineering, Bull. 34, p. 89–109.

KRUMBEIN, W. C., 1955, Experimental design in the earth sciences: Am. Geophys. Union Trans., v. 36, p. 1–11.

LANGBEIN, W. B., and SCHUMM, S. A., 1958, Yield of sediment in relation to mean annual precipitation: Am. Geophys. Union Trans., v. 39, p. 1076–1084.

LEOPOLD, L. B., and MADDOCK, THOMAS, JR., 1953, The hydralic geometry of stream channels and some physiographic implications: U.S. Geol. Survey Prof. Paper 252.

——— and WOLMAN, M. G., 1956, River channel patterns: braided, meandering and straight: U.S. Geol. Survey Prof. Paper 282-B, p. 39–85.

LOVE, S. K., 1936, Suspended material in small streams: Am. Geophys. Union Trans., pt. 2, p. 268–281.

MACKIN, J. H., 1948, The concept of the graded river: Geol. Soc. America Bull., v. 59. p. 463–512.

MALINA, F. J., 1941, Recent developments in the dynamics of wind-erosion: Am. Geophys. Union Trans., pt. 2, p. 262–287.

MELTON, F. A., 1936, An empirical classification of flood-plain streams: Geog. Rev., v. 26, p. 593–610.

MILLER, J. P., 1958, High mountain streams: effects of geology on channel characteristics and bed material: New Mexico Bur. Mines Mem. 4.

NICHOLS, R. L., and MARSTON, A. F., 1939, Shoreline changes in Rhode Island produced by hurricane of September 21, 1938: Geol. Soc. America Bull., v. 50, p. 1357–1370.

O'BRIEN, M. P., 1951, Wave measurements of the Columbia River light vessel, 1933–1936: Am. Geophys. Union Trans., v. 32, p. 875–877.

——— and RINDLAUB, B. D., 1934, The transportation of bed-load by streams: Am. Geophys. Union Trans., p. 593–602.

RECTOR, R. L., 1954, Laboratory study of equilibrium profiles of beaches: Beach Erosion Board, Corps of Eng., Tech. Mem. 41.

ROSENAN, E., 1953, The surface movement of blown sand in relation to meteorology: Desert Res., Research Council of Israel Spec. Pub. 2, p. 94.

——— 1954, The direction of seif dunes and wind direction in Sinai and Negev, 4 pp. (Hebrew).

SMITH, D. D., and WISCHMEIER, W. H., 1957, Factors affecting sheef and rill erosion: Am. Geophys. Union Trans., v. 38, p. 889–896.

THOM, H. C. S., 1954, Frequency of maximum wind speeds: Proc. Am. Soc. Civ. Eng., v. 80, Sept. No. 539.

UNITED STATES ARMY, 1950, South shore, state of Rhode Island, beach erosion control study: Letter from Sec. Army, 81st Cong., 2d Sess., House Doc. 490.

U.S. NAVY HYDROGRAPHIC OFFICE, 1951, Techniques for forecasting wind waves and swell: H.O. Pub. 604.

WOLMAN, M. G., 1959, Factors influencing erosion of a cohesive riverbank: Am. Jour. Sci., v. 257, p. 204–216.

——— and EILER, J. P., 1958, Reconnaissance study of erosion and deposition produced by the flood of August, 1955, in Connecticut: Am. Geophys. Union Trans., v. 39, p. 1–14.

——— and LEOPOLD, L. B., 1956, River flood plains: some observations on their formation: U.S. Geol. Survey Prof. Paper 282-C, p. 87–109.

WOOLLEY, R. R., 1946, Cloudburst floods in Utah, 1850–1938: U.S. Geol. Survey Water Supply Paper 994.

ZINGG, A. W., 1949, A study of the movement of surface wind: Agr. Eng., v. 30, p. 11–13.

2
DISCONTINUITIES IN EROSION PROCESSES
Jean Tricart

This article was translated expressly for this Benchmark volume by M. P. Mosley of the Ministry of Works and Development, New Zealand, from "Les discontinuités dans les phénomènes d'érosion" in Internat. Assoc. Sci. Hydrology Bull. **59**:33–43 (1962), *by permission of the publisher, International Association of Hydrological Sciences.*

The special difficulties encountered in measuring erosion account for the inadequacy of the data available to us. These data are inadequate in two essential aspects.

First, measurements are for only some aspects of the phenomenon and are thus qualitatively incomplete. In particular they involve material transported by rivers but also other modes of evacuation of matter from the basin. Suspended sediment transport is relatively well quantified, in spite of practical difficulties. Measurements of turbidity are numerous and relatively well distributed around the world, which allowed Fournier to attempt statistical analysis. Dissolved matter transport does not present any particular problems, although there are many fewer measurements than for suspended sediment and it is less well known. Moreover, in industrial countries, discharge of pollutants into rivers makes quantitative data completely unreliable.[2] Coarse sediment transport, either in suspension or as bed load, has rarely been measured in its entirety, and the sediment load near the river bed is most often estimated, and rather crudely. Furthermore, it is rare for dissolved, suspended, and coarse bed load transport to be measured systematically and simultaneously at the same point and for the data set to be sufficiently homogeneous to permit a comparative statistical analysis.

Second, the real significance of measurements is not always completely clear, and confusion reigns in the minds of most people who present or use them. Too often, it is implicitly assumed that the material transported past a given cross-section permits measurement of the mean erosion of the basin upstream. A mean depth of removal is calculated, and some authors have gone further and deduced the length of time needed for a given reduction in relief. Such reasoning is rendered still more confusing at the international level by differences in vocabulary. For Soviet, Polish, and Dutch authors, *erosion* is restricted to incision of rivers, while *denudation* refers to removal of material from the interfluves. The two can be grouped together as *ablation*. Because of the paucity of measurements of bed load transport, the rate of reservoir infilling is often used to calculate the mean ablation of the basins upstream. This method is not correct, and the mean thus established does not exactly reflect reality. In fact, part of the suspended load and much of the dissolved load is not retained by a reservoir.[3] Besides denudation often provides much more debris than rivers can carry, and part of the material coming from the slopes is deposited at their base, notably as colluvium. This is also the case with erosion by mountain streams, as evidenced by alluvial cones. Thus measurements of sediment load, including dissolved matter, do not and cannot give an exact picture of the rate of ablation in the basin upstream. That would suppose a model of landscape evolution such that all material coming from the interfluves and mountain streams would be evacuated by the main stream. The measurements provide only a figure for net ablation, which is rather different from that for actual ablation.

Interpretation of measurements of fluvial sediment transport is thus rather difficult and involves transforming data for a phenomenon that is discontinuous in space and time into an equivalent overall value. Thus some understanding of the discontinuities is essential.

TEMPORAL DISCONTINUITIES

The temporal variation of sediment transport rate has been long known. In almost all cases no satisfactory correlation exists between suspended sediment transport or dissolved load and discharge. The processes of stream-flow generation are such that identical flows often have very different concentrations of suspended or dissolved load. Overall the dissolved load is a function of the size of the subsurface water source and is greater when subsurface water has been in contact with the soil and rock for a longer period. Moreover, slow rates of flow allow evaporation to concentrate the dissolved material, modifying its concentration, but they do not increase the solute load. On the contrary, sometimes some of the solutes are precipitated, notably in groundwater bodies in which there is a subterranean flow. Such a phenomenon is important for iron oxides in tropical countries, where it leads to the formation of duricrusts, and also in temperate lands, particularly with limestone, where concretions are formed at or just below the water table. Part of the dissolved material carried by rivers during high water is thus stored in groundwater bodies, going into storage during periods of low flow and being only partially released when discharges increase again. Similarly if no change occurs in flow conditions and the proportion of overland flow and infiltration remains constant, seasonal variations in temperature and plant growth would suffice to prevent deduction of the mean chemical ablation from the solute load of a river for the corresponding time period. Between erosion and evacuation of material exists a period of storage, the importance and duration of which are variable and not directly correlated with river regime. Some of the matter precipitated in the aquifer can be removed after a short time in the following season of high flows, but a portion will remain stored for much longer, on the scale of several years, or on a geologic time scale (for instance, the material that cements alluvium into conglomerate). Additionally, this process of precipitation can interact with complicated and more or less reversible biochemical mechanisms. For example, iron oxides, so important in the hot, humid climates, are generally put into solution in the form of unstable humic complexes, the breakdown of which leads to precipitation of iron. Its subsequent removal from storage in the alluvium is thus made more difficult.

The example that we have considered involves the most mobile material, the dissolved matter. Its destination is the same as that of the water, and it moves at the same velocity, which reduces the temporal discontinuities in its transport. But discontinuities do exist, in the form of periods of temporary (or sometimes permanent) storage, of greater or lesser duration. This means that the discharge of solutes during a given period cannot correspond to the actual ablation in the basin during the same period because of delays in transport, equal in this case to those of the water itself. Thus the discharge of solutes at a given point and for a given period may be less than that which would give the actual ablation of the basin, because part of the dissolved material from the slopes is stored in the aquifers. On the other hand, it may be greater, for it originates in part from solution of alluvium by groundwater in which the groundwater itself is held. Comparison of the solute load of a river and chemical denudation on the slopes of the basin deserves greater interest. It has not been done to our knowledge but would be possible; it would be sufficient to compare actual chemical ablation on the interfluves, using lysimeters, with the solute loss determined by classical methods. Apart from a theoretical interest, such a comparison would have importance for knowledge of the chemical quality of subsurface water in alluvial aquifers.

Time lags are still greater for the less mobile sediments in transport. Rate of movement of material becomes increasingly slow relative to that of water as its size increases. Colloids behave practically like dissolved ions. In contrast, silt and fine sand carried in mechanical suspension by turbulence do not follow the water. During large floods such as that on the Herault in autumn 1958, it is thrown violently into suspension, but that does not stop the particles from being deposited behind obstacles or when turbulence diminishes because of pulsations in the flow. Rates of movement of the particles increase during a flood, and during a flood lasting two or three days particles may travel some tens of kilometers. In June 1957, clay from the Queyras schist reached the confluence of the Rhone and Durance. For a relatively small sediment load, the work expended (in tonne-kilometers) is considerable because it is increased by the large distance covered. Medium and coarse sand, which moves by saltation, travels much more slowly than the water. In June 1957, it was not carried more than forty or fifty km down the Guil, in spite of the quite exceptional flood. Finally, pebbles move still more slowly and advance only in steps. Moreover, the hydraulic conditions that permit entrainment of material, as the classic Hjulström curve shows, occur less and less frequently as one moves from fine sand (0.1 mm) to pebbles and boulders. As particle size increases, particle steps made under favorable conditions become both shorter and more infrequent. The temporal discontinuity of the phenomenon increases.

An interpolation justified for a very mobile material such as an ion in solution is not justified for a less mobile material such as sand, and less so for a class of sediment such as gravel, which has very little mobility. Statistical analysis must therefore be used, bearing in mind the temporal discontinuities in sediment transport. Profiting from the experience in hydrology, where we have moved from consideration of means to that of the range of flows, we must become oriented toward measurement of sediment load not solely by establishing figures for the mean, the value of which is rather doubtful, but by analyzing the frequencies and periodicity of the phenomena. With purely qualitative observations, we can set up the following scale, which may be useful for organizing measurements made in this way:

1. Continuous phenomena, of which the type example is the evacuation of dissolved matter by perennial streams, show numerous variations of different periodicity but are more regular than [seasonal phenomena]. Exceptional floods, for example, will cause only relatively small changes in chemical load, which is more constant than concentration. The problems of statistical analysis are the same as those for the study of floods because the fate of the dissolved matter is the same as that of its solvent.

2. Seasonal phenomena are discontinuous in time and cease for more or less long periods but recur under conditions that exist nearly every year. As for seasonal floods, it is not possible to establish a strict definition of the phenomenon. Examples are the evacuation of dissolved matter by streams that each year cease to flow during the dry season and the evacuation of silt by proglacial channels in middle latitudes, which occurs essentially during the summer melt. The transport of mud often has a similar seasonal character because it is essentially a function of overland flow. In France it occurs particularly in winter when the soil is saturated and poorly protected by vegetation and, notably, when sudden showers coincide with thawing. In summer, even during thunderstorms, it is less important and much more sporadic.

3. Periodic phenomena occur nearly every year but without the regularity of seasonal phenomena. Such is the case with certain forms of solifluction on slopes that are not caused by freezing and that occur only when the soil is sufficiently saturated. In certain cases this solifluction is seasonal, but quite often it is more irregular and only periodic. The transport of pebbles in many streams is similar. It requires powerful rivers with steep slopes, such as the Rhine or middle Rhone, for the flows that set particles in motion to occur several days each year, such that the phenomenon then becomes seasonal. In the mountain streams of the Cevennes, Herault, Gard, and Vidourle, pebbles move only during major floods, which, though they always come between September and April, do not occur every year or can, on the other hand, recur several times during a simple hydrologic year.

4. Sporadic phenomena are less frequent but recur at intervals that permit observation. Ten- or one-hundred-year floods are a good example on the hydrologic level. From the geomorphic point of view, such floods strongly influence erosion, for as they move downstream they cause an abrupt modification of the streambed, followed by a slower recovery, which considerably influences the transport of pebbles and, it seems, finer material. In fact, the material at depth that is stirred up by the flood is washed and made porous. Sedimentologic studies show that the voids are then filled by trapped material; first sand, after the flood and as early as on the falling stage of the flood, then silt and clay, particularly during periods of high water. A large part of the fine sediment may thus be stored in the alluvium, exactly as we have shown happens with dissolved matter, and the diminution of suspended load after large floods may be accounted for partly by this phenomenon. Flow in certain dry valleys also can often have a sporadic character. In the Aigoual, for example, some gorges flow only during exceptional storms, of a type that produce ten-year floods. Then they are more or less flushed clear of the debris with which they are strewn. Sporadic phenomena generally are not susceptible to statistical analysis based on fewer than thirty years of data and tend to show its inadequacy. They therefore merit special attention.

5. Catastrophic phenomena are those that are abnormal and anomalous,

in principle unique. Such was the case with the Guil flood in June 1957.
An exceptional combination of mechanisms reinforced each other and led
to a change in the form of the valley unmatched during postglacial time,
that is in 10,000 years of temperate climate little different from the
present. Major landslides are also catastrophic phenomena. They are
capable of causing profound changes in geomorphic evolution and even
temporarily (on a geologic time scale) reversing a trend. For example,
they can dam a valley and lead to alluviation where normally there would
be incision. They create a more or less lasting disequilibrium, a sort
of "trauma." The Queyras, for example, was affected by the catastrophic
June 1957 flood, and the Guil now carries much more sediment than before. If systematic control work is not carried out, this state will
last for many years and will constitute a danger to Serre-Ponçon. The
cone of the Guil in the Durance Valley is already growing, modifying
subsurface water regimes and tending to divert the main channel, which
has become more unstable and moves its riffles more. Such catastrophic
phenomena introduce a break, which one must take into account in the
establishment of a statistical series.

Temporal discontinuities thus are fundamentally different for the
phenomena involved in ground lowering and sediment transport or for
those with which hydrologists are concerned in flood analysis. But they
are much more accentuated and variable for the successive categories
of material considered. The methods of hydrologic analysis are applicable without any particular precautions to chemical load, but for
suspended load the data series must be longer if one is to maintain
precision. For bed load, a very long series is required because here
the data are almost entirely lacking and indirect and excessive summary methods must be used. This explains how the results of measurement
of solid load are uncertain and how their extrapolation to geologic
time periods can lead to completely fantastic conclusions. For example,
Fournier's work produced a worldwide ground lowering of 1 m per 2500
years. For the Quaternary, which lasted one million years, mean lowering
would thus be 400 m, which is contrary to all geomorphic observation.
Such evaluations are made impossible by temporal discontinuities and
also by spatial discontinuities.[4]

SPATIAL DISCONTINUITIES

The temporal discontinuities are closely linked to the spatial
discontinuities because material in transport stops and is stored in
locations dictated by geomorphic evolution. The time lag between the
variations in chemical ablation of a river with a porous bed is the
consequence of storage of part of the dissolved matter coming from the
interfluves in the alluvial aquifer in the bottom of the valley. The
banks of sand or pebbles that fill the riverbeds are a form of storage
of coarse material that is awaiting a flood before moving downstream
again. The duration of these periods of storage is rather variable. It
is thus impossible to infer from the sediment transport rate at a point
a mean ablation in the drainage basin upstream. One can successfully
measure ablation in a region by means of accumulation rates in a reservoir only if this storage effect is eliminated, which is possible only
on a geologic time scale. In fact, over a million years, for example,
an aquifer in a valley bottom formed at one time can be destroyed by
erosion, and the material that was stored for several hundred million

years can be restored to the system. This is why, for example, Gabert, in a remarkable study, could deduce the denudation undergone by the Italian flank of the Alps from the Plioquaternary accumulation of sediment in the plain of the Po during the corresponding period. The geologic viewpoint, by adopting another time scale at which effects of momentary storage disappear, allows calculation of a mean value of ablation and as an intermediary a total sediment discharge that is correct for a given surface during a given time period.

On the human time scale, the interruptions that are the rule in sediment transport prevent a correlation between ablation on the interfluves and the solid load, however perfectly calculated, at a given point on a river. Measurements of solid load can only have strictly local value, as we will show for various categories of process and types of material.

1. Diffuse flow on hill slopes largely takes the form of discontinuous runoff. During storms, on a less permeable soil surface, often a bare surface pounded by raindrops, a film of water begins to flow and becomes concentrated in threads that entrain fine debris, silt, and sand. It can continue if there are no obstacles and, growing little by little, can cut a rill. This is the initial phase of concentrated runoff, a first-order channel. But more often things are not so simple, and the large roughness of the soil surface, which results from clumps of vegetation, litter, and stones, hinders the runoff. The thread of water divides around obstacles and spreads out; it then drops the material it had entrained and partly infiltrates into the more permeable soil. This material must wait for the formation of another thread of water to pass over the obstacle before being picked up again. It is rare for that to happen during the same storm, for rainfall intensity must be increasing. More often the material will be picked up again only during another storm weeks, months, or even years later, after which the microtopography will have changed and the position of obstacles will no longer cause the division of threads of water at the same places. On a geomorphic time scale, the whole slope is swept by discontinuous runoff, which gives it a characteristic regular form, but on the scale of hydrologic observation there is discontinuous erosion. A stake implanted at any point on the slope can register an accumulation and then a lowering over several months or years. On the very short time scale at which our observations are made, this double variability in time and space requires a large number of measurements.

2. The foot of the slope is a critical zone for sediment transport because it corresponds by definition to a sudden and large reduction in slope angle in most cases. Rarely are there gorges whose walls fall directly to the main channel of a torrent. Usually the slope ends, exceptionally above a major river, above a high terrace, glacis surface, or pediment. The diffuse flow from the slope abruptly loses its energy when slope angle is reduced. It also generally drops its sediment load, which forms a continuous accumulation that is spread out along the foot of the slope, giving it a characteristic concave profile. This is colluviation, a characteristic phenomenon on all slopes from which debris is not evacuated at the same rate as it is produced. It is very widespread across extremely diverse climates and results from both overland flow and solifluction in its various forms, creep included. The concavity at the foot of a slope, particularly an irregular concavity, with local mounds that are small, alluvial cones at the mouths of gullies or swales on the hillside, is a useful indicator (of course,

that must be checked by examination of the material of which it is composed). Colluviation is extremely significant for the geomorphic balance. It reveals a net ablation on the slope greater than that which can be evacuated by the river. In certain climates, the transport rate of debris from the slopes is such that it can inhibit downcutting and can cause a buildup in the valley bottom. This was the case during the Quaternary in the Paris Basin, during the cold periods, which caused "climatic" river terraces, because those beds of sediment were then incised by streams when transport from the slopes decreased (during the interglacial and at the present day). Wherever colluviation occurs, values of ablation deduced from fluvial sediment transport must be inexact. Their comparison with values of ablation measured on the slopes themselves and with the volume of colluvium obtained from the change in ground level nevertheless would allow a closer estimate of the morphogenetic balance.

3. Alluvial cones bear the same relation to concentrated runoff as does colluvium to the phenomena that produce the overall form of the hill slopes. Mountain streams with steep slopes reach a piedmont area and abruptly drop part of their load, which is deposited as a cone in an intermountain basin or more simply in a flat-bottomed valley. The accumulation smoothes the angle of junction between the stream's longitudinal profile and the flatter area, producing a regular concave form. Other processes of slope formation similarly produce terminal deposits under the same conditions as these mountain streams: avalanches, mud slides, and scree chutes. Their activity is very discontinuous, seasonal or periodic, and sometimes sporadic. Material arrives suddenly and in large volumes but for a short time. Some mountain streams cut down into their alluvial cones during the period between the floods that increase their size so that a certain quantity of material is reentrained. But this is always a weak, and often a very weak, tendency, for their competence is much less so that the floods that would cause rapid downcutting are prevented from doing so by the formation of an armored bed. Meanwhile, just after a flood, the recovery of a channel can be explained by a period of decreasing debris transport from the cone into the main stream, a sort of queue of moving sediment. Storage is no less dominant. In this as in the preceding case, it prevents calculation of the true ablation in a basin where its tributaries build alluvial cones consisting of the material carried by the main stream. A detailed geomorphic map of these forms of storage is necessary to define their significance for sediment discharge figures.

4. The main rivers are affected by similar types of phenomena. Natural sediment traps of very diverse nature recur along their courses. In a large, orderly valley, storage is on the groundwater of the alluvium and in the main riverbed. During floods, silt is trapped by vegetation on the floodplain, and sometimes banks of sand and even gravel are left there where the current was stronger. It must remain for a period that can be very long, sometimes thousands of years, before being reentrained, either by incision during a flood of a new channel bypassing the main channel (a meander cutoff for example) or by bank collapse (on concave banks). This material moves a more or less long distance depending on its grain-size distribution. Pebbles are generally dropped a little farther down in midchannel or point bars and wait to be reentrained before moving farther. But fine material is put into suspension; it is abundant along well-vegetated plains with a not-too-steep slope, for it is trapped during floods by

a filtering effect. As a result, the suspended sediment load measured downstream has only a remote relationship with ablation on the interfluves. In rivers in alluvial valleys, eroding banks quite often supply most of the suspended load. Part of the dissolved load also comes from solution of particles in transport, notably from pebbles that are exposed on the bed to the flowing water and that can change. Floods also cause minor changes in the minor streambeds. It is well known that they have a general tendency to accentuate differences between riffles and pools. Afterward the riffles are incised and the pools built up, and the long profile becomes less irregular. It is certain that most pebbles taken from a riffle are deposited in the following pools. But particularly on the flood recession, fine material is often deposited within the gravel, and this, moving farther and modifying the sediment load, has an effect like that of downcutting into alluvial cones. Such changes in the streambed can also occur during small floods; thus we have some sediment load that is not related to simultaneous ablation on the slopes. In less orderly valleys we find the same phenomena, but they are more pronounced. The absence of constrictions plays an important role. Gorges are cleaned out by the floods, but widenings are favorable places for deposition. The bed often divides into several channels, sometimes anastomosing, between banks of alluvium. The debris arriving from upstream is stored there and, by a steplike process that has been studied on the Guil in particular, is reentrained after a very variable period of time by the development of new channels. Large changes in sediment load can occur during a single short, violent flood.

This series of mechanisms shows that it is illusory to correlate ablation on the slopes and sediment load in the rivers on the human time scale. The numerous breaks in the movement of debris and the enormous differences in rate of movement, which are a function of particle size, make the phenomenon extremely complex. Analysis is valuable only when many precise observations are available, which is not generally the case. This is why Fournier's estimates of denudation, based on figures for solid load and limited to the most mobile particles in mechanical transport, produce unacceptable results. In contrast, recourse to a much longer time scale, on the order of millions of years, removes this temporal and spatial heterogeneity and permits the satisfactory estimates Gabert made.

Is this to say that we should give up the measurements that are already available, which are still too few in number? Certainly not. They have value--for example, for predicting rates of infilling of reservoirs. But the methods used at present have been too misleading. Inadequate statistical analysis has been carried as a result of an inadequate qualitative understanding of the phenomena. Advances in modern geomorphology shed new light on the problem. This results in one consequence of practical importance: the solid load varies along the course of the river as a function of factors that in large part are local. For calculation of the rate of infilling, it is thus necessary to use measurements of sediment load at the same location as the planned work if one wishes to get as close as possible to reality.

Similarly another error will be avoidable only with difficulty; that which arises from changes in the geomorphic behavior of a river that arises from its own adjustment. Bank collapses and bed scouring, if important, can strongly modify bed load and suspended load. Aggradation of alluvial banks, levees, and fans may also have a modifying influence but in the opposite direction.

At the level of basic research, our results and thoughts encourage prudence in the interpretation of sediment transport data. They do not have the wide value that one has wished to give them, but they have a local significance that deserves consideration if one approaches it with an intimate knowledge of the basin. They must be backed up by measurements of ablation carried out directly on the slopes and systematically executed at a sufficiently large number of stations. The techniques are simple and cheap, but the work must be directed by a highly qualified specialist in geomorphology, because spatial and temporal discontinuities make the measurements difficult. Rougerie and Rapp have provided some examples of the application of these methods, but their use still remains exceptional. Elsewhere the geomorphic map, which has been developed in several countries and which constitutes one of the bases of applied geography, provides a description of the processes acting in a drainage basin, which is indispensable for deciding what to measure. Detailed geomorphic maps at the scale of 1:5000 to 1:50,000, depending on type of area, show types of slope form, riverbeds, alluvial or colluvial accumulation, eroding banks, stream sources, areas of discontinuous overland flow, elemental channels, the degree of resistance of the vegetation cover, lithology, and other information. They distinguish relict features inherited from another time characterized by other processes. They have become a sure instrument that allows geomorphologists to advise engineers on site selection. Also they are able to provide the necessary information for a thorough analysis of sediment load, which permits a distinction between that coming directly from slopes and that liberated by reentrainment of material stored in various ways, and for widely varying durations. From the practical point of view, geomorphic maps reemphasize a fact already demonstrated by Henin, Michon, and Gobillot: the material in danger of filling reservoirs and estuaries and causing channel instability generally has a restricted source and comes from only a small part of the basin. Control works that are appropriate and often of relatively modest cost thus permit a significant improvement.

Thus, thanks to a team effort by hydrologists, engineers, and geomorphologists, we will obtain more precise results on the subject of the erosion and geomorphic balance of drainage basins.[5]

REFERENCES

Devderiani, A. S., 1956, Méthodes de mesure directe de l'épaisseur de la couche sédimentée, *Izv. Ak. Nauk. SSSR, Ser. Geogr.*, pp. 103-113 (Tran. SIG du BRGM, 2835).

Fournier, F., 1960, *Climat et érosion. La relation entre l'érosion du sol par l'eau et les précipitations atmosphériques*, Paris, P.U.F., 201p.

Gabert, P., 1960, Une tentative d'évaluation du travail de l'érosion sur les massifs montagneux qui dominent la Plaine du Po, *Rev. de Géogr. Alpine* 48:593-605.

Jackli, H., 1957, Gegenwartgeologie des bündnerischen Rheingebietes, *Beitr. zur Geol. der Schweiz, Geotechn. Series, 36*, 136p.

Kunholtz-Lordat, G., 1953, Une équation de l'érosion, *Rev. Gén. Sc.* 40:288-291.

Rapp, A., 1960, Recent development of mountain slopes in Kärkevagge and surroundings, Northern Scandinavia, *Geogr. Annaler* 42:73-200.

Henin, S., X. Michon, et T. Godillot, 1954, Etude de l'érosion des vallées de Haute-Durance et du Haut Drac, *A.I.H.S., Public* 36:158-171.

Remenieras, G., et G. Braudeau, 1951, Quelques observations sur l'alluvionnement dans les réservoirs français, *Congrés Grands Barrages, 50*, New-Delhi.

Taillefer, F., 1950, Projets d'une carte de l'érosion dans les Pyrénées, *1st Congrès International*, Pirenaista, Zaragosa.

Tixeront, J., et E. Berkaloff, 1954, Méthodes d'études et d'évaluation de l'érosion en Tunisie, *A.I.H.S., Public* 36:172-177.

Tricart, J., 1953, Un nouvel instrument au service de l'agronome, les cartes géo-morphologiques, *Sols Africains* 4:67-102.

Tricart, J., 1960, Les modalités de la morphogenèse dans le lit du Guil au cours de la crue de la mi-juin 1957, *A.I.H.S., Public* 53:65-73.

Tricart, J., 1960, Les modalités du transport des alluvions dans les rivières ceve noles, *Bull. A.I.H.S.* 20:75-84.

Tricart, J., 1960, L'evolution du lit du Guil au cours de la crue de juin 1957, *Bull. Sect. Geographie Comité Trav. Hist. et Scient. Min. Education nationale* 72:169-403.

Tricart, J., 1962, L'épiderme de la terre, ésquisse d'une géomorphologie appliquée, *Coll. Evolution des Sciences*, Masson, Paris 21:1-167.

NOTES

1. The editors thank Professor Tricart for his comments on the translation, which are added here as footnotes.
2. This is also true for fertilizers and pesticides washed from agricultural watersheds.
3. Where evaporation rates are high, reservoirs (particularly when constructed to control runoff and flooding) may be empty for much of the time and hence may be able to store much of the runoff and the suspended and dissolved matter it carries.
4. A long series of recordings is desirable for statistical analysis, but this leads to a more likely inhomogeneity of the data because of cyclic variations or progressive change in the climate.
5. This approach has been used, for example, in the Soumman watershed, Algeria, in the planning for a soil and water conservation program and irrigation project.

3

Reprinted from *Geog. Annaler* **42**:184–187 (1960)

RECENT DEVELOPMENT OF MOUNTAIN SLOPES IN KÄRKEVAGGE AND SURROUNDINGS, NORTHERN SCANDINAVIA

Anders Rapp

[*Editors' Note:* In the original, material precedes this excerpt.]

CONCLUDING DISCUSSION ON THE TOTAL DENUDATION OF SLOPES IN KÄRKEVAGGE

HOW TO COMPARE THE PROCESSES QUANTITATIVELY?

It is not easy to compare and evaluate the various slope processes discussed in the previous Chapters 4–10. Some processes are rapid, others slow, some are active on steep slopes, others on gentle.

A helpful simplification to obtain fairly comparable quantitative dimensions is to calculate the "exogene mass transfer" (German "geologische Massenverlagerung") as defined by JÄCKLI (1957, p. 28) in tons × metres.

It seems to be three simple possibilities of expressing mass × movement, viz. either (a) mass × vertical component or (b) mass × horizontal component or (c) mass × resultant or "inclined" component. Of these JÄCKLI has calculated both (a) and (b). As we restrict our discussion to processes acting on the slopes from the water divides down to the valley bottom, and exclude the further transportation by rivers, only the vertical component is calculated here.

The mass transfer expressed in ton-metres/year makes it possible to evaluate the quantitative importance of different processes within a given area. But it is not suitable for comparisons with other areas of different sizes. For this purpose it seems convenient to add what we here preliminarily call the *relative mass transfer* = the quantity of material moved within or removed from a unit area of 1 km² (tons/km²/year). The whole drainage area of Kärkevagge is about 18 km². From this we exclude section B and the top plateau of Mt. Vassitjåkko, which are in part greatly influenced by glaciers and which have not been examined continuously. The remaining area of Kärkevagge considered in Table 32 occupies 15 km². We restrict our discussion to pro-cesses acting from the water divides down to the valley bottom and exclude the further transportation by rivers.

EXAMINATION OF TABLE 32

The letters a and b in Table 32 indicate the accuracy of the calculations. a = Direct measurement in Kärkevagge. b = Extrapolation from direct measurements in Kärkevagge.

The values in the column "Average movement" refer to the "inclined" component. The pebble-falls are supposed to move 90 metres, which is half of the average height of the rock-walls. The boulders are considered to fall the same distance as the pebbles and to continue to the base of the talus slopes (90 + 135 m). In the case of dissolved salts the average transport way is estimated at 700 m, which is half of the average distance from the water divides to the valley bottom.

An examination of the mass transfer in tons/km²/year and in ton-metres/year listed in Table 32 gives the following ranking list of transporting slope processes in Kärkevagge 1952–1960.

Transportation of dissolved salts

The most important transport process is that of dissolved salts, which removes a quantity of 26 tons/km²/year. This value is based on analyses of the water in Lake Rissajaure and in streams from Mt. Kärketjårro (p. 165). Morphological evidence of the chemical weathering are rust-coloured weatering crusts on mica-schist rocks and debris, poisonous effect on vegetation (p. 95), and white crusts of lime on rockwall and talus, especially in sections K and L (p. 165). The corresponding average denudation due to chemical solution is 0.01 mm/year or 70 mm during the postglacial period.

The continuous character of this process, its wide area of activity (100 % of the valley) and its long transport increase its importance. The main quantity of salts consists of sulphates, probably essentially dissolved from the mica-schist till and other loose debris on the slopes (cf. p. 167) but also from joints in bedrock. Calcium carbonate contributes to the salt content, which is probably considerably higher than the average chemical denudation in the mountains of Lappland (p. 168). The annual supply of atmospheric salts, reduced in the calculation, is estimated at 9 tons/km² (p. 167).

Earth-slides and mudflows

The second transport process is the earth-slides, which are due to the extreme rainstorm of October, 1959. This event very likely represents a centennial, or probably even millennial maximum (pp. 150, 157), raising the figures considerably over the "normal" means.

The relative mass transfer due to sheet-slides was 23 tons/km²/year and a further 18 tons/km²/year consisted of sheet-slides which continued their movement as mudflows or stream wash. The bowl-slides comprised a transfer of 20 tons/km²/year but a very low value in ton-metres, due to their short average movement (0.5 m).

The earth-slides and mudflows etc. mainly caused a removal of till from the upper slopes of Mt. Kärketjårro and a re-deposition as alluvial fans or sheets on the lower parts of sections J, K and L. Some material was also deposited on the upper part of sheltered talus slopes. On other places the talus slopes were eroded by gullies or slides (p. 152) continuing as mudflows. This transport together with mudflows in other years gave 8.4 tons/km²/year.

Unlike the chemical solution the slides and mudflows are only local transfers that do not (or only to a minor extent) reach the valley bottom and the main stream. This kind of slope processes probably has a still greater importance in climates with more frequent and heavier torrential rains (p. 164).

Table 32. Denudation of slopes in Kärkevagge 1952—1960 given in quantities per year. The average gradient is roughly indicated in 45°, 30° or 15°. The transfer by sheet-slides has been calculated for each case (12—420 m and 70—600 m respectively). For further explanation, see comments in the text.

Process	Volume, m³	Density	Tons (t)	Tons per km²	Average movement m	Average gradient	Ton-metres (vertical)
Rockfalls							
Pebble-falls..........	5 b	2.6	13	1	90 a	45°	845
Small boulder-falls....	10 b	2.6	26	1.7	225 a	45°	4,160
Big boulder-falls......	35 a	2.6	91	6	225 a	45°	14,560
Avalanches							
Small avalanches......	8 b	2.6	21	1.4	100 b	30°	1,050
Big avalanches							
(Slushers)...........	80 a	2.6	208	14	200 b	30°	20,800
Earth-slides etc.							
Bowl-slides..........	170 a	1.8	300	20	0.5 a	30°	75
Sheet-slides..........	190 a	1.8	340	23	12—420 a	30°	20,000
Sheet-slides + mud-							
flows...............	150 a	1.8	270	18	70—600 a	30°	70,000
Other mudflows......	70 b	1.8	126	8.4	100 b	30°	6,300
Creep							
Talus-creep..........	300,000 b	1.8	—	—	0.01 b	30°	2,700d
Solifluction..........	550,000 b	1.8	—	—	0.02 b	15°	5,300e
Running water							
Dissolved salts........	150 b	2.6	390	26	700 b	30°	136,500
Slope wash..........	?			?			?

d Horizontal component of talus-creep = 4,700.
e Horizontal component of solifluction = 19,800.

46

Dirty avalanches

Dirty avalanches are separated into two types, big and small. The big ones are the three cases of slush avalanches which occurred in 1956 and in 1958 (p. 138). They are sporadic processes probably typical of mountains in high latitudes. The morphology of the avalanche tracks shows that such big events as the three slushers are rare but not at all unique. The 14 tons/km²/year removed by slush avalanches is therefore considered as a high average value. Together with the more continuous, small avalanches we get the figure of 15.4 tons.[1]

Rockfalls and frost-weathering

The mass transfer by rockfalls is 8.7 tons/km²/ year, mainly in the form of big boulder-falls (p. 114). The rockfalls are of a special interest as they show the denudation of bedrock on steep slopes. The average annual retreat of the rockwalls in Kärkevagge was 0.04–0.15 mm/ year, indicating a probable postglacial "continuous" retreat of about one meter (cf. pp. 115, 122).

The rockfalls are mainly released by thawing after frost-bursting (p. 105). Thus the retreat of rockwalls indicates a maximum value of frost-weathering on more gentle bedrock surfaces. There the penetration of frost-shattering is probably slower due to insulating loose debris upon the rock. The water content may in places reverse this supposed relation of weathering in rockwalls contra more gentle rock slopes. An average frost-shattering of 0.04 mm (the minimum value of rockwall retreat) corresponds to an annual production of rock waste amounting to roughly 100 tons/km² of rockwall surface in Kärkevagge.

Solifluction and talus-creep

The continuous mass-movements of solifluction and talus-creep are difficult to compare with the momentary mass-movements. Solifluction is active on an area of about 2.2 km² and the

movement is estimated at 2 cm/year of a layer 25 cm thick (p. 182). Talus-creep is active on an area of 1.5 km² with an estimated movement of 1 cm/year in a layer 20 cm thick (p. 175).

As regards these slow processes, the values in ton-metres tell more than the quantities in tons/km². The mass transfer of solifluction is calculated at 20,000 ton-metres and that of talus-creep at 5,000. The talus-creep seems to decrease towards the base, indicating that it functions as a shifting process on growing talus slopes, which are later affected by momentary removing processes, such as slides, or gullying and mudflows.

The solifluction is the dominant transporting process in certain areas, marked by lobes etc. One function of solifluction is that it delivers material to runnels and slope wash for further transportation.

Other processes

Creep due to *needle-ice* has been noted by the author but it is considered to be of small importance in Kärkevagge, where a large part of the slopes are either grass-covered or consist of talus debris, too coarse for formation of needle-ice.

Wind erosion has been observed (p. 110) but its quantitative importance is believed to be very small in Kärkevagge.

Another factor not considered in Table 32 is *slope wash*. It has not been measured in Kärkevagge, only observed a few times (p. 158). The few recordings made in other periglacial areas do not permit a comparison. The opinion of the author, based on indirect evidence (p. 160), is that slope wash is a process of minor importance on the grass-covered slopes and the naked talus slopes in Kärkevagge.

Summary

The quantitative analysis summarized in Table 32 indicated the following order of the transporting processes acting on slopes in this environment (steep and moderate mountain slopes mainly in the "tundra zone", mica-schist and limestone bedrock and till, maritime, arctic climate).

1. Transportation of salts in running water
2. Earth-slides and mudflows

---

[1] Jäckli (1957, p. 126) estimated the mass transfer by avalanches in Graubünden at 450,000 tons/year, a quantity which corresponds to 105 tons/km²/year. A comparison with the 15 tons in Kärkevagge supports our view that Jäckli's estimation of avalanche removal is too high (cf. p. 146).

3. Dirty avalanches
4. Rockfalls
5. Solifluction
6. Talus-creep.

Where slope wash should be placed in this list is not possible to say.

In other valleys with other types of slopes, bedrock, soils etc. the processes may have quite another order.

Frost-bursting is not included in the list as it is not a transporting process and as it is very difficult to measure directly. But the annual production of rock waste by frost-bursting on the rockwalls is calculated to be 100–400 tons/ km² of wall surface. If frost-bursting were included in the list above, it would possibly be the leading one.

Final remarks[1]

The figures given in Table 32 are more or less

---

[1] See also general summary and conclusions in: RAPP, A.: Studies of the postglacial development of mountain slopes. Meddelanden från Uppsala Universitets Geografiska Institution, A: 159. 10 pp. 1961.

approximate and should be looked upon as an attempt to evaluate the order of magnitude of the processes acting in the selected area. The figures thus can be checked and corrected in many respects, both in Kärkevagge and by comparative studies in other mountain areas. One of the first complementary studies that should be made is measurements of slope wash, for instance by the methods used by JAHN (1961) in Spitsbergen.

Table 32 may serve as a summary of the quantitative measurements made in Kärkevagge and also as a hypothesis for future work, both in the mountains of Scandinavia and other latitudes. Even if we here in many respects have emphasized the importance of direct recordings of slope processes, the geomorphological analysis of slope forms may not be forgotten. In this connection two other methods of quantitative slope studies can be mentioned, viz. comparisons of old and new photographs, and volume measurements of rock waste in talus cones, demonstrated by the author in a previous work.

ERRATUM

Page 184, line 7 from the bottom in the right hand column should read: "are rust-coloured weathering. . . ."

4

Reprinted from *Nature* **224**:851–852 (1969)

Present Rate of Land Erosion

by

ANTHONY YOUNG

School of Environmental Sciences,
University of East Anglia

A fresh look at available estimates of the rates of erosion suggests that landforms have taken longer to evolve than has been thought.

KNOWLEDGE of the approximate time required for the evolution of a landform fundamentally affects any consideration of its origin. A valley initiated some 100,000 years ago would have been shaped largely by geomorphological processes active in the climatic conditions of the last glaciation; this would no longer be the case if rates of erosion were found to be such that the valley could have been cut in less than 10,000 years.

Evidence on which to base estimates of the rate of erosion comes from quantitative studies of geomorphological processes, which have only become frequent during the past 15 years. I have grouped together the various types of evidence relating to the rate of lowering of the land surface by erosion now and in the recent past, and shall discuss here the orders of magnitude obtained.

In the history of ideas about rates of geomorphological change there has been a progressive shortening in the conceptual time scale. In this respect there is a contrast with geology, in which recent evidence has tended to increase the absolute lengths of time attributed to the periods of the geological column. Early geomorphological thinking assigned to landforms ages comparable with those of the rocks from which they were formed; thus early Tertiary and even Triassic dates were attributed to erosion surfaces. W. M. Davis estimated the time required for an erosion cycle to proceed to old age (that is, for the reduction of an uplifted dissected land surface to a gently undulating plain) to be 20–200 million years. Since the late 1930s, new evidence has indicated that landforms are substantially younger. An important reference point was the dating of the 200 m erosion surface in southern England as late Pliocene (revised in 1947 to early Pleistocene); the complex series of landform changes known to post-date this surface must therefore have taken place in 2 million years. This change of approach was reviewed by Linton[1]. In 1954 Thornbury[2] stated as a fundamental concept of geomorphology: "Little of the earth's topography is older than Tertiary, and most of it no older than Pleistocene".

Four types of evidence are available for estimating the rate of loss of material from the land. First, there is a method based on estimates of the suspended and dissolved material transported by rivers, obtained by sampling the load and comparing it with discharge records[3-18]. Most previous estimates of erosion rates on a world scale have used evidence of this type[19]. The second method involves measurement of the sediment accumulated in reservoirs[8,20-24], as illustrated in Fig. 1; catchment areas known to be affected by accelerated erosion of agricultural origin are excluded. The third line of evidence involves measurements of surface processes on slopes, including rates of soil creep, surface wash and landslides[21,25-31]. The fourth class of evidence is the comparison of known geological or radiocarbon dates with landform changes identified as subsequent to them[32-37]; for example, Caine and Jennings[37] estimated that a 10 m high basalt scarp in New South Wales retreated 33 m in 35,000 years. The third and, in part, the fourth types of evidence relate only to surface processes on slopes; the rest of the evidence includes the erosion effects of rivers.

Differences between environmental conditions are such that no two of the reported measurements are rigorously comparable; the principal variables affecting each result are rock type, climate, vegetation, and the nature of the landforms affected, including basin area, relative relief and steepness of slope. A further source of variability is differences in the techniques used. In relation to the number of variables, the data are insufficient for a multivariate analysis; previous attempts to isolate the effect of climate have reached mutually inconsistent conclu-

Fig. 1. Sediment in Strines Reservoir, Sheffield, constructed in 1869 and drained for repairs in 1956. In 87 years, 85,000 m³ of sediment accumulated (seen dissected by stream flow following drainage). Excluding undissected moorland, the estimated mean rate of lowering of the land surface for steeply sloping, uncultivated, valleys of the catchment is 1,016 mm/1,000 years[24].

sions[7,8,19]. With respect to rock type, there are no marked differences in rates of erosion between igneous and metamorphic rocks, siliceous sedimentaries and limestones; unconsolidated rocks are eroded at rates of the order of 10–1,000 times faster than consolidated (for example, 4,600 mm/1,000 years[38]), and data from such rocks have been excluded. I have given equal standing to each result reported, and examined the data for agreement between orders of magnitude.

A clear grouping emerges if the results are divided into two classes of relief: normal relief, including plains, moderately dissected areas, and gentle to moderate slopes; and steep relief, including mountainous areas and individual steep slopes. The results are shown in Fig. 2. The data have been converted to mm/1,000 years and are shown on a logarithmic scale. When a range of values is reported, the extremes are shown. The individual values reported are of unequal standing, but cannot be given relative weights because of the diverse nature of the evidence; in such circumstances the median and inter-quartile ranges are appropriate measures of central tendency. These are, in mm/1,000 years, normal relief 46 (20–81), steep relief 500 (92–970).   Expressed as years required for the removal of a layer of ground 1 m thick, the median values are: normal relief 22,000 years, steep relief 2,000 years. These results are comparable in magnitude with previous studies based chiefly on river load; for plains and mountains respectively, Schumm[8] gave values of 72 and 915, and Corbel[10], 22 and 206 mm/1,000 years. For suspended sediment discharge Holeman[18] found a mean value for all continents except Asia equivalent to 19 mm/1,000 years. Thus the loss of ground from mountainous areas and steep slopes is of the order of ten times faster than from other landforms.

If the median values are extrapolated in time, they indicate that in 10,000 years a steep slope retreats by 5 m, and a gentle slope by less than 0·5 m; a gully 50 m deep could be formed in 100,000 years. In 2 million years a mountain range is reduced in altitude by 1,000 m, and a gently sloping landscape by 100 m. A possible estimate for the reduction of a major erosion surface is the lower limit of the inter-quartile range for normal relief; this gives a lowering of 200 m in 10 million years. Of the many reservations concerning such extrapolations, the most important are the effects of agriculture on present erosion, and of the Pleistocene glaciations on erosion in the past. Comparison of cultivated and uncultivated catchments has shown that sediment yields estimated from partially cultivated river basins may overestimate geological rates

of erosion by at least a factor of 2 (refs. 39 and 40). It is widely held that surface processes in polar climates are more efficacious than in temperate conditions, and that the present landforms of temperate latitudes are largely relict from periglacial conditions; this is unproven but, if true, extrapolation through the Pleistocene is affected.

These results confirm that the revised ideas about rates of landform evolution put forward since 1940 are basically correct. They suggest, however, that the shortening of the conceptual time-scale for geomorphological change may have been carried to excess; that is, landforms have taken somewhat longer to evolve than is sometimes thought to have been the case.   Slopes are unlikely to have been greatly modified during the past 20,000 years; therefore interpretations of the origin of the detailed form of individual slopes, as for example in studies based on slope profile analysis, must necessarily take into account climatic changes during and since the last glaciation.   Minor erosional landforms, such as V-shaped valleys 20–50 m deep, are correctly attributed to the late Pleistocene. Relief forms of intermediate scale are appropriately discussed in the context of the whole of the Pleistocene period; this applies, for example, to studies of erosion surface remnants, valley formation and denudation chronology in regions of the British Isles, or other areas of comparable size.   The origin of erosion surfaces of continental extent, such as those of the African continent, may reasonably be attributed to events dating substantially back into Tertiary time or even earlier; thus King's controversial suggestion[41] that the earliest "Gondwana" surface of Africa originated in the Jurassic is not contradicted by the orders of magnitude for rates of erosion I have obtained.

[1] Linton, D. L., *Adv. Sci.*, 14, 58 (1957).
[2] Thornbury, W. D., *Principles of Geomorphology*, 26 (Wiley, New York, 1954).
[3] Dole, R. B., and Stabler, H., *Water Supply Paper US Geol. Survey*, 234, 78 (1909).
[4] Wundt, W., *Erdkunde*, 6, 40 (1952).
[5] Cavaille, A., *Rev. Géomorphol. Dyn.*, 4, 57 (1953).
[6] Corbel, J., *Rev. Géomorphol. Dyn.*, 8, 4 (1957).
[7] Corbel, J., *Z. Geomorphol.*, 3, 1 (1959).
[8] Schumm, S. A., *Prof. Paper US Geol. Surv.*, 454 (1963).
[9] Williams, P. W., *Irish Geog.*, 4, 432 (1963).
[10] Corbel, J., *Anales Géog.*, 398, 385 (1964).
[11] Douglas, I., *Z. Geomorphol.*, 8, 452 (1964).
[12] Gilluly, J., *Bull. Geol. Soc. Amer.*, 75, 483 (1964).
[13] Judson, S., and Ritter, D. F., *J. Geophys. Res.*, 69, 3395 (1964).
[14] Ritter, D. F., *J. Geol. Education*, 15, 154 (1967).
[15] Pitty, A. F., *Proc. Geol. Assoc.*, 79, 153 (1968).
[16] Khosla, N. A., *Cen. Board Irrig. Power (India), Pub.*, 51 (1953).
[17] Wegman, E., *Rev. Géog. Phys. Géol. Dyn.*, 1, 3 (1957).
[18] Holeman, J. N., *Water Resources Res.*, 4, 737 (1968).
[19] Stoddart, D. R., in *Water, Earth and Man* (edit. by Chorley, R. J.), 43 (Methuen, London, 1969).
[20] Journaux, A., *Premier Rapport de la Commission pour l'étude des Versants* 133 (Union Geog. Intern., Amsterdam, 1956).
[21] Starkel, L., *Czas. Géog.*, 33, 459 (1962).
[22] Cummins, W. A., and Potter, H. R., *Mercian Geol.*, 2, 31 (1967).
[23] Jahn, A., *Czas. Géog.*, 39, 117 (1968).
[24] Young, A., *Proc. Yorks. Geol. Soc.*, 31, 149 (1958).
[25] Gerlach, T., *L'Evolution des Versants*, 129 (Univ. Liége, 1967).
[26] Lamarch, V. C., *Prof. Paper US Geol. Surv.*, 352-I, 341 (1968).
[27] Williams, M. A. J., thesis, Australian National Univ. (1969).
[28] Rapp, A., *Skr. Norsk. Polarinst.*, 119 (1960).
[29] Bauer, F., *Erdkunde*, 18, 95 (1964).
[30] Eardley, A. J., *Bull. Geol. Soc. Amer.*, 77, 777 (1966).
[31] Soons, J. M., and Rayner, J. N., *Geog. Annlr.*, 50, 1 (1968).
[32] King, L. C., *Trans. Geol. Soc. S. Africa*, 43, 153 (1940).
[33] Steiner, A., *Geographica Helv.*, 8, 226 (1953).
[34] Bout, P., Derruau, M., and Fell, A., *Z. Geomorphol.*, Suppl. 1, 133 (1960).
[35] King, L. C., *South African Scenery*, third ed., 29 (1963).
[36] Voronov, P. S., *Geogr. Obsch. SSSR, Izv.*, 96, 3 (1964).
[37] Caine, N., and Jennings, J. N., *J. Proc. Roy. Soc. NSW*, 101, 93 (1968).
[38] Leopold, L. B., Emmett, W. W., and Myrick, R. M., *Prof. Paper US Geol. Surv.*, 352-G, 193 (1966).
[39] Douglas, I., *Nature*, 215, 925 (1967).
[40] Meade, R. H., *Bull. Geol. Soc. Amer.*, 80, 1265 (1969).
[41] King, L. C., *Morphology of the Earth*, 242 (Oliver and Boyd, Edinburgh and London, 1962).

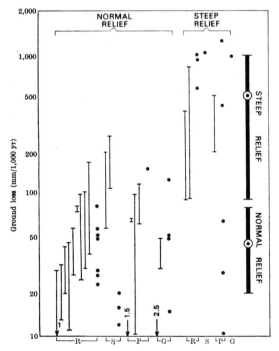

Fig. 2.   Reported rates of lowering of the land surface. To the right are shown the median values and inter-quartile ranges for normal and steep relief. Types of evidence: R, river load; S, sedimentation in reservoirs; P, surface process measurements; G, geological evidence.

5

Reprinted from *U.S. Geol. Survey Prof. Paper 454-H*, 1963, pp. H1–H13

THE DISPARITY BETWEEN PRESENT RATES OF DENUDATION AND OROGENY

By S. A. Schumm

ABSTRACT

Denudation rates are calculated for drainage basins, which average 1,500 square miles in area and which are underlain predominantly by sedimentary and metamorphic rocks. Average denudation rates range from 0.1 to 0.3 feet per 1,000 years, whereas, an average of maximum denudation rates is about 3 feet per 1,000 years. Denudation rates are an exponential function of drainage-basin relief, indicating that denudation rates increase rapidly with uplift.

Modern rates of orogeny are about 25 feet per 1,000 years, or about eight times greater than the average maximum denudation rate. As a result of this disparity between rates of orogeny and denudation, it is concluded that hillslope form is more a function of the difference between the rates of hillslope erosion and stream incision than of rates of uplift. This difference also makes it unlikely that a balance between rates of uplift and denudation will yield time-independent landforms.

Calculations of the time required for planation of 5,000 feet of relief suggest that ample time exists between orogenic periods for the development of peneplains.

Rapid orogeny accompanied and followed by relatively slower denudation should cause epicycles of recurrent uplift owing to isostatic adjustment to denudation. These phases of recurrent uplift after the cessation of tectonism may partly explain the existence of multiple erosion surfaces and terraces as well as the isostatic anomalies associated with old mountain ranges.

INTRODUCTION

The statement is often made that the rate of denudation for the entire United States approximates 1 foot in 10,000 years. It is difficult to grasp the significance of such a pronouncement, for when one considers the topographic and geologic variability of the continent, it is sure that no one rate of denudation is applicable to the whole. In addition, most geologic problems involve areas of less than continental extent, perhaps a mountain range or a section of a physiographic province. New data on sediment-yield rates from drainage basins of about 1,500 square miles in area or smaller, which are characterized by a variety of topographic, geologic, and climatic characteristics, permit calculation of denudation rates for relatively small areas.

Available data on recent rates of uplift afford an opportunity for consideration of the disparity between rates of mountain building and destruction. The extension of present rates of denudation and orogeny into the past and into the future is valid only if the principle of uniformitarianism can be extended to rates as well as processes. At any given locality, denudation rates must have varied widely during the past; however, each time a mountain range existed in a given area, rates of denudation would have been high, and this in general should be true of the past as it is of the present. Gilluly (1955, p. 15) suggests that present rates of erosion are perhaps no greater than the average during the Cenozoic, and Menard (1961, p. 160) shows that denudation rates vary greatly, but "if mountain-building is random in time the rate of erosion of the whole world may be relatively constant when averaged for periods of 10^7–10^8 years." It is necessary to assume that present rates of denudation and orogeny are representative of past rates if there is to be any consideration of geologic and geomorphic problems with regard to the rates presented here.

ACKNOWLEDGMENTS

This paper was read by several people at one or another stage of its preparation. R. F. Hadley, G. C. Lusby, R. B. Raup, Jr., James Gilluly, and H. E. Malde of the U.S. Geological Survey and R. J. Chorley of Cambridge University all made suggestions which resulted in an improved manuscript. Dr. Gilluly's encouragement at a crucial stage in the preparation is deeply appreciated. The author also wishes to express his thanks to M. N. Christensen of the University of California at Berkeley for the time spent on a thorough reading and criticism of an early draft.

DENUDATION RATES

LARGE DRAINAGE BASINS

In contrast to sediment yield rates, which are the weight or volume of sediment eroded from a unit area, denudation rates are generally expressed as a uniform lowering of the land surface in feet per 1,000 years or years per foot of denudation. Obviously, no surface is lowered uniformly in this manner, but it is a convenient way to treat the data.

The denudation rates given in recent geologic publications generally are those for large areas. For example, the rate for the Missouri River drainage basin is about 1 inch in 650 years or 1 foot in 7,800 years (Gilluly, Waters, and Woodford, 1951, p. 135). The same authors state that the rate of denudation for the entire United States is about 1 foot in 9,000 years, and that if this rate could be maintained without isostatic compensation all the landmass would be eroded to sea level in about 23 million years.

Russell (1909, p. 81–84) presents some data on 19th century attempts to calculate denudation rates. Those rates were uniformly high, for no consideration was given to the increase in volume involved in the conversion of rock to sediment.

Early in this century, Dole and Stabler (1909) presented calculations of rates of denudation for major river systems and geographic areas of the United States. Their rates of denudation for a few major rivers are as follows:

River	Drainage area (square miles)	Denudation rate (feet per 1,000 years)
Mississippi	1,265,000	0.17
Missouri	528,000	.13
Colorado	230,000	.19
Ohio	214,000	.17
Potomac	14,300	.09
Susquehanna	27,400	.10

These rates are calculated from data obtained on the dissolved and suspended loads of streams; bedload is not included in the calculations. Dole and Stabler considered that each 165 pounds of sediment in the stream was equivalent to the erosion of 1 cubic foot of rock; in this way they adjusted for the difference in weight and volume between rock and sediment. Again these values, though of considerable interest, were derived from large areas throughout which the topography, geology, and climate varied greatly. Their average for the entire United States is 8,800 years per foot or, if it is assumed that the closed drainage basins of the Great Basin yield no sediment, 9,120 years per foot. These data are probably the source of the denudation rates of 9,000 to 10,000 years per foot that have been so widely quoted.

More recently the existing data have been reviewed, and two papers appeared almost simultaneously that afford new information on erosion from smaller areas (Langbein and Schumm, 1958; Corbel, 1959). Corbel has made a worldwide compilation of erosion rates, which for the most part include the total sediment load of streams. His data were summarized on the basis of climate (Corbel, 1959, p. 15), and indicate that the highest rates of erosion occur in glacial and periglacial regions (2 feet per 1,000 years). Erosion was high also in the high mountain chains having Mediterranean climate (1.5 feet per 1,000 years). In fact, one of the highest rates recorded in nonglacial regions was for the Durance River in southeastern France (1.7 feet per 1,000 years). Corbel's data reveal that denudation rates can be considerably in excess of those reported by Dole and Stabler (1909); this is true also of the Langbein–Schumm data for sediment yields within the United States. These data were averaged for certain ranges of effective precipitation, that is, precipitation adjusted to that yielding equivalent runoff in regions having mean annual temperature of 50°F. The data show that sediment-yield rates are highest in semiarid regions (Langbein and Schumm, 1958, figs. 2 and 3).

Average values for two sets of sediment-yield data are presented in table 1. The first set includes data from about 100 sediment-measuring stations maintained by the Geological Survey. A variety of climatic, topographic, and geologic types are represented, and the drainage areas average 1,500 square miles. The measurements of sediment load at these stations include the dissolved and suspended load but not the bedload. The second set of data are sediment yields from drainage basins averaging 30 square miles in area. As the sediment yield was obtained by measuring sediment accumulation in small reservoirs, these data represent an approach to total load, although small amounts of the suspended and dissolved material are probably lost through spillage of the reservoir.

TABLE 1.—*Denudation rates of drainage basins within the United States*

[From Langbein-Schumm data]

1	2	3	4	5
Effective precipitation (inches)	Mean sediment yield (tons per square mile)	Mean denudation (feet per 1,000 years)	Mean denudation (years per foot)	Time required for planation of 5,000 feet of relief [1] (millions of years)
Gaging-station data				
10	670	0.29	3,400	85
10–15	780	.34	2,900	75
15–20	550	.24	4,200	105
20–30	550	.24	4,200	105
30–40	400	.17	5,900	150
40–60	220	.10	10,000	250
Reservoir data				
8–9	1,400	0.61	1,600	40
10	1,180	.51	2,000	50
11	1,500	.65	1,500	40
14–25	1,130	.49	2,000	50
25–30	1,430	.62	1,600	40
30–38	790	.34	2,900	75
38–40	560	.24	4,200	105
40–55	470	.21	4,800	120
55–100	440	.19	5,300	135

[1] Assuming a 5×isostatic adjustment, reduction of altitude by 1,000 feet requires 5,000 feet of denudation, that is, denudation rate of column 4 multiplied by 25,000.

To convert the sediment-yield rates in tons per square mile to a denudation rate in years required to remove 1 foot of material, it is first necessary to convert the sediment tonnage to cubic feet of rock. An average density of sediment was assumed to be 2.64, and the sediment yield in pounds per square mile was divided by 165 to yield the volume of the surface of the earth removed per square mile per year. Thus, the erosion of 1 ton of sediment per square mile equals the removal of 12.1 cubic feet of rock which is equivalent to 4.34×10^{-7} foot of denudation per year. The average denudation rates for the two sets of Langbein–Schumm data are listed in table 1. The rates are higher than those presented by Dole and Stabler (1909) and are similar to some of the average values presented by Corbel (1959) but do not approach his maximum values, for the sediment yield rates presented in table 1 are average values based on effective precipitation, and the range of individual values comprising each mean is great.

SMALL DRAINAGE BASINS

Even higher denudation rates occur in drainage basins smaller than 30 square miles; for example, a small drainage basin eroded into the sediments of the White River Group of Oligocene age in Nebraska yields 32 acre-feet of sediment per square mile annually (Schumm and Hadley, 1961, fig. 1). Denudation is occurring in this small basin at a rate of 24 feet per 1,000 years. The Halls Debris Basin in the San Gabriel Mountains of California trapped 29.95 tons of sediment per acre of drainage basin per year, between 1935 and 1954 (Flaxman and High, 1955). This material is eroded from an area of 1.06 square miles and at a rate of 8.5 feet per 1,000 years. The maximum sediment-yield rate recorded by the Federal Interagency River Basin Committee (1953, p. 14) is 97,740 tons per square mile for a small drainage basin located in the Loess Hills area of Iowa. The drainage area of this basin is 0.13 square mile. With a sediment yield of this magnitude, denudation will progress at a rate of 42 feet per 1,000 years. This high rate, however, is due to gullying in loess rather than erosion of bedrock. Obviously, the above rates are extreme, but one objective of this report is to indicate that, in the light of recent studies of sediment-yield rates, our concepts of the time required for denudation of less than continental areas should be revised downward.

MAXIMUM DENUDATION RATES

Denudation in large areas requires more time, and in general the sediment yield per unit area decreases at about the -0.15 power of drainage-basin area (Brune,

1948, figs. 6 and 7; Langbein and Schumm, 1958, p. 1079). Sediment from the upper part of the larger basins may be deposited, eroded, and redeposited several times before reaching the basin mouth, although very high denudation rates may pertain to headwater areas. In the smaller basins, steeper slopes allow rapid and generally efficient transport of sediment through and out of the system. For example, on the basis of Brune's (1948) relation between drainage area and sediment yield, it can be calculated that the maximum denudation rate from the Loess Hills area should decrease from 24 to 10 feet per 1,000 years as the size of the drainage basin increases from 0.13 to 1500 square miles. In addition, the denudation rate of 8.5 feet per 1,000 years for the Halls Debris Basin, which has a drainage area of 1.06 square miles, would decrease to about 5 feet per 1,000 years in a 30-square-mile drainage basin, and it would decrease to about 2.8 feet per 1,000 years in a 1,500-square-mile basin. This rate approaches that of the Durance River mentioned by Corbel and other large drainage basins in mountainous areas.

Can this denudation rate of about 3 feet per 1,000 years be considered an average maximum rate for drainage basins on the order of 1,500 square miles in area? It is instructive to compare this rate with rates of erosion in some major mountain ranges. Wegman (1957, p. 6) refers to some earlier work to show that the northern Alps are being lowered at a rate of about 2 feet per 1,000 years. Khosla (1953, p. 111) reports on the suspended sediment yield from the Kosi River above Barakshetra, Bihar, India, which has a drainage basin of 23,000 square miles. Within this basin lie the highest mountain peaks in the world, Mount Everest and Mount Kanchenjunga. The annual suspended sediment yield from this basin is 4.1 acre-feet per square mile which, when converted to a denudation rate and adjusted for change in volume, equals a denudation rate of 3.2 feet per 1,000 years. These limited data indicate that 3 feet per 1,000 years approximates an average maximum rate of denudation.

The denudation rates as calculated here may be extremely high in comparison to those of the geologic past, for man's activities are known to have increased erosion rates many times in certain areas. Yet Gilluly (1949) estimates that 3 miles of denudation occurred in the Rocky Mountains during the Late Cretaceous. If the duration of the Late Cretaceous was 27 million years (Kulp, 1961), denudation occurred at a rate of 0.59 feet per 1,000 years which is almost twice the rate for the large drainage basins (table 1). Gilluly (1949, p. 570–571) also indicates that 5,000 feet of sediment were eroded from the Ventura Avenue anticline in about 1 million years. This is a denudation rate of 5 feet per 1,000 years, a rate higher than the maximums calculated above. Thus, a denudation rate of 3 feet per 1,000 years may not be excessive during the early stages of the erosion cycle when relief is high.

THE EFFECT OF CHANGING RELIEF ON DENUDATION

The denudation rates presented in table 1 are average values which may be used to calculate the time required for peneplanation. It is well known, however, that the rates of denudation change with uplift or during an erosion cycle. Other factors remaining constant, denudation rates will be dependent on the relief of a drainage basin. Data are available on the sediment yields from drainage basins of about 1 square mile in area that are underlain by sandstone and shale in semiarid regions of the western United States. When these data are plotted against the average slope or relative relief of the drainage basin (relief of basin divided by basin length) in figure 1, sediment-yield rates are found to be an exponential function of this relief-length ratio (Schumm and Hadley, 1961). In log form the equation for the regression line of figure 1 is

$$\log S = 27.35R - 1.1870, \tag{1}$$

where S is sediment yield in acre-feet per square mile and R is relief-length ratio. This relation is a straight line when plotted on semi-log paper. When plotted on arithmetic paper, the resulting curve shows clearly the rapid increase in sediment yield or denudation rates with the increase in relief-length ratio. This relation is shown on figure 2 (curve 2), where the sediment-yield rates have been converted to denudation rates for drainage basins 1,500 square miles in area and are plotted against relief-length ratio. Because basin length is constant, figure 2 can also show the increase in denudation rates as the relief of a drainage basin is increased to 30,000 feet. The equation for curve 2 in figure 2 is

$$\log D = 26.866H - 1.7238$$

where D is denudation in feet per 1,000 years and H is relief-length ratio.

To return to the problem of maximum denudation rates for mountainous areas, it is now possible to adjust the data of Table 1 for an increase in relief. The data for the gaging stations on table 1 indicate that the maximum average rate occurs in a semiarid climate and is 0.34 feet per 1,000 years. As the slope or relief of the basins is increased, however, as it would with uplift, the sediment-yield rates will increase greatly. For example, when the average slope of a drainage basin or the ratio of relief to length increases from 0.005 to

about 0.05 during orogeny then, according to the relation between relief-length ratio and sediment yield (fig. 1), such an increase in basin mean slope would increase sediment yield rates roughly tenfold. When this relationship is applied to the data of table 1, the gaging-station data yield a maximum rate of 3.4 feet per 1,000 years. The maximum rate for the reservoir data is 6.7 feet per 1,000 years, but when this value for the small basins is extended to a 1,500-square-mile basin, it becomes 3.6 feet per 1,000 years. Therefore, as suggested previously, a value of about 3 feet per 1,000 years may approach the average maximum rate of denudation for mountainous areas.

When this average maximum rate of denudation for drainage basins having an average relief-length ratio of 0.05 is plotted on figure 2, the point falls far to the left of curve 2. A curve drawn through this point (curve 1, fig. 2) may represent the maximum denudation rates for a given relief-length ratio. Undoubtedly a family of curves would be required to show the relation between denudation rates and relief-length ratio for varying lithology and climate. As rock resistance and vegetative cover increase, the curve should shift to the right.

Figure 2 may be viewed in another manner. If the curves can be considered the locus of points occupied by one drainage basin during a cycle of erosion (high relief is analogous to the geomorphic stage of youth), then the decrease in denudation rates with time during the erosion cycle is illustrated.

In all the preceding discussion it should be remembered that the data on which figures 1 and 2 are based are representative of small drainage basins under a semiarid climate and underlain by sedimentary rocks. Therefore, only the general shape of the curves and the form of equation 1 may be considered as an approximation to large-scale denudation.

RATES OF OROGENY

Before one can discuss the implications of the denudation rates presented here, consideration should be given to rates of orogeny. To be comparable to modern rates of denudation, the rates of orogeny should be modern measured rates rather than rates based on a study of the geologic history of an area. For example, Zeuner (1958, p. 360) presents data showing that uplift occurs at rates of a fraction of a millimeter per year when the age of a formation exposed in the Alps or Himalayas is divided by its present altitude. The rates obtained are only minimum values for the actual rate of uplift, which probably occurred during a relatively short time.

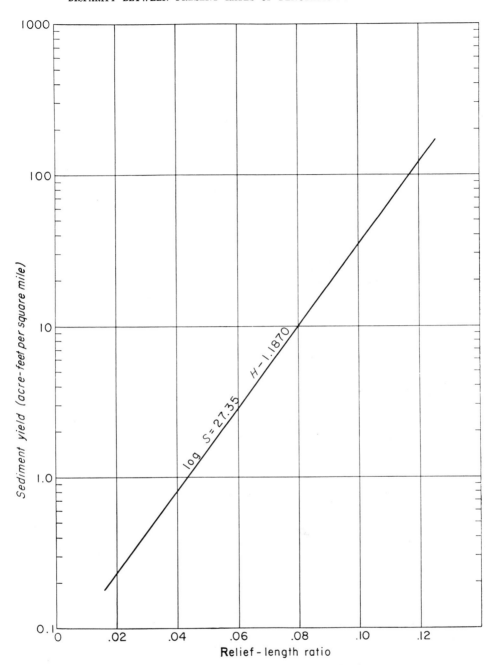

FIGURE 1.—Relation of sediment-yield rates to relief-length ratio.

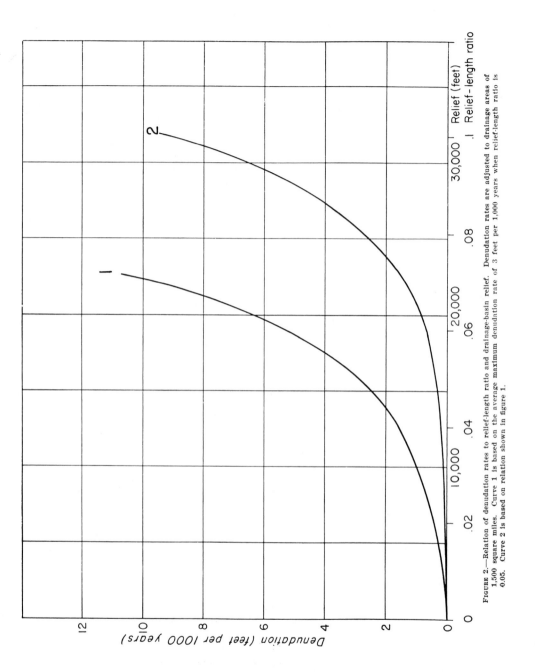

FIGURE 2.—Relation of denudation rates to relief-length ratio and drainage-basin relief. Denudation rates are adjusted to drainage areas of 1,500 square miles. Curve 1 is based on the average maximum denudation rate of 3 feet per 1,000 years when relief-length ratio is 0.05. Curve 2 is based on relation shown in figure 1.

Rates of orogeny being measured at the present instant of geologic time are far in excess of the minimum values obtained by geologic studies. Gilluly (1949) lists some rates of uplift for areas in California in an effort to demonstrate that orogeny is an important aspect of the present. The rates listed are as follows:

Location	Uplift (feet per 1,000 years)
San Antonio Peak	17
Buena Vista Hills	42
Cajon Station	20
Baldwin Hills	29
Alamitos Plain	16

The average of these values is 25 feet per 1,000 years. All these rates of uplift are greatly in excess of the highest rates of denudation presented in table 1, but they are of the same order of magnitude as maximum rates of denudation for very small areas underlain by easily erodible materials, loess and shale.

A recent report on rates of orogeny (Stone, 1961) from results of repeated precise leveling in the Los Angeles area reveal that some stations on the flanks of the Santa Monica Mountains are rising at a rate of 13 feet per 1,000 years. This rate applies also to some stations on the south flank of the San Jose Hills. Stations on the south flank of the San Gabriel Mountains are rising at a rate of 20 feet per 1,000 years.

Gutenberg (1941) has compiled information on postglacial uplift in Fennoscandia and North America. Maximum uplift at the head of the Gulf of Bothnia is 110 centimeters per 100 years or 36 feet per 1,000 years. Geologic evidence indicates the total uplift during the past 7,000 years to be 100 meters or a rate of 48 feet per 1,000 years. Present rates of uplift in North America are 50 centimeters per 100 years to the north of Lake Superior, a rate of 16 feet per 1,000 years. Gutenberg (1941) estimates that present rates of uplift average about half the average for the past 4,000 years. Uplift of about 500 meters has occurred and an additional uplift of 200 meters is required to establish equilibrium. Because over two-thirds of the adjustment has been accomplished, the rates of uplift are slowing but still are high in comparison to rates of denudation.

In contrast to uplift occurring as a result of decrease of load, Lake Mead provides an example of rapid subsidence caused by the addition of 40 billion tons of water and 2 billion tons of sediment (Gould, 1960) to an area of 232 square miles (147 pounds per square inch). Subsidence is occurring at a rate of 40 feet per 1,000 years although total subsidence is expected to be only 10 inches (Longwell, 1960).

Average rates of uplift measured on the eastern European plain of Russia range from 6.5 to 13 feet per 1,000 years (Mescheryakov, 1958); maximum rates are 33 feet per 1,000 years. Data presented by Tsuboi (1933) show an average rate of uplift for leveling stations in Japan of 15 feet per 1,000 years and a range from 250 to 2.6 feet per 1,000 years. Lees (1955) reports rates of uplift in the Persian Gulf area at 33 and 10 feet per 1,000 years.

The above rapid rates of uplift were measured in tectonically active areas or those adjusting to increase or decrease of load. On the other hand, data obtained on epeirogenic uplift along seacoasts (Cailleux, 1952) show that uplift is occurring at an average rate of only 3.2 feet per 1,000 years. The range is from 12 to 0.3 feet per 1,000 years. It is to be expected that slow rates of uplift will be measured even in tectonic areas, for the beginnings and end of orogeny are probably slow.

Some data on present rates of orogeny indicate that these rates are very rapid. Although it is not certain that uplift will continue at these rates, it seems probable that the formation of mountain ranges occur at rates comparable to the rates measured in existing mountains. That is at a rate approaching 25 feet per 1,000 years.

SUMMARY

Calculations based on the best available recent data indicate that denudation will occur at the average maximum rate of 3 feet per 1,000 years in the early stage of the erosion cycle and that an average rate of denudation will be about 0.25 feet per 1,000 years when effective precipitation is less than 40 inches and drainage area is 1,500 square miles.

The relation of relief–length ratio to sediment yield shows that denudation rates will be an exponential function of drainage-basin slope or the relief of a drainage basin of constant size as it is raised by uplift or lowered by denudation. Modern rates of orogeny are about 25 feet per 1,000 years, or about 8 times greater than the average maximum rate of denudation. Existing data show a marked unbalance between the recent rates of orogeny and denudation.

DEDUCTIONS FROM THE COMPARISON OF RATES OF OROGENY AND DENUDATION

Based on the preceding conclusion that rates of orogeny are rapid in relation to maximum rates of denudation, it is possible to reconsider some classic geologic and geomorphic problems relative to hillslopes, peneplains, and the effects of isostastic adjustment and denudation on landforms.

THE EROSION CYCLE

Davis' assumption of rapid uplift of mountain ranges, which allows little erosional modification of the area before the cessation of uplift, is supported to some extent by the disparity between rates of uplift and denudation. Depending on rock type, the rapidly uplifted block may either be little modified by stream incision or significantly modified by it. Nonetheless, when uplift stops, channel incision should be occurring at a maximum rate. The uplifted area, therefore, would be in a youthful stage of geomorphic development or in the beginning of the Davisian cycle. However, if there is isostatic adjustment during the cycle of erosion, uplift will reoccur, perhaps intermittently, throughout much of the cycle. This will extend the duration of and complicate the cycle, but the basic principle of the evolution of landforms from a youthful topography is not changed.

HILLSLOPES

If channel incision is occurring at a rapid rate when uplift ceases, a hillslope profile will not reflect varying rates of diastrophism as suggested by Penck (1953), for long before the slope form could change from convex to concave in response to a waning of the endogenetic forces, uplift would have ceased.

In spite of the fact that most rates of orogeny exceed maximum denudation rates, it is possible to conceive of extremely slow, perhaps epeirogenic, uplift. When uplift is slow in an area of high relief, denudation would be at a maximum, and the landforms would change in relation to the dominant factor of rapid denudation rather than from the effects of slow uplift. If the uplift were long-continued, however, the decrease in the rate of denudation with reduction of relief might continue until it equalled the rate of slow uplift. It is difficult to visualize what effect this balance might have on landforms, but it has been suggested that when uplift and denudation are equal an equilibrium or time-independent landform develops (Penck, 1953, Wegman, 1957, p. 5). If an area of low relief is slowly uplifted at rates equal to or less than the denudation rates, it would seem that little change in the landscape would occur and that again equilibrium landforms, whatever these might be, would be maintained.

This conclusion is, I believe, incorrect and results from the fallacy of assuming that because denudation rates are calculated as a uniform lowering of a land surface that denudation, in fact, occurs in this manner. Obviously it does not, for the forces of denudation are composed of two parts, hillslope erosion and stream channel erosion. Channel erosion may be rapid and in some cases approach the rate of uplift, but hillslope erosion is much less effective in lowering interfluve areas. A theoretically perfect balance between rates of uplift and denudation will, therefore, manifest itself by channel incision and extension of the drainage pattern. Mescheryakov (1959), for example, attributes recent channel erosion in the south Russian steppes to contemporary uplift.

Attempts to relate hillslope form to the interaction of rates of uplift and denudation (Penck, 1953) seem misguided in view of the importance of factors other than orogeny in determining hillslope form. Variations in hillslope form in areas of homogenous materials can best be explained as the result of available relief (Glock, 1932; Schumm, 1956a) or the erosion process (Schumm, 1956b).

The data presented here show that uplift in orogenic areas will be rapid. In fact, figure 2 (curve 2) shows that maximum denudation rates do not reach the average rate of uplift, 25 feet per 1,000 years, until relief is well above 20,000 feet. If so, differences in hillslope form should not be a result of the disparity between uplift and denudation, but probably are the result of the difference between the two components of denudation, rates of channel incision and hillslope erosion. For example, when rocks are very resistant, channel incision will be relatively much greater than hillslope erosion and a narrow canyon is formed. When the rock is less resistant, weathering and erosion on the upper slope form a convex profile. More easily eroded material will pass from an initially convex to a straight and then possibly to a concave profile in relatively rapid succession. This evolution has been noted in badlands (Schumm, 1956a, p. 635), where erosion in the channels and on the slopes proceeds at a rapid rate. Time-independent or equilibrium landforms probably cannot result from a theoretical balance between uplift and denudation. If time-independent forms did develop, it would be as a result of rapid channel incision in response to rapid uplift. For example, if uplift is rapid and of a large amount, initial channel incision will form convex hillslopes, but as channel deepening continues and if the rock is easily eroded, steep straight slopes will result which may be maintained at an angle typical of this material (Strahler, 1950). Again such slopes have been observed in badland areas where channel incision and slope erosion is rapid (Schumm, 1956b).

In conclusion, although slow uplift cannot be neglected entirely, rapid uplift probably is the rule in orogenic areas, as the existence of mountains attest. If rates of uplift in orogenic regions always exceed denudation rates, then hillslope form will not reflect the relation between uplift and denudation. Further, a theoretical balance between rates of uplift and denudation would be reflected in channel incision rather than by equilibrium landforms.

PENEPLANATION

Some questions have been raised not only with regard to the cycle of erosion but also with regard to the peneplain itself. For example, Gilluly's (1949) conclusion that diastrophism has not been periodic but was almost continuous through time has been used as evidence against the uninterrupted evolution of landforms through a cycle of erosion as deduced by Davis. This objection has also been leveled at the concept of peneplanation (Thornbury, 1954, p. 189). However, Gilluly qualified the above statement to indicate that the location of diastrophic movements has continually changed. This shift in location is the crucial point in this connection, for when a period of stability occurs in a given tectonic area, a pediment or peneplain may form if the period of stability is long enough.

Davis (1925) estimated 20 to 200 million years was required for the planation of fault-block mountains in Utah. If, as some assume (de Sitter, 1956, p. 471), there are periods of about 200 million years of "relative quiescence" following shorter periods of diastrophism, then the cycle of erosion must run its course and peneplanation must occur within 200 million years. The upper estimate of Davis is close to the maximum allowable time, and the need for such long periods of stability introduces an element of the implausible.

Recent estimates of the time involved in peneplanation involve much shorter periods; without isostatic readjustment, the continental United States could be reduced to base level in about 10 million years (Gilluly, 1955, p. 15). This figure is only one-twentieth of the allowable time, but no correction for isostatic readjustment during erosion of this mass of rock has been made. This is an extremely important factor which may increase the time required for peneplanation by a factor of five or much more. For example, Holmes (1945, p. 190) considers that to reduce a land surface by 1,000 feet, erosion of 4,000 feet of material would be required; that is, there would be 3,000 feet of isostatic adjustment to erosion. Gilluly (1955, p. 14) in his calculation of the time required for the planation of mountain areas, areas more than 0.2 kilometers in elevation, allows for the erosion of 5.5 times the volume of existing mountain areas owing to isostatic uplift. Considering this isostatic factor, Gilluly (1955) concludes that 33 million years are necessary for the planation.

To use the denudation rates presented in table 1 and figure 2 to calculate time required for peneplanation, it is necessary to consider the 1,500 square mile areas as components of a larger area. It is possible to visualize a series of such basins alined and forming a mountain range. Assuming our model mountain range is so composed and has 15,000 feet of relief then it

should be possible to calculate the time required for peneplanation using the changing rates of denudation shown in figure 2. However, it is difficult to know what amount of denudation is required for peneplanation. It appears that if a mountain block were uplifted 15,000 feet, denudation may need only remove about 5,000 feet of rock, for the adjacent lowlands are being built up by deposition of sediment as the mountains are eroded. This is a complication not shown in figure 2.

Let us assume that 5,000 feet of rock must be eroded for peneplanation. If rates of denudation for the 1,500-square-mile drainage basins are average, all except the basins within the 40–60 inch range of effective precipitation fall within the 200-million-year time limit for peneplanation (table 1, column 5). If less than 5,000 feet of denudation were required for peneplanation, then all would fall within the time limit. The smaller drainage basins (30 square miles) require much less time for planation as expected (table 1). From figure 2 (curve 1) it is possible to obtain denudation rates during the lowering of the mountain range 5,000 feet. At 15,000 feet on the maximum curve, denudation is 2.6 feet per 1,000 years; at 10,000 feet denudation is 1.1 feet per 1,000 years. If the denudation rate at 12,500 feet (1.6 feet per 1,000 years) is the average of these rates, 3 million years are required for the reduction of 5,000 feet of relief. If isostatic compensation requires a 5-fold adjustment, then peneplanation will occur in 15 million years.

On the curve obtained from sediment yields of small drainage basins (fig. 2, curve 2), the average denudation rate between 15,000 and 10,000 feet is 0.23 feet per 1,000 years. At this rate peneplanation would require about 110 million years. These periods of time seem short enough to make peneplanation a distinct possibility in the geologic past.

Much has been ignored in the above analysis, for example, the effects of uplift on the climate of the area and vegetative changes during geologic time. Nevertheless, if denudation occurred at the maximum rate, (3 feet per 1,000 years), an area with 5,000 feet of relief would be reduced to base level in 10 million years.

DENUDATION AND ISOSTASY

Isostatic adjustment to erosion will occur continuously with denudation only if the earth's crust behaves as a fluid. The crust has considerable strength locally (Gunn, 1949, p. 267); before isostatic adjustment can occur, this strength should be exceeded by the removal of rock by denudation. Indeed, even when the strength is exceeded there may be a lag before isostatic adjustment occurs as that which allowed submergence of glaciated lands following the melting of the Pleistocene ice sheets (Charlesworth, 1957, p. 1361). In addition, isostatic adjustment to the retreat of the Pleistocene ice sheets was episodic (Lougee, 1953). This adjustment may be explained most simply by assuming that isostatic adjustment accompanied periods of rapid melting of the ice and that no uplift occurred during pauses in deglaciation. However, one may object to this simple explanation on the grounds that pulses of rapid isostatic adjustment followed rapid melting with a lag of about 2,000 years (Charlesworth, 1957, p. 1345).

Many field studies indicate that discontinuous episodes of uplift occurred in most mountain ranges. For example, Wahlstrom (1947, p. 568) states that uplift of the Front Range in Colorado "to its present elevation was not the result of a single upheaval. The presence of more or less poorly developed terraces in the canyons * * * and well-developed terraces in the valleys east of the mountains suggests intermittent uplifts."

In general, a discussion of these multiple terraces and stepped erosion surfaces, when it is assumed that they have been formed by uplift alone, raises the subject of the role of isostatic adjustment to denudation as a factor in their formation. Whether isostatic adjustment to denudation will occur in a given area depends on the local strength of the earth's crust. Any one of three conditions for adjustment may prevail in a given area: mechanical equilibrium whereby a rigid crust is capable of supporting uncompensated loads; isobaric equilibrium whereby there is a regional compensation for loading or unloading; and isostatic equilibrium whereby a local compensation to loading or unloading occurs along fractures in the crust (Hsu, 1958). Compensation for deglaciation is isobaric, whereas compensation for denudation will generally be isostatic.

The episodic nature of isostatic adjustment to deglaciation and the disparity between rates of uplift and denudation in orogenic areas suggest that isostatic adjustment to denudation will also be episodic. When initial diastrophism occurs, uplift will be relatively rapid until an equilibrium is approached. Orogeny will then cease, and denudation will proceed at a slower rate until the strength of the crust is exceeded, when rapid isostatic adjustment should occur. This relation is shown diagrammatically in figure 3. The orogeny raises the area 15,000 feet; during and following the orogeny, denudation rates increase to a maximum, to be followed by a decline as relief is lowered. This orderly sequence of events is interrupted by a short period of isostatic adjustment, during and following which denudation rates again increase to a maximum. If the result solely of isostasy, the succeeding uplifts will not reach the altitude of the initial uplift due to

mountain-building processes. If such a relation between denudation and isostasy exists, the topographic form of many mountain ranges will be due to episodes of isostatic uplift as well as to the postorogenic uplift discussed by Pannekoek (1961). Depending on the strength of the crust in a given area, the recurrent isostatic adjustment may be long delayed and large, or more frequent and of smaller magnitude.

Renewed tectonism can interrupt these epicycles of denudation and uplift but, if tectonism ceased after initial uplift, the sequence of erosion and isostatic adjustment may be considered analogous to a positive feedback system. Initial uplift increases denudation rates which in turn increase the tendency for further uplift; when this removal of material per unit area is such that it exceeds the strength of the crust, isostatic adjustment occurs and the cycle begins again. These epicycles within the cycle of erosion occur because the components of the system, denudation and isostatic adjustment, operate at greatly different rates. These epicycles, if real, may partly explain the occurrence of multiple terraces, multiple or warped erosion surfaces, and piedmontreppen. For example, King (1955) and Pugh (1955) have attributed the multiple scarps and erosion surfaces formed in western and southern Africa to periods of isostatic adjustment caused by the retreat of major scarps over long distances from the sea coast.

With the preceding hypothesis in mind, it is instructive to compare the condition of isostatic adjustment in young and old mountain ranges. Most of the younger mountain ranges (Rockies, Alps, Andes, Himalayas) have a deficiency of mass at depth, which causes a Bouguer anomaly of about −300 milligals (mgal). This deficiency compensates for the mass of the mountains above sea level, and the young mountains are in isostatic balance. The old, eroded mountain ranges (Appalachians), however, have a smaller deficiency of mass, and although the Bouguer anomalies may be only 0 to 100 mgal, the mass deficiency is larger in many cases than is required for isostatic compensation. "As a result, isostatic anomalies of approximately −50 mgal may be obtained, indicating that erosion and reduction in elevation in these areas has proceeded faster than the readjustment of the compensating mass at depth" (Jacobs, Russell and Wilson, 1959, p. 100).

It would be possible to conclude from this quotation that denudation works faster than uplift, but the data just reviewed disprove this, even if (as figure 3 shows), during long periods of geologic time, denudation does continue while uplift is dormant. The isostatic anomalies in old, eroded mountain ranges, therefore, seem

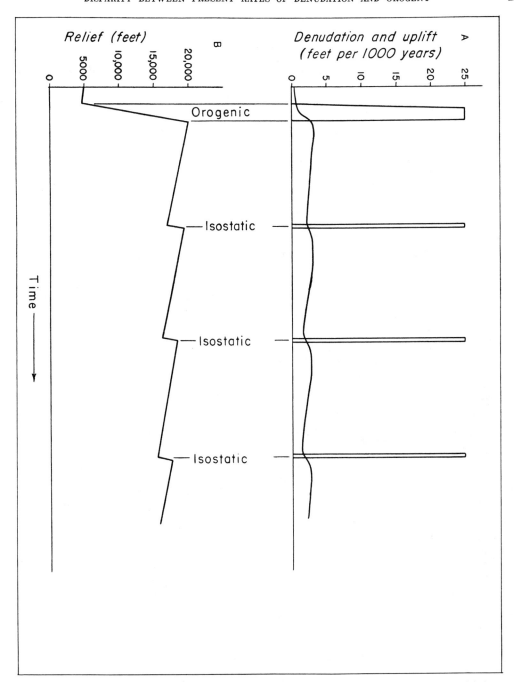

FIGURE 3.—A, Hypothetical relation of rates of uplift and denudation (solid line) to time.   B, Hypothetical relation of drainage-basin relief to time as a function of uplift and denudation shown in A.

to indicate that erosion, although considerable, is not enough to trigger another isostatic adjustment. The young mountain ranges, however, may not be significantly affected by denudation since the last isostatic adjustment, or these adjustments may occur more frequently in the younger ranges and prevent any major isostatic anomaly. Rapid denudation in the younger ranges of high relief make it likely that frequent adjustments keep the young ranges almost in isostatic balance.

Evidence of these epicycles should appear in the stratigraphic record. For example, in sediments derived from an area subjected to cyclic isostatic uplift, gravels might recur through a great thickness of sediment. An example of this as cited by Gilluly (1945) is the Sespe Formation of Eocene and Oligocene age which represents a depositional period of 12 to 15 million years. The recurrent uplift proposed by Gilluly to explain the persistence of coarse sediments in the Sespe Formation may be partly the result of isostatic adjustment to denudation as discussed above.

CONCLUSIONS

Tentative conclusions based on present rates of denudation and uplift are presented as follows:

1. Rates of denudation for areas of about 1,500 square miles average 0.25 feet per 1,000 years and reach a maximum of 3 feet per 1,000 years. These rates are relatively rapid and are representative of areas for the most part underlain by sedimentary rocks in a semiarid climate. Denudation rates in humid regions are about four times slower.

2. Present rates of orogeny exceed rates of denudation significantly. An average maximum rate of orogeny is about 25 feet per 1,000 years.

3. The rapid rates of orogeny in contrast to denudation and valley cutting make it unlikely that hillslope form can be used to decipher the earth's recent diastrophic history. Rather the form of a hillslope profile in an area of high relief probably reflects the difference in rates of channel incision and hillslope erosion.

4. Because denudation has two components, channel and hillslope erosion, which operate at much different rates, a balance between rates of denudation and uplift will not yield time-independent or equilibrium landforms.

5. Relatively rapid rates of denudation make peneplanation a very likely event under conditions which were probably common in the geologic past. Planation of 5,000 feet of relief may require perhaps 15 to 110 million years.

6. An erosion cycle will be interrupted by periods of rapid isostatic adjustment separated by longer stable periods of denudation.

7. The episodic recurrence of isostatic adjustment may partly explain the existence of multiple or warped erosion surfaces, the recurrence of coarse sediments through a thick sedimentary deposit, and isostatic anomalies in old mountain ranges.

REFERENCES CITED

Brune, G., 1948, Rates of sediment production in midwestern United States: Soil Conservation Service Tech. Pub. 65, 40 p.

Cailleux, A., 1952, Récentes variations du niveau des mers et des terres: Geol. Soc. France, Bull. 6, v. 2, p. 135–144.

Charlesworth, J. K., 1957, The Quaternary era: London, Edward Arnold, 2 v., 1700 p.

Corbel, J., 1959, Vitesse de l'Érosion: Zeitschr. Geomorph., v. 3, p. 1–28.

Davis, W. M., 1925, The basin range problem: [U.S.] Natl. Acad. Sci. Proc., v. 11, p. 387–392.

Dole, R. B., and Stabler, H., 1909, Denudation: U. S. Geol. Survey Water-Supply Paper 234, p. 78–93.

Federal Inter-Agency River Basin Comm., 1953, Summary of reservoir sedimentation surveys for the United States through 1950: Subcomm. on Sedimentation, Sedimentation Bull. 5, 31 p.

Flaxman, E. M., and High, R. D., 1955, Sedimentation in drainage basins of the Pacific Coast States: Soil Conservation Service, Portland [mimeographed].

Gilluly, James, 1949, Distribution of mountain building in geologic time: Geol. Soc. America Bull., v. 60, p. 561–590.

——— 1955, Geologic contrasts between continents and ocean basins: Geol. Soc. America Special Paper 62, p. 7–18.

Gilluly, J., Waters, A. C., and Woodford, A. O., 1951, Principles of geology: San Francisco, W. H. Freeman & Co., 631 p.

Gould, H. R., 1961, Amount of sediment, in Smith, W. O., Vetter, C. P., Cummings, G. B., and others, Comprehensive survey of sedimentation in Lake Mead, 1948–49; U.S. Geol. Survey Prof. Paper 295, p. 195–200.

Glock, W. S., 1932, Available relief as a factor of control in the profile of a landform: Jour. Geology, v. 40, p. 74–83.

Gunn, R., 1949, Isostasy—extended: Jour. Geol., v. 57, p. 263–279.

Gutenberg, Beno, 1941, Changes in sea level, postglacial uplift, and mobility of the earth's interior: Geol. Soc. America Bull., v. 52, p. 721–772.

Holmes, Arthur, 1945, Principles of physical geology: New York, Ronald Press, 532 p.

Hsu, K. J., 1958, Isostasy and a theory for the origin of geosynclines: Am. Jour. Sci., v. 256, p. 305–327.

Jacobs, J. A., Russell, R. D. and Wilson, J. T., 1959, Physics and geology: New York, McGraw-Hill Inc., 424 p.

Khosla, A. N., 1953, Silting of reservoirs: Central Board of Irrigation and Power [India] Pub. 51, 203 p.

King, L. C., 1956, Pediplanation and isostasy: An example from South Africa: Geol. Soc. London Quart. Jour., v. 111, p. 353–359.

Kulp, J. L., 1961, Geologic time scale: Science, v. 133, p. 1105–1114.

Langbein, W. B., and Schumm, S. A., 1958, Yield of sediment in relation to mean annual precipitation: Am. Geophys. Union Trans., v. 39, p. 1076–1084.

Lees, G. M., 1955, Recent earth movements in the Middle East: Geol. Rundschau, v. 43, p. 221–226.

Longwell, C. R., 1960, Interpretation of the leveling data: U.S. Geol. Survey Prof. Paper 295, p. 33–38.

Lougee, R. J., 1953, A chronology of postglacial time in eastern North America: Scientific Monthly, v. 76, p. 259–276.

Menard, H. W., 1961, Some rates of regional erosion: Jour. Geol., v. 69, p. 154–161.

Meshcheryakov, Y. A., 1959, Contemporary movements in the earth's crust: Internat. Geol. Rev., v. 1, p. 40–51 (*Translated by* A. Navon *from* Sovremennyye dvizheniya zemnoy kory: Priroda, 1958, no. 9, p. 15–24).

Pannekoek, A. J., 1961, Post-orgenic history of mountain ranges: Geol. Rundschau, v. 50, p. 259–273.

Penck, W., 1953, Morphological analysis of landforms: (Translated) London, Macmillan, 429 p.

Pugh, J. C., 1955, Isostatic readjustment in the theory of pediplanation: Geol. Soc. London Quart. Jour., v. 111, p. 361–369.

Russell, I. C., 1909, Rivers of North America: New York, G. P. Putnam's Sons, 327 p.

Schumm, S. A., 1956a, Evolution of drainage systems and slopes in badlands at Perth Amboy, New Jersey: Geol. Soc. America Bull., v. 67, p. 597–646.

—————— 1956b, The role of creep and rainwash on the retreat of badland slopes: Am. Jour. Sci., v. 254, p. 693–706.

Schumm, S. A., and Hadley, R. F., 1961, Progress in the application of landform analysis in studies of semiarid erosion: U.S. Geol. Survey Circular 437, 14 p.

Sitter, L. U. de, 1956, Structural geology: New York, McGraw-Hill, Inc., 552 p.

Stone, Robert, 1961, Geologic and engineering significance of changes in elevation revealed by precise leveling, Los Angeles area, California [abs.]: Geol. Soc. America Spec. paper 68, p. 57–58.

Strahler, A. N., 1950, Equilibrium theory of erosional slopes approached by frequency distribution analysis: Am. Jour. Sci., v. 248, p. 673–696, 800–814.

Thornbury, W. D., 1954, Principles of geomorphology: New York, John Wiley & Sons, 618 p.

Tsuboi, C., 1933, Investigation on the deformation of the Earth's crust found by precise geodetic means: Japanese Jour. Astronomy and Geophysics: v. 10, p. 93–248.

Wahlstrom, E. E., 1947, Cenozoic physiographic history of the Front Range, Colorado: Geol. Soc. Am. Bull., v. 58, p. 551–572.

Wegman, E., 1957, Tectonique vivante, dénudation et phénomènes connexes: Rev. Géog. Physique et Géol. Dynamique, pt. 2, v. 1, p. 3–15.

Zeuner, F. E., 1958, Dating the past: 4th ed., London, Methuen & Co., 516 p.

6

Reprinted from *Geol. Soc. America Bull.* **75**:483–492 (1964), courtesy of the
Geological Society of America

Atlantic Sediments, Erosion Rates, and the Evolution of the Continental Shelf: Some Speculations

JAMES GILLULY   *U. S. Geological Survey, Denver, Colo.*

Abstract: The volume of Triassic and younger sediment on and offshore from the Atlantic coast between Virginia and Nova Scotia can be estimated from the isopach map by Drake, Ewing, and Sutton (1959) of Atlantic coastal and offshore sediments. This volume is compared with that which would have been derived from the probable source area at present rates of erosion. It is found that the average rate of erosion in Triassic and later time was probably not less than three fourths and perhaps equal to the present rate.

The arrangement of the sedimentary troughs identified by Drake, Ewing, and Sutton suggests that the Continental Shelf was formed in part by isostatic sinking of the offshore crust beneath the sedimentary load supplied by the rivers. But the presence of a nearly continuous median ridge in the sedimentary basin suggests that isostatic sinking

is not alone responsible for the downwarp and in fact could not suffice to bring it about. The density of the sediment must be less than two thirds that of the subcrustal material displaced. If the basement had been originally horizontal or sloping uniformly seaward, the basin landward of the median ridge which would have sunk by a mechanism other than isostatic depression beneath an additional 5000 feet of sediment, for the seafloor is roughly at the same depth over both ridge and inner basin. This differential depression of the basement, as well as the sinking of a former land area to form such basins, requires thinning of the crust; it is here suggested that subcrustal erosion by currents beneath the M discontinuity caused the thinning. Possibly the median ridge and the basins are due partly to instabilities produced by such currents, although they seem too large to be attributed to drag.

CONTENTS

General statement 483
Acknowledgments 484
Volume of sediments 484
Source area 485
Present erosion rate 485
Comparison of present erosion rate with average of
    Mesozoic and Cenozoic time 486
Neglected factors whose omission decreases the
    computed rate 487
    Introduction 487
    Reworking of Coastal Plain sediments 487
    Possible diversion of sediment from postulated
        source area to other deposition sites 488
    Omission of offshore sediments 488
Neglected factors whose omission increases the
    computed rate 488

Introduction 488
Possible presence of carbonate sediments 488
Coastal erosion 488
Possible inclusion of pre-Triassic sediments . . . 489
Clues to depth of erosion of the source area . . . 489
Discussion 489
References cited 491

Figure
1. Isopach map of sediments of the Atlantic Coastal
    Plain, Continental Shelf, and Continental
    Slope. 484
2. Assumed source area for sediments of the Coastal
    Plain, Continental Shelf, and Continental
    Slope between Chesapeake Bay and Nova
    Scotia 486

GENERAL STATEMENT

The isopach map of the Atlantic Continental Shelf and Slope by Drake, Ewing, and Sutton (1959) supplies data fundamental to many geological problems. It permits an estimate of the volume of offshore sediment and shows a

rough approximation of the form of the offshore basins and the presence of a nearly continuous median ridge separating an inner group of deep basins from an outer group.

Seismic evidence suggests that the basement on which the sediments lie is continuous from the Piedmont seaward. It is assumed, therefore, that the sediment is chiefly younger than Triassic, but inasmuch as several troughs of Triassic sediment are overlapped by the formations of the Coastal Plain, a larger proportion of offshore sediment may be of Triassic age than would be inferred from maps of the Piedmont province.

The map thus furnishes a basis for comparing the average rate of Mesozoic and Cenozoic

(near the Virginia–North Carolina border), and a line trending southeast from a point on the Nova Scotia coast near lat. 45° N. As the seaward limit, I have taken an extrapolation of the 5000-foot isopach north of lat. 40° N., and the 10,000-foot isopach and its extrapolation on strike to the south of that parallel (Fig. 1). Thus at the cutoff on the seaward side, sediment thicknesses are still great. The omitted

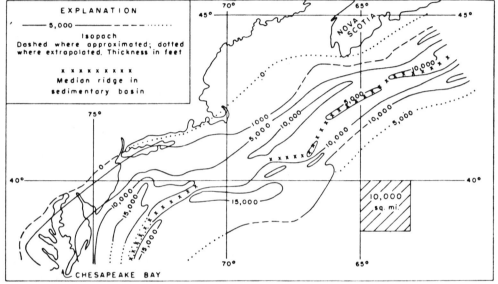

Figure 1.   Isopach map of sediments of the Atlantic Coastal Plain, Continental Shelf, and Continental Slope; slightly modified and somewhat extrapolated from Figure 29 of Drake, Ewing, and Sutton (1959)

erosion with that of the present. It also gives clues as to the origin of the Continental Shelf and the Mesozoic and Cenozoic history of the Atlantic Ocean.

ACKNOWLEDGMENTS

The patience of several friends in talking over these speculations is much appreciated. S. A. Schumm, Henry Berryhill, V. E. Swanson, J. R. Gill, L. C. Craig, H. A. Tourtelot, and T. A. Hendricks were all most helpful in reading and discussing earlier drafts. The very pertinent critique of Gordon Rittenhouse is especially appreciated.

VOLUME OF SEDIMENTS

Very slight extrapolation of the trends of the isopachs determined by Drake, Ewing, and Sutton (1959, Fig. 29) makes possible a rough estimate of the volume of sediments on the Shelf and Slope between lat. 37° 30′ N.

seaward extensions, if included, would notably increase the volume of sediment measured and accordingly the rate of erosion here deduced; the present estimates are thus very conservative.

By planimetry, the mapped area of Triassic and younger sedimentary rocks on the Coastal Plain, Continental Shelf and Slope is about 190,000 mi². The volume as computed from the 0-, 1000-, 5000-, 10,000-, and 15,000-foot isopachs is approximately 280,000 mi³. The porosity of the sediment is greater than that of the source rock, which included both more compacted older sediment and a considerable volume of crystalline rock. A mean density of 2.3 for the present sediments seems reasonable; if the source bedrock had a mean density of 2.65, the present volume of sediments represents about 245,000 mi³ of bedrock.

Because the interface between sediments and basement is continuous from that exposed at the Fall Line to the Shelf edge throughout the map area of Drake, Ewing, and Sutton (1959), it seems reasonable to conclude, as these authors imply, that nearly all this sediment is of Triassic and younger age.

Although there are a few marls in the exposed rocks of the Coastal Plain, there are no true limestones in the latitudes here considered; probably the offshore sediments are likewise predominantly noncalcareous. Thus the sediments were supplied primarily by the suspended and traction loads, rather than the dissolved load, of the streams emptying into the Atlantic, with minor contributions from shore erosion.

SOURCE AREA

To estimate the average rate of erosion implied by the measured volume of sediment, we must make assumptions as to both time of accumulation and source area. The time we can assume to be about 225 m. y., the usually accepted estimate for the span since the beginning of the Triassic (Holmes, 1960). The source area is considerably more uncertain. Because the general trend of the isopachs determined by Drake, Ewing, and Sutton (1959) is roughly parallel to the coast, and the width of the shelf is roughly constant both northeast and southwest of the map area, no very great convergence or divergence of sediment seems to have occurred during its accumulation. Therefore the segment of the continent directly inshore from the mapped area probably supplied most of the sediment. .

The landward limit of the source area is highly uncertain. The volume of Mesozoic and Cenozoic sediment in the Gulf of Mexico implies nearly as great a rate of supply during Mesozoic and Cenozoic time (mainly via the Mississippi River) as that of the present (Gilluly, 1955). The Rocky Mountain Cretaceous geosynclinal sea extended eastward almost to the Mississippi River. The Mississippi now drains more country than it did prior to the glacial diversion of the Missouri and Ohio rivers and thus the Atlantic drainage in the area of the St. Lawrence divide probably did not head much farther inland than it does today. Clearly, much of the headwater extension of the St. Lawrence drainage was due to Pleistocene diversions. I therefore assume that the ancestral St. Lawrence headed no farther west than Lake Erie and thus arbitrarily delimit the hypothetical source area as shown on Figure 2, by lines roughly normal to the coast at either end of the mapped area.

This area includes Virginia and all the coastal States to the northeast, most of New Brunswick, and parts of Ontario, Quebec, and Nova Scotia. The boundary extends along the zero isopach of Figure 1, and inland along the south border of Virginia, west border of West Virginia, Pennsylvania, and New York, the divide between the drainage of the St. Lawrence and that of James Bay as far as the Ungava Bay divide, from which point it trends southeastward to the coast in central Nova Scotia. The submarine contours of Figure 2 show that the northeastern boundary assigns relatively much greater segments of the Shelf off Labrador, Cape Breton Island, and Newfoundland to a relatively much smaller source area in Quebec, Labrador, and Newfoundland. From the standpoint of estimation of erosion rates in the area of our concern, the boundary is conservative. The St. Lawrence drainage has probably emptied about where it now does since well back in the Tertiary, so that the sedimentation area of our concern has probably not received its full quota of St. Lawrence sediment. The rate of denudation arrived at in this paper is thus probably lower than the true one.

The source area thus arbitrarily arrived at is about 510,000 mi^2.

PRESENT EROSION RATE

The hypothetical source area is now drained mainly by the St. Lawrence and by the rivers included by Dole and Stabler (1909) in their North Atlantic Group. Dole and Stabler estimated (p. 84) that the North Atlantic streams are now carrying enough suspended sediment to lower their drainage basins 0.000, 20 inch per year. The St. Lawrence now carries so little suspended load that its drainage basin is lowered only 0.000,005 inch per year on this account, although clearly the present regimen of the St. Lawrence is highly abnormal because of the settling basins of the Great Lakes. I have therefore considered the rate estimated by Dole and Stabler for the "North Atlantic" streams the one to be tested for the entire source area.

Dole and Stabler made no effort to estimate the traction load of the streams, and even now few data are available. Some data gathered by Corbel (1959) show traction loads of several streams to range from 2 to 300 per cent of the suspended load; except in alpine environments he found the traction load to be generally 5–15 per cent of the suspended. For my rough estimates I assume a traction load 10 per cent of the suspended, or one equivalent to 0.000,020 inch of denudation per year.

The dissolved load, which I assume did not contribute to the offshore sediments appreciably, but of course did aid in denudation of the source area, is such as to lower the drainage basins of the North Atlantic Group 0.000,68 inch per year and the St. Lawrence basin 0.000,60 inch per year (Dole and Stabler, 1909); I assume 0.000,64 inch per year as the average.

COMPARISON OF PRESENT EROSION RATE WITH AVERAGE OF MESOZOIC AND CENOZOIC TIME

If the present rate of denudation were applicable to the assumed source area of 510,000 mi² for 225 m. y., the clastic sediment transported would aggregate 390,000 mi³ of material of the density of the source rock. Allow-

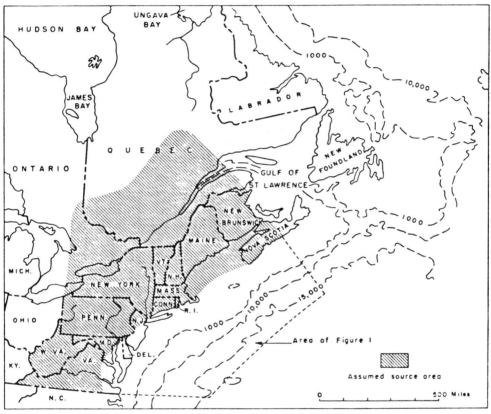

Figure 2.   Assumed source area for sediments of the Coastal Plain, Continental Shelf, and Continental Slope between Chesapeake Bay and Nova Scotia

In summary, the approximate rate of denudation of the assumed source area at present is:

By suspended load	0.000,20 in/yr
By traction load	0.000,02 in/yr
Total clastic sediments	0.000,22 in/yr
By dissolved load (assumed lost to general oceanic circulation)	0.000,64 in/yr

Total erosion is thus equivalent to 0.000,86 inches per year.

ing for increased porosity, the resulting, less-consolidated sediment would occupy perhaps 450,000 mi³. This volume is to be compared with the 280,000 mi³ actually found in the adjacent Coastal Plain, Continental Shelf, and Continental Slope. The ostensible average rate of erosion from the beginning of the Triassic to the present would thus be 62 per cent of the present rate, or 0.000,136 inch per year.

Actually, I think the inferred average rate is highly conservative and that the true rate may have been fully equal to that of the present. The computation neglects several probably significant factors that would increase the inferred rate; neglected factors of the opposite tendency do not seem very weighty.

NEGLECTED FACTORS WHOSE OMISSION DECREASES THE COMPUTED RATE

Introduction

The arbitrary assumptions that were used in the preceding estimates neglect several noteworthy items:
(1) Reworking of Coastal Plain sediments;
(2) Probable overestimate of source area;
(3) Omission of the wedge of sediment offshore from the mapped area in computing the volume of erosion products.

Reworking of Coastal Plain Sediments

Without question, there has been considerable reworking of the Coastal Plain sediments; some of the measured sediment has been reworked, perhaps more than once. The areas now covered by Triassic, Cretaceous, or Miocene strata obviously did not supply sediment to the Triassic, Cretaceous, or Miocene deposits respectively; instead, they and areas adjacent to them (now denuded) were recipients of sediments for part of the time we are considering. In other words, for long periods the source area was considerably smaller than I have assumed, and much of the sediment measured in the offshore area has been transported more than once. Both factors lead to an underestimate of the average rate of past erosion.

The Triassic is still represented on the Coastal Plain and Piedmont by deposits aggregating nearly 20,000 mi³, as deduced from the Paleotectonic folio (McKee and others, 1959). These deposits have been included in the sediment volumes just considered. Certainly Triassic rocks were formerly much more extensive and voluminous. Sanders (1963) has suggested that the now-separated basins of New Jersey and Connecticut were originally parts of a single fault trough, and that other large areas of New York and Massachusetts once contained Triassic sediments. Presumably this implies comparable expansions of the Triassic sediments of Nova Scotia and of the Piedmont of Pennsylvania, Maryland, and Virginia and thus the source area during perhaps 15 m. y. of Late Triassic time would have been much restricted, perhaps by as much as 50,000 mi², and

the volume of Triassic sediments since reworked might be as great as 100,000 mi³. Correction for the first factor, the restriction of the source area, would only result in perhaps a 2 per cent adjustment in the rate of erosion deduced and may be neglected in computations so rough as these. But the correction implied by the volume of reworked Triassic sediments might be as great as 30 per cent and even if the volumes implied by Sanders are too high, the correction must still be considerable. If we assume that only 50,000 mi³ of Triassic rocks was reworked and add this amount to the 280,000 mi³ measured on the Coastal Plain and offshore prism the average rate of denudation would be not 62 per cent, but 73 per cent of the present rate. If Sanders' reconstruction were accepted the average rate deduced would be still higher.

Similar corrections may be needed for Jurassic modifications of source area and for reworking of Jurassic sediments, but there is no basis for evaluating them.

Cretaceous history, however, does imply the need of some corrections. An extreme view of Cretaceous paleogeography is that of Johnson (1931) who thought that the Cretaceous Coastal Plain sediments at one time covered most of the folded Appalachians. Were this true, the volume of reworked sediment (and the area necessarily eliminated as a source during much of Cretaceous time) would be very great. Johnson's theory would demand deposition and later stripping of sediment from a belt 125–200 miles wide—surely more than enough to accommodate the difference between the present rate of erosion and the average deduced for the Mesozoic and Cenozoic. In other words, if Johnson had been correct, we would have to infer erosion rates for most of this time notably higher than those of the present.

Johnson's theory seems, however, to be purely deductive; no direct evidence of so wide a former extent of the Coastal Plain formations has been presented. Groot (1955) has shown that some, at least, of the Cretaceous formations were supplied from nearby Piedmont sources and could never have extended very much farther northwest than they now do.

The figures derived do require a correction for reworking of Cretaceous rocks, but far less correction than Johnson's theory would demand. As a rough guess, a wedge 20 miles wide and 500 feet thick at the butt (an average thickness of exposed Coastal Plain formations of Cretaceous and later ages) along the coast might approximate the volume of the reworked sediments of the Coastal Plain. Such a volume would add nearly 1000 mi³ to the aggregate sediment and diminish the source area by some part of, perhaps 20,000 mi²; but these corrections would be only a very small fraction of that computed. Any reasonable estimate seems to me only a few times as large. In view of the crudeness of all these estimates, I believe the correction for reworked Cretaceous and Tertiary sediments would probably be less than 5 per cent of the computed average.

By thus considering restrictions of the source area at various times and reworking of sediments, we would arrive at an average rate for Mesozoic and Cenozoic erosion between 80 and 90 per cent of the present rate as computed by Dole and Stabler (1909).

Possible Diversion of Sediment From Postulated Source Area to Other Deposition Sites

Nearly one third of the hypothetical source area lies in the present drainage basin of the St. Lawrence River. Absence of Mesozoic and Cenozoic sediments, glacial drainage modifications, and isostatic rebound make it impossible to evaluate the persistence or position of the Atlantic-Arctic drainage divide during pre-Recent time. Continued rebound should enlarge St. Lawrence drainage at the expense of the James Bay and Ungava Bay tributaries; perhaps this relation was the normal one in nonglacial times. The great volume of post-Triassic sediments of the Mississippi Embayment and the eastward extent of the Rocky Mountain Cretaceous into Minnesota and Iowa make it highly improbable that the source area assumed for the Atlantic sediments should be extended significantly into the present Mississippi drainage.

The St. Lawrence now discharges about 200 miles to the northeast of the Shelf section considered, and almost certainly did so for much of Tertiary and perhaps even Cretaceous time. The great width and volume of the shelf northeast, east, south, and southwest of Newfoundland (Fig. 2) suggest a tremendous contribution of sediment from the St. Lawrence to that area—much of the St. Lawrence sediment must have accumulated there. Nevertheless, some St. Lawrence sediment has probably been carried into our area by longshore and other currents, and thus, for my rough estimates (and to be conservative as to past erosion rates) I have drawn the end boundaries of the "source area" about normal to the coast.

My computations may, indeed, have given too much weight to the part of the area in the St. Lawrence drainage and thus may have erred on the side of conservatism. Perhaps, instead of supplying 30 or 35 per cent of the computed volume of sediment, as implied by Figure 2, the St. Lawrence drainage supplied only 20 or 25 per cent. Correcting for such an overestimate of source area would surely bring the hypothetical average erosion rate to within 90 per cent of the rate computed by Dole and Stabler, perhaps even to 95 per cent.

Omission of Offshore Sediments

The volume of sediment computed from the planimetry is highly conservative: it neglects a voluminous wedge of sediment offshore from the mapped area. For half the length of the mapped strip this wedge has a base height of 10,000 feet and for the other half, 5000 feet; although it feathers out, it is unlikely to do so in less than 100 miles. If we include such a wedge, the volume to be added to that measured is nearly 50,000 mi³, a correction of another 10 per cent.

NEGLECTED FACTORS WHOSE OMISSION INCREASES THE COMPUTED RATE

Introduction

Among the possible factors whose omission tends to increase the computed rate of denudation are:
(1) Possible carbonates in the coastal and shelf sediments;
(2) Coastal erosion by the sea;
(3) Possible inclusion of pre-Triassic sediments in the measured volume of deposits.

Possible Presence of Carbonate Sediments

In the area here considered, the Coastal Plain sediments include a few marl units, but they are so subordinate that their omission does not affect the crude computations the data permit. Possibly in the presumably milder climate of Cretaceous and early Tertiary time effective limestone deposition occurred farther north than at present, but a correction for such additions seems of a second order.

Coastal Erosion

Coastal erosion by waves and currents contributes steadily to the sediment. The amount contributed is difficult to evaluate from Atlantic coastal data, but Kuenen (1950, p. 234) has estimated that the global average contribution of sediment by coastal erosion is less than 1 per cent of that supplied by the rivers. For our rough estimates, such a correction is negligible.

Possible Inclusion of Pre-Triassic Sediments

The whole thesis of this paper is based on the postulate that all the sediment overlying the apparently continuous basement is of Triassic or younger age. This assumption seems to me safe for the area as far offshore as the median ridge in the basement; beyond that it is increasingly doubtful. The uniformity of seismic properties throughout the map area seem to me, however, to justify the assumption as a working hypothesis; the reader who rejects the hypothesis will of course reject the entire argument.

CLUES TO DEPTH OF EROSION OF THE SOURCE AREA

Our estimates can be checked by determining whether the depths of subaerial erosion here inferred from the sedimentary record are at all consonant with the geology of the postulated source area. To the depth of erosion deduced by extrapolating the present suspended and traction loads of the streams (3300 feet) it is necessary to add the depth corresponding to the solution load. At the present rate of 0.000,64 inch per year, the denudation by solution during 225 m.y. would be 11,600 feet. The total of traction, suspension, and solution loads at present rates would be 16,300 feet or almost 3.1 miles.

An estimate of the depth of erosion permitted by the geology observable at present is difficult. Chamberlin (1910), in his study of the folded Appalachians of Pennsylvania, drew sections of restored folds and by measuring the area between the base of the Pottsville Formation and the "Kittatinny" surface deduced an average denudation of 3 miles between Tyrone and Harrisburg, in postfolding (Permian?) time. If we include the volumes of Pottsville and younger rocks, and the large areas reduced below the "Kittatinny" surface, we see that in this one area the theoretical erosion during 225 m.y. at present rates and the actual depth of denudation suggested by the geology agree remarkably.

Erosion would almost surely have been slower over much of the crystalline areas; on the other hand many of these were unquestion-ably covered by greater or lesser thicknesses of Paleozoic sedimentary rocks at the close of the Appalachian revolution. With so many uncertainties, it seems reasonable to me to accept an average rate of denudation for Mesozoic and Cenozoic time not very different from that of the present.

DISCUSSION

The data and computations presented are of course speculative. Nevertheless when combined with the more definite information from the Gulf of Mexico (Weaver, 1950; 1955; Murray and others, 1952; Gilluly, 1955) they strongly suggest that present rates of erosion are not 10 times higher than the average of those of the geologic past (Barrell, 1917; Kuenen, 1950, p. 168); I think it most unlikely that the present rates are even twice as high—the most conservative figure suggested by Kuenen (1946, p. 571). I believe the evidence suggests that the average rate of denudation for Triassic and younger time along the Atlantic seaboard and the rate of denudation deduced by Dole and Stabler for the present are not significantly different: the present rate is probably only a few per cent higher. When data comparable to those of Drake, Ewing, and Sutton for Atlantic coastal and offshore sediments become available for the wide Continental Shelf off Newfoundland a much more definite estimate will be possible.

The facts that the shelf and slope retain a continental crust beneath the sediment, even though this crust thins seaward as Drake, Ewing, and Sutton show, and that the surface of the basement was in Triassic and earlier time one of subaerial erosion, go far to prove the reality of a former "Appalachia"—a landmass it has become increasingly popular to discredit. The shelf, without doubt, was variably emergent during parts of Paleozoic time and a source of sediment for the Appalachian geosyncline.

These facts throw some light on the origin of the Continental Shelf. Inasmuch as no sediment is likely to be more than two thirds as dense as the subcrustal material displaced, patently no amount of accumulating sediment can depress a surface of active sedimentation below its original level by purely isostatic forces. The old surface of subaerial erosion of Appalachia was thus not changed into a submarine surface of deposition merely by being loaded with sediment. Some subcrustal process must have brought about the submergence;

isostatic adjustment to a growing sedimentary load would of course operate to continue the sinking.

The Continental Shelf, therefore, is not due primarily to subsidence of an offshore area under a load of fluvial sediments discharged at the shore, as conceived by Kuenen (1950). The subsidence preceded much of the sedimentary loading. The loading must indeed have induced further subsidence, but it cannot account for the 5000 feet of differential sinking of the landward troughs with respect to the median rise, on the seaward side of which even deeper basins occur. Five thousand feet of sediment cannot account for the difference in basement elevation between basin and rise when both columns are overlain by the same depths of water. The still greater subsidence of the troughs seaward of the median ridge is even harder to attribute to isostasy, for here the upper surface of the sediment in the troughs is in deep water, well down the Continental Slope.

Possibly fluctuations in the position of the shore line because of eustatic shifts, both Pleistocene and older, have modified the relations of inner and outer troughs and median ridge. But these structures must owe their primary forms to subcrustal processes, with sedimentation adding only secondary modifications. The seismic speeds presented by Drake, Ewing, and Sutton do not suggest that the outer troughs contain much basalt. It seems unlikely that the offshore troughs are strongly eugeosynclinal, even though several seamounts there are doubtless volcanoes.

For the former Appalachia to have subsided the sialic crust beneath it must have been thinned. Inasmuch as subaerial erosion can only reduce the crust to sea level, the thinning must have been subcrustal. The sial thins steadily southeastward and most abruptly beneath the Continental Slope (Drake, Ewing, and Sutton, 1959, Fig. 28), although there is no very marked change in its seismic properties and the presedimentary surface appears continuous with that of the Piedmont. The continuity of the upper surface of the basement and the general similarity of the seismic properties of the crust from the Piedmont to the Continental Slope suggest that the whole crust is sial. The M discontinuity here separating this sial from the mantle is almost surely not a phase transition, but a compositional boundary. This conclusion is also in accord with considerable experimental work (i.e., Yoder and Tilley, 1962). The crustal

thinning must then be due to subcrustal processes that involve not phase transformations but mass transfer. I think this is a strong argument for subcrustal flowage and erosional thinning of the sial by movements at its base. That such processes exist is suggested by many other phenomena described by Vening-Meinesz (1934; 1948), Ampferer (1941; 1944), Kraus (1951), Gilluly (1955; 1963), Hess (1962), Wilson (1963), and Kaitera (1963).

The origin of currents in the outer mantle is perhaps convectional, but a consideration of scale (Hubbert, 1937; Maillet and Parans de Ceccaty, 1937) shows that during millions of years any large-volume density differences, however generated, will inevitably tend to equalize by flow. Here it is important to reiterate a fact pointed out by Lawson (1932) and recently elaborated by Kaitera (1963): Once a coast has been established isostasy necessitates a subcrustal flow from ocean basin to beneath the continent to compensate for the mass eroded from the continent and deposited in the sea. Of course such a motion might be too deep to influence the crust directly, although the considerable variation of immediately sub-Moho seismic velocities from place to place suggests otherwise. The coastal zone as a whole is a zone of torque about a horizontal axis, with (1) the continent tending to flow out over the ocean floor, (2) the floor near the coast being depressed by both this continental spreading and by the increasing load of sediment, and (3) a continent-trending flow in the mantle compensating for the mass transferred. The folded basement of the Atlantic sediments suggests that the continent-trending flow is close enough beneath the sial to deform it and to crowd the sedimentary prism against the continental mass. The basins and swells are too large to have been caused directly by drag; they may have been produced by instabilities generated or maintained by the postulated landward flow. According to this interpretation, the Continental Shelf is due to a complex of factors—sedimentation, isostatic response to loading, and subcrustal flow. I suggest that the thinning of the crust toward the foot of the Continental Slope is a result of subcrustal erosion by the mantle current, localized by the sedimentary load.

The relationships discussed do not seem to require either a drastic change in the volume of the sea in Cretaceous time, as has been suggested to account for the Pacific guyots, or a late date for the birth of the Atlantic. The postulated westward drift of America from the Mid-Atlantic Ridge (Ampferer, 1941; Runcorn, 1962; Hess, 1962; Wilson, 1963) seems to be permitted but not compelled by the evidence. A critical point here is the volume of pelagic sediment in the North Atlantic on both sides of the Mid-Atlantic Ridge. After the volume of these sediments has been as well determined as that in the area covered by Drake, Ewing, and Sutton in their important paper, we may be able to give a firmly based interpretation as to the age of the ocean and of its sediments and as to the erosion rates of the geologic past.

REFERENCES CITED

Ampferer, Otto, 1941, Gedanken über das Bewegungsbild des atlantischen Raumes: Akad. der Wissenschaften in Wien, Math.-Naturw. Kl., Sitzungsber., Abt. I, v. 150, p. 19–36

—— 1944, Vergleich der tektonischen Wirksamkeit von Kontraktion und Unterstromung: Mitt. Geol. Gesell. Wien, v. 35, p. 107–123

Barrell, Joseph, 1917, Rhythms and the measurements of geologic time: Geol. Soc. America Bull., v. 28, pt. 3, p. 745–904

Chamberlin, R. T., 1910, The Appalachian folds of central Pennsylvania: Jour. Geology, v. 18, p. 228–251

Corbel, J., 1959, Vitesse de l'erosion: Zeitschr. Geomorphologie, n. f. Bd. 3, H. 1, p. 1–28

Dole, R. B., and Stabler, Herman, 1909, Denudation, p. 78–93 in Papers on the conservation of water: U. S. Geol. Survey Water-Supply Paper 234, 96 p.

Drake, C. L., Ewing, Maurice, and Sutton, G. H., 1959, Continental margins and geosynclines—the east coast of North America north of Cape Hatteras, p. 110–198 in Ahrens, L. H., and others, Editors, Physics and chemistry of the earth: New York and London, Pergamon Press, v. 3, 464 p.

Gilluly, James, 1955, Geologic contrasts between continents and ocean basins, p. 7–18 in Poldervaart, A., Editor, Crust of the earth: Geol. Soc. America Special Paper 62, 762 p.

—— 1963, The tectonic evolution of the western United States: Geol. Soc. London Quart. Jour., v. 119, p. 133–174

Groot, J. J., 1955, Sedimentary petrology of the Cretaceous sediments of northern Delaware in relation to paleogeographic problems: Delaware Geol. Survey Bull., no. 5, 157 p.

Hess, H. H., 1962, History of ocean basins, p. 599–620 in Engel, A. E. J., James, H. L., and Leonard, B. F., Editors, Petrologic studies: a volume in honor of A. F. Buddington: Geol. Soc. America, 660 p.

Holmes, Arthur, 1960, A revised geological time scale: Edinburgh Geol. Soc. Trans., v. 17, pt. 3, p. 183

Hubbert, M. K., 1937, Theory of scale models applied to the study of geologic structures: Geol. Soc. America Bull., v. 48, p. 1459–1519

Johnson, D. W., 1931, A theory of Appalachian geomorphic evolution: Jour. Geology, v. 39, p. 497–508

Kaitera, Pentti, 1963, Sea pressure as a factor shaping the Earth's crust: Terra, Year 75, no. 4, p. 342–347

Kraus, Ernst, 1951, Die Baugeschichte der Alpen: Berlin, Akademie-Verlag, v. 2 (Neozoikum), 489 p.

Kuenen, P. H., 1946, Rate and mass of deep-sea sedimentation: Am. Jour. Sci., v. 244, p. 563–572

—— 1950, Marine geology: New York, John Wiley and Sons, 568 p.

Lawson, A. C., 1932, Insular arcs, foredeeps, and geosynclinal seas of the Asiatic coast: Geol. Soc. America Bull., v. 43, p. 353–381

Maillet, R., and Parans de Ceccaty, R., 1937, Le physicien devant la tectonique: Paris, 2nd World Petroleum Congress

McKee, E. D., Oriel, S. S., Ketner, K. B., MacLachlan, M. E., Goldsmith, J. W., MacLachlan, J. C., and Mudge, M. R., 1959, Paleotectonic maps of the Triassic System: U. S. Geol. Survey Misc. Geol. Inv. Map I-300

Murray, G. E., and others, 1952, Sedimentary volumes in Gulf Coastal Plain of United States and Mexico: Geol. Soc. America Bull., v. 63, p. 1157–1228

Runcorn, S. K., 1962, Paleomagnetic evidence for continental drift and its geophysical cause, p. 1–40 in Runcorn, S. K., Editor, Continental drift: New York, Academic Press, Internat. Geophysics Ser., v. 3, 338 p.

Sanders, J. E., 1963, Late Triassic tectonic history of northeastern United States: Am. Jour. Sci., v. 261, p. 501–524

Vening-Meinesz, F. A., 1934, Report of the gravity expedition in the Atlantic of 1932 and the interpretation of the results: Delft, Netherlands Geod. Comm., v. 2, 208 p.

—— 1948, Major tectonic phenomena and the hypothesis of convection currents in the earth: Geol. Soc. London Quart. Jour., v. 103, pt. 3, p. 191–207

Weaver, Paul, 1950, Variations in history of continental shelves: Am. Assoc. Petroleum Geologists Bull., v. 34, p. 351–360

—— 1955, Gulf of Mexico, p. 269–278 *in* Poldervaart, A., *Editor*, Crust of the earth: Geol. Soc. America Special Paper 62, 464 p.

Wilson, J. T., 1963, Hypothesis of earth's behavior: Nature, v. 198, no. 4884, p. 925–929

Yoder, H. S., Jr., and Tilley, C. E., 1962, Origin of basalt magmas: an experimental study of natural and synthetic rock systems: Jour. Petrology, v. 3, p. 342–564

Part II

PROCESSES OF EROSION AND SEDIMENT TRANSPORT

Editors' Comments
on Papers 7 Through 11

7 WOLLAST
*Kinetics of the Alteration of K-Feldspar in Buffered Solutions
at Low Temperature*

8 ELLISON
Studies of Raindrop Erosion

9 CHEPIL
*Dynamics of Wind Erosion: I. Nature of Movement of Soil by
Wind*

10 EINSTEIN
Formulas for the Transportation of Bed Load

11 GRISSINGER
Resistance of Selected Clay Systems to Erosion by Water

Study of the processes rather than the causes or effects of erosion and sediment transport presents major problems. For example, it is relatively simple to measure the suspended sediment load of a stream or the amount of sand being blown across a field and to relate these quantities to such factors as water discharge or wind velocity. It is less easy to explain precisely how the sediment is initially entrained into the moving fluid or how it is maintained in transport, and Raudkivi (1976), in a very lucid chapter on initiation of sediment transport, commented that "there is no generally accepted theory by which the forces acting on the grain in the bed and in the bed vicinity can be determined."

Transport mechanisms such as fluvial or mass wasting processes that involve large distances of transport have been studied more intensively and more quantitatively than the in situ processes that involve physical and chemical changes of rock and soil. Ollier (1969) has stated that "it is not possible to treat real weathering with any chemical vigour," but this belief is shown to be excessively pessimistic by work such as that by Wollast (Paper 7).

Weathering is the first, indispensable, stage in the denudation of a land mass. Much of the material carried to the sea in solution or suspension would not be available for transport without initial preparation by weathering processes. Weathering traditionally has been subdivided into physical and chemical processes, but in fact many of the physical processes, such as insolation and salt weathering, occur primarily in the presence of water, which activates chemical reactions that involve aqueous solutions (Barton, 1916; Blackwelder, 1933; Cooke, 1979). Freeze-thaw, rock unloading, animal burrowing, and root wedging are physical processes that may be locally important, but biochemical processes are extremely important wherever flora and fauna are present. For example, Schatz et al. (1954) have discussed the role of plants in mobilizing insoluble residues by chelating agents, Boyle, Voigt, and Sawhney (1974) have examined the processes whereby organic acids affect the weathering of biotite, and Marshall and Patnaik (1953) showed that the weathering potential of plants is primarily due to the high pH surrounding rootlets, a high cation-exchange capacity condition that enables the breaking of strong silica-alumina bonds.

The most common chemical processes are dissolution, carbonation, hydrolysis, and redox reactions (Keller, 1968). We include here a paper by Wollast that considers the hydrolysis of silicates as a representative of modern research into the processes of weathering (Paper 7). There are three basic reasons for this focus on silicate weathering: silicate minerals and their derivatives cover over 92 percent of the area of the earth's crust; the hydrolysis of silicates is necessary for the alteration of igneous and metamorphic rocks to sediments and soils; and dissolution is a relatively simple process, carbonation involves chemical reactions that can be accurately determined for equilibrium conditions, and redox reactions are slow even in geological terms.

Tamm (1930) and Correns (1940) conducted the first accurate experimental studies of silicate weathering. Once the products of weathering of various silicate minerals were identified, it remained essential to determine from kinetic information the mechanism proper—that is, the reaction paths involved. The first such modern study was undertaken by Lagache, Wyart, and Sabatier (1961), but it dealt with elevated temperatures irrelevant to most weathering at the earth's surface. Wollast has studied the alternation of potassium feldspar at 25°C; we include his study because of its pioneering significance and because this perceptive kinetic investigation has triggered the interest of many other researchers. Wollast suggested a hypothesis (the amorphous precipitate hypothesis) concerning alteration mechanisms and the limiting factors that control rates of alteration, but other hypotheses of feldspar alteration have also been proposed. These include the crystalline precipitate hypothesis (Helgeson, 1971); the surface reaction hypothesis (Lagache, Wyart, and Sabatier, 1961), and the leached layer hypothesis (Paces, 1973). Only the hypotheses of Wollast and Holgeson attribute rate control in incongruent dissolution to diffusion of feldspar components by a precipitate layer. Although modern techniques of X-ray photoelectron microscopy may be used to determine whether precipitate layers do indeed occur, the debate started by Wollast has yet to be resolved.

Much research on the processess of erosion has been designed to solve practical problems, and a detailed understanding of the processes has often been a research objective only insofar as it facilitated control measures. For example, much of the early work on fluvial sediment transport was by hydraulic engineers charged with the task of maintaining navigable waterways, and even until recently, the development of relationships between rate of sediment movement and such simple parameters as mean flow velocity or mean shear stress has been acceptable. On the other hand, agricultural engineers, particularly in North America, have found that a detailed understanding of the processes of land erosion by wind and water has been indispensable to the development of effective control measures on agricultural lands, and they have led the way in detailed studies of erosion processes.

E. Wollny, who has been called the Father of Soil Conservation Research, carried out in the late nineteenth century a number of experiments that pointed to the effects of raindrop impact, soil structure, and erosion and the modifying influence of vegetative canopy, soil type, slope angle, and aspect upon soil surface erosion rates. His work, published in German in a series of long papers, has been summarized by Stallings (1957). A major coordinated research effort began in the United States in 1917, and a number of state and federal government experiment stations were established in localities where erosion presented problems on agricultural lands. Data on runoff and soil loss from experimental plots with a range of inclinations, lengths, vegetative covers, and soil types was collected, and an increasing number of reports appeared, particularly during the 1930s.

Ellison's contribution is fundamental and oriented toward elucidation of the processes involved rather than merely to the presentation of empirical relationships (Paper 8). Ellison explicitly considered raindrop impact as the dominant agent; Stallings (1957) believed that "he was the first to realise that the falling raindrop was a complete erosive agent within itself and that little or no erosion occurred when the ground surface was protected by ample cover." Stallings viewed this discovery as the dawning of a new era in soil conservation research; the paper by Ellison illustrates this new dawn.

Ellison was an employee of the Soil Conservation Service, U.S. Department of Agriculture, and during the decades since his results were published, others in the Soil Conservation Service and Agricultural Research Service have built upon his work. Later studies have considered in some detail the processes of rainsplash, overland flow, and rill erosion (Foster and Meyer, 1972; Free, 1952; Meyer, Foster, and Nikolov, 1975; Meyer and Monke, 1965), but the main objective of this research has been to develop predictive models for facilitating soil conservation measures. Perhaps the best-known of these, the Universal Soil-Loss Equation, is in some respects an outgrowth of Zingg's equation relating soil loss to slope length and inclination (see Paper 12). Using over 10,000 plot-years of data, Wischmeier and Smith (1965) developed a regression model relating soil loss to rainfall erosivity, slope length and inclination, soil erodibility, cropping system, and management practices. More recently, attempts have been made to move from a statistical approach to one based on physical principles, using the type of information obtained by Ellison (for example, Meyer and Onstad, 1977). Nevertheless a significant empirical component remains in these physically based models.

The serious soil erosion problems that beset agricultural regions in North America in the 1920s and 1930s also prompted an increase in research into wind erosion processes. Wind erosion and its geomorphic effects, particularly in the arid regions of the Middle East and Africa, had already attracted the attention of European geographers and earth scientists; the classic work of Brigadier Bagnold (1943) represents perhaps a peak of achievement. In North America, Chepil published a long series of articles on the dynamics of wind erosion (1945–1946) and the properties of soils that influence wind erosion (1953–1955). These papers represent the greatest contribution by a single researcher to our knowledge of the processes and effects of sediment transport by wind, and we include here one of the series, which treats the nature of movement of soil by wind (Paper 9). Chepil paid considerable attention in this paper to the actual processes involved in the initiation of motion of sediment particles. He noted that grains tended to leave the soil surface in a near-vertical direction and speculated that the forces required were generated by the steep velocity gradient near the ground and by spinning of the grains, which produces a pressure gradient between the top and bottom of the grains. More recently, detailed observations have demonstrated (Bisal and Nielsen, 1962; Lyles and Kraus, 1971) that particles vibrate with increasing intensity as wind velocity (and water velocity; see Cheng and Clyde, 1972) increases and then leave the surface instantaneously, as if ejected. Although much more research is required, the observations of Chepil and later workers suggest that turbulent bursting at the air-soil boundary may be an important mechanism for initiation of motion of grains, as suggested by Sutherland (1967) for sediment in flowing water.

Sediment transport by stream flow has been the subject of thousands of publications and has attracted the attention of some of the world's outstanding scientists. Progress in understanding of the complex phenomena involved has generally been by small steps and has achieved successively better approximations to reality so that identification of Benchmark Papers in the field is difficult. The work of one of the leaders in this field, H. A. Einstein, is represented in this volume by Paper 10. We have already noted the major contribution of scientists of the U.S. Department of Agriculture in the field of erosion research. Einstein was also with the Soil Conservation Service at the time he wrote this landmark paper on bedload transport, working on problems of stable channel design for irrigation and drainage schemes. His 1942 paper was further developed (Einstein, 1950) into a technique for computing bed material discharge (including material from the bed carried in suspension), and with various additions and refinements (Colby and Hembree, 1955; Colby and Hubbell, 1961) is currently in wide use by U.S. government agencies for shallow sand bed channels.

Bed materials transport had been studied for almost two centuries before the publication of Einstein's paper; Graf (1971) provides a useful review. In 1753 Brahms had introduced the "sixth-power law",

$$U_{cr} = kW^{1/6}, \tag{1}$$

in which U_{cr} is the critical bottom velocity at which incipient sediment motion occurs, W is the immersed weight of a particle, and k is a constant. Du Buat in 1816 tabulated values of critical velocity for entrainment for a range of sediment sizes, and in 1879 Du Boys extended Du Buat's work on the shear-resistance concept to suggest a quantitative expression for bedload transport,

$$q = \chi \, F \, (F - F_{cr}), \tag{2}$$

in which q is transport rate, F is the force of entrainment under the given flow conditions, F_{cr} is the critical force at the beginning of motion, and χ is a characteristic sediment coefficient. This type of approach, in which initiation of motion or the rate of transport of bed material is related to the extent to which critical values of force (or bed shear stress, velocity, discharge, or steam power) are exceeded, underlies many scour criteria and bed load discharge formulas (Schoklitsch, 1934; Shields, 1936; Kalinske, 1947; Meyer-Peter and Müller, 1948; all reviewed by Graf, 1971, and Yalin, 1972).

Equations based on the notion of a critical condition at which sediment motion commences imply an ability to discern this condition. In fact, it is impossible to discern accurately or predict incipient motion, and Einstien's model of sediment transport is particularly noteworthy because it is based on the notion that the fluid forces acting on the bed materials fluctuate randomly due to the turbulent structure of the flow. Einstein's original application of probability concepts to sediment transport was published in 1937. Even earlier (Einstien, 1934) he had introduced the idea that the energy expended by a river may be apportioned into that which induces sediment movement (approximately 3 percent according to Rubey, 1933) and that which is expended in overcoming resistance to flow and is dissipated as heat. Einstein's model is closer to reality than many others, but even as modified it is by no means universally applicable. For example, his assumption that the average particle step length is constant, so that the time interval between steps is the decisive control on bed load discharge, has been shown (Hung and Shen, 1979) to be incorrect. Furthermore, entrainment is controlled not only by hydraulic conditions but also by antecedent conditions and the texture and structural arrangement of the stream bed (Laronne and Carson, 1976). This drawback applies to all bed load transport equations, of course. Thus, although the most important formulas have been used and been calibrated against Gilbert's (1914) data, they do not give the same predictions for sediment transport under the same conditions (Vanoni, 1975, figs. 2.113 and 2.114). Continuing research is producing further innovation in this field, based upon such concepts as the tendency for channels to minimize their rate of energy expenditure (Yang, 1973) and the dependence of sediment transport rate on stream power (Bagnold, 1977).

Much of the work on sediment transport by flowing water has considered noncohesive sediments, but erosion and transport of fine cohesive sediments have also attracted the attention of many researchers, particularly those concerned with the design of stable channels and canals flowing in a variety of materials. The behavior of fine sediments, those for which surface forces are equal to or greater than body forces, is complex and depends on many factors, including the electrochemical characteristics of the sediment. Much of the work on erosion of cohesive sediments has been concerned with mechanical properties (Dunn, 1959; Flaxman, 1963), but several workers, notably Partheniades and Grissinger, have stressed the importance of mineralogy and chemical factors. The decreased erodibility of surficial materials due to cohesion was first envisaged by Hjulström (1935). Hjulström's curve was later amended by Sundborg, who suggested that the cohesive force is proportional to vane shear strength. However, Grissinger demonstrated the difficulty of using simple parameters such as Atterberg limits as control indexes for erosion studies (Paper 11), while in an earlier study of erosion of a cohesive river bank, Wolman (1957) pointed to the multitude of variables such as antecedent moisture content and occurrence of freeze-thaw that control the erosion of cohesive sediments. There have been a number of experimental studies of the behavior of cohesive sediments in flowing water (reviewed by Partheniades, 1971), but we include here Grissinger's work on the resistance of selected clay systems to erosion by water as a pioneering attempt representative of this research. Once entrained, clay-sized particles are maintained almost indefinitely in suspension in turbulent flow unless flocculation (or coagulation) occurs. Partheniades (1971) has reviewed the processes involved; they are clearly of significance to any consideration of erosion and sediment yield of fine sediment.

REFERENCES

Bagnold, R. A., 1943, *The Physics of Blown Sand and Desert Dunes*, Methuen, London.

Bagnold, R. A., 1977, Bed Load Transport by Natural Rivers, *Water Resources Research* **13**:303–312.

Barton, D. C., 1916, Notes on the Disintegration of Granite in Egypt, *Jour. Geology* **24**:382–393.

Bisal, F., and K. F. Nielsen, 1962, Movement of Soil Particles in Saltation, *Canadian Jour. Soil. Sci.* **42**:81–86.

Blackwelder, E., 1933, The Insolation Hypothesis of Rock Weathering, *Am. Jour. Sci.* **26**:98–113.

Boyle, J. R., G. K. Voigt, and B. L. Sawhney, 1974, Chemical Weathering of Biotite by Organic Acids, *Soil Sci.* **117**:42–45.

Cheng, E. D. H., and C. G. Clyde, 1972, Instantaneous Hydrodynamic Lift and Drag Forces on Large Roughness Elements in Turbulent Open Channel Flow, in *Sedimentation*, ed. H. W. Shen, Water Resources Pub., Fort Collins, pp. 3.1–3.20.

Chepil, W. S., 1945, Dynamics of Wind Erosion, Part I: Nature of Movement of Soil by Wind; Part II: Initiation of Soil Movement; Part III: The Transport Capacity of the Wind, *Soil Sci.* **60**:305–320, 397–411, 475–480.

Chepil, W. S., 1946, Dynamics of Wind Erosion, Part IV: The Translocating and Abrasive Action of the Wind; Part V: Cumulative Intensity of Soil Drifting Across Eroding Fields; Part VI: Sorting of Soil Material by the Wind, *Soil Sci.* **61**:167–178, 257–263, 331–340.

Chepil, W. S., 1953a, Factors That Influence Clod Structure and Erodibility of Soil by Wind, Part I: Soil Texture, *Soil Sci.* **75**:473–483.

Chepil, W. S., 1953b, Factors That Influence Clod Structure and Erodibility of Soil by Wind, Part II: Water Table Structure, *Soil Sci.* **76**:389–399.

Chepil, W. S., 1954, Factors That Influence Clod Structure and Erodibility of Soil by Wind, Part III: Calcium Carbonate and Decomposed Organic Matter, *Soil Sci.* **77**:473–480.

Chepil, W. S., 1955, Factors That Influence Clod Structure and Erodibility of Soil by Wind, Part IV: Sand, Silt and Clay; Part V: Organic Matter at Various Stages of Decomposition, *Soil Sci.* **80**:155–162, 413–421.

Colby, B. R., and C. H. Hembree, 1955, Computations of Total Sediment Discharge, Niobrara River near Cody, Nebraska, *U.S. Geol. Survey Water-Supply Paper 1357*, 187p.

Colby, B. R., and D. W. Hubbell, 1961, Simplified Methods for Computing Total Sediment Discharge with the Modified Einstein Procedure, *U.S. Geol. Survey Water-Supply Paper 1593*, 17p.

Cooke, R. U., 1979, Laboratory Simulation of Salt Weathering Processes in Arid Environments, *Earth Surf. Proc.* **4**:347–360.

Correns, C. W., 1940, Die chemische Verwitterung der Silikats, *Naturwissenschaften* **28**:369–373.

Dunn, I. S., 1959, Traction Resistance of Cohesive Channels, *Am. Soc. Civil Engineers Proc., Jour. Soil Mechanics and Found. Div.* SM3 **85**:1–24.

Einstein, H. A., 1934, Der hydraulische oder Profilradius, *Schweizerische Bauzeitung* **103**:89–91.

Einstein, H. A., 1937, *Der Geschiebetrieb als Wahrscheinlichkeitsproblem,* Mitteilung der Versuchsanstalt fur Wasserbau am der Eidgenossische tech. Hochschule, Zurich, 110p. (Bed Load Transport as a Probability Problem, in *Sedimentation*, ed. H. W. Shen, trans, W. W. Sayre, Water Resources Pub., Fort Collins, 1972, pp. C.1–C.105.)

Einstein, H. A., 1950, The Bed Load Function for Sediment Transportation in Open Channels, *U.S. Dept. Agriculture Tech. Bull. 1026*, 78p.

Flaxman, E. M., 1963, Channel Stability in Undisturbed Cohesive Soils, *Am. Soc. Civil Engineers Proc., Jour. Hydraulics Div.* HY2 **89**:87–96.

Foster, G. R., and L. D. Meyer, 1972, Transport of Soil Particles by Shallow Flow, *Am. Soc. Agricultural Engineers Trans.* **15**:99–102.

Foster, G. R., L. D. Meyer, and C. A. Onstad, 1977, An Erosion Equation

Derived from Basic Erosion Principles, *Am. Soc. Agricultural Engineers Trans.* **20**:678–682.

Free, G. R., 1952, Soil Movement by Raindrops, *Agricultural Engineering* **33**:491–495.

Gilbert, G. K., 1914, The Transportation of Debris by Running Water, *U.S. Geol. Survey Prof. Paper 86,* 263p.

Graf, W. H., 1971, *Hydraulics of Sediment Transport,* McGraw-Hill, New York.

Helgeson, H. C., 1971, Kinetics of Mass Transfer Among Silicates and Aqueous Solutions, *Geochim. et Cosmochim. Acta* **35**:421–469.

Hjulström, F., 1935, Studies of the Morphological Activity of Rivers as Illustrated by the River Fyris, *Geol. Inst. Uppsala Bull.* **25**:221–527.

Hung, C. S., and H. W. Shen, 1979, Statistical Analysis of Sediment Motions on Dunes, *Am. Soc. Civil Engineers Proc. Jour. Hydraulics Div.* **105**: 213–227.

Keller, W. D., 1968, *Principles of Chemical Weathering,* Lucas Bros. Pub., Columbia, Missouri.

Lagache, M., J. Wyart, and G. Sabatier, 1961, Mécanisme de la dissolution des fedspathes alcalins dans l'eau pure on chargée de CO_2 a 200°C, *Acad. Sci. Comptes Rendus* **253**:2296–2299.

Laronne, J. B., and M. A. Carson, 1976, Interrelationships Between Bed Morphology and Bed Material Transport for a Small Gravel Bed Channel, *Sedimentology* **23**:67–85.

Lyles, L., and R. K. Kraus, 1971, Threshold Velocities and Initial Particle Motion as Influenced by Air Turbulence, *Am. Soc. Agricultural Engineers Trans.* **14**:563–566.

Marshall, C. E., and N. Patnaik, 1953, Ionisation of Soil and Soil Colloids: IV. Humic and Hymatomelanic Acids and Their Salts, *Soil Sci.* **75**: 153–165.

Meyer, L. D., and E. J. Monke, 1965, Mechanics of Soil Erosion by Rainfall and Overland Flow, *Am. Soc. Agricultural Engineers Trans.* **8**: 572–577, 580.

Meyer, L. D., G. R. Foster, and S. Nikolov, 1975, Effect of Flow Rate and Canopy on Rill Erosion, *Am. Soc. Agricultural Engineers Trans.* **18**: 905–911.

Ollier, C. D., 1969, *Weathering,* Oliver and Boyd, Edinburgh.

Paces, T., 1973, Steady-State Kinetics and Equilibrium Between Ground Water and Granitic Rock, *Geochim. et Cosmochim. Acta* **37**:2641–2663.

Partheniades, E., 1971, Erosion and Deposition of Cohesive Materials, in *River Mechanics,* ed. H. W. Shen, Shen, Fort Collins, pp.25-1-25-91.

Raudkivi, A. J., 1976, *Loose Boundary Hydraulics,* Pergamon Press, Oxford.

Rubey, W. W., 1933, Equilibrium Conditions in Debris-Laden Streams, *Am. Geophys. Union Trans.* **14**:497–505.

Schatz, A., N. D. Cheronis, V. Schatz, and G. S. Trelawny, 1954, Chelation (Sequestration) as a Biological Weathering Factor in Pedogenesis, *Pennsylvania Acad. Sci. Proc.* **28**:44–51.

Stallings, J. H., 1957, *Soil Conservation,* Prentice-Hall, Englewood Cliffs, New Jersey.

Sutherland, A. J., 1967, Proposed Mechanism for Sediment Entrainment by Turbulent Flows, *Jour. Geophys. Research* **72**:6183–6194.

Tamm, C., 1930, Experimentelle Studien uber die Verwitterung und Tenbildung von Feldspathen, *Chem. Erde* **4**:420–430.

Vanoni, V. A., ed., 1975, *Sedimentation Engineering,* Am. Soc. Civil Engineers, New York.

Wischmeier, W. H., and D. D. Smith, 1965, Predicting Rainfall Erosion Losses from Cropland East of the Rocky Mountains, *U.S. Dept. Agriculture, Agriculture Handb. 28,* pp. 1–47.

Wolman, M. G., 1957, Factors Affecting Erosion of a Cohesive River Bank, *Am. Jour. Sci.* **257**:204–216.

Yalin, M. S., 1972, *Mechanics of Sediment Transport,* Pergamon Press, Oxford, England.

Yang, C. T., 1973, Incipient Motion and Sediment Transport, *Am. Soc. Civil Engineers Proc., Jour Hydraulics Div. HY10* **99**:1676–1704.

7

Kinetics of the alteration of K-feldspar in buffered solutions at low temperature

R. Wollast

Université Libre de Bruxelles, Department of Solid State Chemistry, Belgium

Abstract—A study has been made of the release of Si and Al to solution from the alteration of a potassic feldspar in solutions buffered at pH values between 4 and 10. Release of both Si and Al is consistent with diffusion from an altered layer, presumably formed by rapid initial hydration and exchange of H^+ for K^+. In a limited volume of solution diffusion ceases when the Al concentration in the external solution reaches a fixed value at each pH; this value is reasonably consistent with the solubility of $Al(OH)_3$. The Si concentration tends to reach a maximum at each pH.

The interpretation is made that the maximum in Si concentration corresponds to a balance between Si diffusion into the solution, and Si removed from the solution by reaction with $Al(OH)_3$ to form a hydrated silicate. The calculated equilibrium value for the reaction $Al(OH)_3 + SiO_{2aq} = Al\text{-silicate}$ is 5 ppm SiO_2.

Introduction

In TWO earlier studies of the weathering of feldspars (Wollast, 1961, 1963), emphasis was placed upon determining the conditions controlling the composition of the reaction products and particularly in differentiating the processes that lead to formation of aluminum oxide hydrates from those that lead to kaolinite. As a continuation of this program, the present study is directed toward an analysis of the effects of various experimental conditions on the rate of alteration of potassic feldspar. The experiments were carried out at room temperature and pressure, so that the rates of reaction observed would resemble nature more closely than any preceding investigations at elevated temperatures and pressures.

Correns and von Engelhardt (1938, 1940) demonstrated decomposition of feldspar at low temperature by continually eliminating the ions placed in solution by means of a dialysis cell and ultrafilters. The results show that the various constituents of feldspars go into solution at different rates, leaving a slightly soluble residual layer at the surface. The composition of the layer was shown to be dependent on pH. In all cases the ratio of SiO_2 to Al_2O_3 in the residual layer was greater than that of kaolinite ($>2/1$). Consequently, they concluded that kaolinite could form only by reactions among dissolved substances. The rate of the alteration reaction was controlled by the rate of diffusion of ions through the residual layer. The diffusion rate was found to be very slow, approximately 10^{-19} cm²/sec (10^{-14} cm²/day).

On the other hand, Lagache, Wyart and Sabatier (1961a, 1961b) observed that at 200°C and 5 bars pressure the dissolution rate of feldspar was not controlled by a residual layer. Instead the feldspar dissolved continuously to yield precipitates and dissolved materials; but the reaction rate depended on the concentration of silica and alumina in the solution.

In this work an attempt is made to explain the markedly different behavior found by these two sets of investigators.

EXPERIMENTAL

Procedure

The concentrations of silica and alumina liberated at room temperature from suspensions of finely ground feldspar were determined in solutions buffered at constant pH, in the range pH 4–10.* The suspensions were continuously agitated to keep the solid in suspension and to homogenize the liquid phase. At given intervals, a few cm³ of the suspension were removed, centrifuged, and alumina and silica were determined in the clear liquid by colorimetric analysis (SHAPIRO and BRANNOCK, 1956). The feldspar used was an orthoclase containing about 5% of quartz as an impurity. No analysis was made for the Na/K ratio of the feldspar.

Results

Experimental results are given in Table 1 (a) and (b), which shows silica in solution as functions of time, pH, and wt.% suspended feldspar. Results are also shown graphically: Figs 1 and 2 indicate the release of silica as a function of time for various pH and Fig. 3 the alumina dissolved at pH 4 and 5.

At the higher pH values, the amount of alumina in solution is always less than 1 mg/l. and it is difficult to measure the change of this constituent during a given time interval. Table 2 gives the results obtained in a particular experiment during which concentrations of Al were measured with fair accuracy. Table 3 summarizes the results obtained after 25 days indicating the maximum concentrations realized during the alteration of the feldspar.

These experimental results show that when a potassium feldspar is ground and placed in buffered aqueous solutions between pH values of 4 and 10, the alumina content of the solution quickly reaches a low and constant value in all but strong acid solutions.

The silica content of the solution rises over a much longer period of time, and reaches a maximum that is nearly independent of the weight ratio of feldspar to solution. Nevertheless, it must be emphasized that the *rate* of dissolution is greatly affected by the wt.% of feldspar in suspension as shown by Table 1(b). Finally, the dissolution of feldspar is incongruent whatever the pH value.

Interpretation

It has already been pointed out, in the work of CORRENS, that K^+ is quickly released during alteration of K feldspar. The appearance of K^+ in solution is in

---

* The buffers used had the following compositions:

pH	0·1 M K biphthalate (ml)	0·1 M NaOH (ml)	0·1 M H₃BO₃ (ml)
4	50	0·40	
5	50	23·85	
6	50	45·45	
8	—	3·97	50
9	—	21·30	50
10	—	43·90	50
	(adjusted to 100 ml)		

Table 1(a). Silica concentration (mg/l.) in solution at various pH
(5% solid in suspension)

t (hr)	pH 4 C (mg/l.)	pH 6 C (mg/l.)	pH 8 C (mg/l.)	pH 10 C (mg/l.)
0·25	2·9	0·525	0·55	—
2	7·75	1·55	—	0·05
6	9·1	1·65	0·95	0·78
11	10·0	2·17	1·04	1·03
24	12·4	3·45	3·18	3·44
48	14·2	3·70	3·97	5·50
72	16·6	5·50	4·56	6·75
95	17·7	—	4·86	6·66
144	20·4	7·31	5·38	7·09
168	21·8	7·61	6·09	7·91
200	24·4	9·12	6·18	8·12
320	26·8	10·4	5·99	—
576	31·1	11·5	6·30	—

Table 1(b). Silica concentration (mg/l.) in solution as a function of time and
% suspended solids, at pH 4

Time Wt.% suspension	8 hr C (mg/l.)	24 hr C (mg/l.)	48 hr C (mg/l.)
2	3·15	3·53	4·72
5	6·9	8·3	9·7
10	13·0	15·0	19·1
25	20·1	33·0	36·6
50	38·0	56·0	69·3

Table 2. Alumina concentration in solution at pH 4·2
(10% solid in suspension)

t (hr)	C (mg/l.)
2	4·36
4	7·99
6	9·05
8	11·8
24	16·75
48	21·5

Table 3. Concentrations of silica and alumina in the presence of solid
feldspar after 25 days

pH	SiO_2 (mg/l.)	Al_2O_3 (mg/l.)
4	87·8	139
5	44·3	25·4
6·8	33·1	0·45
8	17·9	0·65
9	15·9	0·10
10	22·5	1·6

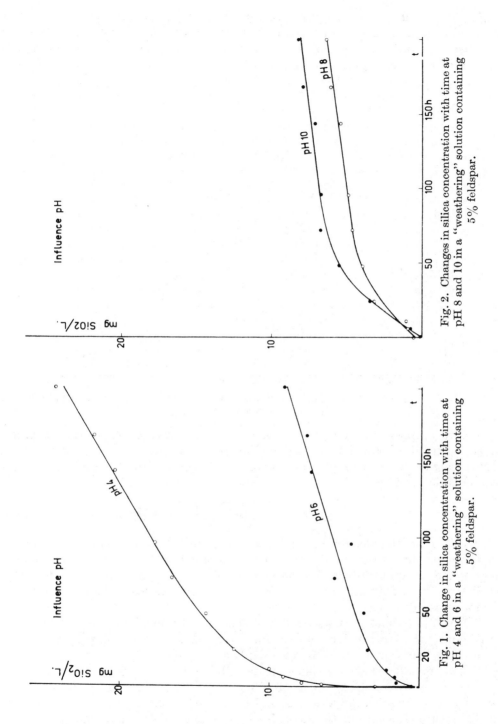

Fig. 1. Change in silica concentration with time at pH 4 and 6 in a "weathering" solution containing 5% feldspar.

Fig. 2. Changes in silica concentration with time at pH 8 and 10 in a "weathering" solution containing 5% feldspar.

fact accompanied by the disappearance of H$^+$ ions: the pH of a suspension of feldspar in distilled water increases during the early stages of the reaction from 6·8 to 9·4. The first reaction step was analyzed by determining the rate of dissolution of silica and alumina for time intervals between 15 min and 2 hr.

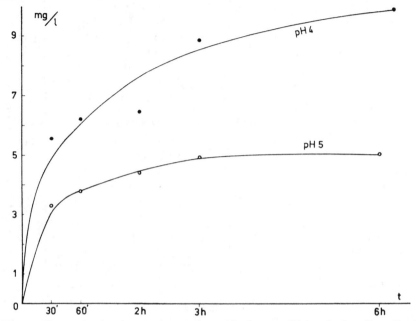

Fig. 3. Change in alumina concentration with time at pH 4 and 5 in a "weathering" solution containing 10% feldspar.

Figure 4 represents the logarithm of silica and alumina concentrations expressed in mg/l., found after 30 min in various buffered solutions. It can be seen that the initial reaction rate is a simple function of pH. In terms of molar concentrations the rate equation can be written

$$v_1 = \frac{d[\text{Al}_2\text{O}_3]_{\text{sol}}}{dt} = \frac{1}{1·75}\frac{d[\text{SiO}_2]_{\text{sol}}}{dt} = k_1[\text{H}^+]^{1/3}$$

This fractional order for H$^+$ and the increase of pH in distilled water indicates that the first step of the reaction is probably exchange of H$^+$ for K$^+$ with decomposition of the structure into a thin surface layer of amorphous Al(OH)$_3$ and SiO$_2$ or H$_4$SiO$_4$. Loss of alumina and silica from this sheath increases the content of these components in the solution.

The solution quickly saturates with respect to alumina. As shown in Fig. 5, the maximum concentration of alumina into solution is in good agreement with the solubility of amorphous Al(OH)$_3$ at low pH and is too high as compared with solubilities of crystallized species like gibbsite or bayerite.

Because of the relatively high solubility of H$_4$SiO$_4$ ($\sim$115 mg/l.), silica continues to diffuse from the sheath but at a diminishing rate because of the increasing distance of diffusion of silica from the fresh feldspar to the solution through the alumina-enriched outer portion of the sheath.

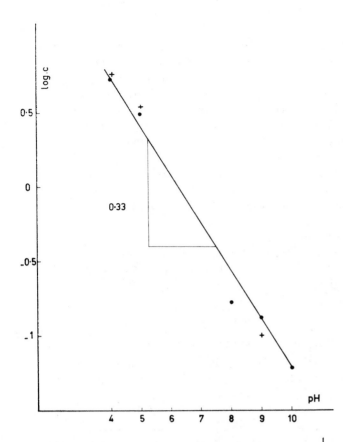

Fig. 4. Logarithm of silica (filled circles) and alumina (crosses) concentrations in mg/l. as a function of pH after 30 min. Solutions contained 5 wt% feldspar.

Fig. 5. Maximum concentration of alumina in solution ($\oplus$) compared with calculated solubilities of amorphous $Al(OH)_3$ (open circles) and gibbsite.

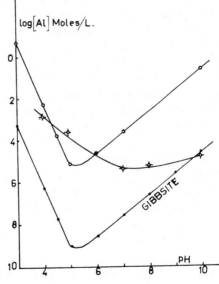

91

Furthermore, SiO_2 reaches a series of maxima at various pH values; these values are not in accord with the solubility of any known species. This leads to the argument that the silica values represent an increase toward 115 ppm by reason of diffusion and a decrease because of some reaction that tends to remove silica. The maximum value of silica is consistent with a rate equilibrium between the two processes. As suggested by R. M. Garrels, the removal reaction may be the formation of an amorphous alumina-silicate, perhaps a hydrated kaolinite-like material $H_4Al_2Si_2O_9 \cdot nH_2O$, by reaction of dissolved silica with the aluminous outer portion of the altered feldspar grains. It will be shown further that the kinetics data are in good agreement with this interpretation.

Let us imagine a grain of feldspar placed in a finite volume of pure water. It is well established that the silica, alumina and alkalies that dissolve are not in the proportions found in feldspar, and that around the grain a residual layer of slightly soluble substances is formed.

The dissolution rate can depend on
the rate of the chemical reaction itself, between the feldspar and the water,
the diffusion rate of the reacting substances through the residual layer.

In the beginning the fresh feldspar is in contact with pure water and the chemical reaction determines the dissolution rate following

$$v_1 = k_1[H^+]^{1/3}$$

The grain of feldspar is then partially altered and surrounded by a residual layer. If the rate of the chemical reaction is much higher than the diffusion rate of the ions through the residual layer, the rate determining step of the weathering reaction is mass transfer through this layer. One may then consider that at the boundary between fresh feldspar and residual layer, the concentrations of dissolved species are controlled by the equilibrium (stable or metastable) of the chemical reaction.

To approach quantitatively the phenomenon of diffusion through the residual layer it is necessary to put down simplifying hypotheses, We assume that the solution is perfectly agitated and that the concentrations are uniform throughout. Let us consider a feldspar grain of radius R partially altered and surrounded by a residual layer of thickness l. If l is much smaller than R, one may consider that contact area Ω of feldspar with the solution does not vary and that diffusion takes place only at right angles to this surface. Also in order to simplify the diffusion equations we will consider that at any time, a quasi-stationary diffusion is realized.

Let us analyse the diffusion of H_4SiO_4 formed during the weathering of feldspar. At time t, the thickness of the residual layer is l_t. The concentration of silicic acid on the surface of the non-altered feldspar is C_s, while in aqueous solution, it is C at time t.

The amount of silicic acid brought into solution during the interval dt is then given by the diffusion equation

$$\frac{dq}{dt} = Dn\Omega \frac{C_s - C}{l} \tag{1}$$

wherein n is the number of particles of feldspar considered. If C_0 represents the amount of silica present initially in each unit of volume of fresh feldspar (moles

SiO_2/cm^3 of feldspar) and q the amount of silica eliminated at time t, one gets

$$q = C_0 n l t \tag{2}$$

Now, if V represents the total volume of solvent and if $C = 0$ at $t = 0$

$$C = \frac{q}{V} \tag{3}$$

Substituting (2) and (3) in (1), the result is

$$\frac{dC}{dt} = \frac{n^2 D \Omega^2}{V^2} C_0 \frac{C_s - C}{l} \tag{4}$$

or

$$\frac{C dC}{(C_s - C)} = n^2 D \frac{\Omega^2}{V^2} C_0 \, dt \tag{5}$$

and integrating

$$-C - C_S \ln \frac{C_S - C}{C_S} = \frac{D n^2 \Omega^2}{V^2} C_0 t \tag{6}$$

When C is much smaller than C_S, the relation becomes, after expanding the series

$$C = \frac{n \Omega}{V} \sqrt{(2 C_0 C_S)} \cdot \sqrt{(Dt)} \tag{7}$$

Equations (6) and (7) only give the amount of silica brought into solution by diffusion from the feldspar. We must now subtract the amount of silica lost from the solution by the formation of an alumino-silicate, to find the actual concentration of silica.

If dissolved silica reacts with amorphous $Al(OH)_3$ on the exterior of feldspar grains, the chemical reaction is

$$H_4SiO_{4 \text{ solution}} + Al(OH)_{3 \text{ amorph}} = \tfrac{1}{2}(H_4Al_2Si_2O_9)_{\text{amorph}} + 5/2 \, H_2O$$

The rate of the reaction might be expected to be proportional to H_4SiO_4 concentration above the equilibrium level (C_e). This concentration represents the maximum solubility of the amorphous-silicate and the reaction takes place only when the concentration of H_4SiO_4 is superior to this value. We have then

$$\frac{dC}{dt} = -k_2(C - C_e) \tag{8}$$

If C is much smaller than C_s (as is borne out by the experimental values except at pH 4), equation (7) holds. After differentiation it becomes

$$\frac{dC}{dt} = k_3 t^{-1/2} \tag{9}$$

Combining equations (8) and (9), one obtains the material volume of silica into solution:

$$\frac{dC}{dt} = k_3 t^{1/2} - k_2(C - C_e)$$

This expression is not easily integrated to obtain an algebraic relation, so that graphical integration was employed to obtain the calculated curves of Figs. 6–9.

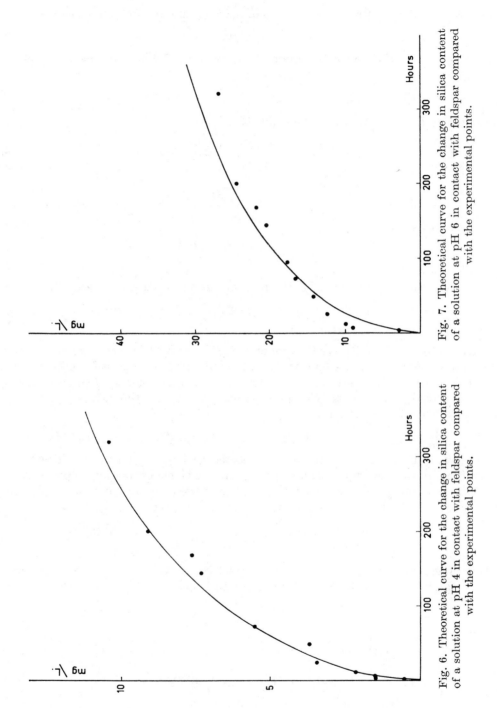

Fig. 6. Theoretical curve for the change in silica content of a solution at pH 4 in contact with feldspar compared with the experimental points.

Fig. 7. Theoretical curve for the change in silica content of a solution at pH 6 in contact with feldspar compared with the experimental points.

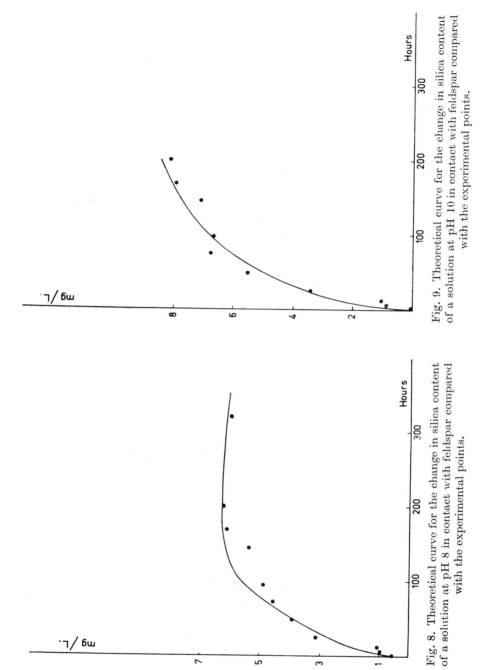

Fig. 9. Theoretical curve for the change in silica content of a solution at pH 10 in contact with feldspar compared with the experimental points.

Fig. 8. Theoretical curve for the change in silica content of a solution at pH 8 in contact with feldspar compared with the experimental points.

Their correspondence with observed relations is good. In fitting the experimental relations, it was found that the best value for C_e is about 5 mg/l. as SiO_2. The silica concentration for equilibrium between crystalline kaolinite and crystalline gibbsite is 1–2 mg/l. (POLZER, 1962) so that we can consider our value as a reasonable value for equilibrium between the amorphous equivalents.

It is possible to determine an approximate specific rate constant/cm^2 of feldspar surface/cm^3 of solution. If the rate is expressed in mg/l-hr the constants are

pH	k_2	k_3
4	$3 \cdot 0 \times 10^{-6}$	$2 \cdot 6 \times 10^{-4}$
6	$3 \cdot 8 \times 10^{-6}$	$6 \cdot 4 \times 10^{-6}$
8	$5 \cdot 6 \times 10^{-5}$	$5 \cdot 6 \times 10^{-5}$
10	$1 \cdot 5 \times 10^{-5}$	$7 \cdot 4 \times 10^{-5}$

These values suggest that at the beginning of the reaction the silica content of the solution is only determined by diffusion from the fresh feldspar especially at low pH. Further, the rate of dissolution decreases rapidly and the second term of the rate equation, i.e. the synthesis of alumino-silicate becomes predominant until the equilibrium concentration C_e is reached.

The order of magnitude of the diffusion coefficient of silicic acid, calculated from the value of k_3 is 10^{-14} cm^2/sec, decreasing from low to high pH.

As shown before (equation (7)) the rate of diffusion of silica is proportional to the surface area $n\Omega/V$; the rate of the counter-reaction also would be expected to be proportional to the surface area, as it takes place on the exterior surface of the feldspar grain. These relations are in accord with the observation that the maximum silica content observed at a given pH is nearly independent of the weight of feldspar (surface area) used in a given volume of solution, since both the addition and removal reaction rates would be changed proportionally.

We studied the influence of the amount of feldspar in a suspension in great detail, varying the weight from 0·1 to 50%.

Let us first consider the experimental results wherein the concentration of silica is negligible, a condition realized for dilute suspensions and, in the case of concentrated suspensions in the beginning of the weathering reaction. In this case, relation (7) shows that the silica concentration in solution varies proportionately to the amount of feldspar in suspension represented by n/V. This relation is verified over a long period of n/V in Fig. 10 where the log of the concentration of silica liberated after 2 hr is represented as a function of the log of n/V. The experimental points fall well along a straight line with a 45° slope.

We also calculated the concentrations of silicic acid in solution after various times, in feldspar suspensions where the wt.% ranged from 1 to 50% of solid. The theoretical concentrations thus obtained, compared with the experimental concentrations are shown in Table 4. Here too, the agreement is satisfactory.

It is interesting to compare the thickness of the residual layer formed around the feldspar grains in these various suspensions. From equations (2) and (3)

$$l_t = \frac{CV}{n\Omega C_0}$$

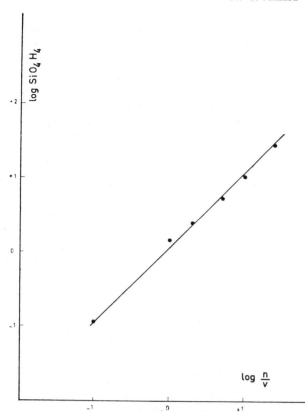

Fig. 10. Log of concentration of silica released after two hours as a function of surface area (n/V).

Table 4. Comparison of the theoretical and experimental concentration of silica (mg/l.) at pH 4 for various suspensions and various times

Time	8 hr		24 hr		48 hr	
% Solid in suspension	C_{calc}	C_{obs}	C_{calc}	C_{obs}	C_{calc}	C_{obs}
2	1·8	3·15	3·1	3·53	4·0	4·72
5	4·5	6·9	7·8	8·30	10·0	9·7
10	8·7	13·0	14·4	15·0	19·3	19·1
25	20·8	20·1	33·1	33·0	40·0	36·6
50	37·2	38·0	56·2	56·0	67·5	69·3

Table 5. Calculated thickness of the residual layer formed after 48 hr in a solution at pH 4

% Solid in suspension	C_t (mg/l.)	l_t (cm)
0·01	0·022	$1·5 \times 10^{-6}$
0·1	0·22	$1·5 \times 10^{-6}$
1	2·40	$1·6 \times 10^{-6}$
10	19·1	$1·45 \times 10^{-6}$
25	36·6	$1·15 \times 10^{-6}$
50	69·3	$0·95 \times 10^{-6}$

97

Table 5 indicates the thickness of the residual layer evaluated from the experimental results obtained by weathering of various suspensions of feldspar in a solution at pH 4, after 48 hr. The thickness of the layer formed is independent of the concentration of solid as long as the ratio C/C_S is inferior to approximately 0·25. For higher values of the ratio, the growing rate of the residual layer is slowed owing to the influence of the substances in solution and the increase of silica into solution is limited by the counter reaction to form silicate.

DISCUSSION

The weathering of feldspar under natural conditions can be described by a diffusion mechanism of H_4SiO_4 through a residual layer, constituted by slightly

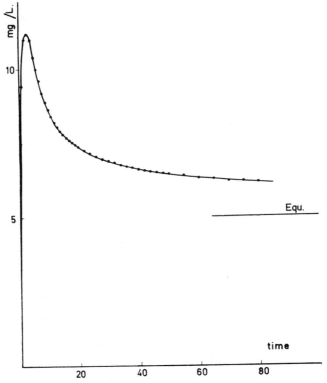

Fig. 11. Theoretical curve for the change in silica content of a solution in contact with feldspar initially saturated with $Al(OH)_3$.

soluble $Al(OH)_3$ and subsequent reaction of these two substances to form a hydrated alumino-silicate. The mechanism proposed shows that there is no contradiction between the results of Correns and Wyart. If the concentration of the substances in the solution bathing the feldspar is weak, the reaction rate is limited by the thickness of the residual layer as postulated by Correns.

The reaction rate is only slowed by the presence of dissolved substances, as in the results of Wyart, when their concentration approaches the saturation concentration or when a counter-reaction can occur.

It is perhaps worthwhile to speculate briefly on the events that should occur after long time intervals. Figure 11 shows a theoretical curve for the change in silica content of a very concentrated suspension of feldspar (75% by weight). This curve would correspond to the alteration of feldspar into kaolinite, with amorphous $Al(OH)_3$ as intermediate product. Continuous kaolinization of the feldspar should take place, but perhaps at a slowly increasing rate as the thickness of the diffusion layer diminishes by the kaolinization reaction and the silica concentration again increases because of the excess of silica in the feldspar over that required to make kaolinite. However further studies are required to determine, for instance, whether the initial alumino-silicate will be converted to a new species higher in silica (montmorillonite?) as H_4SiO_4 in solution increases.

The theoretical and experimental results emphasize the importance of an open system in the bauxitization of feldspar. Only so long as the concentration of silica in the external solution is maintained below about 5 mg/l. can the counter-reaction to form kaolinite be avoided. The natural conditions suggested are good drainage and rains of high intensity and frequency, but of short duration, to avoid waterlogging. Also, the role of pH is somewhat minimized, because of the demonstrated independence of reaction rates.

At pH values less than 5, where the solubility of $Al(OH)_3$ is high, no protective sheath is formed, because silica and alumina dissolve at very nearly the same rate in solutions undersaturated with both. Under such conditions, the feldspar dissolves, and formation of kaolinite would be through the uninvestigated reactions in homogeneous media:

$$H_2O + 2Al^{3+} + 2H_4SiO_4 = H_4Al_2Si_2O_9 + (6H^+)$$

or

$$2H^+ + 2AlO_2^- + 2H_4SiO_4 = H_4Al_2Si_2O_9 + 3H_2O$$

However, the present study shows that these reactions must be slow relative to the reaction between H_4SiO_4 in solution and "amorphous" $Al(OH)_3$ on grain surfaces.

Acknowledgment—The author wishes to thank the CEMUBAC, which sponsored this study, and Prof. W. L. DE KEYSER who directed his activities.

The author expresses his sincere thanks to Prof. R. M. GARRELS for his interest in this study, and for his suggestion concerning the formation of an alumino-silicate at the feldspar surface.

REFERENCES

CORRENS C. W. and VON ENGELHARDT W. (1938) Neue untersuchungen über die Verwitterung des Kalifeldspates. *Chem. d. Erde* **12,** 1.

CORRENS C. W. and VON ENGELHARDT W. (1940) Die chemische Verwitterung der Silicate. *Naturwissen.* **28,** 369.

LAGACHE M., WYART J. and SABATIER G. (1961a) Dissolution des feldspaths alcalins dans l'eau pure ou chargee de CO_2 a 200°C. *C.R. Acad. Sci. Paris* **253,** 2019.

LAGACHE M., WYART J. and SABATIER G. (1961b) Mecanisme de la dissolution des feldspaths alcalins dans l'eau pure ou chargee de CO_2 a 200°C. *C.R. Acad. Sci. Paris* **253,** 2296.

POLZER W. (1962) Personal communication.

SHAPIRO L. and BRANNOCK W. W. (1956) Rapid analysis of silicate rocks. *U.S. Geol. Survey Bull.* 1036 C.

WOLLAST R. (1961) Aspect chimique du mode de formation des bauxites dans le Bas-Congo. *Bull. Acad. Roy. Sci. d'Outre-Mer (Belgique)* **7,** 468.

WOLLAST R. (1963) Aspect chimique du mode de formation des bauxites dans le Bas-Congo, II. *Bull. Acad. Roy. Sci. d'Outre-Mer (Belgique)* **9,** 392.

8

Copyright © 1944 by the American Society of Agricultural Engineers
Reprinted from pages 131–136 and 181–182 of *Agricultural Engineering*
25:131–182 (1944)

Studies of Raindrop Erosion

By W. D. Ellison

R ECENT experimental studies at Coshocton, Ohio, have shown that in addition to the soil moved downslope by flowing surface water, some soil is carried downhill by the splash from raindrops. When falling raindrops strike either the soil surface or a thin film of water covering the surface, soil and water particles are splashed into the air. If the raindrops fall vertically onto the surface of a hillside, they tend to strike glancing blows and the splashed soil is mostly thrown downhill. Another cause for the splash moving soil toward the base of the hill is that the particles moving downslope travel greater horizontal distances than those directed upslope.

Other actions in the raindrop erosion process, and of prime importance in problems of soil and water relations, have been found to include breaking down soil aggregates and muddying surface waters before infiltration takes place. The pore openings in the soil surface, partially reduced by the initial breakdown of the aggregates, are still further closed by deposits from the infiltration which contains small fragments of the aggregates and some silt and clay particles.

A broad grouping of the factors affecting raindrop erosion processes would include (1) variables of rainfall, (2) slope of the land, (3) soil characteristics, and (4) protection of the soil against, or its exposure to, raindrop impact. The experiments reported in this paper were designed for studying relationships of rainfall intensity, drop size and drop velocity to raindrop erosion processes. Only one soil type, one slope, and a bare, loose, air-dry soil were used for all tests.

In some of the exploratory experiments used for observing splash phenomena, soil was placed in small cans so as to reduce the runoff and to assist in isolating the factor of raindrop erosion. In other experiments small plots were used, but only one soil (Muskingum silt loam with No. 4 erosion) was tested.

To measure the soil transported by raindrop splash, some new devices were needed. These were developed during the course of the investigations and they are described in another article[4]*.

With the rainfall applicator used[5], the raindrop velocity, drop size, and rainfall intensity could be varied independently. While controlling these factors the effects of changes in each variable on raindrop erosion were studied. For these studies, velocities of raindrops were limited to a maximum of 19.2 fps (feet per second) and to a minimum of 12.0 fps. Drop sizes of 3.5 mm and 5.1 mm were applied and rainfall intensities included rates of 4.8, 6.6, 8.1, and 14.8 iph (inches per hour).

Most of the measurements and samples taken during this experimental work pertained to raindrop splash, but in order to study some of the effects of raindrop impact and the associated splash on soils not found in the splash samples, soils contained in the surface flow on the plot as well as that in the runoff at the bottom of the plot were tested. In one series of experiments samples were taken from the soil surface following rainfall. Analyses of samples included quantity determinations as well as determinations of aggregate sizes, and soil-particle size distributions.

Since these investigations represent a type of erosion study different in both technique and objective from that of measuring soil loss in the runoff waters, a clarification of terms becomes desirable. For this clarification the following terms are described:

[1] *Loss of soil through surface runoff.* This is the soil carried off an area by the runoff waters. The amounts carried are largely controlled by the concentration of flow, the density of channels, the hydraulic efficiency of channels, the gullying which takes place, and by the erosional activity over the entire surface of the land.

2 *Loss of soil through raindrop splash.* This is the soil carried off a given area by raindrop splash. Major meteorological factors affecting this loss are sizes of raindrops, intensity of rainfall, raindrop velocity, and wind direction and velocity.

3 *Erosional activity.* This is a process of soil movement and it refers to displacement of soil particles from initial positions regardless of whether or not they are carried off the field or watershed. The displacement may be caused by either raindrops or by flowing surface water. Through erosional activity thin soils can be made thinner and thick soils can be made thicker; top soils may be buried under subsoils, either in valley bottoms or on hillsides; soil structure may be destroyed through breaking down aggregates, and this breakdown may increase rates of water and soil losses.

4 *Erosional damage.* This refers to damages resulting from both soil loss and erosional activity. On some watersheds the soil lost in runoff waters represents only a small part of all the erosional damages. Other damages are caused by actions described as occurring under erosional activity; for example, such as breaking down soil structure, depleting soils of their fertility elements, making thin soils thinner, thick soils thicker, and the burying of topsoils under subsoils.

There are no reports showing that the soils transported by raindrop splash have been measured or analyzed by others. Inadequacy of equipment has delayed studies of splash phenomena, and many of the actions involved have been difficult to detect and identify with the specific results they produce. The process of erosion by splash is usually not apparent to the eye because of high velocities of particle movement, and also because of unfavorable conditions of light and background. So long as the process is not recognized, the results it produces are usually attributed to surface flow, which, like the splash, removes soils from the hilltops and deposits them at points farther down slope.

Although previous measurements have not been made of splash erosion, investigators have long recognized that raindrops striking the soil surface cause erosion. As early as 1877, Wollny described the effect of a beating rain breaking down the soil, in washing fine particles into the tiny crevices and pores, in sealing the soil surface and thereby decreasing porosity, and that of a cover in decreasing just such effects[11]. Research since that time has tended to confirm Wollny's observations and to treat aspects of the problem which he did not stress.

Lowdermilk, in studying the effects of forest litter, concluded that the formation of a fine-textured layer at the surface of a bare soil by the filtering of suspended particles from percolating muddy water was the decisive factor in increasing the surficial runoff from bare surfaces[9]. Other experiments have helped confirm the facts leading to this conclusion[6]. Duley and Kelly described and photographed the formation of a compact surface layer which greatly reduces the infiltration rate and showed the effect of mulch in preventing its formation[3]. They regarded the surface conditions as having a larger influence on the infiltration capacity than soil type, initial moisture content, and rainfall intensity combined.

Borst and Woodburn[1] working with plots 6x29 ft compared the soil losses resulting from the application of artificial rainfall to a bare soil, to a soil covered by a mulch and to a soil having a mulch supported an inch above its surface, and concluded that the elimination of raindrop impact with its destructive effect on the soil surface, rather than the reduction of overland flow velocity, appeared to be the major contribution of the mulch in reducing soil loss.

Laws discovered that the rate of infiltration after 3/4 hr of rainfall decreased with an increase in the kinetic energy of the drops falling per unit area. He also found "that the erosional losses, which were measured in terms of the concentration of the soil in the runoff water, increased by as much as 1200 per cent", as the drop sizes were increased[8].

The physical properties of raindrops have been mentioned or stressed in other experiments. Cook believed that raindrop velocity was one of the variables in the water erosion process[2]. Neal and Baver described a device for measuring drop impact, mentioning the momentum per unit area as a property of rainfall[10]. Horton in considering the effects of rainfall intensity and drop size on infiltration regarded energy per inch of rain as the important property[7].

This paper was prepared expressly for AGRICULTURAL ENGINEERING.
W. D. ELLISON is project supervisor, North Appalachian Experimental Watershed, Soil Conservation Service, U. S. Department of Agriculture.
*Superscript numbers indicate the references appended to this paper.

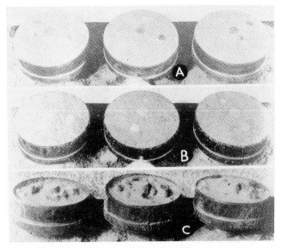

Fig. 1 (A) Soil samples prepared with coins on surface, before rainfall was applied. (B) Samples 45 sec after beginning of rainfall. (C) Samples 1 hr and 15 min after beginning of rainfall

EXPLORATORY EXPERIMENTS FOR OBSERVING RAINDROP EROSION

Exploratory experiments were carried on for purposes of observing splash phenomena and identifying specific erosion processes involved. While this part of the experimental work was in progress, some of the equipment was developed and final plans for the studies were largely completed.

For the first of these experiments, the large stones were removed from an air-dry soil by passing it through a screen with four openings per square inch. The soil was then placed in the tops of standard rain gages, as shown in Fig. 1, and artificial rainfall was applied. Intensity of application was 6.6 iph, drop size 5.1 mm, and drop velocity 19.2 fps. A in Fig. 1 shows three soil samples with coins placed on the surface, before rainfall was applied; B shows these same samples 45 sec after rainfall started (note soil splashed on the white backboard), and C shows these samples 1 hr and 15 min later (large quantities of soil splashed by the raindrops can be seen on the backboard). Since the rainfall produced very little surface flow, the greater part of the erosional loss shown in C may be ascribed to raindrop splash.

Fig. 2 shows a plot severely eroded by raindrop action. These pictures of the soil surface were taken after 5 hr of rainfall, applied at the same intensity, drop size, and drop velocity as described in the preceding paragraph. None of the stones had been screened out of this soil, and after small amounts of erosion had occurred many stones were exposed on the surface and these tended to retard the erosional action of the raindrops. These surface stones were especially effective during the latter part of the experiment.

A in Fig. 2 shows the upper end of the plot and here more soil splashed down slope than was splashed back up to replace the loss. Also, there were some additional losses through splash which carried soil materials out over the plot borders.

B in Fig. 2 includes the entire plot of which a section is shown in A. In this picture erosional deposit can be seen near the base of the slope and particularly on the lower third of the plot length.

To continue with the exploratory observations of raindrop splash, a plot was prepared in a darkened room and a mirror was used to reflect a sunbeam across the soil surface so as to obtain a photograph of the splash while applying artificial rainfall (Figs. 3, 4, and 5). The photograph of Fig. 3 was made with an exposure of 1/25 sec, Fig. 4 is an enlargement of a portion of Fig. 3, and Fig. 5 is an enlargement from a picture taken with 1/100-sec exposure. The vertical streaks in these photographs indicate paths of falling raindrops, while the parabolic curves in Figs. 3 and 4, and the short lines which tend toward the horizontal in Fig. 5, indicate paths of soil and water particles contained in the raindrop splash. In these experiments, where the raindrop splash

was photographed, the rainfall intensity, drop size, and drop velocity were 6.6 iph, 5.1 mm, and 19.2 fps, respectively.

Where paths of soil and water particles describe parabolic curves such as are seen in Figs. 3, 4, and 5, it would be expected that maximum distances of particle travel would be about four times the height of rise. An experiment was set up to check these distances and to observe how far the different particle sizes of soil would be carried by the splash. A sample of soil was prepared as shown in Fig. 1 and set on a white sheet, and artificial rainfall was applied. When using a drop size of 5.1 mm and a drop velocity of 18 fps, the maximum distance of splash was found to be 5 ft. Some stone fragments as large as 4 mm were splashed 8 in and soil aggregates and particles of 2 mm were carried as far as 16 in.

In a second experiment, drop size was reduced to 3.5 mm, and this change reduced the drop velocity to about 17 fps. Under this rainfall the maximum distance particles were splashed was about 3.5 ft, and some soil aggregates and stone fragments of 2 mm were carried as far as 8 in.

It was observed that raindrop impact, under certain conditions, would move stones as large as 10-mm diameter when they were partially or wholly submerged in water. When raindrops would strike these submerged stones, the stones would rise and frequently they would move some distance down slope. Where there was surface flow, this would assist the downhill motion even though the surface flow acting alone would not move them.

Following an exploratory test of 1½ hr duration, with rainfall at 6.6 iph, drop size 3.5 mm, and drop velocity at 18 fps, three 10-lb. samples of soil were scraped from the surface cutting to about ½-in depth. One sample was taken from the upper end of the slope, one from the center of the slope, and one from the lower end of the slope. All clods and large aggregates were broken down using a rubber pestle in a mortar, and the samples were then screened, first through the 8-mm screen and then through the 2-mm screen. Results of these screening tests seemed to confirm previous observations by showing greater percentages of large stone on lower portions of the plot, thus indicating a downhill movement of stone fragments as large as 8-mm diameter. The results are summarized as follows:

Sample from top of slope		Sample from center of slope		Sample from bottom of slope	
81.5%	< 2 mm	74.0%	< 2 mm	70.2%	< 2 mm
14.0%	2 - 8 mm	20.5%	2 - 8 mm	21.5%	2 - 8 mm
4.5%	> 8 mm	5.5%	> 8 mm	8.3%	> 8 mm

Fig. 2 (A) This shows upper portion of plot 5 hr after beginning of rainfall; note stones supported by soil columns as in (C) Fig. 1. (B) This shows entire plot of which a section is shown in (A). There is some soil deposit on extreme lower portion of the slope

CONTROLLED EXPERIMENTS FOR STUDYING VARIABLES

The exploratory experiments previously described indicated that considerable amounts of soil materials were carried by raindrop splash, and following their completion plans were made for analyzing samples of this splash to determine physical properties of the soils, the amounts carried, and relationships of these factors to the intensity, size and velocity of raindrops. In some of this work, in addition to studying the raindrop splash, both the soil carried by the surface runoff and that remaining on the surface following rainfall were tested to determine the disposition of the different particle and aggregate sizes contained in the original soils. Experiments were of 30-min duration and different values of raindrop velocity, drop size, and rainfall intensity were used.

The plot was that shown equipped with splash samplers in Fig. 6. A flow sampler[4] can be seen on the lower end of the plot, at the left. Splash samplers were set so that the lip would be ½ in above the soil, and splash plates were set perpendicular to the soil surface. The plot was prepared by spreading approximately 1½ in of top soil over fine gravel about 14 in deep. The top soil was air-dried and all stones and clods larger than ½-in diameter were removed by screening. Six-inch tile were laid in the gravel and good ·outlets were provided to afford drainage which would prevent backing up of subsurface water into the surface soils. The plot was 5 ft long by 6 ft wide with a slope of 10 per cent. Fresh soils were spread over the surface for each separate experiment.

Soil Particle Sizes in Raindrop Splash. Soil particle sizes contained in the samples of raindrop splash, as well as those contained in the original soils, are shown in Table 1. Each value tabulated represents an average of three determinations, except for those marked with an asterisk, and these represent averages of two determinations each. Averages of all values included in Table 1 indicate that the samples of splash contained a greater percentage of sand and gravel than did the original soils. This increase was from about 25.3 per cent for the original soils to 31.2 per cent for the samples of raindrop splash.

Soil Particle Sizes in Runoff. After it was found that the samples of splash contained greater percentages of sand and gravel than were contained in the original soils, samples of the surface flow were collected from the plot, shown in Fig. 6, to determine how the percentages of sand and gravel carried by the flow compared with those of the original soils. A flow sampler was used to collect these samples. One sample was taken from near each

corner of the plot, one from the bottom, and one from the top of the slope about half way from side to side. These six small samples were combined into one large sample before making the laboratory determinations.

TABLE 1. MECHANICAL ANALYSES† OF SOIL MATERIALS CONTAINED IN RAINDROP SPLASH
ERODED MUSKINGUM SILT LOAM

Gravel, 2 to 1 mm Per cent	Sand, 1 to 0.5 mm Per cent	Sand, 0.5 to 0.25 mm Per cent	Sand, 0.25 to 0.125 mm Per cent	Sand, 0.125 to 0.05 mm Per cent	Total sand and gravel, 2 to 0.05 mm Per cent	Total silt and clay <0.05 mm Per cent
Raindrop size, 5.1 mm						
Raindrop velocity, 19.2 fps; height of raindrop fall, 6.7 ft						
3.83	2.58	2.90	3.84	17.28	30.44	69.56
2.34*	3.27*	2.53*	3.70*	21.57*	33.41*	66.59*
Raindrop velocity, 15 fps; height of raindrop fall, 3.9 ft						
1.87	2.87	3.37	4.27	18.01	30.39	69.61
1.52*	2.97*	2.70*	3.82*	22.92*	33.93*	66.07*
Raindrop velocity, 12 fps; height of raindrop fall, 2.3 ft						
1.91	2.97	3.45	4.26	17.06	29.64	70.36
1.20*	2.04*	2.45*	3.88*	26.06*	35.63*	64.37*
Raindrop size, 3.1 mm						
Raindrop velocity, 18.0 fps; height of raindrop fall, 6.7 ft						
1.07	2.18	3.43	3.60	17.00	27.83	72.17
Raindrop velocity, 14.5 fps; height of raindrop fall, 3.9 ft						
0.62	2.18	3.60	4.73	17.75	28.88	71.12
Raindrop velocity, 12 fps; height of raindrop fall, 2.3 ft						
1.00	2.50	4.13	5.11	18.35	31.09	68.91
Analyses of original soil						
0.55	1.81	2.27	3.49	17.16	25.28	74.72

†Values indicated by asterisk (*) are averages of two experiments; all others are averages of three.

Particle-size determinations made from these samples indicated that soil materials in the surface flow were comprised of only about 5 per cent sand and gravel and 95 per cent silt and clay. Results of these determinations are summarized as follows:

Particle Sizes in Runoff		Per Cent of Sample
2.0	- 1.0 mm	0.25
1.0	- 0.5	0.36
0.5	- 0.25	0.49
0.25	- 0.125	0.67
0.125	- 0.05	3.20

Total sand and gravel, 4.97 % > 0.05 mm
Total silt and clay, 95.03% <0.05 mm

Soil Aggregates in Raindrop Splash. Soils collected in the splash samplers were analyzed to determine aggregate size distribution. Results of the analyses are shown in Table 2. These data indicate that the samples of splash contained a greater percentage of aggregates smaller than 0.105 mm than did the original soils. This increase was from about 53.5 per cent for the original soils to 72.2 per cent for the samples of raindrop splash. Such increases could be caused by either (1) breaking down of large aggregates under the impact of raindrops, (2) the large aggregates being carried away by surface flow, or (3) the large aggregates being left on the surface of the plot and not carried by either raindrops or surface flow. In some of the experiments that follow, checks were made to determine the disposition of the large aggregates and the possible causes for the increase in small aggregates contained in the raindrop splash.

Soil Aggregates in the Surface Flow. For aggregate analyses of soils contained in the surface flow, samples were collected in the same way they were for studies of soil-particle sizes. The results of the laboratory determinations are shown in Table 3, and these (like the splash) indicate an increase in the percentage of aggregates smaller than 0.105 mm, over and above the percentage found in the original soils. This increase was from about 53.5 to 90.2 per cent. These results indicated that large aggregates not

Fig. 3 (Upper left) This photograph was made by directing a sunbeam onto the plot surface during artificial rainfall. Rainfall intensity, 6.6 iph; drop size, 5.1 mm; drop velocity, 19.2 fps; exposure, 1/25 sec. Vertical marks indicate paths of falling raindrops; parabolic curves indicate trajectories of soil and water particles which splash from the soil surface as part of the reaction to the impact of the falling drops ● Fig. 4 (Right) This is an enlargement of a section of Fig. 3 ● Fig. 5 (Lower left) This is an enlargement of a section of a photograph made in the same manner as Fig. 3, except exposure was 1/100 sec. Long vertical lines indicate falling raindrops, and those lines not vertical indicate particles of soil and water contained in raindrop splash

found in the raindrop splash were not being carried away by the surface flow, and in the following experiments the surface soils were tested to determine if they were left on the plot.

TABLE 2. AGGREGATE* ANALYSES OF SOIL MATERIALS CONTAINED IN RAINDROP SPLASH
ERODED MUSKINGUM SILT LOAM

> 2 mm Per cent	2 to 1 mm Per cent	1 to 0.5 mm Per cent	0.5 to 0.25 mm Per cent	0.25 to 0.105 mm Per cent	< 0.105 mm Per cent
Raindrop size, 5.1 mm					
Raindrop velocity 19.2 fps; height of raindrop fall, 6.7 ft					
2.40	3.89	5.07	5.24	11.38	71.98
Raindrop velocity, 15 fps; height of raindrop fall, 3.9 ft					
1.41	3.68	4.98	5.12	11.53	73.25
Raindrop velocity, 12 fps; height of raindrop fall, 2.3 ft					
1.04	2.87	5.74	6.56	12.11	71.65
Raindrop size, 3.5 mm					
Raindrop velocity, 18.0 fps; height of raindrop fall, 6.7 ft					
0.74	2.87	5.24	6.65	10.83	73.66
Raindrop velocity, 14.5 fps; height of raindrop fall, 3.9 ft					
0.94	2.81	5.66	5.85	10.81	73.91
Raindrop velocity, 12 fps; height of raindrop fall, 2.3 ft					
0.44	2.31	6.25	8.22	13.86	68.89
Analyses of samples of original soils					
13.34	6.33	6.69	8.01	12.05	53.55

*Aggregates include stone fragments.

Soil Aggregates on the Surface Following Rainfall. Following application of rainfall and after runoff had ceased, soil was scraped from the plot surface and determinations were made of aggregate-size distribution. A safety razor blade was used for scraping up these samples from the soil surface, and the depth of scraping did not exceed 1/16 in. Three samples were taken at the bottom, three at the middle, and three at the top of the slope. Each sample was taken on about the quarter point, across the width of the plot.

TABLE 3. AGGREGATE* ANALYSES OF SOIL MATERIALS CONTAINED IN THE RUNOFF†
ERODED MUSKINGUM SILT LOAM

> 2 mm Per cent	2 to 1 mm Per cent	1 to 0.5 mm Per cent	0.5 to 0.25 mm Per cent	0.25 to 0.105 mm Per cent	< 0.105 mm Per cent
Raindrop size, 5.1 mm					
Raindrop velocity, 19.2 fps; height of raindrop fall, 6.7 ft					
0.40	1.68	2.39	2.85	5.27	87.26
Raindrop velocity, 15 fps; height of raindrop fall, 3.9 ft					
0.41	0.91	1.71	2.25	6.25	88.45
Raindrop velocity, 12 fps; height of raindrop fall, 2.3 ft					
0.61	0.56	1.27	1.55	3.95	92.04
Raindrop size, 3.5 mm					
Raindrop velocity, 18 fps; height of raindrop fall, 6.7 ft					
0.47	1.13	1.99	2.14	6.03	88.21
Raindrop velocity, 14.5 fps; height of raindrop fall, 3.9 ft					
0.08	0.66	1.35	1.91	3.02	92.88
Raindrop velocity, 12 fps; height of raindrop fall, 2.3 ft					
0.23	0.80	1.47	1.41	3.64	92.43

*Aggregates include stone fragments.
†Analyses of original soil shown in Table 2.

Aggregate sizes are summarized in Table 4. These samples, like the samples of splash and surface flow, contained higher percentages of aggregates smaller than 0.105 mm than did the original soils. The increase was from about 53.5 to 57.1 per cent.

Upon completion of these aggregate studies it was concluded that some of the larger soil aggregates were broken down by the

TABLE 4. AGGREGATE* ANALYSES OF SOIL MATERIALS SCRAPED FROM THE SURFACE OF THE PLOT FOLLOWING RAINFALL†

> 2 mm Per cent	2 to 1 mm Per cent	1 to 0.5 mm Per cent	0.5 to 0.25 mm Per cent	0.25 to 0.105 mm Per cent	< 0.105 mm Per cent
Raindrop size, 5.1 mm					
Raindrop velocity, 19.2 fps; height of raindrop fall, 6.7 ft					
19.85	4.38	4.47	4.09	12.62	54.56
Raindrop velocity, 15 fps; height of raindrop fall, 3.9 ft					
22.13	5.52	4.33	3.47	7.89	56.62
Raindrop velocity, 12 fps; height of raindrop fall, 2.3 ft					
18.91	4.53	4.62	3.92	7.61	60.37
Raindrop size, 3.5 mm					
Raindrop velocity, 18 fps; height of raindrop fall, 6.7 ft					
18.32	4.77	4.32	4.17	10.23	58.16
Raindrop velocity, 14.5 fps; height of raindrop fall, 3.9					
19.45	5.11	4.63	3.89	10.07	56.36
Raindrop velocity, 12 fps; height of raindrop fall, 2.3 ft					
25.29	4.15	3.39	3.29	7.12	56.73

*Aggregates include stone fragments.
†Analyses of original soil shown in Table 2.

Fig. 6 Splash samplers installed in preparation for an experiment. A surface flow sampler can be seen at the lower end of the plot

rainfall during the experiment. This conclusion was based on aggregate analyses of the raindrop splash, the surface flow, and the surface soils following rainfall, all showing higher percentages of small aggregates than did the original soils.

Quantities of Soil in Samples of Raindrop Splash. Fifty-nine experiments were run for purposes of studying quantities of soil contained in the samples of raindrop splash. For these tests, splash samplers were installed on the plot as shown in Fig. 6.

Values of E shown in Table 5 represent the total number of grams of soil caught in the 8 splash samplers during experiments of 30-min duration. These results indicate that quantities of soil in the samples of raindrop splash were increased each time either drop size, drop velocity, or rainfall intensity was increased.

Fig. 6 shows that two of the samplers located midway between the top and bottom of the slope are installed with the splash plates set perpendicular to the length of the plot. Soils caught in upslope sides of the samplers represent materials splashed downhill, while the catch in the downslope sides represents materials splashed uphill. Splash plates were set perpendicular to the soil surface, not vertical.

TABLE 5. QUANTITIES OF SOIL CONTAINED IN RAINDROP SPLASH USING DIFFERENT VALUES OF d, I, AND V

Drop size, 3.5 mm						Drop size, 5.1 mm				
Exp. No.	Distance of drop fall, ft	V, fps	I, iph	E, g		Exp. No.	Distance of drop fall, ft	V, fps	I, iph	E, g
A	6.7	18.0	4.8	267		M	6.7	19.2	4.8	308
A1	6.7	18.0	4.8	176		M1	6.7	19.2	4.8	526
A2	6.7	18.0	4.8	227		M2	6.7	19.2	4.8	504
			Avg.	223					Avg.	446
B	6.7	18.0	6.6	283		N	6.7	19.2	6.6	597
B1	6.7	18.0	6.6	173		N1	6.7	19.2	6.6	505
B2	6.7	18.0	6.6	281		N2	6.7	19.2	6.6	528
			Avg.	245					Avg.	543
C	6.7	18.0	8.1	368		P	6.7	19.2	8.1	690
D	6.7	18.0	14.8	440		Q	6.7	19.2	14.8	679
D1	6.7	18.0	14.8	614		Q1	6.7	19.2	14.8	789
D2	6.7	18.0	14.8	424		Q2	6.7	19.2	14.8	890
			Avg.	492					Avg.	786
E	3.9	14.5	4.8	81.5		R	3.9	15.0	4.8	210
E1	3.9	14.5	4.8	51.9		R1	3.9	15.0	4.8	209
E2	3.9	14.5	4.8	68.0		R2	3.9	15.0	4.8	189
			Avg.	67.1					Avg.	203
F	3.9	14.5	6.6	116		S	3.9	15.0	6.6	242
F1	3.9	14.5	6.6	71.8		S1	3.9	15.0	6.6	171
F2	3.9	14.5	6.6	101		S2	3.9	15.0	6.6	286
			Avg.	96.3					Avg.	233
G	3.9	14.5	8.1	138		T	3.9	15.0	8.1	295
H	3.9	14.5	14.8	206		U	3.9	15.0	14.8	307
H1	3.9	14.5	14.8	190		U1	3.9	15.0	14.8	352
H2	3.9	14.5	14.8	300						
			Avg.	232					Avg.	329
I	2.3	12.0	4.8	17.0		V	2.3	12.0	4.8	51.3
I1	2.3	12.0	4.8	13.6		V1	2.3	12.0	4.8	28.0
						V2	2.3	12.0	4.8	27.8
			Avg.	15.3					Avg.	35.7
J	2.3	12.0	6.6	26.9		W	2.3	12.0	6.6	97.3
J1	2.3	12.0	6.6	11.2		W1	2.3	12.0	6.6	35.1
J2	2.3	12.0	6.6	25.7		W2	2.3	12.0	6.6	52.8
J3	2.3	12.0	6.6	18.4						
			Avg.	20.5					Avg.	61.7
K	2.3	12.0	8.1	33.2		X	2.3	12.0	8.1	67.3
L	2.3	12.0	14.8	49.5		Y	2.3	12.0	14.8	168
L1	2.3	12.0	14.8	49.0		Y1	2.3	12.0	14.8	133
L2	2.3	12.0	14.8	45.0		Y2	2.3	12.0	14.8	170
			Avg.	47.8					Avg.	157

Tables 6, 7, and 8 have been prepared for making comparisons of the relationships between quantities of soils caught in the uphill sides of the samplers and designated by "EU", and those caught in the downhill sides and designated by "ED", when using different drop sizes, rainfall intensities, and drop velocities.

The average ratio of EU to ED, for the 59 experiments included in Tables 6, 7, and 8, is 3.1, and this would indicate that considerably more soil is splashed downslope than is splashed back upslope. Part of this greater downhill movement can be ascribed to the fact that those particles moving downhill travel greater horizontal distances than those moving upslope. Another consideration is the angle at which raindrops strike the surface. The still photographs (Figs. 3, 4 and 5) do not show the greater part of the splash moving downhill, but movies made at the time these photographs were taken show that raindrops falling on this 10 per cent slope tend to strike glancing blows and that most of the splash moves out in downhill directions.

Soils in Raindrop Splash Referenced to Time. In this phase of the work both quantities and particle sizes of soils contained in raindrop splash were referenced to time.

In Fig. 7 the quantities of soil contained in raindrop splash are plotted against time for a 30-min period of rainfall, and rainfall characteristics are indicated in the figure. The samples of splash used in this work included those intercepted during one-minute periods for the first five minutes of rainfall, and those intercepted during 5-min periods throughout the next 25 min. Note that all rainfall intensities, drop sizes, and velocities used produced maximum rates of soil splash at between two and three minutes after beginning of rainfall. As in all previous experiments, soils were air-dry at the beginning of the tests.

Observations made during the course of these experiments indicated that the soil surface was mostly sealed against infiltration at the time maximum rates of soil splash were measured. Surface conditions on the plot, during the time of these experiments, were undergoing changes which seemed to exert considerable influence on results obtained.

TABLE 6. QUANTITIES OF SOIL CAUGHT IN UPSLOPE AND DOWNSLOPE SIDES OF SPLASH SAMPLERS, EU AND ED, RESPECTIVELY, WHILE USING DIFFERENT DROP SIZES

Drop size, 3.5 mm		Drop size, 5.1 mm	
EU g	ED g	EU g	ED g
962	299	1700	547
$\frac{EU}{ED} = 3.2$		$\frac{EU}{ED} = 3.1$	

TABLE 7. QUANTITIES OF SOIL CAUGHT IN UPSLOPE AND DOWNSLOPE SIDES OF SPLASH SAMPLERS, EU AND ED, RESPECTIVELY, WHILE USING DIFFERENT RAINFALL INTENSITIES

I = 4.8 iph		I = 6.6 iph		I = 8.1 iph		I = 14.8 iph	
EU g	ED g	EU g	ED g	EU g	ED g	EU g	ED g
504	172	637	242	301	71.5	1220	361
$\frac{EU}{ED} = 2.9$		$\frac{EU}{ED} = 2.6$		$\frac{EU}{ED} = 4.2$		$\frac{EU}{ED} = 3.4$	

TABLE 8. QUANTITIES OF SOIL CAUGHT IN UPSLOPE AND DOWNSLOPE SIDES OF SPLASH SAMPLERS, EU AND ED, RESPECTIVELY, WHILE USING DIFFERENT RAINDROP VELOCITIES

V = 12 fps		V = 14.5 fps		V = 15 fps		V = 18 fps		V = 19.2 fps	
EU g	ED g	EU g	ED g	EU g	ED g	EU g	ED g	EU g	ED g
202	70.9	208	74.9	374	122	713	207	1165	372
$\frac{EU}{ED} = 2.8$		$\frac{EU}{ED} = 2.8$		$\frac{EU}{ED} = 3.1$		$\frac{EU}{ED} = 3.4$		$\frac{EU}{ED} = 3.1$	

TABLE 9. SOIL PARTICLE SIZES IN RAINDROP SPLASH FOR DIFFERENT PERIODS OF TIME DURING RAINFALL
ERODED MUSKINGUM SILT LOAM
Raindrop size, 5.1 mm
Mechanical analysis

| | Sand and gravel | | | | | Totals | |
Time interval sample taken*	2 to 1 mm Per cent	1 to 0.5 mm Per cent	0.5 to 0.25 mm Per cent	0.25 to 0.125 mm Per cent	0.125 to 0.05 mm Per cent	Sand and gravel 2 to 0.05 mm Per cent	Silt and clay < 0.05 mm Per cent
			Raindrop velocity, 19.2 fps				
0-2	2.11	2.51	2.82	3.87	16.09	27.40	72.60
6-8	5.01	2.76	2.97	3.76	18.08	32.58	67.42
13-15	4.36	2.47	2.92	3.90	17.68	31.33	68.67
			Raindrop velocity, 18 fps				
0-2	1.91	2.65	2.75	3.89	16.36	27.56	72.44
6-8	2.79	2.63	3.22	4.75	17.76	31.15	69.85
13-15	1.43	2.46	3.42	4.77	19.69	31.77	68.23
			Raindrop velocity, 15 fps				
0-2	1.20	2.83	3.28	4.27	16.46	28.04	71.96
6-8	2.63	3.49	3.69	4.50	19.55	33.86	66.14
13-15	1.79	2.30	3.13	4.03	18.03	29.28	70.72
			Raindrop velocity, 12 fps				
0-2	1.01	2.22	2.87	4.19	16.14	26.43	73.57
6-8	2.29	3.06	3.69	4.22	17.68	30.94	69.06
13-15	2.44	3.62	3.79	4.37	17.33	32.55	68.45
Before rainfall	0.55	*Analysis of original soil* 1.81	2.27	3.49	17.16	25.28	74.72

*0-2, 6-8, and 13-15 indicate time intervals that samples were taken, expressed in minutes after beginning of experiment.

The first raindrops striking the soil tended to break down aggregates and release fine particles of silt and clay. The splash carried some of these particles and as it returned to the surface it was no longer clear water as was the original raindrop; instead, it

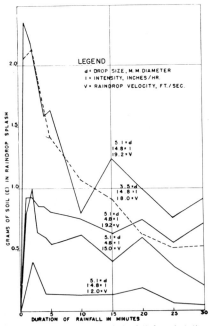

LEGEND
d = DROP SIZE, M.M DIAMETER
I = INTENSITY, INCHES / HR.
V = RAINDROP VELOCITY, FT./SEC.

GRAMS OF SOIL (E) IN RAINDROP SPLASH

DURATION OF RAINFALL IN MINUTES

Fig. 7 Soil in raindrop splash plotted against time

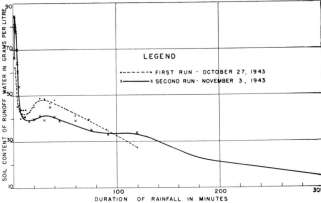

LEGEND
•------• FIRST RUN · OCTOBER 27, 1943
x——x SECOND RUN · NOVEMBER 3, 1943

SOIL CONTENT OF RUNOFF WATER IN GRAMS PER LITRE

DURATION OF RAINFALL IN MINUTES

Fig. 8 Soil loss plotted against time. Rainfall = 6.6 iph; runoff = 6.5 + iph; drop size = 5.1 mm diameter; velocity = 19.2 fps; plot size = 5x6 ft; soil, eroded Muskingum silt loam

represented a mixture of water, clay, silt, and sand. At the time this mixture infiltrated, the sand was largely deposited on the surface, and the clay and silt particles were mostly filtered out of the mixture while percolating through the soil mass. The eroded Muskingum silt loam that was used in these experiments seemed to reach an advanced stage of surface sealing within 2 to 3 min after beginning of rainfall.

As the surface-sealing process developed, more and more water was accumulated on the surface, and this accumulation caused the soil materials to be partially submerged, just as a spoonful of soil would be submerged if it were dumped into a large shallow plate containing water. In this analogy, the plane where the soil is sealed compares with the bottom of the plate and the loose soil above this plane compares with the spoonful of soil. These loose materials above the plane of sealing were easily picked up by the splash, and reductions in the maximum rates of soil splash were associated with reductions in amounts of these loose materials.

Particle sizes of sand and gravel contained in the raindrop splash were also referenced to time. After it was found that the splash contained higher percentages of sand and gravel than were contained in the original soils, particle-size determinations were made for the sand and gravel, using three 2-min samples. These 2-min samples were for periods of 0 to 2 min, 6 to 8 min, and 13 to 15 min following beginning of rainfall. The results are shown in Table 9.

Soil Loss in Runoff Referenced to Time. It was thought desirable to obtain curves of soil loss in surface runoff water, as these may be helpful in later studies of the data. Each sample was collected in a quart jar by intercepting for a very short period of time all of the runoff passing off the lower end of the plot. Samples were taken every minute for the first 5 min and at irregular intervals thereafter for a period of 5 hr. Results of these studies are shown in Fig. 8. A smooth curve was drawn through the points, and it will be seen the points define an irregular recession graph.

REFERENCES

1 Borst, H. L., and Woodburn, Russell. 1942. The effect of mulching and methods of cultivation on runoff and erosion from muskingum silt loam, Agr. Engr. 23: 19-22.

2 Cook, Howard L. 1936. The nature and controlling variables of the water erosion process, Soil Sci. Soc. Amer. Proc. 1: 487-494.

3 Duley, F. L., and Kelly, L. L. 1939. Effect of soil type, slope and surface conditions and intake of water, Nebr. Agr. Exp. Sta. Res. Bul. 112. 16pp.

4 Ellison, W. D. 1943. Two devices for measuring soil erosion, Agr. Engr. vol. 25, no. 2, February, 1944.

5 Ellison, W. D., and Pomerene, W. H. 1943. A rainfall applicator, unpublished.

6 Hendrickson, B. H. 1934. The choking of pore-space in the soil and its relation to runoff and erosion, Trans. Amer. Geophys. Union, pp. 500-505.

7 Horton, Robert H. 1940. An approach toward a physical interpretation of infiltration capacity, Soil Sci. Soc. Amer. Proc., 5: 399-417.

8 Laws, J. Otis. 1940. Recent studies in raindrops and erosion, Agr. Engr., vol. 21, no. 11, (November) pp. 431-433.

9 Lowdermilk, W. C. 1930. Influences of forest litter on runoff, percolation, and erosion, Jour. Forestry, 28: 474-491.

10 Neal, J. H., and Baver, L. D. 1937. Measuring the impact of raindrops, Jour. Amer. Soc. Agron., 29: 708-709.

11 Wollny, Ewald. 1877. Der einfluss der pflanzendecke und beschattung auf die physikalischen eigenschaften und die fruchtbarkeit des bodens, Berlin, pp. 171-174.

[*Editor's Note:* This article was continued in a later issue of the journal and is reprinted on the following pages.]

(Continued from the April issue)

MATHEMATICAL ANALYSIS OF DATA

The only data treated mathematically were those relating to the effects of rainfall variables on quantities of soil intercepted with the samples of raindrop splash. This analysis was made by H. W. Alexander and statements taken from his report are quoted as follows:

"*Regression Analysis of Splash Experiment Data.* The following analysis is based upon the data from the 59 experimental runs shown in Table 5.

"Preliminary graphing suggested an exponential relationship as the appropriate one. Hence it was decided to fit a relationship of the form

$$E = k \ V^a d^b I^c \qquad [1]$$

where $E =$ grams of soil carried by the samples of splash in a 30-min run

$V =$ drop velocity in feet per second

$d =$ diameter of drops in millimeters

$I =$ intensity in inches per hour

and k, a, b, and c are the constants to be determined. In logarithmic form, formula [1] is written

$$\log E = a \log V + b \log d + c \log I + \log k \qquad [2]$$

"The process used was that of fitting a relationship of the form 1 by the method of least squares. This is equivalent to fitting a relationship of the form 2 by the method of least squares, provided that each of the original observations is given a weight proportional to E^2*. The loss of accuracy is not great if an average weight is assigned to a group of points. In the computations that follow, an average weight was assigned to each of the three sets of experimental values corresponding to the three distances of drop fall.

"The F-test was used to evaluate the significance of the whole relationship, and the t-values for each of the coefficients a, b, and c were computed to evaluate their significance. These values are shown in Table 10.

This paper was prepared expressly for AGRICULTURAL ENGINEERING. W. D. ELLISON is project supervisor, North Appalachian Experimental Watershed, Soil Conservation Service, U. S. Department of Agriculture.

*Leland. O. M. Practical Least Squares. 1921. Page 136.

TABLE 10

	a	b	c
Coefficients	4.33	1.07	0.65
Standard error	0.31	0.16	0.066
Value of t	13.98	6.72	9.85

"For 55 degrees of freedom, all of the t-values are highly significant. The F-value for this relationship is 143.17 and it is also highly significant.

"The preliminary graphing suggested that the relationship between E and V might be sufficiently curvilinear to justify including another term in V. Hence the relation

$$E = k \ V^a d^b I^c \ 10^{eV}$$

or in logarithmic form

$$\log E = a \log V + b \log d + c \log I + eV + \log k$$

was fitted. The coefficients, their standard errors, and their t-values are given in Table 11.

TABLE 11

	a	b	c	e
Coefficient	8.8	0.95	0.53	1.39
Standard error	4.2	0.16	0.048	1.11
Value of t	2.12	6.01	10.96	1.26

The F-value here is 121.02. Comparing the t-values in Table 11 with those in Table 10, as well as the corresponding F-values, it appears that there is no gain in including the additional factor 10^{eV}.

"The best fitting relationship as established by the analysis is thus the one given by equation 1 with the values shown in Table 10, or

$$E = .00007661 \ V^{4.33} \ d^{1.07} \ I^{.65} \qquad [3]$$

Curves for this equation are shown in Figs. 9 and 10."

SUMMARY AND CONCLUSIONS

It may be helpful to review the problem of raindrop erosion and consider what part of the whole is covered by these studies before summarizing findings and attempting to draw conclusions. For purposes of review the problem has been divided into four parts: (1) the storm, (2) the soil, (3) the slope of the surface, and (4) the vegetal canopies, mulches and other materials which impede the fall of raindrops and the movement of raindrop splash.

Since only one soil type, one slope, and no vegetal covers were tested, studies of relationships of the variables must be limited to (1) above (the storm). Nor were studies of the first part of this problem (the storm) completed. The storm variables include both rainfall and wind. In reference to the rainfall, a wider range of drop velocities, drop sizes, and rainfall intensities must be tested. And concerning the effects of wind which may be anticipated include its influences on the angle of fall and strike of the raindrops, and transporting of splashed materials. The wind may also cause velocity changes and these must be studied, but the effects of these velocity changes are shown in formul [3].

Returning now to the specific results obtained, three distinct actions in the raindrop erosion processes, which may be harmful to the land, were found to be associated with raindrop impact and splash. These are (1) displacement and transportation of the soils, (2) changing the clear rainfall water to a mixture of water and soil materials preceding infiltration, and (3) breaking down soil aggregates.

Specific amounts of soil materials displaced and transported by splash on each unit of surface area have not been studied, but some

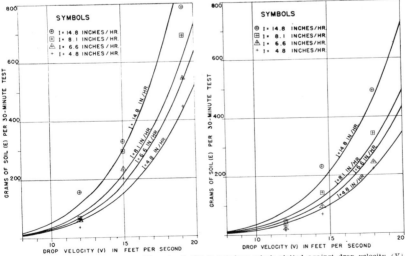

Fig. 9 (Left) Average observed quantities of soil (E) in raindrop splash plotted against drop velocity (V), and curves of $E = 7.66(10)^{-5} V^{4.33} d^{1.07} I^{0.65}$ for drops of 5.1 mm diameter ● Fig. 10 (Right) Average observed quantities of soil (E) in raindrop splash plotted against drop velocity (V), and curves of $E = 7.66 (10)^{-5} V^{4.33} d^{1.07} I^{.65}$ for drops of 3.5 mm diameter

results of this action are shown in Figs. 1 and 2. Figs. 3, 4, and 5 show raindrops falling and particles of soil and water splashing up from the soil surface. The relative effects of rainfall intensity (I), drop size (d), and drop velocity (V), in causing soil to splash are shown in Table 5, and mathematical relationships of these data are shown in formula [3]. Formula [3] indicates that only small changes in velocities of raindrops may cause large differences in quantities of soil carried by the splash.

Fig. 7 indicates that greater amounts of soil are carried by the splash after the surface becomes covered with a film of water. This seems partly due to a reduction of the cohesive forces working between the soil particles, and it is perhaps due in part to changes in splash phenomena which occur with changes in the depth of the water film covering the soil. This figure also indicates a reduction occurs in the rate of soil splash after the surface has become sealed and the loose particles above the plane of sealing have been carried away by erosional processes.

Tables 6, 7, and 8 show the amounts of soil caught in upper and lower sides of splash samplers when the sampler was installed with the splash plates perpendicular to the direction of slope. About three times as much soil was caught in the uphill side of samplers as was caught in the downhill side and this would seem to indicate that three parts of the soil splash were moved downhill, while only one part was being moved uphill on this 10 per cent slope. The probable effects of such action become more apparent when this part of the study is referenced to Fig. 1 (C); had one-half the soil lost from this surface been transported down slope, downhill movement caused by raindrop splash would have been very appreciable.

In reference to raindrop impact and splash which changes the rainfall from clear water to mixtures of water and soil materials, data relating to the soil content of these mixtures are shown in Tables 1, 2, 5, and 9. Rates of infiltration may be very sensitive to changes in the soil content shown in these tables. Also, the soil content of surface flow, especially in prechannel stages, may be proportional to the soil content of the raindrop splash. This is shown to be the case when comparing Figs. 7 and 8. Note similarity of recession curves and that they both break at about 10 minutes after beginning of rainfall.

Breaking down of the soil aggregates under raindrop impact is indicated by the data in Tables 2, 3, and 4. Under this impact many of the aggregates too large to be carried by raindrop splash were broken down. This breakdown made more small aggregates of sizes readily carried by the raindrop splash and transported by the surface flow.

The data in Tables 2, 3, and 4 indicate that the degree of breakdown of a soil aggregate is not sensitive to changes in drop size and velocity, at least not under the conditions of these tests. This information supports a conclusion that the total aggregate breakdown for a particular soil will be proportional to the total number of aggregates involved in the erosional processes. These totals may be represented by E in formula [3]. Therefore, the total aggregate breakdown is proportional to $0.00007661\ V^{4.33}\ d^{1.07}\ I^{.65}$. The structure of soil pores may be changed by a breakdown of the aggregates, and these changes are certain to have an effect on infiltration and runoff.

Experimental studies of the soil aggregate and particle sizes contained in the raindrop splash and in the surface flow indicate that the raindrop splash carries higher percentages of large particles and aggregates than does surface flow in the prechannel stages as represented on this plot.

Experimental results indicate that in addition to measures of rates and quantities of precipitation that are now being made in most watershed studies, there is an outstanding need for measuring drop sizes and drop velocities of natural rainfall.

Fig. 1 (C) indicates that particles of splash rising from the surface do not have sufficient size and velocity to cause erosion when they again strike the soil. If they were capable of this, they would have destroyed the vertical sides of soil columns contained under the coins.

Since infiltration depends on both (a) properties of the soil through which the water must filter and (b) properties of the water that is to be filtered, raindrop erosion which changes properties of both is certain to be a most important factor affecting the infiltration process.

Results of these studies suggest a need for infiltration experiments where both quantities and physical properties of soil materials contained in raindrop splash are manipulated and studied as independent variables.

It is recognized that high concentration of soil in the runoff water may be caused by either (a) excessive erosion in a few localized gullies and channels or (b) high rates of erosional activity over the entire watershed surface. If it is caused by the former, it will probably bear little relationship to infiltration and runoff on the entire watershed. If it is caused by the latter, it may indicate that the entire surface of the watershed is covered by a mixture of muddy water and this should have considerable bearing on rates of infiltration and runoff. Before attempting to analyze runoff and associated erosional data on a watershed basis it would seem desirable and even necessary to study erosional activity on the entire watershed. Where this is not done, the relationships of erosion, infiltration and runoff cannot be determined.

Since erosional activity reacts so sensitively to changes in raindrop velocities, it will be important to study methods of controlling these velocities under field conditions. Considerable work is now being done with mulches which partially control the delivery of raindrops to the soil surface. However, most mulches that are of practical use are probably not as effective as many vegetal canopies. Perhaps the effectiveness of the canopies may be improved by developing lower growing plants and through better use and improved distribution of canopies in the land use plans.

Vegetal canopies in addition to reducing high raindrop velocities approaching the soil surface also impede the movement of soils in the splash process and thereby curtail soil loss from the sloping areas; they may also prevent winds from carrying the splash any considerable distances.

Since the splash and not the surface runoff seems to carry the larger particle and aggregate sizes, and since erosional deposits along the bases of many slopes are often composed of coarse textures and large aggregates, the question is asked as to whether the downhill splash of raindrops does not account for some of the erosional deposits found near the bottoms of many hillsides. Results of these experiments indicate that it probably does, and that this question should be given further study. Determinations of particle and aggregate sizes carried by raindrop splash and surface flow under field conditions may make it possible to identify the erosional deposits caused by each.

The results of raindrop erosion are most apparent at the tops of hummocks and hills, where return splash does not balance outgoing splash. It is the author's belief that most of the severe sheet erosion found at these locations is largely caused by raindrop splash.

Erosion by surface flow is usually most apparent near the base of the slope where greatest amounts of surface flow concentrate.

The many detrimental processes involved in raindrop erosion may be more damaging on some types of soils than are gullying and other forms of erosion caused principally by runoff. Since raindrop erosion may cause infiltration curves to make a sharp downward break at the beginning of the rainstorm, and the runoff curves to make a sharp upward break, it is apparent that infiltration and runoff are dependent on raindrop erosion and that prevention of raindrop erosion through use of canopies and mulches represents a basic approach to the broader problems of soil and water conservation and floods. Research work necessary to achieve most effective use of these covers must be based on many considerations. A study of raindrop splash and the erosional activity associated with this splash seems to represent a step in the right direction, and provides a means for making better evaluations of soil and water conservation practices which may be applied to the lands.

ACKNOWLEDGMENTS

The author expresses sincere thanks to the following persons:

C. E. MacQuigg, dean of engineering, Ohio State University, for university equipment and for his advice and consultation which helped to guide the work.

R. W. Powell, professor of mechanics, Ohio State University, for assisting with formulation of plans and with analyses.

R. E. Yoder, chief in agronomy, Ohio State Agricultural Experiment Station, for suggestions and for use of station equipment.

F. R. Dreibelbis, soils technologist, Coshocton Project, for supervision of mechanical and aggregate analyses of samples, and for suggestions.

W. H. Pomerene, agricultural engineer, Coshocton Project, for assisting with design of rainfall applicator used, and for suggestions.

H. W. Black, soil conservationist, Coshocton Project, for suggestions.

D. A. Parsons, project supervisor, U. S. Soil Conservation Service, Auburn, Ala., for review and criticism of preliminary manuscript.

Members of C. P. S. Camp No. 23, for suggestions, for assisting with the experiments, and for work on the mathematical, mechanical and aggregate analyses.

Reprinted by permission from *Soil Sci.* **60**:305–320 (1945)

DYNAMICS OF WIND EROSION: I. NATURE OF MOVEMENT OF SOIL BY WIND[1]

W. S. CHEPIL[2]

Canada Department of Agriculture

Received for publication January 27, 1945

The energy relationships between the air currents, commonly known as wind, and the soil are of great importance in the problem of wind erosion and its control. But, unfortunately, much of the information available in the past on the nature of air currents near the ground and on the importance of wind as a geologic agent in soil formation is too fragmentary and inadequate for wide application to the practical aspect of wind erosion control. Consequently, when research work on wind erosion of soil was begun at this laboratory, little aid was obtainable from the records of previous work, and it was found necessary to investigate the fundamental aspects of the problem, namely, the energy relationship between wind and soil.

The factors influencing wind erosion are numerous and hence add considerably to the complexity of the problem. The most important of these factors may be listed as follows:

I. Air
  1. Velocity
  2. Turbulence
  3. Density, affected by
     a. Temperature
     b. Pressure
     c. Humidity
  4. Viscosity

II. Ground
  1. Roughness
  2. Cover
  3. Obstructions
  4. Temperature
  5. Topographic features

III. Soil
  1. Structure, affected by
     a. Organic matter
     b. Lime content
     c. Texture
  2. Specific gravity
  3. Moisture content

It is evident that the wind erosion problem depends on the mutual relationship of a combination of many factors. The influence of any factor that is involved in any condition may be negative or positive with regard to erosion, and in fact, one factor may counteract the influence of another in virtually any situation. For example, although it has been found (4) that wind turbulence increases erosion, yet the degree of erosion of a rough surface, where turbulence is more developed, is much less than that of a smooth surface over which the mean velocity of wind is greater. In both cases it is the net effect of the two opposing trends which determines the actual amount of erosion of the soil. Thus, in order that any condition may be understood and properly interpreted, it is essential that the individual factors involved be known and their relative significance accurately evaluated.

---

[1] Contribution from the Experimental Farms Service (P.F.R.A.), Dominion Department of Agriculture, Ottawa, Canada.

[2] Agricultural scientist, Soil Research Laboratory, Dominion Experimental Station, Swift Current, Sask.

Briefly, the immediate problems for the soil conservationist are to determine the relative significance of the various factors influencing the movement of the soil material by the wind and to appreciate the physical nature of wind translocation. In order to answer some of the elementary questions connected with these problems, it was necessary to run a large series of experiments. The first of these had to do with the physical nature of soil drifting, the results of which are herewith reported.

REVIEW OF LITERATURE

The movement of soil particles by the wind has been studied by a number of investigators. Free (6) asserted in his review of the literature on the problem up to 1911 that the largest proportion of the soil carried by the wind is moved in a series of short bounces called "saltation." He reported that sand never bounces high above the ground and in the desert cannot be felt by a person mounted on a camel. The smaller the soil particles, the greater is the influence of the wind upon it and the closer the approach of the path of saltation to a line parallel with the direction of the wind. The smallest quartz particles carried in saltation are about 0.1 mm. in diameter. Particles smaller than this have a velocity of fall lower than the upward velocity of the turbulent wind. Such particles are carried more or less parallel with the general direction of the wind, and form what has been termed a "suspension movement." They may be carried through the atmosphere for long periods of time and will fall to earth only with rain or after the wind has slackened considerably. Fine dust is thus often carried great distances from its original location.

In addition to flow in saltation and suspension, there is still another type of movement. Udden (7) observed that quartz grains larger than about 0.5 mm. in diameter and smaller than 1 mm. are too heavy to be transported through the air, but roll and slide along the surface of the ground. This type of movement is termed, by Bagnold (1), "surface creep." On the other hand, grains greater than 1 mm. in diameter are too large to be moved by ordinary erosive winds (5).

Although the manner of the transport of sand by wind was understood in a general way many years ago, it was not until recently that a comprehensive study was made by Bagnold (1). He asserted that dune sand is carried by wind mainly in saltation and surface creep and to a minor degree in true suspension. Bagnold concluded from theoretical calculations that with average dune sand the suspension flow, even under a relatively strong wind, does not exceed one twentieth of the flow in saltation and surface creep.

For arable soils, the only information available is a preliminary report from this laboratory (2), concerning the proportion of soil carried in different types of movement. Chepil and Milne (3) made measurements of the relative amounts of soil carried by wind over cultivated fields. The soil catchers used at that time were satisfactory for measuring rate of flow in saltation and surface creep, but were not entirely effective in trapping fine dust carried in suspension. It was observed that much soil was moved in saltation, but a very substantial proportion constituting the finest fraction was carried in true suspension.

109

In addition to information on the general mechanism of transport of soil by wind, data are herewith presented on the relative proportion of different types of movement on different soils with varying degrees of roughness of surface, and on the relative nature and intensity of soil movement as influenced by some of the major types of tillage treatments.

EXPERIMENTAL PROCEDURE

In the development of experimental technique to study the quantity of soil transported by wind, it was found necessary to recognize the fundamental differences that exist between the transport of particles by saltation and surface creep and by movement in suspension. A fine grain of quartz dune sand when shot into the air at a speed of 9 miles per hour travels 20 cm. before its speed is reduced 50 per cent. Because of this continuing and directed motion, the rate of the flow of particles by saltation is determined easily by trapping the particles in narrow containers, open to windward. On the other hand, the measurement of the movement of fine dust is rather more complicated, for such particles are seriously affected by the wind's internal movements, that is, by the

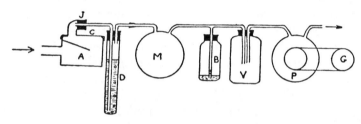

Fig. 1. Diagrammatic Representation of Apparatus Used for Measuring Quantities of Soil Carried in Saltation and Suspension

continued and instantaneous changes in its velocity and direction. Consequently, small particles do not enter the volume of still air contained in the trap, but instead are deflected from it and carried along with moving air. The procedure for measuring the rate of suspension flow is therefore much more complicated. In the first place, the volume of air in which dust is suspended must be determined accurately, and in the second, the dust itself must be filtered off completely and removed into a receptacle in which it may be weighed.

The method used in measuring the quantity of soil moving in saltation and suspension is indicated diagrammatically in figure 1. Soil material carried by the wind enters a narrow box A through a rectangular nozzle, $\frac{1}{2}$ inch wide and 2 inches high, facing into the wind. The particles carried in saltation are trapped in the box and may be removed and weighed after detachment of A from the rest of the apparatus at a stoppered joint J and removal of a tight-fitting cover C. The particles in suspension, however, are carried into the trap by suction created by a vacuum pump P and collected in a 6-inch column of distilled water in cylinder D. Air intake through the apparatus is measured with a gas meter M and its volume corrected on the basis of atmospheric pressure with the aid of a mercury barometer B. More than one dust-catching unit can be connected by rubber tubing to a vacuum flask V. This type of arrangement facilitates simultaneous measurement of soil flow at various positions. The vacuum pump was driven by a small gasoline motor at G.

The amount of dust trapped was determined by evaporating the water and weighing the residue. The relative concentration of particles carried in suspension at any height was determined by dividing the weight of the residue by the corrected volume of filtered air. Suspension flow was determined by multiplying concentration per unit volume by velocity of the wind. This was done on the assumption that the velocity of particles carried in suspension was equal to that of the wind. Photographs and indirect measurements substantiated this assumption.

In addition to measurements of soil flow at different heights, the total amounts of soil carried in different types of movement were determined. Total saltation and surface creep were determined by the method of Bagnold (1), and the suspension flow was ascertained by subtracting the flow in saltation and surface creep from the total flow as determined by the difference in the weight of thoroughly air-dry soil before and after exposure to an erosive wind. These measurements were made both in the open field and in a portable field tunnel, described previously (2), and on different types of soil with varying degrees of roughness of surface.

In order to determine the nature of soil movement under different conditions of soil and wind, photographs were taken of the paths of moving soil particles. Sunlight admitted vertically through a lens in the ceiling of a darkened wind tunnel was used for illumination. The lens was 25 cm. long and 12 cm. wide and had a focal length of approximately 2 feet. It produced a very intense vertical beam of light, which illuminated an area, 25 cm. long and 1 cm. wide, parallel with the direction of the wind. Particles flying through the illuminated space reflected light distinctly, and their paths appeared on a photographic plate as silvery threads, or as a series of dots or dashes against a black background. Photographic exposures of $\frac{1}{25}$ and $\frac{1}{10}$ of a second were made for the purpose of indicating the shape and length of grain paths and $\frac{1}{100}$ and $\frac{1}{80}$ of a second for determining the speed of the grains through the air.

RESULTS

Nature of wind translocation

Photographs and direct visual observation indicated that all soils were carried by wind in the three types of movement already mentioned. The relative proportion of each type of movement varied greatly for different soils.

The greatest proportion of the movement in all cases examined was by particles in saltation. After being rolled by the wind, the particles suddenly leaped almost vertically to form the initial stage of the movement in saltation. Some grains rose only a short distance, others leaped 1 foot or more, depending directly on the initial velocity of rise from the ground. They also gained considerable forward momentum from the pressure of the wind acting upon them, and acceleration of horizontal velocity continued from the time grains began to rise to the time they struck the ground. In spite of this acceleration, the grains descended in almost a straight line invariably at an angle between 6 and 12 degrees from the horizontal. On striking the surface they either rebounded and continued their movement in saltation, or lost most of their energy by striking other grains, causing these to rise upward and themselves sinking into the surface or forming part of the movement in surface creep. Irrespective of whether the movement was initiated by impact of descending particles or by impact of rolling grains, the initial rise of a grain in saltation was generally in a vertical direction.

The cause of this vertical rise was not at all apparent. It was found that soil grains jumped vertically off a smooth surface, such as a wooden floor, after rolling for as short a distance as 2 cm., although there were neither soil grains nor other obstructions against which they could strike and rebound into the air. The probable cause of the vertical rise was thought to be a direct impact of a facet of an irregularly shaped grain against the tunnel floor. From a theoretical point of view, however, if no other impact forces are involved, the angle with which a descending grain would rebound from a smooth horizontal surface should be equal to the angle of descent, which was 6 to 12 degrees. Actually, the angle of ascent was between 75 and 90 degrees in the majority of cases. This seemed incredible unless it was presumed that the vertical rise was due to some force other than the force of impact of the grain against the surface.

The only logical explanation of the vertical rise of particles in saltation appears be by the theory of the Bernoulli effect. This effect is apparently due to two causes, the spinning of the grains and the steep velocity gradient near the ground. Actual photographs, some of which are shown in figures 2 and 3, indicate that grains carried in saltation rotate at a speed of 200 to 1,000 revolutions per second. The photographs show clearly that about 50 per cent of the grains carried in saltation are spinning, while another 25 per cent exhibit relatively indistinct rotation. It is possible, however, that more than 75 per cent of the grains were rotating but that this could not be indicated without greater intensity of illumination. Nor was it possible to indicate rotation for grains approaching a spherical shape, since photographs merely show the variation in the intensity with which the reflecting facets of the grain were illuminated. The more angular the grain, the more distinctly rotation appeared on the plate.

On account of a rapid clockwise spinning of the grain, the air near the grain surface is carried around it. On the lower side, assuming the wind to be traveling from left to right, this air is moving against the direction of the wind; on the upper side, it moves with it. The velocity of the air at any point near the grain is thus made up of two components, one due to the wind and the other to the spinning of the grain. On the upper side these components have the same direction, whereas below they have opposite directions. Thus, the velocity is greater at the top surface than at the bottom, and according to the Bernoulli theorem, the pressure is decreased at the top and increased at the bottom, and so the grain tends to rise.

The Bernoulli effect is further intensified by virtue of the fact that a steep velocity gradient exists near the ground. At a threshold wind velocity, that is,

FIG. 2. PHOTOGRAPHS OF PATHS OF WIND-BORNE SOIL PARTICLES, INDICATING MOVEMENT IN SALTATION AND SUSPENSION OVER A LEVEL SURFACE OF (TOP) SCEPTRE HEAVY CLAY AND (BOTTOM) HATTON FINE SANDY LOAM

a velocity just high enough to initiate the movement in saltation, the following is a typical case of average velocities encountered near a smooth ground surface:

Height, mm	0.02	0.1	0.2	0.3	0.6	0.9	1.2	1.5	1.8
Velocity, cm./sec.	0	104	152	189	234	254	272	297	304

The variation in air velocity near the ground causes a substantially higher rate of air flow at the upper than the lower surface of the grain of, say, 0.2 mm.

in diameter, and this difference in velocity may be expected to produce a similar and additive effect to that caused by the spinning of the grain in a current of air. Consequently, if the total difference in the pressure between the upper and the lower surfaces is greater than the force of gravity acting downward, the grain will rise in a vertical direction.

FIG. 3. PHOTOGRAPHS OF PATHS OF WIND-BORNE SOIL PARTICLES, INDICATING (TOP) MOVEMENT IN SALTATION AND SUSPENSION OVER A LEVEL SURFACE OF HAVERHILL LOAM, (BOTTOM) MOVEMENT IN SUSPENSION OVER A RIDGED SURFACE

As no experimental data are available concerning a sphere the size of a sand grain spinning in an air current near the ground, it is impossible to confirm the existence of a vertical component of wind force as postulated above. Experiments are now being undertaken to obtain definite information on this problem.

After being shot into the air, the grains rose to various heights and, because of force of gravity, fell at an accelerating velocity, There was at the time a horizontal acceleration of the falling particle due to the forward pressure of the wind upon it. Photographs indicate that the downward and the forward accelerations were approximately equal, and the inclined path of the falling grain was therefore almost a straight line. Only a slight curvature downward was observed in most cases. The angle of descent varied but little and, as already pointed out, was between 6 and 12 degrees from the horizontal. Smaller grains descended at a somewhat smaller angle than the larger grains. The angle of descent did not vary greatly with wind velocity, but the higher the wind the greater the height to which some grains rose in the air, and hence the corresponding longer average path.

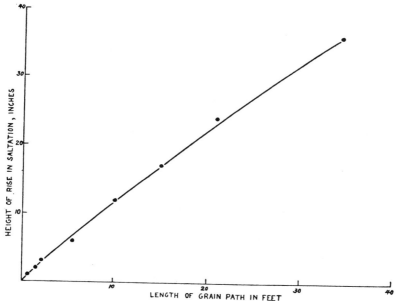

FIG. 4. RELATION OF HEIGHT OF RISE IN SALTATION TO LENGTH OF GRAIN PATH (HORIZONTAL PLANE)

On the whole, the horizontal distance through which the grain continued to rise was about one fifth to one fourth of the total horizontal length of a single leap in saltation.

On a smooth surface consisting only of erosive grains, there was a remarkable consistency in the shape of grain paths. The shape of the paths was much less regular over a rough surface, but the general character of the movement in saltation remained the same. Irrespective of the degree of surface roughness, a close relationship existed between the average height to which the grains rose in the air and the horizontal length of grain leap. This relation is indicated in figure 4. The ratio of height of rise to the horizontal equivalent of length of grain leap was about 1:7 for a rise up to 2 inches in height, 1:8 for a rise of 2 to 4 inches, 1:9 for 4 to 6 inches, and 1:10 for heights above 6 inches. The results were the same for all soils used in the investigation.

The surface creep of soil grains could not be recorded photographically but was plainly visible to the naked eye. The grains in surface creep were too heavy to be moved by the direct pressure of the wind, but derived their kinetic energy from the impacts of smaller grains moving in saltation. Grains in saltation, on the other hand, received most of their impact energy from the direct pressure of the wind. It is evident that the movement of soil by wind is dependent, not so much on the force of the wind acting on the surface of the ground, as on the velocity distribution to such height as the grains rise in saltation. This height is definitely limited, and it may be concluded, therefore, that wind erosion is mainly a surface phenomenon and is not directly dependent on the condition of the wind above that restricted distance.

Movement of dust in suspension

The results of measurements made in the open field showed that the mechanism of transport of soil by wind is very similar to that of dune sand described by Bagnold (1), except that in addition to movement in saltation and surface creep there was, in some cases, a substantial proportion of the soil carried in true suspension. The presence of fine dust, even in very appreciable amounts, did not seem to affect the nature of the movement in saltation or surface creep, but it greatly influenced both the threshold wind velocity and the intensity of erosion for a given wind (5).

Once lifted off the ground, the particles in suspension were completely borne up by the wind. In the open country they usually reach great heights and do not fall to the ground except with rain or after the wind has slackened considerably.

The mechanism by which fine dust is lifted off the ground is entirely different from that of saltation. A previous study (5) showed that samples of soil composed only of fine dust particles were extremely resistant to erosion by wind. In fact, quartz particles less than 0.05 mm. in diameter could not be moved by wind velocities as high as 37 miles per hour at a 6-inch height. In mixtures with coarser grains ranging up to 0.5 mm., however, these particles moved readily, and the threshold velocity of the mixtures was lowered very considerably. It may be said, therefore, that movement of fine dust in an air current is mainly the result of movement of grains in saltation; hence, without saltation movement, dust clouds would not arise, except on a relatively limited scale as a result of disturbance by moving vehicles, animals, etc.

The relative quantities of fine dust particles blown off different soils are indicated in figures 2 and 3. The suspension flow, though clearly visible to the naked eye, is indicated by very faint lines on a photographic plate. The difficulty is attributed to the relatively small diameter of the particles and to their high speed, which is approximately that of the wind. A very sensitive film and light much brighter than sunlight are essential for a clear indication of suspension flow.

In spite of these difficulties, the relative concentration of suspended dust is plainly indicated for different types of soil. Photographs indicate almost no suspension flow over Sceptre heavy clay, an appreciable concentration over

Hatton fine sandy loam, and a concentration so dense over Haverhill loam as to mask the appearance of movement in saltation. Wind velocity in all cases was about 17 miles per hour at a 12-inch height.

In contrast to the movement of grains in saltation, the movement of fine dust in suspension, after it has been lifted off the ground, is completely governed by the characteristic movement of the wind. The influence of an eddy on suspended dust is indicated clearly in the lower photograph of figure 3, which was taken at the time of a gentle movement of air following a stronger wind that initiated the movement of saltation. The back-eddy of air currents is shown

TABLE 1
Relative quantities of soil carried in surface creep and saltation under different conditions of surface roughness and wind

WIND VELOC-ITY AT 12-INCH HEIGHT	SOIL TYPE	SMOOTH SURFACE							RIDGED SURFACE*						
		Surface creep	Saltation carried below the height of						Surface creep	Saltation carried below the height of					
			1 inch	3 inches	6 inches	12 inches	24 inches	36 inches		1 inch	3 inches	6 inches	12 inches	24 inches	36 inches
m.p.h.		mgm./cm. width/sec.	per cent	per cent	per cent	per cent	per cent	per cent	mgm./cm. width/sec.	per cent	per cent	per cent	per cent	per cent	per cent
17	Sceptre heavy clay	15.9	27	66	91	98	....	....	1.9	20	54	74	96	99+	....
	Haverhill loam	3.1	24	63	86	97	99+	....	0.4	18	47	72	95	99+	....
	Hatton fine sandy loam	2.8	30	65	88	98	99+	....	0.5	21	50	75	97	99+	....
25	Sceptre heavy clay	142.1	28	56	77	91	99	99+	4.5	25	55	76	93	98	99+
	Haverhill loam	9.2	24	54	75	90	97	99+	1.2	18	45	72	95	98	99+
	Hatton fine sandy loam	3.3	29	63	83	95	99	99+	0.6	20	47	73	95	99	99+

* Ridges were 2.5 inches high, 9 inches wide, running at right angles to the wind.

plainly by the characteristic paths of suspended particles and almost duplicates a diagrammatic representation of wind structure previously recorded over the same type of surface with the aid of sensitive oscillating plates (4).

Concentration of wind-borne particles at different heights

Measurements of the concentration of wind-borne particles in a portable field tunnel and also in the open field indicated that most of the soil movement in saltation was carried below the height of 2 or 3 feet. In fact, over 90 per cent of the soil was transported below the height of 12 inches, and this was found to be true for several widely different soils chosen for investigation.

The results of this study obtained in the open field under various conditions of soil, wind velocity, and surface roughness are given in table 1. In general, the movement in saltation was the same on all soil types investigated, but some variation was found in the proportion of grains carried at different heights.

Coarsely granulated soils, such as Sceptre heavy clay, drifted closer to the ground than the more pulverized Haverhill loam, but the difference was not appreciable.

There were large differences in the rate of surface creep, which was particularly high on coarsely granulated Sceptre heavy clay and least on finely granulated Hatton fine sandy loam. Roughly, the amount of surface creep depended on the quantity of erosive grains greater than 0.5 mm. in diameter for cultivated soils and over 0.25 mm. for dune sand.

The relative concentration of wind-borne particles at different heights above a rough surface differed widely from that above a smooth one. Figures 5 and 6 show typical differences in relative concentration of drifting soil over the two

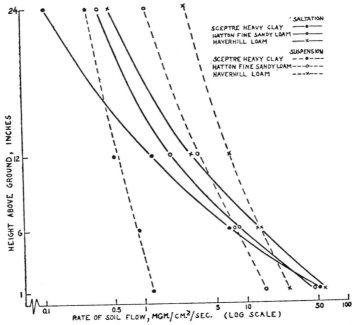

FIG. 5. DISTRIBUTION OF WIND-BORNE PARTICLES CARRIED IN SALTATION AND SUSPENSION AT DIFFERENT HEIGHTS ABOVE A SMOOTH SURFACE

types of surface—wind velocity in both cases being the same at 12-inch height. The ridges had a marked effect on lowering the total rate of soil flow and virtually eliminated surface creep. Furthermore, many of the coarser granules that moved in saltation over a smooth surface were apparently trapped by the ridges, thus lowering appreciably the relative concentration of particles near the ground.

The relative concentration of soil particles at different heights remained the same under a wide range of wind velocity; hence the graphs presented in figures 5 and 6 give only averages for a range of wind velocity from 13 to 30 miles per hour at 12-inch height. The data show that the ratio of saltation to suspension decreased rapidly with height above the ground, as would be exacted. As Grains in saltation do not generally rise higher than several feet above the ground, the soil carried above this height is in true suspension and capable of being carried to great heights and over long distances from the original location.

Figures 5 and 6 indicate further that virtually a straight-line relationship exists between height and the logarithm of the rate of flow in saltation and suspension over all soils investigated. There seems to be some functional relationship between the rate of soil flow and height, for nearly all of the experimental values fall very closely to curves of the same characteristic shape.

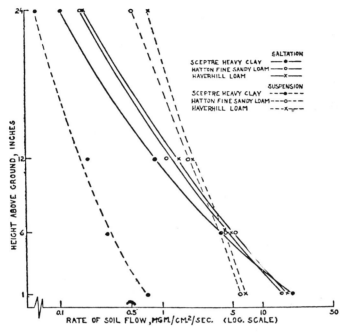

FIG. 6. DISTRIBUTION OF WIND-BORNE PARTICLES CARRIED IN SALTATION AND SUSPENSION AT DIFFERENT HEIGHTS ABOVE A RIDGED SURFACE

Proportion of different types of flow on different soils

The relative rates of soil transport in each of three types of movement varied widely on different soils (table 2). The proportion moved in surface creep constituted 7 to 25 per cent of the total flow, the lowest rate being on highly pulverized Haverhill loam, the highest on coarsely granulated Sceptre heavy clay. Flow in saltation varied much the same way as surface creep and composed 55 to 72 per cent of the total flow, depending on soil type. The proportion of flow in suspension constituted about 3 per cent of the total flow on Sceptre heavy clay and as much as 38 per cent on typically dusty Haverhill loam and was approximately equal to the proportion of particles smaller than 0.1 mm. found in the soil. The data in table 2 indicate that some particles larger than 0.1 mm. must have been carried in suspension, probably those of lowest specific gravity. As the percentage of particles smaller than 0.15 mm. in all soils was substantially greater than the percentage carried in suspension, it is evident that only a small proportion of the size above 0.1 mm. was carried in true suspension.

There was a constant proportion of the three types of flow throughout the wide range of wind velocity used in the investigation, and on this account the values in table 2 indicate only the average results obtained for many individual cases. Measurements were made on a level surface on which the surface projections in the form of clods and surface ripples did not exceed 0.75 inch in height.

TABLE 2

*Relative proportion of three types of flow on different soils**

SOIL TYPE	SOIL REMOVED IN		
	Saltation	Suspension	Surface creep
	per cent	*per cent*	*per cent*
Sceptre heavy clay.	71.9	3.2	24.9
Haverhill loam.	54.5	38.1	7.4
Hatton fine sandy loam.	54.7	32.6	12.7
Fine dune sand.	67.7	16.6	15.7

* The size distribution of erosive grains in the soils was as follows:

SIZE OF PARTICLES. *mm.*	DISTRIBUTION OF PARTICLES				
	0.83–0.42	0.42–0.25	0.25–0.15	0.15–0.1	0.1
	per cent	*per cent*	*per cent*	*per cent*	*per cent*
Sceptre heavy clay.	33.5	46.1	14.9	4.0	1.5
Haverhill loam.	13.3	17.3	15.1	22.3	32.0
Hatton fine sandy loam.	1.1	6.0	26.4	40.5	26.0
Fine dune sand.	0.1	2.1	54.2	35.6	8.0

Increasing the roughness of surface caused a proportional reduction in the rate of movement in surface creep. A rough surface, such as that composed of cultivator ridges extending at right angles to the direction of wind, trapped most of the surface creep but failed to reduce the movement in saltation and suspension to quite the same degree.

Relative efficiency of cultural treatments in trapping drifting soil

The data in table 2 give some idea of the relative amounts of soil that are carried away by the winds. Dust in suspension is transported far and wide; hence the regions in which it is deposited benefit but little in the way of additional soil, but the much more limited eroded area loses a great deal. In saltation and surface creep, on the other hand, soil is not usually carried far and is deposited in or near the vicinity of the affected area. Many cultural and cropping methods are devised to trap the grains in saltation and surface creep to prevent the spread of erosion to surrounding unaffected areas. It has been pointed out that surface creep may be almost eliminated and saltation greatly reduced as a result of ridging a highly erosive soil. The higher the ridges the more effective they are in stopping surface creep and saltation, but as movement in suspension and surface creep is dependent wholly on movement in saltation, the elimination of saltation will eliminate all other forms of movement. The whole program of wind erosion control is therefore based on reduction or elimination of movement in saltation.

Soil ridges may be used to reduce or eliminate saltation. It is often preferable to ridge the whole of the affected area, but this treatment is not always possible, and ridging narrow strips at regular intervals across the field is often resorted to. Stubble and crop strips may be used for a similar purpose.

The width of trap strip that may be required depends partly on the length of jump of grain in saltation and partly on the trapping capacity or receptiveness of the surface. Standing grain stubble is probably the most effective form of trap, for it will trap all the soil moving into it in saltation. It can therefore be considered as fully receptive. The minimum width of stubble strip, however, would have to be not less than the maximum horizontal length of a single leap of the grain in saltation. The data in table 1 and figure 4 combined supply complete information on the approximate percentage of soil grains that may be trapped by any width of a wholly receptive trap strip.

Thus, supposing the width of a strip of short stubble to be 10 feet, figure 4 indicates that to jump this distance the grain in saltation would have to rise to a height of approximately 12 inches. Furthermore, it is indicated in table 1 that between 90 and 98 per cent of the flow in saltation is below this height; hence a 10-foot strip of short stubble may be expected to trap between 90 and 98 per cent of the flow in saltation, depending on soil type and wind velocity. It can be found in like manner that a 2-foot stubble strip will trap about 50 per cent of the flow in saltation and a 30-foot strip over 99 per cent. These values corroborate the results obtained from actual practice in the field.

The data indicate further that the effectiveness of a trap varies somewhat with wind velocity. Thus, it is shown that in order to trap 99 per cent of the movement in saltation a 20-foot strip of totally receptive surface would be required for a 17-mile-per-hour wind, and a 35-foot strip for a 25-mile-per-hour wind.

There are other angles to be considered in deciding on the width of trap strip that would be most effective. The next of these considerations is the height of the stubble. A 3-foot width of 6-inch stubble, for example, will trap on an average about 85 per cent of the total flow in saltation, but a 3-inch stubble will trap about 60 per cent.

The minimum width of trap required depends also on the capacity to hold the blown soil. Long stubble, in addition to being more effective in trapping the encroaching drift, has a greater holding capacity. Other factors being equal, the holding capacity varies directly with the height of the stubble. The effectiveness of the trap is reduced to zero as soon as the trap has reached its holding capacity. Hence, to be fully effective, the minimum width should be that required to store the encroaching drift, plus that width to leeward that will remain relatively free to act as an effective trap. The width that may be required to store the encroaching drift cannot be estimated with any degree of accuracy, for the amount of erosion is dependent to a large measure on the conditions of the weather, which cannot be predicted. Hence, a considerable margin over and above the minimum requirement must be allowed.

Ridges were found to have a lower trapping capacity than the standing grain stubble, for many of the grains bounced off the ridges one or more times before they were finally trapped in the furrows. The trapping capacity of ridges depends on their size. Ridges 2.5 inches high and 9 inches wide were about 50 per cent as effective as a 6-inch wheat stubble, whereas cultivator ridges 5 inches high and 18 inches wide and lister furrows 12 inches deep were 77 to 83 per cent and 85 to 92 per cent as effective, respectively.

The foregoing results were obtained on a highly nonerosive clay soil that was exceedingly resistant to the grinding action of flying grains. Many soils do not exhibit such marked resistance, for the ridges may wear down rapidly and lose much of their sheltering effect. The trapping capacity of the grain stubble, on the other hand, is not at all affected by abrasion.

CONCLUSIONS

Because of rapid spinning of the grains moving in saltation and a steep velocity gradient, there appears to be a considerable vertical component of wind force near the ground. On account of these effects the grains rise steeply and descend very obliquely toward the surface. As the downward acceleration, due to gravity, and the forward acceleration, due to wind pressure, are approximately equal, the grains fall in almost a straight line. They strike the surface at an angle of 6 to 12 degrees.

Movement in suspension and in surface creep is a result of movement in saltation. The whole program of wind erosion control should therefore depend on methods designed to reduce or eliminate saltation.

The intensity of soil movement depends not so much on the force of the wind acting on the ground, as on its pressure against the grains as they leap in saltation. Soil movement is therefore dependent not on velocity at any fixed height but on the velocity distribution to the height of saltation.

Dust in suspension does not affect the general character of the movement in saltation or in surface creep, but the presence of dust in the soil increases the minimum velocity required to initiate erosion and decreases the intensity of erosion for a given erosive wind. Once lifted off the ground, fine dust is carried to great heights and distances from its original location and thus may be considered a complete loss to the eroding area. The soil moved in saltation and surface creep, on the other hand, usually remains within the eroding area, especially when the erosive winds are from different directions. The maximum diameter of soil particles carried in suspension is on an average slightly greater than 0.1 mm.

The proportion of the three types of movement varies widely for different soils. In the cases examined, between 55 and 72 per cent of the weight of the soil was carried in saltation, 3 to 38 per cent in suspension, and 7 to 25 per cent in surface creep. Coarsely granulated soils erode mainly in saltation, and finely pulverized soils, in saltation and suspension.

The trapping capacity of stubble or ridged strips depends on the relative receptiveness of the surface and the length of jump of particles in saltation. The data presented supply information on the approximate percentage of soil grains that may be caught by trap strips of different widths.

The effectiveness of a particular type of trap depends also on the height and density of the obstructions and on the resistance of these obstructions to the abrasive action of wind-borne grains. Soils vary greatly in resistance to abrasion, but grain stubble is virtually unaffected.

REFERENCES

(1) BAGNOLD, R. A. 1941 The Physics of Blown Sand and Desert Dunes. Methuen & Co. Ltd., London.

(2) Canada Department of Agriculture, Soil Research Laboratory 1943 Report of Investigations. Swift Current, Sask.

(3) CHEPIL, W. S., AND MILNE, R. A. 1939 Comparative study of soil drifting in the field and in a wind tunnel. *Sci. Agr.* 19: 249–257.

(4) CHEPIL, W. S., AND MILNE, R. A. 1941 Wind erosion of soil in relation to roughness of surface. *Soil Sci.* 52: 417–431.

(5) CHEPIL, W. S. 1941 Relation of wind erosion to the dry aggregate structure of a soil. *Sci. Agr.* 21: 488–507.

(6) FREE, E. E. 1911 The movement of soil material by the wind. U. S. Dept. Agr. Bur. Soils Bul. 68.

(7) UDDEN, J. A. 1894 Erosion, transportation, and sedimentation performed by the atmosphere. *Jour. Geol.* 2: 318–331.

10

FORMULAS FOR THE TRANSPORTATION
OF BED LOAD

By H. A. Einstein,[1] Assoc. M. Am. Soc. C. E.

With Discussion by Messrs. Joe W. Johnson, A. A. Kalinske, O. G. Haywood, Jr., Samuel Shulits, John S. McNown, and H. A. Einstein.

Synopsis

A method for the representation of bed-load data is given in this paper. The method is based on the conception that bed-load transportation is the movement of bed particles, as governed by the laws of probability. By means of this method an equation is obtained, which describes a great number of experiments in channels with uniform beds. A group of experiments conducted on sand mixtures provides material for describing another application of the method.

Introduction

In the past the problem of bed-load transportation has been studied mostly by empirical methods. More recently, there has been a tendency to base transportation formulas on the new theories of turbulence.[2,3] It is the writer's belief that an approach to the problem of transportation can be made by a combination of the empirical and rational methods and that the results can be expressed by dimensionless plots.

Before proceeding with the development of these formulas, it is necessary to discuss briefly two important considerations: (1) The difficulty or impossibility of defining, accurately, the so-called "critical" values; and (2) the possibility of correlating bed-load movement with local fluctuations in water velocity along the bed.

(1) Attempts have been made in the past to derive an expression for the "initial movement" that is governed by certain definable "critical" conditions to be used as the first step toward the solution of the transportation problem.

Note.—Published in March, 1941, *Proceedings*.

[1] Hydr. Engr., SCS, U. S. Dept. of Agriculture, Greenville, S. C.

[2] "Anwendung der Aehnlichkeitsmechanik und der Turbulenzforschung auf die Geschiebebewegung," by A. Shields, *Mitteilungen der Preussischen Versuchsanstalt für Wasserbau und Schiffbau*, Heft 26, Berlin, 1936.

[3] "An Analysis of Sediment Transportation in the Light of Fluid Turbulence," by Hunter Rouse, Assoc. M. Am. Soc. C. E., SCS, Sedimentation Div., SCS-TP-25, Washington, D. C., July, 1939 (mimeographed).

In interpreting the results of many experiments on bed-load movement, and in comparing them with those obtained by other experimenters, the writer has concluded that a distinct condition for the beginning of transportation does not seem to exist. It is just as impossible to determine the limit of initial movement as to determine the maximum possible flood of a river. Just as the engineer is able to predict the probable maximum flood of a river to be expected within a given range of years, however, so is he able to define the hydraulic conditions in a stream that will produce any given small rate of movement, which might be called the limit. This value can be chosen without any restriction. It is difficult to believe that the hydraulic conditions that will produce such movement could have any special meaning in the problem of transportation as a whole. Therefore, the writer will not use the conception of critical tractive force, or any other critical value, when the term "critical" pertains to the flow where transportation begins.

(2) In general, transportation of bed load has been described as follows:[4] A particle of the bed moves when the pushing force or lifting force of the water overcomes the weight of the particle. This push or "lift" is expressed in terms of the average flow. The usual conception is that transportation begins when the velocity increases enough to overcome the weight, and that, with further increasing velocity of the water, the rate of transportation will also increase, following a certain law that is found empirically. To prove that this conception is misleading, assume that the force acting on a particle could be described by means of the average flow alone. If the velocity of the water is increased gradually to a point at which the first particle would just be moving, the force acting on all the other particles of the same kind and size would move those too. Therefore, in a uniform bed where all particles have the same size and shape, all would start moving together; they would be unable to settle again because at all points the water velocity is just sufficient to start movement. If this were true, there could be no law governing the rate of transportation, but only a critical velocity. At all undercritical velocities, there would be no transportation, whereas, at all supercritical velocities, the rate of transportation would be limited only by the number of particles available. Therefore, it cannot be presumed that the rate of bed-load transportation is a function of the average flow. Instead it is proposed to express it in terms of the fluctuations of the water velocity near the bed.

Results of previous studies[5] describing the movement of a bed-load particle by means of statistical methods are to be used in an attempt to coordinate the rate of transportation with the fluctuations of the water velocity near the bed. The results of these studies can be summarized as follows:

(a) These flume studies dealt with the movement of rather coarse particles along a bed consisting of the same kind of grains. Being coarser than 1 in. these particles always remained near the bed, rolling, sliding and, sometimes, in saltation according to the normal description of bed-load movement. It

[4] "The Force Required to Move Particles on a Stream Bed," by William W. Rubey, *Professional Paper No. 189-E*, U. S. Geological Survey, Washington, D. C., 1938.
[5] "Der Geschiebetrieb als Wahrscheinlichkeitsproblem," von H. A. Einstein, *Mitteilung der Versuchsanstalt für Wasserbau an der Eidgenössische technische Hochschule in Zürich*, Verlag Rascher & Co., Zürich 1937.

was found that the moving bed load and the bed on which it was moving formed a unit, inasmuch as there was a steady and intensive exchange of particles between the two. Thus it is concluded that all the particles of the bed, down to a certain depth, take equal part in the movement, alternatively moving and returning into the bed.

(b) Bed-load movement is to be considered as the motion of bed particles in quick steps with comparatively long intermediate periods of rest. Thus bed-load movement is a slow downstream motion of a certain top-layer of the bed.

(c) The average step of a certain particle seems always to be the same even if the hydraulic conditions or the composition of the bed changes; and

(d) Different rates of transportation are produced by a change in average time between two steps and by a change in the thickness of the moving layer.

These concepts permit the development of a formula in general terms. The rate of transportation will be described by means of this average "step."

NOTATION

The letter symbols in this paper are defined where they first appear and are assembled for convenience of reference in Appendix I.

DERIVATIONS

This paper will treat only the bed-load movement of uniform sediment and mixtures acting like uniform sediment. In both cases it is possible to describe the sediment by a representative diameter D and its density ρ_s. The expression "acting like uniform material" means that both bed material and moving material have the same composition and, therefore, the same representative diameters. It is possible in this case to describe transportation at a certain point of the bed by one symbol; namely, the rate of transportation q_s. Two special cases of transportation, both of which have been observed and described in engineering literature, may be excluded from treatment in this paper: (1) Bed material moving in suspension; and (2) bed-load movements, in which the composition of the bed is essentially different from the composition of the transported material.

A bed-load formula is an equation linking the rate of bed-load transportation with properties of the grain and of the flow causing the movement. A formula of this kind can be derived by expressing in an equation the fact that all particles passing the unit width of a section as bed load are just on the way to perform one of these steps of the constant length $L = \lambda_0 D$. Fig. 1 shows the cross section and the rectangular area with the length L and with unit width, where all particles start the steps that together form the rate of transportation q_s. Eq. 1 expresses the condition that the total volume passing the unit width of the section per second (q_s divided by the specific gravity of the particles, both under water)

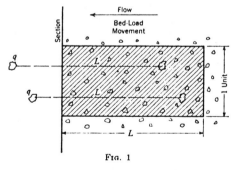

Fig. 1

equals the total volume of all the particles starting a step during a second in the aforementioned rectangular area. This volume is obtained by multiplying the number of particles in the surface of the area by the probability that a particle in the bed surface will start moving during a given instant and with the volume of a given particle. Eq. 1 follows:

$$\frac{q_s}{(\rho_s - \rho_f)\, g} = \frac{L}{A_1 D^2}\, p_s\, A_2\, D^3 = \frac{A_2}{A_1}\, \lambda_0\, p_s\, D^2 \dots\dots\dots\dots (1)$$

in which: q_s = the rate of transportation, in weight (under water), per unit of width, per second; ρ_s and ρ_f = density of particle and fluid, respectively; g = acceleration due to gravity; D = representative diameter of the particles; A_1 and A_2 = dimensionless ratios such that $A_1 D^2$ = the area that the grain covers in the bed and $A_2 D^3$ = the volume of the particle; p_s = the probability that a particle will start moving in any given second; and λ_0 = the dimensionless measure for the length of the single step.[5] It must be kept in mind, however, that λ_0 may or may not be a constant; that is, it has not been proved to be constant.

If A and D are transposed to the left side of Eq. 1, that side will include all terms pertaining to the grain, whereas the right side $\lambda_0 p_s$ is still an unknown function of the flow—that is:

$$\frac{q_s}{(\rho_s - \rho_f)\, g\, D^2} \frac{A_1}{A_2} = \lambda_0\, p_s \dots\dots\dots\dots (2)$$

In this equation A_1, A_2, and λ_0 are dimensionless, but p_s has the dimension sec^{-1}. In order to make the right side of Eq. 2 dimensionless, p_s must be multiplied by a given time. The most reasonable time to use is the average time, t, required for the water to remove one particle from the bed. If $p = t\, p_s$, then p is dimensionless and gives the number of steps that start from any given place during the time it takes to remove one particle. The maximum value of p is 1, and indicates that at all times and at all points particle by particle starts to move. The minimum value of p is zero; therefore, p expresses the probability that a step is about to begin at a given place. These steps start everywhere on the bed. Therefore, p can be interpreted as the probable part of the bed area in which steps are starting. A step is started only at a point where the hydraulic lift of the water is able to overcome the weight of the particle. Therefore, p expresses the probability that the hydraulic lift on any particle along the bed is about to overcome the weight of the particle.

Unfortunately there is no method of expressing or measuring the time t required for the lifting force to pick up a particle. It is assumed to be proportional to some other characteristic time of the particle in the water. The time that the particle requires to settle in water, a distance equal to its own diameter D, is chosen for this characteristic. The reason for choosing this time was the fact that it is the only expression with the dimension of a time, which is representative for the behavior of the particle in the liquid without including any characteristics of the flow. This time can be expressed as

$$\frac{D}{v_f} = \frac{1}{F}\sqrt{\frac{D\,\rho_f}{g\,(\rho_s - \rho_f)}} \cdot \dots\dots\dots\dots (3)$$

in which: v_f = the velocity of a particle settling in water; and F = a parameter for settling velocity. In Eq. 3, $F = 0.816$ for particles greater than 1 mm, settling in water of normal temperature. Fig. 2 shows the values of F for smaller grain sizes. Characteristics of the materials in Fig. 2 are: Kinematic viscosity $\nu = \dfrac{\mu}{\rho_f}$ equals, for water, 0.012, and, for air, 0.16, cm² per sec; and specific densities, $\dfrac{\rho_s - \rho_f}{\rho_f}$, are as follows—

Material	Specific density
Barite in water	3.22
Gravel in water	1.65
Coal in water	0.25
Gravel in air	2,210.0

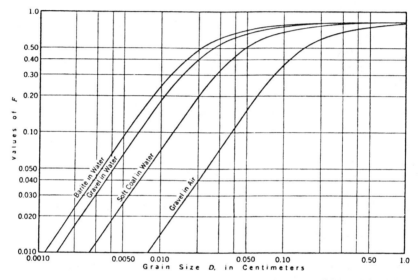

FIG. 2.—PARAMETER F FOR DETERMINING THE SETTLING VELOCITY OF VARIOUS MATERIALS

In Fig. 2 the following equation for the settling velocity derived by William W. Rubey[6] has been used for the determination of F:

$$v_f = \sqrt{\frac{2}{3} g \frac{\rho_s - \rho_f}{\rho_f} D + \frac{36 \mu^2}{\rho_f^2 D^2}} - \frac{6 \mu}{\rho_f D} = F \sqrt{D g \frac{\rho_s - \rho_f}{\rho_f}} \quad \ldots \ldots (4)$$

In this formula all terms are measured in centimeter-gram-second units. Hence, F will be

$$F = \sqrt{\frac{2}{3} + \frac{36 \mu^2}{g D^3 \rho_f (\rho_s - \rho_f)}} - \sqrt{\frac{36 \mu^2}{g D^3 \rho_f (\rho_s - \rho_f)}} \quad \ldots \ldots (5)$$

6 "Settling Velocities of Gravel, Sand, and Silt," by William W. Rubey, *American Journal of Science,* Vol. 25, No. 148, April, 1933.

The time t required to remove a particle from its place in the bed then will be

$$t = \frac{A_3}{F} \sqrt{\frac{D\,\rho_f}{g\,(\rho_s - \rho_f)}} = \frac{p}{p_s} \dots\dots\dots\dots (6)$$

in which A_3 is still an unknown constant. Eq. 1 can now be changed to the form

$$p = \frac{A_1\,A_3}{\lambda_0\,A_2} \left[\frac{1}{F} \frac{q_s}{(\rho_s - \rho_f)\,g} \sqrt{\frac{\rho_f}{\rho_s - \rho_f}} \frac{1}{g^{0.5}\,D^{1.5}} \right] \dots\dots\dots (7)$$

In an attempt to express p as the probability of the local hydraulic lift[4] to overcome the weight of the particle, p refers to the part of the bed in which locally (at a certain moment) the lifting force is greater than the weight of the particle. It can be stated that p refers to the part of the bed in which the ratio of the local lift to the average lift is greater than the ratio of the weight of the particle to the average lift. In mathematical terms this is

$$p = f\left(\frac{\text{Weight of the particle}}{\text{Average lift of the particle}} \right) \dots\dots\dots\dots (8)$$

in which f is an unknown function. The weight of the particle under water is $A_2\,D^3\,(\rho_s - \rho_f)\,g$, and the average lift is

$$\text{Lift} = A_4\,D^2\,v^2\,\rho_f \dots\dots\dots\dots\dots (9)$$

v being a local velocity at some still unknown distance from the bed. An approximate value for v is:

$$v = 11.6 \sqrt{\frac{\tau}{\rho_f}} \dots\dots\dots\dots\dots (10)$$

Eq. 10 may need to be corrected in the future. It defines the velocity[3] at the edge of the laminar boundary layer if the wall is smooth, or the velocity at the distance D if the wall is rough, in which D is a measure of average roughness. The shearing stress τ along the wall is

$$\tau = S\,R\,\rho_f\,g \dots\dots\dots\dots\dots (11)$$

in which S is the slope and R is the hydraulic radius; therefore

$$v = 11.6 \sqrt{S\,R\,g} \dots\dots\dots\dots\dots (12)$$

Eq. 8 can now be written

$$p = f\left[\frac{A_2\,D^3\,(\rho_s - \rho_f)\,g}{(A_4\,D^2\,\rho_f)\,(135\,S\,R\,g)} \right] = f\left[\frac{A_2}{135\,A_4} \times \frac{(\rho_s - \rho_f)\,D}{\rho_f\,S\,R} \right] \dots (13)$$

By assuming that Eq. 10 gives the correct value for the velocity, Eqs. 7 and 13 can be combined and a new transportation formula formed:

$$A \left\{ \frac{1}{F} \left[\frac{q_s}{(\rho_s - \rho_f)\,g} \right] \sqrt{\frac{\rho_f}{\rho_s - \rho_f}} \frac{1}{g^{0.5}\,D^{1.5}} \right\} = f\left[B\left(\frac{\rho_s - \rho_f}{\rho_f} \frac{D}{S\,R} \right) \right] = p. (14)$$

in which

$$A = \frac{A_1 A_3}{\lambda_0 A_2}; \dotfill (15a)$$

and

$$B = \frac{A_2}{135 A_4} \dotfill (15b)$$

are constants which, however, may vary with different shapes of the particles. Whether λ_0 and A_4 really are constant under all conditions must be determined later. Introducing

$$\phi = \frac{1}{F} \frac{q_s}{(\rho_s - \rho_f) g} \sqrt{\frac{\rho_f}{\rho_s - \rho_f}} \frac{1}{g^{0.5} D^{1.5}} \dotfill (16a)$$

and

$$\psi = \frac{\rho_s - \rho_f}{\rho_f} \frac{D}{S R}. \dotfill (16b)$$

Eq. 14 can be written in the short form:

$$A \phi = f(B \psi) = p. \dotfill (17)$$

The function f as well as the two constants A and B must be determined empirically. Data from a great number of measurements using various materials have been analyzed, and values of ψ and ϕ computed. A semilogarithmic plot of these values is shown in Fig. 3(a). The grain sizes range from 0.315 to 28.65 mm in diameter, the water depth from 18 to 1,100 mm, and the specific gravity of the particles from 1.25 to 4.22. All these experiments are performed in flumes with uniform sediment. The experiments marked "Zürich" are described briefly in a paper by E. Meyer-Peter[7] whereas the remaining experiments are taken from the generally known paper by G. K. Gilbert.[8] It seems that all these points follow, reasonably, a single curve. It should be noted that from these data not a single experiment has been omitted even when the measurement appeared questionable. It might be mentioned also that the hydraulic radius R is computed by a method[9] that eliminates the effect of side-wall friction and gives results comparable to a channel of infinite width (see Appendix II).

In Fig. 3(a) all the points with values of ϕ less than 0.4 seem to follow the straight line, curve (1), the equation of which is

$$0.465 \phi = e^{-0.391 \psi} \dotfill (18)$$

If Eq. 18 is assumed to represent the law of transportation:

$$\left.\begin{array}{l} A = 0.465 \\ B = 0.391 \\ f(x) = e^{-x} \end{array}\right\} \dotfill (19)$$

---

[7] "Neuere Versuchsresultate über den Geschiebetrieb," by E. Meyer-Peter, H. Favre, and A. Einstein, *Schweizerische Bauzeitung*, Vol. 103, No. 13, March, 1934.

[8] "The Transportation of Debris by Running Water," by Grove Karl Gilbert, *Professional Paper No. 86*, U. S. Geological Survey, Washington, D. C., 1914.

[9] "Der hydraulische oder Profil-Radius," by H. A. Einstein, *Schweizerische Bauzeitung*, Vol. 103, No. 8, February 24, 1934; see also Appendix II.

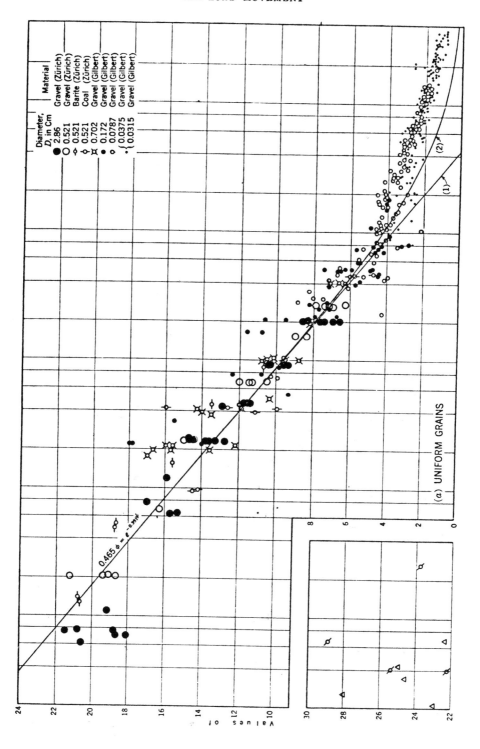

(a) UNIFORM GRAINS

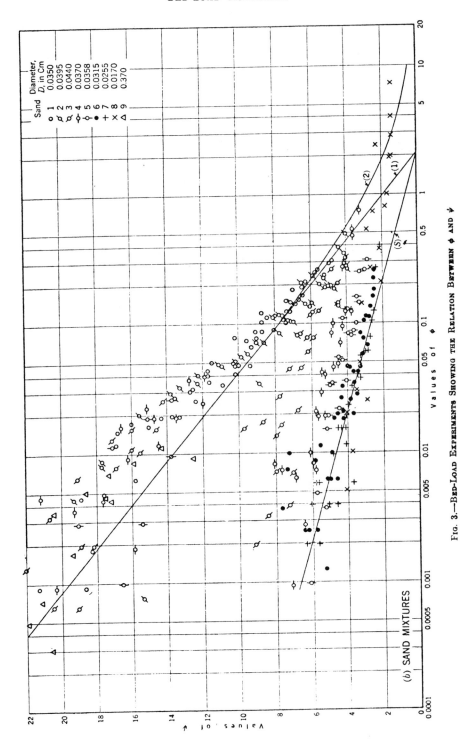

FIG. 3.—Bed-Load Experiments Showing the Relation Between ϕ and ψ

it remains only to explain why the points $\phi > 0.4$ seem to be too high. If Eq. 18 is accepted as a general law, the value $\phi > 2.15$ would not be possible because p cannot exceed 1. Therefore, values of $\phi > 2.15$ are possible only if A becomes smaller (A consists of the constants A_1, A_2, A_3, and λ_0).

The constants A_1, A_2, and A_3 are not likely to change with increasing values of p, but λ_0 does. The distance λ_0 has been found to be constant for small values of p—that is, when the hydraulic lift seldom exceeds the weight. As p increases, it more often happens that, in the very spot where the step would have ended, there exists a local lift strong enough to keep the particle from settling. The oftener this happens the more λ_0 seems to increase on the average. The symbol p expresses the probability that the lift exceeds the weight of the particle for every point on the bed. Therefore, only $(1 - p)$ particles of the unit will be able to settle after a step λ_0. The other p particles will start for another λ_0, and out of these $(1 - p)\,p$ will settle after the second λ_0, and so on. The average distance traversed by the unit, therefore, is:

$$\lambda = \sum_{m=0}^{\infty} (1 - p)\, p^{m-1}\, m\, \lambda_0 = \frac{\lambda_0}{1 - p} \dots\dots\dots\dots (20)$$

in which $m = $ a whole positive number. Introducing λ instead of λ_0 yields curve (2) instead of curve (1). This new curve follows the plotted points more closely than curve (1).

The constant A_4 for the lift in Eq. 9 is introduced without further discussion. The question arises whether the deviation of the points from curve (2) could be due to a change in the constant A_4. Constant A_4 could change only with the Reynolds number of the flow around the particle, or with the value of $\frac{D}{\delta}$, the ratio of the grain size to the thickness of the laminar layer. Eq. 10 gives the local velocity of the water:

$$v = 11.6 \sqrt{\frac{\tau_0}{\rho_f}} = 11.6 \sqrt{S\,R\,g} \dots\dots\dots\dots (21)$$

The Reynolds number of the local flow is

$$R = \frac{D\,v}{\nu} = \frac{11.6\,D\,\sqrt{S\,R\,g}}{\nu} \dots\dots\dots\dots (22)$$

and the thickness of the laminar layer is

$$\delta = \nu\,\frac{D\,v}{\tau_0/\rho_f} = \frac{11.6\,\nu}{\sqrt{S\,R\,g}} \dots\dots\dots\dots (23)$$

or

$$\frac{D}{\delta} = \frac{D\,\sqrt{S\,R\,g}}{11.6\,\nu} = \frac{R_D}{134} \dots\dots\dots\dots (24)$$

As $\frac{D}{\delta}$ differs from R_D only by a constant factor, it is sufficient to study the influence of only one of them. The deviation of the measured points from curve (2) plotted against $\frac{D}{\delta}$ failed to disclose any satisfactory relationship. Each grain size appears to follow a separate curve; therefore, it appears much more probable that A_4 is a constant, but that the exponential law for p does not extend down to $\psi = 0$. Another explanation for the deviation of the points may be that part of the grains have been transported in suspension. In this case those experiments would be outside the field of application of Eq. 17. More experiments, however, will be required in this range to determine the exact shape of the curve. It may be emphasized that in most American rivers the bed load is largely transported under conditions pertaining to this part of the curve, and, therefore, further study in this direction is most urgent.

The title of this paper was chosen specifically to avoid the impression that any attempt was being made to discover "the law of bed-load transportation," because it is the writer's belief that such universal law does not exist in a simple mathematical form. Just as it is necessary to distinguish between friction in smooth and rough pipes or channels, so it is necessary to distinguish between different kinds of transportation. Nevertheless, the distinction between friction along rough, wavy, and smooth walls was only possible on the discovery of the general method of plotting the friction factor against Reynolds' number. An attempt is made in this paper to determine a corresponding method of presenting transportation data by introducing the quantities ψ and ϕ, both of which are derived by pure speculation, using only generally known facts.

As an example, the method is used to discuss the results of experiments with sand mixtures, conducted at the U. S. Waterways Experiment Station, at Vicksburg, Miss.[10] Fig. 3(b) gives the results of these experiments as a ϕ-ψ graph. The method described in Appendix II has been used to determine the effective hydraulic radius. Curves (1) and (2) are transferred from Fig. 3(a), and all the various sand mixtures have been assigned different symbols.

The first problem was to determine the effective diameter of the mixtures—that is, the value of D that would represent the mixture in the formulas. Experience gained in previous studies has convinced the writer that the most usable value for this effective diameter is the grain size of which 35% to 45% of the material is finer. This value is readily obtained from the cumulative size-frequency curve of the mixture. The 40% value was used for Fig. 3(b), although it is realized that the use of a 35% value would tend to bring the high points closer to curves (1) and (2).

The distribution of the points in Fig. 3(b) is very interesting to note. At first glance one observes that the points for sands 1, 2, and 9 distinctly follow curves (1) and (2). Sands 3, 4, and 5 follow the curves in the upper part only. Sands 6, 7, and 8 fall below the curves at all points, but a distinct grouping of points along a line curve (S) is noticed.

The two curves, (2) and (S), seem to represent limits of maximum and minimum transportation for a given value of ψ. In searching for an explanation of this the three following questions naturally arise: (1) Is there any relationship between the position of the points and friction loss? (2) Is there any relationship between the position of the points and the condition of the bed?

---

[10] "Studies of River Bed Materials and Their Movement, with Special Reference to the Lower Mississippi River," *Paper No. 17*, U. S. Waterways Experiment Station, Vicksburg, Miss., January, 1935.

(3) Would a similar distribution be possible also in experiments with uniform material, or is it characteristic only of mixtures?

With regard to question (1), it was found that Manning's n, without exception, increased suddenly when the points leave curves (1) and (2). This means that the bed becomes rougher than the original material as soon as the rate of transportation decreases below that shown by curves (1) and (2). The reverse is also true—that is, the rate of transportation will decrease as soon as the roughness of the bed increases.

The reason for this increased roughness is of interest. As a rule, riffles begin to form precisely at the place where the points depart from curve (1). If riffles are the reason for the increase in Manning's n, this value must always increase when riffles are formed. Sand 1 does not show this increase and sand 2 only very slightly—although these sands develop general riffles like all the other mixtures. Therefore, the riffles are not the reason for the increased roughness, but merely happen to develop simultaneously. This answers question (2) in the negative.

Question (3) suggests that perhaps some kind of sorting of the grains is the reason for the deviation from curve (1) and for the increase of roughness at the same time. It is the writer's belief that this is true, but unfortunately it is not subject to direct proof. This sorting would be caused by the lack of material in the upper end of the flume. If the sand is fed in at a smaller rate than the stream is able to transport it, the bed starts to build some kind of a protective layer of coarse grains on the surface and buries all the finer grains beneath. The average grain size in this coarse layer is much greater than the average grain size in the bed, and scour is either reduced or completely prevented. For this reason, it is impossible, during an experiment, to detect a lack of feeding merely by watching the position of the bed. Curve (1) gives the results obtained when the highest quantity of sand is fed in that can be transported without deposition, and curve (S) gives the smallest amount that will be transported without scour on the protecting layer. If D is the effective diameter of the original bed material, it is reasonable to believe that these two limits coincide for uniform material and that curve (S) falls more and more below curve (1) as the material decreases in uniformity.

These are merely some suggested methods of studying bed-load transportation. It would be very easy to determine, by experiment, the validity of such reasoning. If the interpretations are correct, it should be possible to determine all points between the two limiting curves by merely changing the rate of sand feed. It would also be very instructive to conduct a similar group of experiments in the opposite sequence—that is, by beginning with high discharges and progressively decreasing the discharge and load. If the explanation is correct, one will probably not return to the same curve obtained with increasing flow. This would also answer question (3). This ϕ-ψ method is offered as a new procedure for studying bed-load problems. It may be possible to refine the method by introducing corrections for the velocity v, and various constants; but as a whole it seems to be satisfactory in its present form.

Summary

In concluding it may be stated that the treatment of transportation problems by means of statistical methods, made possible by the use of large-scale experiments, led to the proposed method of representation:

Two dimensionless functions ψ and ϕ have been developed theoretically, ψ as the ratio of the forces acting on the particle, and ϕ including the rate of transportation and the size and settling velocity of the particle. The interrelation between these two functions expresses the law of transportation, and at the same time expresses the statistical distribution p of the velocity of the liquid close to the laminar boundary layer. The transportation law is derived by use of a great many experiments with uniform sediment, and is then used in discussing published results of experiments with sand mixtures.

Acknowledgments

The writer is greatly indebted to Prof. E. Meyer-Peter of Zürich, Switzerland, for having permitted the use of valuable data, even though the results of the original work have not as yet been published in full.

[*Editors' Note:* Appendixes I and II and the discussion have been omitted.]

11

Reprinted from *Water Resources Research* 2:131–138 (1966)

Resistance of Selected Clay Systems to Erosion by Water

EARL H. GRISSINGER

Sedimentation Laboratory, Soil and Water Conservation Research Division
U. S. Agricultural Research Service, Oxford, Mississippi

Abstract. Rates of erosion of selected clay systems were determined by subjecting molded samples to a uniform erosive force. Increased concentrations of clay minerals generally induced greater stability. Increased bulk densities, however, had little influence on stability. The influence of clay particle orientation and antecedent water content (water content at the start of the test) was not consistent but varied depending upon the clay mineral mixture. Stability increased with increased antecedent water for the Grenada silt loam and for illitic, montmorillonitic, and oriented kaolinitic samples. Stability decreased with increased antecedent water for unoriented kaolinitic samples. Increased orientation of the Grenada silt loam reduced stability but did not change the influence of antecedent water. Greater erosion rates occurred as the temperature of the eroding water was increased. (Key words: Erosion; soil; clay minerals)

INTRODUCTION

Stability of stream channels is a function of both streamflow dynamics and the resistance of channel material to erosion. This study is limited to the determination of properties of certain clay systems that are conducive to erosion resistance. It represents preliminary findings of current research at the U. S. Department of Agriculture Sedimentation Laboratory, Oxford, Mississippi, concerning the stability of stream channels in cohesive materials. The ultimate goal of this work is the development of practical tests for use in channel design.

The influence of selected soil properties on the stability of cohesive materials was evaluated qualitatively by subjecting samples to a constant erosive force produced in a small flume. Remolded samples were employed to achieve good control of the variables and to fabricate a known range of soil properties. The soil properties related to erodibility include bulk density, antecedent water, type of clay mineral, percentage of clay, and orientation of the clay minerals.

Dunn [1959], *Smerdon and Beasley* [1959], *Lane* [1955], *Moore and Masch* [1962], and others have used critical tractive force values as stability measurements. For this study, values for rate of erosion expressed as grams of material eroded per minute were used. Changes in

area and in roughness of the eroding surface during erosion were considered functions of the erodibility of the material.

PROCEDURES

Material. A Grenada silt loam soil was used as stock material to which various clay minerals were added to simulate a wide range of natural conditions. The Grenada soil is loessial. The sample used as stock material was composed of 20% clay, 74% silt, and 6% fine sand, computed on the basis that the material was free of organic matter. The organic matter content was low (0.6%). Moisture percentages for the liquid limit and plastic limit were 31 and 20, and the plasticity index was 11. The clay fraction was composed of kaolin, montmorillonite, vermiculite, illite, and some amorphous material. Clay minerals were identified by standard X-ray diffraction techniques.

Clay mineral additives were kaolinite, purified Fithian illite, and two montmorillonites. These clays are commercial and possibly differ from naturally occurring soil clay minerals, but they have the advantages of uniformity. The illite sample was obtained from the Illinois Geological Survey.[1] The two montmorillonite sam-

[1] Trade names and company names are included for the benefit of the reader and do not infer any endorsement or preferential treatment of the product listed by the U. S. Department of Agriculture.

ples were a sodium montmorillonite, Volclay, and a calcium montmorillonite, Panther Creek, obtained from American Colloid Company. Differences between these two montmorillonites were not due entirely to the different saturation cations, since the internal compositions also differed.

Three sizes of kaolinite were obtained from Georgia Kaolin Company:

	Average Diameter, Microns	Designation
Hydrite UF	0.2	Fine
Hydrite 121	1.5	Medium
Hydrite MP	9.5	Coarse

For convenience, the three kaolinites were designated as fine, medium, and coarse.

Sample preparation. A bulk sample of the Grenada soil was air dried, ground to pass a No. 18 (1000-micron) sieve, split to appropriate sizes, and stored in plastic bags. Individual samples were thoroughly ground in a ball mill to reduce aggregation. Weighed amounts of the clay minerals were added to the stock material and mixed in the ball mill. The mixed samples were then sprayed with water to the desired weight content. The moist samples were stored overnight to ensure wetting of the clay surfaces.

Samples of wet weights required for desired oven-dried bulk densities were pressed into brass molds until the sample surface was flush with the top surface of the mold. These molds were 5.04 cm wide and 12.5 cm long with a volume of 116 ± 1 cc. The piston of the press moved within a removable collar on top of the mold.

Stainless steel filters having an average pore size of 5 microns were used as bottoms of the molds. These filters permitted water exchange following compaction. The resultant water content, termed antecedent water, was determined prior to testing by sampling a portion of the molded sample not exposed to erosion. Changes in the water content during erosion were considered a property of the material.

Orientation procedures. Two techniques were employed to induce orientation of the clay minerals. These techniques were based on the work of *Lambe* [1958a, b] and *Seed and Chan* [1961].

The first technique involved compaction at water contents approaching saturation for the desired bulk density. Soil systems have an optimum water content for inducing orientation by compaction, i.e., orienting the major plane surface of the clays normal to the compacting effort. For these samples, compaction at about 20% water induced more orientation than was induced by dry compaction, that is, compaction at approximately 10% water. Increased compactive effort at constant water content also resulted in increased orientation. Thus, compaction at 10% water to a 1.6 bulk density produced more orientation than was produced by compaction at 10% water to a 1.3 bulk density.

The second technique was a change in orientation following initial compaction. Compacted samples were allowed to sorb sufficient water through the stainless steel filter to cause swelling. The samples were then mounted on a vacuum rack and subjected to vacuums ranging from 45 cm of water to 60 cm of mercury. Change in volume caused by the withdrawal of sorbed water resulted in realignment of clay particles, the major plane surfaces of the clays orienting perpendicular to the vacuum vector. In all cases reported, vacuum was maintained until the swelled surface was restored to the original position.

The orientation of the clay minerals was measured by X-ray diffraction procedures using the same crystal planes used by *Meade* [1961]. These orientation values were ratios of the intensities of X-ray reflections from two selected crystal planes positioned essentially normal to each other. The ratio of the intensities of the X-ray reflections from these normal planes was minimal for unoriented samples. However, for oriented samples the intensity of one plane, 001, was enhanced in relation to a normal plane, 020. Hence the ratio of intensities of the 001:020 varied directly with orientation. Orientation of the clay minerals in the Grenada silt loam could not be measured because of the masking influence of amorphous material and the small content of crystalline clays. All orientation values are for mixtures of the Grenada with various clay minerals.

Testing procedures. Molded samples were subjected to a constant erosive force in a small flume. The flume bed was essentially horizontal and 5 cm wide. The flow velocity was 55 cm/sec (1.8 ft/sec), and the depth of flow was 2.0 cm. Velocity was checked during all tests by meas-

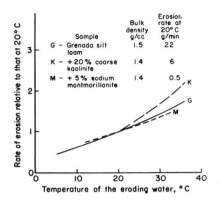

Fig. 1. Influence of temperature on erodibility of Grenada silt loam with admixtures, compacted and tested at 10% water.

uring the depth of flow in the flume and the time of fill of a calibrated barrel. The test material in the mold was mounted on the outside of an abrupt 5° turn, becoming a part of the side wall of the flume. Thus the test material was oriented so that the direction of packing of the material was normal to the general direction of flow. Material remaining in the mold after erosion was transferred to a pan, dried, and weighed. Eroded material was determined by difference in weight.

Controlled variables. The temperature of the eroding water and the wet aging time prior to erosion were found to influence the measured rates of erosion. These variables were investigated to ensure reproducible experimental conditions. They were not incorporated into this preliminary phase of the study of stability of cohesive materials.

The temperature of the eroding water influenced erodibility (Figure 1), although no change in flow velocity or depth was detected. This increase in erodibility with increasing temperature necessitated a recirculating water supply, making it possible to restrict water temperature to 25° ± 2°C.

Stability was also influenced by aging time for samples of given water content (Figure 2). Samples wetted after compaction required at least 4 hours of aging to approach an equilibrium condition.

RESULTS

This study was designed to evaluate qualitatively the soil properties that regulate the stability of cohesive materials against erosion by flowing water. Reproducibility was generally better within days than between days.

The study included only resistance to scour, and it neglected properties such as volume changes associated with freezing and thawing, wetting and drying, etc., which may indirectly affect stability. Values for the rate of erosion were determined for short time tests. The montmorillonite mixture of Figure 1 was eroded for 10 minutes. All other samples were eroded for 3 minutes, or for less than 3 minutes if the rates of erosion were high.

For ease of discussion the four soil-clay systems are considered separately. These four systems are (1) Grenada silt loam, (2) montmorillonite-Grenada mixtures, (3) illite-Grenada mixtures, and (4) kaolinite-Grenada mixtures. The samples are described as stable if the measured rates of erosion are small and as erodible if the rates are high. These descriptions are, of course, relative to the hydraulic conditions.

Grenada silt loam. The stability of the Grenada silt loam was dependent upon antecedent water and upon compaction technique (Figures 3 and 4). Both compaction at approximately 20% water and vacuum treatment were employed to induce orientation greater than that for samples compacted at approximately 10% water. Unfortunately, orientation was not measurable by X-ray diffraction for these samples. The observation was made, however, that 'wet' packed and vacuum treated samples eroded as shale-like fragments, whereas 'dry' packed samples eroded as fairly equidi-

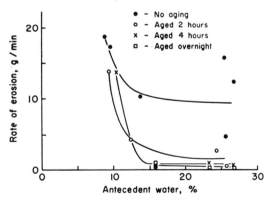

Fig. 2. Effect of aging on the erodibility of a Grenada silt loam, packed at 10% water to a 1.4 bulk density.

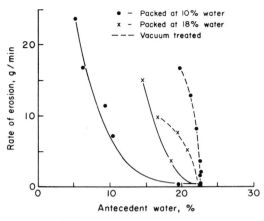

Fig. 3. Influences of packing technique and antecedent water on the erodibility of a Grenada silt loam at a 1.6 bulk density.

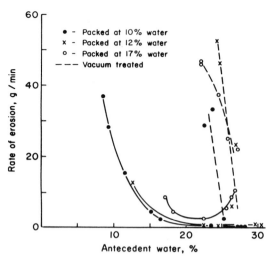

Fig. 4. Influences of packing technique and antecedent water on the erodibility of a Grenada silt loam at a 1.4 bulk density.

mensional particles. The influence of water at time of compaction on orientation was measurable for the 15% coarse kaolinite mixture (Table 1). For these samples orientation, as measured by the 001:020 ratio, increased as the water content at time of compaction increased. These orientation measurements of the kaolinite mixture and the observed platy form of the 'wet' packed and the vacuum treated Grenada silt loam samples indicated that the clay particle orientation was dependent upon sample treatment. For both the 1.4 and the 1.6 bulk density Grenada silt loam samples, the general stability varied inversely with the inferred orientation.

Increased antecedent water in the Grenada silt loam resulted in greater stability except for those water contents greater than 25% for 1.4

bulk density 'wet' packed samples. In general, the 1.6 bulk density samples were more stable than the 1.4 samples. The major differences in erodibility of these two bulk densities occurred for small antecedent water contents.

Dilution of the Grenada silt loam by additions of 10% silt reduced the stability of the material but did not alter the dependency of the stability on antecedent water (Figure 5).

Montmorillonite-Grenada silt loam mixtures. The stability of a calcium montmorillonite-Grenada silt loam mixture increased with increasing antecedent water (Figure 6), as did that of the Grenada silt loam. This increased

TABLE 1.   Orientation Measurements

Rate of Erosion Values	Process Inducing Orientation Change	Material	Bulk Density, g/cc	Per Cent, Clay	Per Cent, Water*	Ratio, 001 : 020
Table 2	Compaction	Ca montmorillonite	1.3	10	10	2.1
Table 2	Compaction	Ca montmorillonite	1.6	10	10	2.4
Figure 8	Moisture	Coarse kaolinite	1.4	15	5	3.4
Figure 8	Moisture	Coarse kaolinite	1.4	15	11	4.2
Figure 8	Moisture	Coarse kaolinite	1.4	15	14	5.2
Figure 8	Moisture	Coarse kaolinite	1.4	15	19	5.4
Table 3	Compaction	Fine kaolinite	1.3	10	10	2.6
Table 3	Compaction	Fine kaolinite	1.6	10	10	3.8
Table 3	Compaction	Coarse kaolinite	1.3	10	10	3.4
Table 3	Compaction	Coarse kaolinite	1.6	10	10	5.9
Table 3	Compaction	Coarse kaolinite	1.3	20	10	4.2
Table 3	Compaction	Coarse kaolinite	1.6	20	10	6.4

* Approximate values for the per cent water at the time of compaction.

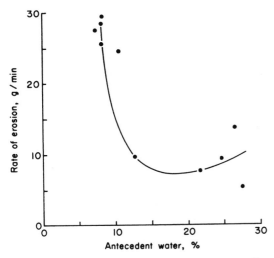

Fig. 5. Influence of antecedent water on the erodibility of a 10% natural silt-Grenada silt loam mixture, packed at 8% water to a 1.4 bulk density.

TABLE 2. Influence of Bulk Density on the Erodibility of Grenada Silt Loam Mixed with Either Montmorillonite or with Illite, Compacted at 10% Water*

Sample	Bulk Density, g/cc	Rate of Erosion, g/min		
		2% Mix	5% Mix	10% Mix
Mixed with Ca montmorillonite	1.3	14	16	21
	1.4	16	17	20
	1.5	17	17	17
	1.6	17	16	14
Mixed with Na montmorillonite	1.3	5	1	<1
	1.4	7	1	<1
	1.5	7	<1	<1
	1.6	7	<1	<1
Mixed with illite	1.3	21	13	6
	1.4	23	15	7
	1.5	23	14	7
	1.6	21	12	6

* Eroded at antecedent water contents of approximately 10%.

stability does not imply that the natural Grenada clay minerals are calcium systems, but only that this antecedent water versus stability relationship is applicable to a range of clay systems. The Grenada clays are probably high aluminum systems, since the soils are naturally acid.

Observation of the influence of bulk density on the stability of various sodium montmorillonite-Grenada mixtures and calcium montmorillonite-Grenada mixtures (Table 2) was confused by the increased orientation of the more dense samples. This increase in orientation was substantiated for calcium montmoril-

lonite-Grenada mixtures by the small but reproducible increase in the ratio of the intensity of the 001:020 reflections for the higher bulk density material (Table 1). Thus the true influence of bulk density on stability was probably somewhat greater than that presented in Table 2.

The complex nature of soil properties is well illustrated in Table 2. Although increased concentrations of calcium montmorillonite induced stability at high bulk densities, they also induced erodibility at low bulk densities. The stability of sodium montmorillonite-Grenada mixtures, however, was independent of bulk density, being controlled by the percentage sodium montmorillonite. The extreme stability of the sodium montmorillonite-Grenada mixtures was undoubtedly related to the minute size and the relatively extreme flexibility of these particles. The rate of erosion values would be greatly reduced with increased antecedent water.

Illite-Grenada silt loam mixtures. The dependencies of stability of illite-Grenada silt loam mixtures on antecedent water, bulk density, and amount of clay (Figure 7 and Table 2) were similar to those of the preceding samples. Stability increased with increasing antece-

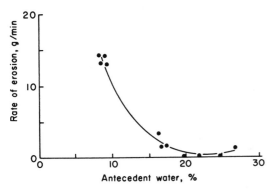

Fig. 6. Influence of antecedent water on the erodibility of a 5% calcium montmorillonite-Grenada silt loam mixture, packed at 8% water to a 1.5 bulk density.

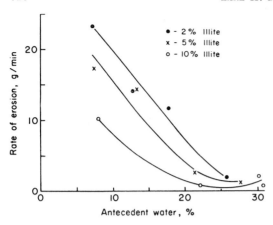

Fig. 7. Influence of antecedent water on the erodibility of illite-Grenada silt loam mixtures, packed at 10% water to a 1.4 bulk density.

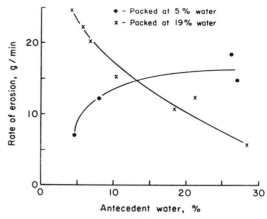

Fig. 9. Influences of packing technique and antecedent water on the erodibility of a 10% coarse kaolinite-Grenada silt loam mixture at a 1.4 bulk density.

dent water and with increasing illite content, but it appeared relatively independent of the compacted bulk density.

Kaolinite-Grenada silt loam mixtures. The relations between stability and various soil properties for kaolinite-Grenada mixtures were different than the relations for the preceding samples. Stability decreased with increasing antecedent water for the unoriented samples compacted 'dry' to a 1.4 bulk density (Figures 8–11). The stability of oriented samples, those compacted at approximately 20% water, increased with increasing antecedent water (Figures 8, 9, and 11). Oriented samples for

Figures 9 and 11 were prepared by compacting the samples at approximately 20% water, saturating selected samples, and air-drying others to varying degrees of water before testing. The antecedent water values for air-dried samples should be considered as average values for individual samples. For these samples, water content was probably variable with depth because of the drying process. The 15% kaolinite-Grenada silt loam samples compacted at 19% water (Figure 8) were relatively stable at 17–23% antecedent water. Stability values were not determined at other antecedent water contents. Differences between the stabilities of these 'wet' packed samples and samples compacted at reduced water contents indicate an abrupt orientation influence on stability, although the orientation as measured by the

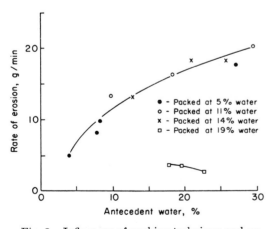

Fig. 8. Influences of packing technique and antecedent water on the erodibility of a 15% coarse kaolinite-Grenada silt loam mixture at a 1.4 bulk density.

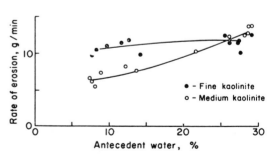

Fig. 10. Influence of antecedent water on the erodibility of 10% kaolinite-Grenada silt loam mixtures packed at 7% water to a 1.4 bulk density.

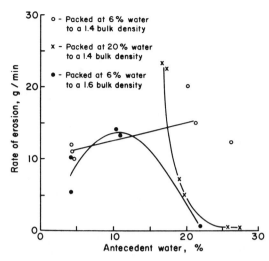

Fig. 11. Influences of packing technique and antecedent water on the erodibility of a 5% coarse kaolinite-Grenada silt loam mixture.

TABLE 3. Influence of Bulk Density on the Erodibility of Grenada Silt Loam Mixed with Kaolinites, Compacted at 10% Water*

Sample	Bulk Density, g/cc	Rate of Erosion, g/min			
		2% Mix	5% Mix	10% Mix	20% Mix
Mixed with fine kaolinite	1.3	15	12	10	
	1.4	17	14	10	
	1.5	18	15	9	
	1.6	18	14	8	
Mixed with coarse kaolinite	1.3	21	19	18	6
	1.4	24	20	15	4
	1.5	22	20	10	3
	1.6	20	17	9	3

* Eroded at antecedent water contents of approximately 10%.

001:020 ratio increased fairly linearly (Table 1).

For these kaolinite-Grenada silt loam mixtures, orientation was also induced by compaction to a high bulk density at low water content (Table 1). The rate of erosion versus antecedent water curve for the 1.6 bulk density sample in Figure 11 is a typical example of a fairly dense, oriented, kaolinitic material.

For unoriented kaolinitic mixtures, the increase of erodibility with increasing antecedent water became smaller as particle size decreased (Figures 9 and 10). Kaolinites typically are coarse clay minerals with a relatively small specific surface. As kaolinite particles of smaller size were used in the mixtures the specific surface was increased, and the surface properties undoubtedly became more important. The erodibility of the fine kaolinite-Grenada mixture was relatively independent of antecedent water.

The erodibility of fine kaolinite-Grenada mixtures was less than that of coarse kaolinite-Grenada mixtures at antecedent water contents of approximately 10% (Table 3). The true effect of bulk density again appears to be masked by the increased orientation with increasing bulk density (Table 1). Increasing the bulk density did increase the stability of the 10% and the 20% kaolinite mixes.

The resistance of kaolinite-Grenada mixtures

to erosion by flowing water is not due entirely to dilution of the Grenada clay minerals by the addition of the kaolinite but is apparently dependent upon the surface properties of kaolinitic clay minerals. Dilution of the Grenada silt loam by 10% natural silt (Figure 5) resulted in decreased stability. Stability, however, increased with increasing antecedent water. This increase is opposite to the observed decrease in stability with increasing antecedent water for unoriented kaolinite mixtures.

Plasticity index. Attempted characterization of these various clay mineral-Grenada silt loam mixtures using the plasticity index as the sole criterion (Table 4) was not fruitful. Similar plasticity index values were obtained for the Grenada silt loam and selected montmorillonite-Grenada mixtures on the one hand and kaolinite-Grenada mixtures on the other hand. Since the effect of antecedent water on stability is different for these materials, erroneous conclusions could be reached by depending solely upon the plasticity index.

CONCLUSIONS

The stability of cohesive materials against the erosive force of flowing water varied with the type and amount of clay minerals, clay mineral orientation, sample bulk density and antecedent water, and temperature of the eroding water. Increasing the temperature of the eroding water decreased stability.

The influence of antecedent water on sta-

TABLE 4. Liquid Limit, Plastic Limit, and Plasticity Index Values

Material	Liquid Limit	Plastic Limit	Plasticity Index
Grenada silt loam	31	20	11
Mixed with			
2% Ca montmorillonite	32	21	11
5% Ca montmorillonite	33	21	12
10% Ca montmorillonite	41	24	17
2% Na montmorillonite	32	21	11
5% Na montmorillonite	40	24	16
10% Na montmorillonite	62	27	35
2% coarse kaolinite	28	21	7
5% coarse kaolinite	29	22	7
10% coarse kaolinite	30	20	10
15% coarse kaolinite	30	20	10
20% coarse kaolinite	32	22	10
2% fine kaolinite	28	21	7
5% fine kaolinite	31	19	12
10% fine kaolinite	29	18	11

bility depended upon clay mineral assemblage and orientation. Stability increased with increasing antecedent water for the Grenada silt loam and for illitic and montmorillonitic mixtures thereof. Stability of the Grenada silt loam decreased with increasing orientation.

For mixtures of coarse kaolinite with the Grenada silt loam, stability decreased with increasing antecedent water for unoriented samples but increased with increasing antecedent water for oriented samples. The influence of antecedent water on the stability of kaolinitic mixtures was, furthermore, dependent upon the size of the kaolinite particles. As the size of the kaolinite was reduced and the specific surface increased, the decrease in stability with increased antecedent water typical of unoriented kaolinitic samples was less extreme.

The extreme stability of sodium montmorillonite mixtures was thought to be due to the minimum size and accompanying maximum specific surface and flexibility of these particles.

Stability increased with increasing clay mineral content at antecedent water optimum for stability. At other water contents stability was reduced for some systems with increasing clay content.

Stability increased slightly with increasing bulk density. The absolute dependency of stability upon bulk density was not clearly defined, owing to the concurrent change in orientation.

These data indicate the complexity of the process of erosion of cohesive materials and the variety of soil properties that determine the resistance to erosion.

Acknowledgments. The indispensable assistance of associates is gratefully acknowledged. Donald Parsons critically reviewed the manuscript and made many valuable comments during the experimentation. Loris Asmussen made most of the X-ray diffraction determinations of clay particle orientation using diffraction equipment of the Department of Geology, University of Mississippi. The close control of all steps in sample production and testing was in large part due to the supervision of Winfred Cook.

This paper is a contribution from the U. S. Department of Agriculture Sedimentation Laboratory in cooperation with the University of Mississippi and the Mississippi Agricultural Experiment Station.

REFERENCES

Dunn, Irving S., Tractive resistance of cohesive channels, *Proc. Am. Soc. Civil Engrs., 85*(Sm 3), Paper 2062, 1959.

Lambe, T. William, The structure of compacted clay, *Proc. Am. Soc. Civil Engrs., 84*(SM 2), Paper 1654, 1958a.

Lambe, T. William, The engineering behavior of compacted clay, *Proc. Am. Soc. Civil Engrs., 84*(SM 2), Paper 1655, 1958b.

Lane, E. W., Design of stable channels, *Trans. Am. Soc. Civil Engrs., 120,* 1234–1260, 1955.

Meade, Robert H., X-ray diffractometer method for measuring preferred orientation in clays, *U. S. Geol. Surv. Prof. Paper 424-B,* 273–276, 1961.

Moore, Walter L., and Frank D. Masch, Jr., Experiments on the scour resistance of cohesive sediments, *J. Geophys. Res., 67*(4), 1437–1446, 1962.

Seed, H. Bolton, and C. K. Chan, Compacted clays: a symposium, structure and strength characteristics, *Trans. Am. Soc. Civil Engrs., 126*(1), 1344–1385, 1961.

Smerdon, E. T., and R. P. Beasley, The tractive force theory applied to stability of open channels in cohesive soils, *Univ. Missouri Agr. Expt. Sta. Res. Bull. 715,* 1959.

(Manuscript received June 21, 1965; revised September 11, 1965.)

Part III

CONTROLS UPON EROSION RATES

Editors' Comments
on Papers 12 Through 19

12 ZINGG
Degree and Length of Land Slope as It Affects Soil Loss in
Runoff

13 ANDERSON
Suspended Sediment Discharge as Related to Streamflow,
Topography, Soil, and Land Use

14 FOURNIER
Climatic Factors in Soil Erosion

15 LANGBEIN and SCHUMM
Yield of Sediment in Relation to Mean Annual Precipitation

16 MILLER
Excerpts from *Solutes in Small Streams Draining Single Rock
Types, Sangre de Cristo Range, New Mexico*

17 GIBBS
*Amazon River: Environmental Factors That Control Its Dis-
solved and Suspended Load*

18 CLEAVES, GODFREY, and BRICKER
Geochemical Balance of a Small Watershed and Its Geomorphic Implications

19 SCHUMM
*Speculations Concerning Paleohydrologic Controls of Ter-
restrial Sedimentation*

Many factors that control erosion rates have been intensively
studied, and several models have been developed to describe the
relationships between erosion and these controlling factors. Perhaps the best known is the Universal Soil-Loss Equation, which
was developed to predict erosion rates in the Midwest of the United

States under various farming systems but which is now used all over the world, and for a variety of purposes. The equation is based on data from many studies conducted by U.S. agricultural scientists; one of the earliest was by A. W. Zingg, who used still-earlier data to present the first comprehensive quantitative view of the relationship between soil erosion and topography (Paper 12).

The Unversal Soil-Loss Equation takes the form

$$E = RKLSCP, \tag{1}$$

in which R is the rainfall erosivity factor, K is the soil erodibility factor, L and S are slope length and steepness factors (Paper 12), C is a cropping-vegetation cover factor, P is a land management factor, and E is soil-loss rate (Wischmeier and Smith, 1965). This conceptual model demonstrates that erosion rate is a function of the relative magnitudes of the resistance of the ground surface to erosion and the erosive forces applied and that topography, vegetation, and human activity can modify this balance. The equation was developed using data from plot studies of raindrop-runoff erosion on agricultural areas (see Paper 8), but the basic concept of a balance between force and resistance may be applied to other agents of erosion and to denudation on a broader scale.

Variability caused by a large range of controlling factors (some of which may be regarded as background noise) is an inherent feature of erosion data, and statistical techniques are an obvious tool for elucidating the basic relationships between erosion rates and the dominant factors. Henry W. Anderson has been very active in introducing statistical analysis, particularly multiple regression and other multivariate analysis techniques, into erosion studies. Paper 13 is an early example of this type of approach, which predates the wide availability of computers and statistical program packages. Although analysis of quantitative data is necessary for a more complete understanding of a phenomenon, it does lead to the problem of selecting simple indexes describing complex factors such as soil type and lithology, which are initially defined qualitatively (Jansen and Painter, 1974). The problem is perhaps less severe in an applied study where future prediction from readily measured indexes is required, but where the objective is a scientific understanding of fundamental relationships, selection of the quantities to be measured is critical.

A decade separates Anderson's study from the first comprehensive presentation of the Universal Soil-Loss Equation (Wischmeier and Smith, 1965), but all the factors that are incorporated in the equation were also found to be significant in Anderson's study. An additional factor, channel processes and particularly bank erosion, was shown to be a major control of suspended sediment yield from watersheds in western Oregon (see also Paper 2; Brune, 1950; Coldwell, 1957).

Fournier's work is an early attempt to use fluvial sediment yield data to quantify the effect of environmental factors upon erosion rates (Paper 14). He concluded that a rainfall seasonality factor p^2/P, in which p is maximum monthly rainfall and P is total annual rainfall, is a simple index of climatic influence but that the relationship between erosion and the rainfall factor is strongly modified by relief. Later Fournier (1960) included a relief factor in a simple statistical model, producing for 104 drainage basins greater than 2000 km² the well-known equation

$$\log DS = 2.65 \log \frac{p^2}{P} + 0.46 \log \frac{H^2}{S} - 1.56, \qquad (2)$$

in which DS is specific sediment loss, H is mean relief (the difference between mean elevation defined by hypsometry and maximum elevation), S is basin area, and p and P are as defined above. Limitations of the approach used by Fournier have been pointed out by Tricart (see Paper 2).

Fournier's work has been extended by Jean Corbel (1959, 1964), who again emphasized the importance of climatic regime and relief as controls on erosion rate, presenting the following mean values for a grouping of 117 rivers:

	Lowlands	*Mountains*
Cold climate	18 m³ km⁻² yr⁻¹	385 m³ km⁻² yr⁻¹
Temperate climate	46	110
Hot, dry climate	11	228
Hot, wet climate	21	92
Mean	22	242 (excluding hot, wet climate)

In another classic study, Langbein and Schumm developed what has come to be known as the Langbein-Schumm rule, relating annual sediment yield to effective precipitation (Paper 15). They placed strong though indirect emphasis on the part played by vegetation in modifying erosion rates, essentially concluding that the peak sediment yield in semiarid areas is due to the sparse vegetative cover coupled with highly erosive rainfall and runoff characteristics. Langbein and Schumm's work has been the subject of considerable subsequent analysis and synthesis, particularly by Douglas (1967) and by Wilson (1973). Douglas reconciled the relationships between erosion rate and annual precipitation presented by Fournier and by Langbein and Schumm, noting, however, that human interference may considerably accentuate the erosion rate peaks in semiarid and tropical areas (see Paper 21). Wilson (1973) maintained that the Langbein-Schumm rule is broadly applicable to areas with a continental climate and that the underlying theory is correct, although its precise form is not. He indicated that different relationships may exist for other climatic regimes and emphasized the importance of land use in producing a large amount of variability about the basic relationships (see also Part IV).

Papers 13, 14, and 15 deal exclusively with suspended sediment load. However, Meybeck (Paper 27) had estimated that the quantity of matter carried to the oceans in solution is $3.25 - 3.5 \times 10^9$ t yr^{-1}, about 20 percent of that carried in suspension. Even taking into account inputs of dissolved matter in precipitation, this corresponds to an annual global chemical erosion rate of 2.4×10^9 t yr^{-1}. Moreover, much of the particulate matter transported to the oceans owes its presence to chemical alteration and disaggregation of the bedrock. In any case, only a portion of the suspended sediment is unaltered residual material, and much is chemically or biogeochemically altered (Curtis, 1975).

Like the processes of physical erosion, chemical weathering is a result of a disequilibrium between two sets of opposing forces, and the concepts of mineral stability and chemical equilibria have been used by geologists such as Goldich (1938). Goldich concluded that there is an inverse relation between mineral persistence in soils and the temperature and pressure of the environment under which they were formed. Curtis (1975) stated the principle as "the more similar the environments of formation and weathering, the less susceptible to weathering the minerals will be" and compiled data (1976) indicating a positive relation between susceptibility to weathering and calculated energy changes in the chemical reactions involved.

The stability series concept may be used to explain differential weathering rates of minerals within a single rock type or of several rock types with different mineralogies. The paper by John P. Miller is a classic comparative study of chemical erosion rates on different rock types (Paper 16). The watersheds underlain by quartzite have the lowest chemical denudation rate; granite watersheds are intermediate, and areas underlain by sandstone have the highest rates. (The sandstones are highly variable in composition, ranging from quartz and quartzite constituents cemented by silica through to shales and mudstones; Miller pointed out the possible significance of highly reactive constituents present in only small quantities.) Reynolds and Johnson (1972) similarly concluded that the chemical composition of runoff in a high-altitude temperate glacial climate in the northern Cascade Mountains is controlled by lithology.

Although lithology is a fundamental control of chemical erosion rates (Davis, 1961; Hembree and Rainwater, 1961) and specifically of the types and quantities of solutes in spring water (for example, Marchand, 1974), the rates are directly controlled by the type and rates of chemical reactions involved and the rate of removal of weathering products by water movement. These in turn are influenced by a variety of factors: climatic factors such as the quantity, form, and distribution of precipitation; topographic factors such as elevation and slope angle; biotic factors such as the type of vegetation present and its nutrient cycling characteristics; and so on. For example, Dunne (1978) showed that differences in solution rates between several admittedly similar rock types for any value of runoff are minor and sometimes contradictory and that mean annual runoff is the dominant variable affecting solute yield. Feth, Roberson, and Polzer (1964) also attributed most of the variation in the quantity and type of solutes in the Sierra Nevada to climatic factors but attached much significance to weathering reactions and to time of residence of groundwater. Dunne's study included the first set of accurate measurements of chemical denudation rates (of watersheds draining mostly single rock types—igneous and metamorphic silicates) in the tropics. These are smaller than total solute yields by an amount equal to rates of input in precipitation and dry fallout. Dunne attached much importance to clay mineral stability relations, which serve to explain the major differences in solute quality (the Na/SiO_2 ratio) in semiarid and tropical regions of Kenya. Miller (Paper 16) concluded that for a given rock type, chemical denudation rates increase with elevation (to which are related precipitation and runoff volumes). Reynolds and Johnson (1972) considered that an abundant supply of well-aerated water accounts for high rates of chemical denudation observed in their study area because water is the reagent, catalyst, and carrier in the weathering reaction. The conclusion that chemical weathering may be most rapid in cool environments receiving large amounts of precipitation is echoed by Gibbs's study of solute and sediment yields from the Amazon watershed (Paper 17). He found that 85 percent of the dissolved matter (and 80 percent of the suspended load) annually discharged by the Amazon is supplied by the Andean mountain environment, which encompasses barely 12 percent of the watershed area. The dominance of the higher elevation areas is not solely due to greater volumes of runoff because solute concentrations are also highest in the headwaters; the high rates of inorganic weathering observed by Reynolds and Johnson presumably are also found in the Andes. Although the immense area of tropical rain forest in the Amazon watershed has a remarkably small effect upon the solute and sediment yield derived from it, the chemical composition of runoff draining the tropical areas differs considerably from that from the mountainous areas, in response to the different weathering processes involved.

Gibbs's data cannot necessarily be taken to indicate that high altitude cool climate weathering rates are higher than those under tropical rain forest conditions. There may be high rates of organic weathering in the rain forest, but the weathering products are rapidly taken up and recycled by the vegetation, so that only minor quantities may be lost from the ecosystem. The role of organic acids in rock weathering has long been recognized (Keller and Frederickson, 1952), but a number of recent studies have considered in detail the quantitative significance of the vegetation cover to the hydrology and denudation of forested watershed. Undoubtedly the best known of these is the Hubbard Brook (New Hampshire) study, the findings of which have been published in a long series of papers and monographs (Likens et al., 1977, and Paper 22). Bormann, Likens, and Eaton (1969) presented some of the early results from Hubbard Brook, demonstrating the role of the biota on losses of particulate and dissolved matter from the watershed. Dissolved and particulate losses were respectively, 14 t km^{-2} y^{-1} (unadjusted for precipitation fallout of 8 t km^{-2} yr^{-1}) and 2.5 t km^{-2} yr^{-1}, of which 30 percent and 52 percent, respectively, were provided solely by biological activity. In a companion paper, Johnson, Likens, and Bormann (1968) estimated a weathering rate of 80 t km^{-2} yr^{-1} for the bedrock-till substrate based on cation export data. They pointed out that the chemical denudation rate is directly proportional to the amount of hydrogen ion flushed through the system, which in turn is a function of throughput of water. This indicates the indirect but important influence of vegetation cover upon denudation rates, an influence that is manifested among others by the high water losses to interception and evapotranspiration (62.4 percent in the Hubbard Brook watershed during the 1965–1967 water years).

Cleaves, Godfrey, and Bricker have provided a detailed

demonstration of the effect of a vegetation cover upon chemical weathering and denudation (Paper 18). Seasonal export of dissolved matter from Pond Branch, a deciduous forested watershed in Maryland, varies in response to the biologic activity of the forest, chemical weathering being most vigorous in the warmer months. In the fall, leaves falling into the stream produce another increase in solute concentrations. Marchand (1971) also studied the implications for chemical denudation of plant uptake, storage, and litter fall of minerals, although measurement difficulties prevented precise quantification of losses.

Bormann, Likens, and Eaton (1969) pointed out the implications of a vegetative cover for mechanical erosion. Intact vegetation tends to reduce flood period runoff and to provide a protective cover for the land surface, and it may actually control the combination of geomorphic processes operating. The data of Cleaves, Godfrey, and Bricker (Paper 18) suggest that chemical weathering, itself controlled to a large extent by the vegetative cover, is the dominant denudation process in this climate. Other forms of biological activity affecting solute and particulate yield, such as burrowing of animals and rock weathering by snails, may be locally important. In fact, Pitty (1971) stated that "ever since 1882 when Charles Darwin's measurements showed that earthworms could move 10 tons per acre per year, attention has been drawn, from time to time, to the tendency for studies of sub-aerial processes to emphasize physical and chemical factors, perhaps at the expense of biological considerations." Notwithstanding the biotic influence, Cleaves, Godfrey, and Bricker believed that over the long term mechanical erosion must be as important as chemical erosion and that high magnitude, low frequency (catastrophic) events are responsible for shaping the valley by periodically flushing the floodplain. If this were not the case, insoluble weathering products (mainly quartz) would have been concentrated as deep valley-fill deposits of sand and gravel, deposits that do not in fact exist in their study area. In a later study, Cleaves, Fisher, and Bricker (1974) concluded that the relative importance of chemical and mechanical denudation in the piedmont of Maryland varies with lithology. Chemical weathering rates in Soldiers Delight, a watershed underlain by serpentinite, were 2–2.5 m/my (meters per million years), in comparison with 1.2–1.3 m/my in Pond Branch, which is underlain by schist. On the other hand, mechanical weathering rates were 0.5–0.03 m/my and 1.3–1.2 m/my respectively. Hence, although total denudation rates are approximately equal in both watersheds, chemical weathering is dominant in the serpentinite terrain but of equal importance to mechanical erosion in the schist. Contrasting topography is a result.

The probable significance of "catastrophic" events in Pond Branch underlines the need to consider erosion on a long time scale, as Tricart (Paper 2) stressed. The final paper in Part III, by S. A. Schumm, considers erosion on the geologic time scale (Paper 19), using his earlier work on the influence of climate, hydrology, and vegetation cover on erosion. Vegetation is viewed as the prime control of sediment yield; it reduces the size of weathered material and the amount of material stored on valley side slopes, increases the stability of channels by stabilizing banks, and decreases the size of material entrained, transported, and deposited by rivers, thereby also affecting channel shape. Schumm stressed the importance of vegetation and also considered the effects of its evolution through geologic time. His paper thus nicely rounds out the section and provides links joining hydrology, climatology, biology, and the broader field of geology.

REFERENCES

Bormann, F. H., G. E. Likens, and J. S. Eaton, 1969, Biotic Regulations of Particulate and Solution Losses from a Forested Ecosystem, *Bioscience* **19**:600-610.

Brune, G. M., 1950, The Dynamic Concept of Sediment Source, *Am. Geophys. Union Trans.* **31**:587-594.

Cleaves, E. T., D. W. Fisher, and O. P. Bricker, 1974, Chemical Weathering of Serpentinite in the Eastern Piedmont of Maryland, *Geol. Soc. America Bull.* **85**:437-444.

Coldwell, A. E., 1957, Importance of Channel Erosion as a Source of Sediment, *Am. Geophys. Union Trans.* **38**:908-912.

Corbel, J., 1959, Vitesse de l'érosion, *Zeitschr. Geomorphologie* **3**:1-28.

Corbel, J., 1964, L'érosion terrèstre, étude quantitative, *Annales Géographie* **73**:385-412.

Curtis, C. D., 1975, Chemistry of Rock Weathering: Fundamental Reactions and Controls, in *Geomorphlogy and Climate*, ed. E. Derbyshire, Wiley, London, pp. 25-51.

Curtis, C. D., 1976, Stability of Minerals in Surface Weathering Reactions, *Earth Surf. Proc.* **1**:63-70.

Davis, G. H., 1961, Geologic Control of Mineral Composition of Stream Waters of the Eastern Slope of the Southern Coast Ranges, California, *U.S. Geol. Survey Water-Supply Paper 1535B*, 30p.

Douglas, I., 1967, Man, Vegetation, and the Sediment Yield of Rivers, *Nature* **215**:925-928.

Dunne, T., 1978, Rates of Chemical Denudation of Silicate Rocks in Tropical Catchments, *Nature* **274**:244-246.

Feth, J. H., C. E. Roberson, and W. L. Polzer, 1964, Sources of Mineral

Constituents in Water from Granitic Rocks, Sierra Nevada, California and Nevada, *U.S. Geol. Survey Water-Supply Paper 1535-I*, 70p.

Fournier, F., 1960, Débit solide des cours d'eau. Essai d'éstimation de la perte en terre subie par l'ensemble du globe terrèstre, *Internat. Assoc. Sci. Hydrology Pub.* **53**:19–22.

Gibbs, R. J., 1971, Water Chemistry of the Amazon River, *Geochim. et Cosmochim. Acta* **36**:1061–1066.

Goldich, S., 1938, A Study in Rock Weathering, *Jour. Geology* **46**:17–58.

Hembree, C. H., and F. H. Rainwater, 1961, Chemical Degradation on Opposite Flanks of the Wind River Range, Wyoming, *U.S. Geol. Survey Water-Supply Paper 1535-E*, 9p.

Jansen, J. M. L., and R. B. Painter, 1974, Predicting Sediment Yield from Climate and Topography, *Jour. Hydrology* **21**:371–380.

Johnson, N. M., G. E. Likens, and F. H. Bormann, 1968, Rate of Chemical Weathering of Silicate Minerals in New Hampshire, *Geochim. et Cosmochim. Acta* **32**:531–545.

Keller, W. D., and A. F. Frederickson, 1952, Role of Plants and Colloidal Acids in the Mechanism of Weathering, *Am. Jour. Sci.* **250**:594–608.

Likens, G. E., F. H. Bormann, R. S. Pierce, J. S. Eaton, and N. M. Johnson, 1977, *Biogeochemistry of a Forested Ecosystem*, Springer-Verlag, New York.

Marchand, D. E., 1971, Rates and Modes of Denudation, White Mountains, Eastern California, *Am. Jour. Sci.* **270**:109–135.

Marchand, D. E., 1974, Chemical Weathering, Soil Development and Geochemical Fractionation in a Part of the White Mountains, Mono and Inyo Counties, California, *U.S. Geol. Survey Prof. Paper 352-J*, pp. 379–424.

Pitty, A. F., 1971, *Introduction to Geomorphology*, Methuen, London.

Reynolds, R. C., and N. M. Johnson, 1972, Chemical Weathering in the Temperate Glacial Environment of the Northern Cascade Mountains, *Geochim. et Cosmochim. Acta* **36**:537–554.

Wilson, L., 1973, Variation in Mean Annual Sediment Yields as a Function of Mean Annual Precipitation, *Am. Jour. Sci.* **273**:335–349.

Wischmeier, W. D., and D. D. Smith, 1965, Predicting Rainfall-Erosion Losses from Cropland East of the Rocky Mountains, *U.S. Dept. Agriculture, Agriculture Handb. 282*, pp. 1–47.

12

Copyright © 1940 by the American Society of Agricultural Engineers

Reprinted from *Agricultural Engineering* 21:59–64 (1940)

Degree and Length of Land Slope as It Affects Soil Loss in Runoff

By Austin W. Zingg

THIS study was initiated to better evaluate the effects of degree and length of land slope upon soil loss, and to serve as a guide for subsequent experimentation which should ultimately lead to precise determination of these effects. The results of research work by various individuals and organizations are grouped and analyzed as a whole to develop a rational equation for soil loss with respect to degree and horizontal length of slope. Results of an original experiment involving several plots with variations of slope and horizontal length of slope are given. It is not assumed that the equation obtained represents absolute values for any specific soil or condition but is merely an average of available data on the subject.

DEGREE OF SLOPE

Few experiments have been conducted which cover sufficient ranges of degree of land slope to permit evaluation of this factor as it is related to soil loss. Most plots studied under field conditions are widely separated in location, or the surface soil has been manipulated to produce varying degrees of slope. Under such conditions the infiltration rates and physical properties of soils in the same experiment can be greatly different, and give results which do not necessarily represent the effect of slope alone.

Research conducted at the Kansas agricultural experiment station[1]* and the agricultural experiment station of the Alabama Polytechnic Institute[2] seem to best approach the requirements necessary for a study which will properly evaluate slope with a degree of freedom from the influences of other variables.

The results of six tests are graphically recorded in Fig. 1. Two are from the data secured by F. L. Duley and O. E. Hays on a Kansas silty clay loam and on Kansas sandy loam soils. The tests on the silty clay loam were conducted in a laboratory upon prepared plots 24 in wide, 28 in deep, and 10 ft long. Plot slope differences were obtained by use of a differential hoist, and water application was by means of the sprinkling can method. The rate of water application was one inch per hour. The studies on Kansas sandy loam were carried out upon field plots 38.85 in wide and 25 ft long. Water application was by means of sprinkling cans at the rate of 2 in in 30 min. The surface soil, which was in a fallow condition, was saturated prior to both series of tests.

The four tests by E. G. Diseker and R. E. Yoder were made on a Cecil clay soil. The plots used were 15 ft wide and 50 ft long. The plots having different slopes were located in adjacent order and separated by concrete walls. Soil was carefully filled in to secure the various degrees of slope. The materials were tamped to their normal volume weight and a surface soil added. Mechanical analysis of the soils showed their texture to be uniform from plot to plot. Water was applied by use of a Skinner type irrigation system and distributed with "catfish" nozzles. Tests on smooth fallow ground and on freshly plowed ground were

made upon a saturated soil with 1.25 in of water applied in 11 min. The results listed in Fig. 1 for a freshly cultivated, and for a compact and crusted surface soil condition, were made on plots planted to cotton in 3-ft contour rows, with a field moisture condition preceding a rate of water application of 2 in in 18 min. While the results obtained from these four runs on different soil conditions are rather erratic, the four of them combined represent soil conditions which occur throughout different periods of the year. Their average is similar to the trends obtained by Duley and Hays.

It is noted that several soils and various sizes of plots have been used and different rainfall intensities and amounts have been applied to them. The soil losses are consequently greatly different in quantity for the various groups of data, and direct comparisons are not possible. The published absolute values of total soil loss have therefore been adjusted by a system of coding for the purpose of comparison.

In Fig. 1 a curve is drawn through the average of the coded total soil loss. Logarithmic plotting of this average of the data shows the coded total soil loss to be proportional to the 1.49 power of slope, or

$$X_c = 0.065 \, S^{1.49},$$

in which X_c = coded total soil loss, and S = land slope in per cent.

A rational equation representing total soil loss for a general condition could therefore be

$$X = C S^m,$$

in which X = total soil loss in weight units, C = a constant of variation, S = land slope in per cent, and superscript "m" = exponent of land slope.

The question arises as to whether results from experiments of the foregoing type are applicable to field conditions, since they do not fully cover the range of intensities and infiltration rates common to field areas throughout the year. All data listed have been obtained from a saturated soil condition where the infiltration rate has been low, or with high rates of water application which have minimized the effect of infiltration. Under field conditions, where a rain is of relatively short duration and has a rate slightly in excess of the infiltration rate, soil loss may be obtained from a steep slope, whereas a relatively moderate slope would give no runoff or soil loss because of greater depression storage. However, since the major part of soil loss over a year's time appears to occur from a relatively few rains of high intensity, and from rainfall on soil having a relatively low infiltration rate, the exponent of slope obtained from the several experiments may be very close to that which may be obtained from field conditions. Unpublished data obtained from degree of slope in continuous corn at the University of Missouri show an exponent for slope only slightly less than the average for the data presented in Fig. 1. These plots have been under observation for a 10-yr period under natural rainfall conditions.

HORIZONTAL LENGTH OF LAND SLOPE

Available data from several midwestern experiment stations of the U. S. Soil Conservation Service have been

The author is assistant agricultural engineer, Soil Conservation Service, U. S. Department of Agriculture, stationed at Bethany (Mo.) soil conservation experiment station.

*Superscript figures refer to literature cited at the end of Mr. Zingg's paper.

grouped for the study. Those grouped are from Tyler, Texas[3], Guthrie, Okla.[4], Clarinda, Iowa[5], Bethany, Mo.[6], and LaCrosse, Wis.[7] The horizontal lengths of plots at the stations are 36.3, 72.6, and 145.2 ft, and all are in continuous intertilled crops of corn or cotton.

The total soil losses in runoff, (tons per plot per year) for each station have been adjusted by a system of coding for purposes of comparison. These coded total yearly soil losses are plotted against horizontal length in Fig. 2. A curve is drawn through the average of the coded soil losses. Logarithmic plotting of this average of the data shows the coded total soil loss to be proportional to the 1.53 power of the horizontal length of slope, or

$$X_c = 0.0025 \, L^{1.53},$$

in which X_c = coded total soil loss, and L = horizontal length of land slope in feet.

It is apparent that there is only a small divergence in the exponential value of the horizontal length of slope for the various stations. When it is considered that the soils are different in formation, the infiltration rates have a large range of values, and the areas receive different quantities and intensities of precipitation, the fact that all data deviate so little from the average is worthy of note.

After averaging the coded data from the relatively short lengths of the control plots, data from other experiments of greater plot length were compared with this average to ascertain whether the same trend of soil loss is operative.

Fig. 3 shows the variation in data from all length of slope studies at the Clarinda, Iowa, and Bethany, Mo., stations. Both stations have plots of greater length than the control plots. Their respective lengths are shown on the abscissa of the graphs.

The Clarinda plots are located on a Marshall silt loam which has an exceedingly high infiltration rate. The data obtained at the Clarinda station during 1934 and 1935 show lesser increases of total soil loss from increased length of plots than the curve representing the general average for all data studied. Soil loss during this time was exceptionally low due to severe drought and high temperatures. An exceptionally high infiltration rate undoubtedly prevailed with these conditions. In 1933, which was the year preceding the drought condition, there was a greater increase in total soil loss for increased length of plots than the average for all stations in three out of

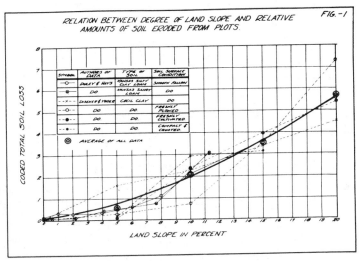

FIG.-1

RELATION BETWEEN DEGREE OF LAND SLOPE AND RELATIVE AMOUNTS OF SOIL ERODED FROM PLOTS

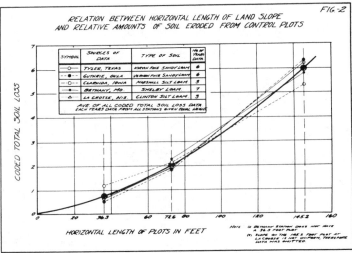

FIG-2

RELATION BETWEEN HORIZONTAL LENGTH OF LAND SLOPE AND RELATIVE AMOUNTS OF SOIL ERODED FROM CONTROL PLOTS

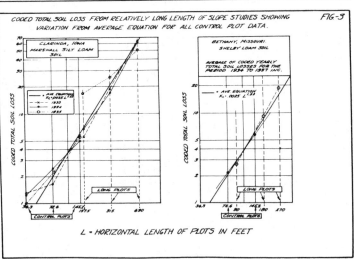

FIG-3

CODED TOTAL SOIL LOSS FROM RELATIVELY LONG LENGTH OF SLOPE STUDIES SHOWING VARIATION FROM AVERAGE EQUATION FOR ALL CONTROL PLOT DATA.

L - HORIZONTAL LENGTH OF PLOTS IN FEET

four comparisons. Musgrave[5] points out the fact that decreased runoff may occur with increased length of slope and expresses the belief that relation of rainfall intensity to infiltration rate is a determining factor. Data reported by Lowdermilk and by Hendrickson show that pore space in the immediate surface soil becomes clogged by small particles from turbid water during periods of intense rainfall. Neal[8], applying artificial rainfall to plots on Putman silt loam, has shown that the open structure of the top layer of soil is destroyed by intense applications of water. With a return to more normal climatic conditions, it is possible that total soil loss may be represented by a greater exponential value for length than occurred in 1934 and 1935. It is to be noted that the coded total soil loss from the group of longer plots follows a similar relationship to the average of coded total soil loss from the control plots.

The graph (Fig. 3) for data from the Bethany station on a Shelby loam soil presents a comparison of coded total soil loss from the control plots and a series of relatively long plots with respect to the average of all data in Fig. 2. The two lengths of slope under study in the control plots (72.6 and 145.2-ft plots) show less increase in relative total soil loss for increased length than the other group (90, 180, and 270-ft plots). The data from the control plots may follow a relatively low exponential value for horizontal length due to the manner in which the plots were prepared. The 145.2-ft plot was filled in with additional surface soil on the upper end to obtain a uniform slope at the beginning of the experiment, thus changing the original surface profile from a convex shape to a uniform slope. The additional depth of topsoil undoubtedly gives the upper portion of the plot a higher infiltration rate than the 72.6-ft plot. As erosion has progressed on the 145.2-ft plot for a period of years, the surface has assumed a concave profile. The concrete wall at the lower end of the plot prevents lowering of the soil surface level at this point. At the present time there are several inches of topsoil adjacent to the concrete wall on the lower end of the plot; on the middle portion of the plot the surface soil is entirely gone and a glacial till is exposed; and the upper portion of the plot, which was originally built up with additional surface soil, has about one foot of surface soil remaining on it. The plots of the longer slope group (90, 180, and 270-ft plots) were located on a virgin soil in 1934. The longer plots have colluvial soil on their lower portion, and it is possible the coded total soil loss for increased length is represented by a greater exponent for length due to the more erosible nature of the colluvial materials. It is also possible that total soil loss is proportional to the horizontal length of the plot to a power which will increase with the horizontal length of slope. This comparison of the two groups of plots is given to show some of the uncontrolled variables which occur in studies of this type and to bring out the fact that the differences obtained on parallel experiments on one soil type are nearly as great as the deviations between all stations. It would also indicate that present information on the subject is too meager to permit more than broad generalizations to be made at the present time.

From the study of the data it is believed a rational equation representing total soil loss for a general condition can be expressed as

$$X = CL^n$$

in which X = total soil loss in weight units, C = a constant of variation, L = horizontal length of land slope in feet, and superscript "n" = exponent of horizontal length of land slope.

In Table 1 all published annual total soil loss data from the experiment stations are tabulated. In this instance the data are presented as ratios produced by doubling the horizontal length of slope. The several length comparisons are specified. The same data are presented in the form of a frequency distribution curve in Fig. 4. The value of the median, or the value selected so that twenty-five, or half the cases, have lesser ratios and an equal number have greater ratios than this value, is 2.84. If this median value is used as the effect of doubling horizontal slope length, the value of (n), the exponent of length, is (1.5). If the value of the arithmetic mean (3.03) is used, the value of (n) will be (1.6).

TABLE 1. SOIL LOSS RATIOS PRODUCED BY DOUBLING LENGTH OF PLOTS

Experiment station	Year	Length of plots compared, feet	Yearly soil loss ratio, long plot/ short plot
BETHANY, MO.	1931	145.2 – 72.6	2.50
	1932	" "	2.15
	1933	" "	2.52
Soil type,	1934	" "	2.28
Shelby loam	1935	" "	2.84
	1936	" "	2.25
	1937	" "	3.80
	1934	180.0 – 90.0	3.33
	1935	" "	4.87
	1936	" "	3.22
	1937	" "	4.16
		Average,	3.08
CLARINDA, IOWA	1933	72.6 – 36.3	1.24
	1934	" "	1.91
	1935	" "	2.20
Soil type,	1933	145.2 – 72.6	3.55
Marshall silt loam	1934	" "	2.02
	1935	" "	2.44
	1933	315.0 – 157.5	3.39
	1934	" "	2.46
	1935	" "	2.39
	1933	630.0 – 315.0	3.45
	1934	" "	3.65
	1935	" "	2.01
		Average,	2.56
LA CROSSE, WIS.	1933	72.6 – 36.3	3.90
Soil type,	1934	" "	2.62
Clinton silt loam	1935	" "	2.60
		Average,	3.04
GUTHRIE, OKLA.	1931	72.6 – 36.3	1.72
	1932	" "	2.92
Soil type,	1933	" "	3.34
Vernon fine	1934	" "	2.23
sandy loam	1935	" "	1.95
	1936	" "	4.20
	1931	145.2 – 72.6	1.65
	1932	" "	4.21
	1933	" "	2.60
	1934	" "	4.62
	1935	" "	4.80
	1936	" "	5.05
		Average,	3.27
TYLER, TEXAS	1931	72.6 – 36.3	3.49
	1932	" "	2.84
Soil type,	1933	" "	2.67
Kirvin fine	1934	" "	2.98
sandy loam	1935	" "	2.84
	1936	" "	3.29
	1931	145.2 – 72.6	2.53
	1932	" "	5.02
	1933	" "	3.66
	1934	" "	3.86
	1935	" "	3.16
	1936	" "	2.21
		Average,	3.21

Average for all ratios, 3.03
Median value of all ratios, 2.84

COMBINED EFFECT OF DEGREE AND LENGTH OF LAND SLOPE

The following symbols will be used in this discussion:

$S=$ degree of land slope

$L=$ horizontal length of land slope

$X=$ total soil loss from a land slope of unit width

$C=$ a constant of variation

$m=$ exponent of degree of land slope

$n=$ exponent of horizontal length of land slope

$A=$ average soil loss per unit area from a land slope of unit width.

From the development of material under the sections on degree of slope and horizontal length of slope, and with the above symbols in mind, it was determined that $X=CS^m$ and $X=CL^n$. It was found, in several experiments where degree of slope was a variable, that the value of (m) appeared to be independent of the lengths of plots used in conducting the experiments; conversely, the length of slope experiments at the several stations were conducted on slopes of various degree, and the values of (n) are not greatly different. It is believed that both degree of slope and horizontal length of slope may, for practical purposes, be considered independent variables in their relation to soil loss. Their effect would thus be expressed as

$$X = CS^m L^n$$

$$A = CS^m L^{n-1}$$

EXPERIMENT IN WHICH THE EFFECT OF DEGREE AND LENGTH OF LAND SLOPE ARE SIMULTANEOUSLY EVALUATED

Procedure. A group of eight plots in which three degrees of slope, 4, 8, and 12 per cent, and two horizontal lengths, 8 and 16 ft, are represented, were prepared for the study. All plots were 42 in wide. The 4, 8, and 12 per cent slope plots were 8 ft in horizontal length. Each slope and length was duplicated at random location to give a record of the variability and to serve as a check on values attained.

The plots were located on an original 8 per cent Shelby loam slope. The original surface soil was entirely removed, the desired plot grades were established on the subsoil, and steel plot divides were placed to serve as plot boundaries. An 8-in depth of surface soil, which had been mixed and sifted through a one-inch mesh screen to produce uniformity, was then added to the plots. Concentrating

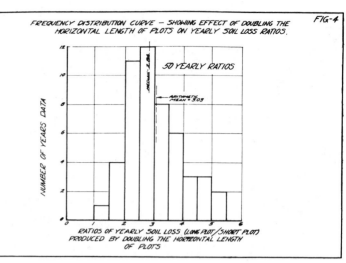

FIG-4

FREQUENCY DISTRIBUTION CURVE – SHOWING EFFECT OF DOUBLING THE HORIZONTAL LENGTH OF PLOTS ON YEARLY SOIL LOSS RATIOS.

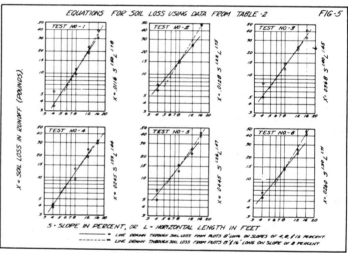

FIG-5

EQUATIONS FOR SOIL LOSS USING DATA FROM TABLE-2

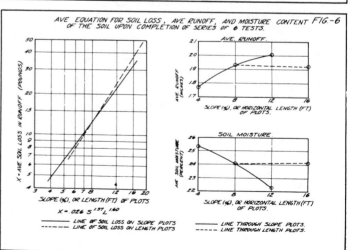

FIG.-6

AVE. EQUATION FOR SOIL LOSS, AVE. RUNOFF, AND MOISTURE CONTENT OF THE SOIL UPON COMPLETION OF SERIES OF 6 TESTS.

TABLE 2.   RESULTS SECURED FROM TESTS

Symbol No.	Plot Length, ft	Slope, per cent	1	2	3	4	5	6	Averages 1-6	
			Test number							
					Pounds of soil loss in runoff					
1-A	8	4	3.53	7.14	7.74	4.20	6.74	5.63	5.83	Soil
1-B	8	4	5.68	5.93	5.05	3.77	5.05	6.44	5.32	moisture
2-A	8	8	11.36	15.55	12.20	9.40	12.75	13.71	12.50	content
2-B	8	8	9.68	14.76	14.57	13.44	15.40	13.85	13.12	in 12-in
3-A	8	12	23.19	25.94	24.30	24.41	26.51	26.42	25.13	depth upon
3-B	8	12	18.91	24.55	23.20	18.77	21.67	23.28	21.73	completion of
4-A	16	8	39.60	48.23	41.57	32.09	43.47	43.15	41.35	six tests,
4-B	16	8	30.65	46.90	33.56	29.72	37.06	46.72	37.44	per cent
			Inches of water loss							
1-A	8	4	1.41	1.82	1.75	1.54	1.87	1.87	1.71	24.9
1-B	8	4	1.75	1.86	1.85	1.78	1.86	1.89	1.83	25.7
2-A	8	8	1.74	1.91	1.91	1.88	1.86	1.89	1.87	24.4
2-B	8	8	1.95	2.05	2.10	1.90	2.00	1.93	1.99	23.5
3-A	8	12	1.92	2.06	2.15	2.13	2.08	2.10	2.07	22.3
3-B	8	12	1.84	2.12	2.06	1.99	2.03	1.94	2.00	23.0
4-A	16	8	1.96	1.99	1.89	1.95	1.95	1.91	1.94	24.7
4-B	16	8	1.85	1.94	1.83	1.82	1.92	2.04	1.90	23.5
Water applied	Amount	In	2.34	2.37	2.37	2.35	2.32	2.36	2.35	
	Rate	In/hr	3.12	3.16	3.16	3.13	3.09	3.15	3.13	

troughs were placed at the lower edge of the plots and downspouting was used to conduct runoff to barrels sunk in the ground.

The rain simulator developed at the Bethany station and described in detail in a preliminary draft of a manuscript by C. M. Woodruff, D. D. Smith, and Darnell M. Whitt[9] was used to apply water to the plots. A rainfall rate of 3.15 in per hour was applied for a 45-min period for all tests made. The one intensity was used due to the fact that drop size will vary with changes in pressure head, for the type of rain simulator used, introducing a variable in addition to variation of the intensity of rainfall. The rate and amount of water applied has a frequency of occurrence of once in 20 yr, based on data by Yarnell[10].

The surface of the plots was cultivated with a rake to a depth of 3 in, preceding each of six tests. The condition of the soil for each run would represent a freshly prepared seedbed, with the surface soil in a moist, loose, granular condition and the subsoil wet. The choice of applying simulated rainfall to this soil condition, in preference to any other, was made because it is representative of one from which a large percentage of soil loss appears to occur in the field. It closely resembles ground in cultivated corn or cotton; and small grain, soybean, or alfalfa seedbed preparation.

The six tests were made on successive days in an attempt to have the plots as near to the same moisture content for each run as possible.

Results. All data collected are listed in tabular form in Table 2. The soil loss in runoff for each of the six tests is plotted on logarithmic scales in Fig. 5. The lines which best fit the data for each are drawn and the equation being followed is given. The equation given is based on the assumption that like values of (m), the exponent of slope in per cent, and (n), the exponent of horizontal length in feet, would be obtained had the experiment been conducted on plots with lengths greater or less than those used. This is not necessarily the case, as will be seen under the heading "Discussion" following.

In Fig. 6 all equations shown in Fig. 5 have been averaged and a graph of the average total soil loss ($X=0.026\ S^{1.37}L^{1.60}$) is shown. Values of S (slope in per cent), and L (horizontal length in feet) may be substituted directly in the equation to obtain pounds of soil loss in runoff. Using the average values of (m) and (n) (the exponents of slope and length, respectively), doubling the degree of slope increased the soil loss 2.61 times, and doubling the hori-

zontal length of slope increased the total soil loss 3.03 times.

Graphs for the total average runoff in surface inches and the per cent soil moisture content for each slope and length studied are illustrated in Fig. 6. Runoff increased appreciably with increases of slope and decreased a very small amount with additional length of slope for the rainfall intensity applied. The soil moisture content of the plots shows an inverse relationship to runoff.

Discussion. Since unequal rates of infiltration have occurred on the different slopes and horizontal lengths of slope studied, it is apparent that any factor which changes this relationship will be reflected in the values of the exponents (m) and (n). A relatively low value of (n) was secured by Neal*. It is therefore probable that an equation for soil loss in relation to both degree and horizontal length of land slope could be expressed as follows:

$$X = [C = f(K)]\ S^{m=f(K,L)}\ L^{n=f(K,S)}$$

in which K represents infiltration rate, physical properties of the soil, intensity and duration of the rain, and other factors.

The purpose of the experiment was to obtain relationships for conditions approximating those from which a large portion of soil loss occurs in the field. For practical purposes the relationship $X = CS^mL^n$, will be of value as a rational equation until more exacting relationships are attained.

It is interesting to note that the relative value of total soil loss, 2.61 (the equivalent of doubling the degree of slope when m = 1.37), is somewhat less than the average of 2.80 (where m = 1.49) obtained by averaging the coded total soil loss from several experiments under the section on degree of slope. The value 3.03 (effect of doubling the horizontal length of slope on soil loss) is equal to the arithmetic mean of the fifty yearly ratios of soil loss listed in Table 1 of the section on length of land slope. Since the effect upon total soil loss by increasing the horizontal length of slope has been found to be the same by the two methods of approach, it follows that the value of (m = 1.37) is probably close to a contemporaneous

*Neal[8], evaluating the effect of degree of slope on a Putman soil in a laboratory plot having a constant length of slope, found the soil loss varied as the 0.7 power of the slope. The amounts of soil loss were obtained after water had been applied for periods of time greater than one hour. The Putman soil has a low minimum infiltration rate, and the runoff from all slopes approached equality.

value for the degree of slope which would be attained as an average of field conditions.

Subsequent use will be made of the rational equation $X = CS^{1.4}L^{1.6}$.

SUMMARY

Available data on soil loss in runoff from degree and horizontal length of land slope experiments were studied by a system of coding. From an average of these coded data it was found that (1) doubling the degree of slope increased the total soil loss 2.80 times, and (2) doubling the horizontal length of slope increased the total soil loss 3.03 times.

An experiment, in which a simulated 20-year frequency rainfall was simultaneously applied to various degrees of slope and horizontal lengths of land slope, was conducted on small plots. For an average of the tests conducted on these plots, the following occurred:

1 Doubling the degree of slope increased the total soil loss in runoff 2.61 times.

2 Doubling the horizontal length of slope increased the total soil loss in runoff 3.03 times.

3 Increasing the degree of slope increased the total runoff.

4 Increasing the length of slope decreased the total runoff.

5 The moisture content of the soil at the completion of the tests showed an inverse relationship to total runoff.

For practical purposes, the rational equation

$X = CS^{1.4}L^{1.6}$ will give a relation between total soil loss, degree of slope, and horizontal length of slope which, as a generalization, will be applicable to field conditions. The equation is based upon a limited amount of data which was not developed for the purpose of this study, and when more satisfactory data are available, refinements will undoubtedly be possible.

LITERATURE CITED

[1]Duley, F. L., and Hays, O. E. The effect of the degree of slope on runoff and soil erosion. Journal of Agricultural Research, Vol. 45, No. 6. 1932.

[2]Diseker, E. G., and Yoder, R. E. Sheet erosion studies on Cecil clay. Bulletin No. 245, Agricultural Experiment Station of the Alabama Polytechnic Institute.

[3]Hill, H. O., Mech, S. J., and Pope, J. B. Progress report, 1931-1936, of soil and water conservation investigations at the soil conservation experiment station, Tyler, Texas.

[4]Hill, H. O., Elwell, H. M., and Slosser, J. W. Progress report, 1930-1935, of soil and water conservation investigations at the soil conservation experiment station, Guthrie, Okla.

[5]Musgrave, G. W., and Norton, R. A. Soil and water conservation investigations at the soil conservation experiment station, Clarinda, Iowa. Technical Bulletin No. 558.

[6]Woodruff, C. M., and Smith, D. D. Progress report, 1930-1935, of soil and water conservation investigations at the soil conservation experiment station, Bethany, Mo.

[7]Hays, O. E., and Palmer, V. J. Progress report, 1932-1935, of soil and water conservation investigations at the soil conservation experiment station, LaCrosse, Wis.

[8]Neal, Jess H. The effect of the degree of slope and rainfall characteristics on runoff and soil erosion, Research Bulletin No. 280, University of Missouri, Columbia, Mo. April 1938.

[9]Woodruff, C. M., Smith, D. D., and Whitt, D. M. Distributed preliminary draft of a manuscript entitled, Results of studies involving the application of rainfall at uniform rates to control plot conditions. Soil and Water Conservation Experiment Station, Bethany, Mo.

[10]Yarnell, D. L. Rainfall intensity-frequency data. U. S. Dept. Agr. Misc. Pub. 204. 1935.

AUTHOR'S NOTE: The author acknowledges the assistance of Dwight D. Smith and Darnell M. Whitt in the collection and interpretation of the data presented.

EDITOR'S NOTE: The study on which the foregoing paper was based, is believed to be the first of its kind, at least of such comprehensive scope. The results yielded are of value, and form the background for a paper, entitled "An Analysis of Degree and Length of Slope Data as Applied to Terracing," which was presented by the author before the fall meeting of the American Society of Agricultural Engineers at Chicago, Ill., December 8, 1939, and which will be published in AGRICULTURAL ENGINEERING for March. The author believes that most of the questions raised at the time the latter paper was presented will be cleared up in the foregoing paper.

13

Reprinted from *Am. Geophys. Union Trans.* **35**:268–281 (1954)

SUSPENDED SEDIMENT DISCHARGE AS RELATED TO STREAMFLOW, TOPOGRAPHY, SOIL, AND LAND USE

Henry W. Anderson

Abstract--The results of suspended-sediment sampling were used to obtain average annual suspended sediment discharge from 29 watersheds of western Oregon by relating sediment-sampling results to streamflow and by using streamflow frequencies. The values of average suspended sediment thus obtained were related by regression analysis to average watershed values of two streamflow variables, two topographic variables, two soil variables, and one channel bank variable. The soil variables were functions of particle size and aggregation determined by analyzing samples of the surface soil taken at standardized locations in the major geologic types. The other variables were functions of data published in maps and other secondary sources. The regression results were used (1) to construct a map of the sediment producing potential of lands in western Oregon under average land use conditions; (2) to estimate how the actual production of sediment would differ from the potential with deviation of land use from average; and (3) to distribute present sediment production to the three major source areas: forest land, agricultural land, and channel banks of the main river.

Introduction

The suspended-sediment load of many streams has been measured in recent years. Estimates of the sediment discharge of the watersheds can be made from these measurements. The rates of sediment discharge thus obtained represent an integrated average of the sediment discharge from all parts of the watershed. The rates from individual parts of the watershed remain unknown. Between watersheds the discharge of sediment varies in response to differences in streamflow, soil, topography, and land use. Within the parts of a single watershed similar responses may be presumed to occur. The purpose of this paper is to report a study of the responses of sediment discharge to the following watershed variables: streamflow, soil, topography, and land use. These responses were used to estimate the contribution to sediment discharge of the individual parts of watershed with different values of the variables.

The locale of the study was the mountain and valley watersheds of western Oregon, from the California border to the Columbia River. Here records of suspended-sediment load were available for 29 streams for a period of one to three years. Information on streamflow, geology, topography, vegetation, and land use has been published, and some additional information on soils was determined by laboratory analysis of soil samples from representative areas of the watersheds.

Specifically, this study included (1) calculation of average annual suspended discharge of 29 watersheds, (2) regression analyses relating sediment discharge to watershed characteristics, (3) estimation from these relationships of the "erosion potential" for lands of western Oregon (erosion potential is defined as the expected average annual suspended sediment production if all lands, both forest and agricultura, were in the same condition--the average 1950 condition for the whole of western Oregon), and (4) an estimation of how much the actual erosion would differ from the erosion potential with deviations of land use from average.

Table 5--Relative discharge frequency, suspended sediment sampling results, and average annual sediment concentration for selected watersheds of western Oregon

Relative discharge[a]				Suspended sediment sampling						Average suspended sediment
				No. of samples	Per cent of days with ppm of					
Class	Average	Frequency	Total		<12.5	12.5-27.5	27.5-72.5	72.5-142.5	>142	
1	2	3	4	5	6	7	8	9	10	11
	pct									ppm
Calapooya River at Albany										
0.01- 0.1	0.05	27.0	0.0135	14	13.50	13.50	....	...	...	13.5
0.1 - 1.0	0.42	45.0	0.1890	19	7.10	37.90	....	...	...	18.0
1.0 - 2.0	1.38	15.0	0.2070	9	....	1.70	13.30	...	...	46.5
2.0 - 4.0	2.62	8.5	0.2227	8	....	....	7.44	...	1.06	65.0
4.0 - 8.0	5.20	3.35	0.1742	6	....	....	1.68	1.67	...	79.0
8.0 -16.0	10.00	0.93	0.0930	1	....	....	0.93	...	...	95.0
>16.0	20.40	0.22	0.0449	0	....	....	....	...	0.22[b]	200.0[c]
Mean	....	....	.....	..	7	20	50	108	170	62.8
Total	....	100	0.9443	57	20.60	53.10	23.35	1.67	1.28	...
Rogue River at Shady Cove										
0.01- 0.1	....	0	.....	0	....	....	....	...	...	...
0.1 - 1.0	0.75	68.00	0.5114	2	68.00	....	....	...	...	5.5
1.0 - 2.0	1.28	26.20	0.3354	9	23.30	2.90	....	...	...	7.1
2.0 - 4.0	2.42	5.26	0.1273	2	....	2.63	2.63	...	...	35.0
4.0 - 8.0	4.80	0.51	0.0246	1	....	....	0.51	...[b]	...	50.0
8.0 -16.0	9.00	0.03	0.0027	0	....	....	....	0.03[b]	...	133.0[c]
>16.0	18.00	0	0	0	....	....	....	...	...	...
Mean	....	....	.....	..	5.5	20	50	133	...	11.3
Total	....	100	1.0014	14	91.30	5.53	3.14	0.03	...	...

[a] Relative discharge is the daily discharge divided by the mean annual discharge.
[b] Value is from frequency of discharge column; location in sediment class was by reference to trend line relationship similar to Figure 11.
[c] Obtained from trend line of a relationship similar to Figure 11.

The results, including a map of erosion potential prepared for western Oregon, should provide
a useful guide to land and water managers in this area. The methods have application to suspended
sediment analysis and erosion and sedimentation studies in other places.

Studies and analysis

Obtaining sediment discharge from sediment load data--The data used in calculating average
annual suspended sediment discharge included depth-integrated samples of the sediment concentration of the streams. These samples were taken between December, 1948, and March, 1952, by
the Corps of Engineers, by the Forest Service, and by the Bureau of Reclamation. The other data
were stream discharge quantities and frequencies. These data were summarized in the form of a
frequency table for each watershed. The form of the frequency table (Table 5) is shown for Calapooya and Rogue watersheds. Similar frequency tables were prepared for all watersheds and are
available upon request. For convenience, in the tables all discharges were expressed as "relatives" by dividing the actual daily discharge by the mean annual discharge.

The steps in calculating average annual suspended sediment discharge were: (1) calculating
the average suspended-sediment concentration of a discharge class; (2) weighting the average by
the volume of water occurring annually in that discharge class; and (3) multiplying this weighted
average concentration by the mean annual runoff. In step (1), the means of each sediment-concen-
tration class (usually the means were 7, 20, 50, 108, and 200 ppm) were weighted by the per cent
of days in a given discharge class. This computation gave the average suspended-sediment concentration of a discharge class (Table 5, Col. 11). When no sediment sample had been taken in a
discharge class, the average suspended sediment was read from the trend line of a relationship
between suspended sediment and relative discharge established for that watershed (Fig. 11). In
step (2), the average suspended-sediment concentration of each discharge class weighted by the
average annual volume of a flow occurring at that discharge (Table 5, Col. 4) gave average suspended-sediment concentration of all water in ppm. In step (3), this weighted average concentration
times the mean annual runoff (Table 6, Col. 2) converted to millions of tons, gave average annual
suspended-sediment discharge of a watershed in tons (Table 6, Col. 16). This was used as the
value of the dependent variable, suspended sediment, for that watershed in the subsequent regression analyses.

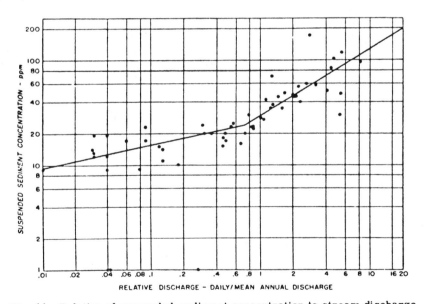

Fig. 11--Relation of suspended-sediment concentration to stream discharge,
Calapooya River near Albany, Oregon, 1949 and 1950

This sediment-discharge-frequency method of obtaining total sediment from sampling results
has three advantages over the sediment-hydrograph method described by the CORPS OF ENGINEERS
[1948]: (1) Average annual sediment delivery is obtained directly, obviating the need to calculate sediment discharge for individual years; (2) not all the discharges during a year of sampling need be
known, only the discharges at the time of sampling; and (3) the method is more rapid than the sediment-hydrograph method.

Table 6--Watershed characteristics and suspended sediment for selected watersheds of western Oregon

Watershed	Watershed characteristics[b]															Sediment discharge[b] SS
	A^a	MA_q	FQ_p	Se	SC	S/A	PF	RC	BC	OC	PC	R	S	EB	SSf	
	1	2	3	4	5	6	7	8	9	10	11	12	13	14	15	16
	sq mi	$\frac{cfs}{sq\ mi}$	$\frac{pct}{100}$	pct	pct	$\frac{cm}{gm\ pct}$	pct	pct	pct	pct	pct	ft/mi	ft	M ton/yr		M ton/yr
Calapooya	362	2.29	3.28	3.1	20.9	0.52	7	5.6	10	27	17	0.60	630	1,600	16	51
Pudding	493	2.26	2.39	2.3	21.7	0.45	18	7.7	22	37	40	0.57	210	0	26	70
Mollala	323	3.21	3.90	3.6	20.4	0.56	24	15.8	4	12	28	0.22	1000	3,400	34	54
McKenzie	1310	3.80	3.27	9.6	19.9	1.04	20	7.1	2	5	22	0.10	1100	16,700	93	149
N. Santiam	665	4.76	3.55	8.6	20.1	0.96	9	5.6	0	4	9	0.12	1500	0	59	190[c]
Long Tom	100	2.52	3.81	3.9	24.2	0.61	36	21.1	3	16	39	0.30	660	0	12	8
Coyote	100	1.40	3.58	3.1	24.4	0.47	6	0	11	16	17	0.45	480	0	3	8
Willamette	991	2.64	2.44	15.4	19.8	1.46	4	1.9	0	1	4	0.11	960	6,800	53	53
Tualatin	568	2.12	3.03	1.7	29.2	0.29	13	2.4	19	48	32	...	...	....	..	53
Santiam	1790	4.10	3.62	5.6	20.3	0.72	10	6.3	3	11	13	0.23	1510	25,500	154	457
Coast Fork	69	2.77	4.23	3.9	20.2	0.59	22	17.1	1	1	23	0.12	480	0	5	5
Marys	155	2.84	3.58	3.7	24.1	0.59	20	1.4	4	26	24	0.38	830	0	12	19
Willamette	7280	3.05	3.24	6.2	21.1	0.76	12	4.3	6	19	18	0.43	1200	204,800	464	1955
Willamette	2030	2.50	3.64	7.3	20.2	0.84	4	1.9	2	6	6	0.17	1290	30,300	103	248
Row	211	2.63	4.27	4.0	20.1	0.59	5	0	1	1	6	0.21	1150	0	12	13
Lucklamute	236	3.74	3.61	3.7	24.1	0.59	8	0	6	44	14	0.40	710	0	14	27
S. Santiam	640	4.28	4.23	3.9	20.2	0.59	8	2.6	1	4	9	0.16	1290	0	60	88
S. Yamhill	502	3.19	3.18	1.8	29.7	0.28	12	5.0	18	32	30	0.53	490	0	42	82
Calapooya	(98)	2.30	3.96	4.2	24.9	0.90	20	5.6	0	4	20	...	...	....	..	23
Elk Creek	(85)	1.53	3.90	11.5	24.8	1.06	4	4.4	0	9	4	...	...	....	..	12
S. Umpqua	454	2.02	4.26	4.6	20.5	0.62	2	0	0	2	2	...	...	....	..	62
S. Umpqua	1640	1.51	4.06	6.8	25.6	0.71	9	1.1	0	4	9	...	...	....	..	182
S. Coquille	169	4.25	3.90	6.0	26.9	0.60	16	6.8	0	3	16	...	...	....	..	86
Illinois	367	2.92	3.90	8.2	28.8	0.68	10	4.9	1	6	11	...	...	....	..	85
Big Butte	(243)	1.04	1.98	17.6	19.1	1.78	12	0	0	5	12	...	...	....	..	3
Elk Creek	133	1.25	4.30	4.0	20.1	0.59	7	0	0	5	7	...	...	....	..	2
Rogue	(1190)	1.96	2.21	16.6	19.6	1.60	7	0	0	5	7	...	...	....	..	26
Wilson	(56)	5.89	3.90	2.0	32.0	0.26	80	30.4	0	0	80	...	...	....	..	21
Wilson	159	7.48	3.90	2.0	32.0	0.26	81	16.3	0	1	81	...	...	....	..	134
Mean	773	2.97	3.56	6.0	23.0	0.73	17	6.0	4	12	21	0.30	910	17,000	68	144

[a]Parentheses indicate area crudely estimated.
[b]Definitions of variables given in text, under the headings relating to variables.
[c]This value thought to be inordinately high due to stream disturbance during construction of Detroit Dam.

Relating sediment discharge to watershed characteristics--Sediment discharge varies with differences in watershed characteristics, which include streamflow, soil, topography, bank conditions, and land-use. Variations in these between watersheds permit evaluation of their individual effects on sediment discharge. The evaluation is possible by regression analysis which relates suspended sediment discharge from many watersheds to quantitative expressions of the characteristics of each watershed [GOTTSCHALK, 1947; ANDERSON, 1949, 1951; and GOTTSCHALK and BRUNE, 1950].

Specific variables were chosen to give quantitative expression to the watershed characteristics. Two requirements were made of a variable: (1) that it be expected to be related to sediment discharge, and (2) that it be expressible in terms applicable to individual land areas within watersheds. The second requirement was imposed because eventually the regression results were to be applied to calculate the contribution of individual land areas to sediment discharge.

The value of a variable for a watershed was obtained by averaging values of the variable as found in the individual land areas within the watershed. Peakedness of streamflow was averaged by weighting according to areal extent and a function of rain and snow-melt frequencies. All other variables were averaged by simply weighting by the areal extent.

The variables selected are listed below together with the methods used in obtaining the variable. In the section on results additional details are given on the methods used in obtaining the slope S and the soil variables SC and S/A. Quantitative values of each of the variables for each watershed are given in Table 6.

Dependent variables (suspended-sediment discharge)--The following were used:

SS   Suspended sediment; the average annual suspended sediment from a watershed under
      1950 watershed condition (obtained from sediment sampling); unit, thousand of tons/yr;
      mean, 144; range, 2 - 1955.

ss   Same as SS except unit is tens of tons/sq mi yr; mean, 18; range, 1 - 84.

Independent streamflow variables--The following were used:

MAq   Mean annual runoff, obtained from U. S. Geological Survey Water Supply Papers; unit,
      cfs/sq mi; mean, 2.97; range, 1.04 - 7.48.

FQp   Discharge peakedness, based on a separate regression analysis which related peak dis-
      charges from watersheds to geologic rock types, to the area receiving rain during
      storms, and to the area in which snow was ripe during the storms (see Flood hydrology
      of forest watersheds, by Henry W. Anderson, California Forest and Range Experiment
      Station, 15 pp typewritten, Jan. 2, 1950); quantitative values of the relative peakedness
      variable may be obtained for western Oregon watersheds from a geologic map and
      Table 7; unitless; mean, 3.56; range, 1.98 - 4.30.

Table 7--Quantitative peakedness variable

Geologic type	Symbol[a]	Relative peakedness	Relative rain area frequency	Relative snow melt frequency	FQ_p[b]
Recent alluvium	Aa	1.0	0.97	1.00	1.0
Younger volcanic	Qv	2.0	0.20	0.30	0.6
Young volcanic	Qpv	2.0	0.53	0.66	1.2
Old volcanic	PMv	5.0	0.82	0.96	4.3
All other	c	4.0	0.96	0.99	3.9

[a]Geologic map of United States, U. S. Geological Survey, 1932.
[b]Relative peakedness times (1.2 relative rain area frequency plus
1.0 relative snow melt frequency) divided by 2.2 gives FQ_p.
[c]Mostly marine sediments, but some very old volcanics included.

Independent topographic variables--The following were used:

A   Area of watershed; unit, sq mi; mean, 773; range, 56 - 7280.

S   Slope of streams of one mile mesh length after method of HORTON [1945] (see deriva-
      tion in section on results); unit, ft/mi; mean, 910; range, 210 - 1510.

Independent soil variables--The following were used:

Se   Soil erosibility; generalized variable of soil characteristics obtained from soil meas-
      urements developed by MIDDLETON [1930]; dispersion ratio times suspension times
      ultimate silt plus clay divided by 100 times the aggregated silt plus clay; unit, pct/100;
      mean, 6.0; range, 1.7 - 17.6.

SC   Silt and clay; fraction of top six inches of soil with particle sizes of less than 0.05 mm
      in diameter; unit, pct; mean, 23.0; range, 19.1 - 32.0.

S/A   Surface aggregation ratio; surface area on soil particles of sand and coarser size
      (>0.05 mm diameter) divided by the aggregated silt plus clay; surface area was ob-
      tained by considering the particles as spheres and assigning mean diameters of 7.5,
      3.5, and 0.9 mm to the >5, 2 to 5, and 0.05 to 2.0 mm particle-size classes, respec-
      tively; aggregated silt plus clay was the ultimate silt plus clay, minus the suspension
      per cent; unit, cm^2/gm pct; mean, 0.73; range, 0.26 - 1.78.

Independent land-use and channel-bank variables--The following were used:

PC   Poor cover; portion of watershed with poorly stocked or non-stocked forest land or land
      in bare cultivation (data from same sources as RC and BC below); unit, pct; mean, 21;
      range, 2 - 81.

PF   Poorly stocked forest; portion of watershed in poorly stocked or non-stocked forest
      land (data from same source as RC below); unit, pct; mean, 17; range, 2 - 81.

R   Roads; portion of watershed area in roads; road lengths measured from National
        Forest maps of the Forest Service and Land and Water Inventory maps of the Soil
        Conservation Service; road widths taken as follows: primary state highways, 50 ft;
        secondary state highways, 40 ft; county roads, 32 ft; forest roads, 20 ft; and railroads,
        15 ft; unit, pct; mean, 0.30, range, 0.05 - 0.60.

RC   Recent cut-over; portion of watershed area cutover in last ten years, read from For-
        est Service Forest Survey maps; unit, pct in 10-year period; mean, 6.0; range, 0 - 30.4.

BC   Bare cultivation; portion of watershed in row crops and small grain, from Soil Conser-
        vation Service Land and Water Inventory Maps; unit, pct; mean, 4; range, 0 - 22.

OC   Other cultivation; portion of watershed area in cultivation other than bare cultivation
        (source of data same as for BC, above); unit, pct; mean, 12; range, 0 - 48.

C   Cultivation; area of watershed in bare and other cultivation; (BC + OC) times A of
        Table 6; unit, sq ml; mean, 8; range, 0 - 67.

EB   Eroding banks; length of eroding main channel banks as given by the CORPS OF ENGI-
        NEERS [1950]; unit, ft; mean, 17,000; range, 0 - 204,800.

Calculated sediment discharge variables--The following were used:

ss3   Calculated sediment from streamflow and area variables, using Eq. (2); unit, tens of
        tons/sq ml yr; mean, 15; range, 2 - 43.

ss7   Calculated sediment from Eq. (3); unit, tens of tons/sq ml yr; mean 18; range, 4 - 102.

SSf   Sediment from forest lands of a watershed, read from the map (Fig. 12) and adjusted
        for deviation of recent cut-over, RC, from average; adjustment factor = antilog [0.0086
        (RC-6.1)] from Eq. (5); unit, thousands of tons/yr; mean, 68; range, 3 - 464.

Regression analyses--The general method of analysis was multiple regression [SNEDECOR,
1946, p. 340]. However, because of the large number of variables which were expected to be important, the analysis was performed in five steps. The first step was to test the relation of sediment discharge SS to mean annual streamflow MAq, and peakedness of streamflow FQp, including
in the analysis generalized expressions of other watershed variables. The generalized variable
for topography was the area A; for soil, the soil erosibility Se; and for land use, the poor cover
PC. Eq. (1) was the result.

The second step was the use of the partial effects of MAq, FQp, and A found in the first test,
with the appropriate constant (2) to calculate sediment discharge ss3 for each watershed based on
these variables alone.

The third step was to test the relation of actual sediment discharge ss, to this calculated sediment discharge ss3 and to some additional variables, soil silt plus clay SC, soil surface-aggregation ratio S/A, bare cultivation BC, other cultivation OC, and poorly stocked forest PF.

In the fourth step, similar to the second step above, the seven variables previously evaluated
as the partial effects of ss3, SC, S/A, BC, and OC (3) were used to obtain another calculated
value of sediment discharge from the watersheds, ss7.

The fifth step, similar to the third step, was to test the relation of actual sediment discharge
ss to this calculated sediment discharge ss7 and to the final additional variables, road area R,
recent cut-over area RC, and slope S. The result was (4), which when the specific variables are
re-inserted for ss7, gave (5).

Eq. (6) was developed from analysis of 17 watersheds in the Willamette Basin for which the
CORPS OF ENGINEERS [1947] had made estimates of the lengths of eroding banks. A test was
made of the relation of the actual sediment discharge, SS, from the 17 watersheds to the length
of eroding banks EB the calculated sediment discharge from forest lands SSf and the area of
cultivated land C with (6) being the result.

Results of regression analyses--The regression equations follow, with logarithms being to
the base 10:

$$\log SS = -2.717 + 1.163 \log MAq + 1.109 \log A + 1.566 \log FQp + 0.211 \log \log e^{PC}$$
$$+ 0.029 \overline{5} \log Se \dots \dots \dots \dots \dots \dots \dots \dots \dots \dots \dots \dots \dots \dots (1)$$

$$\log ss3 = -0.516 + 1.163 \log MAq + 0.109 \log A + 1.566 \log FQp \dots \dots \dots \dots (2)$$

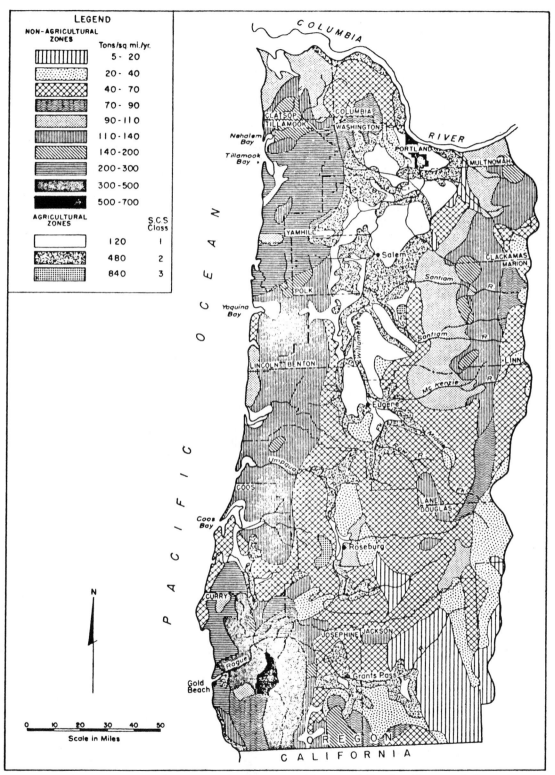

Fig. 12--Erosion potential for land areas of western Oregon, defined as the expected average annual suspended-sediment production if all lands, both forest and agricultural, were in the same condition, the average 1950 condition in the whole of western Oregon

$$\log ss7 = -1.169 + 0.901 \log ss3 + 0.0410 \text{ SC} + 0.406 \text{ S/A} + 0.0236 \text{ BC} - 0.0030 \text{ OC} \\ - 0.0003 \text{ PF} \dots \dots \dots \dots \dots \dots \dots \dots \dots \dots (3)$$

$$\log ss = -1.715 + 1.186 \log ss7 + 0.942 \text{ R} + 0.0085 \text{ RC} + 0.401 \log \text{S} \dots \dots \dots \dots (4)$$

$$\log ss = -3.721 + 0.116 \log \text{A} + 1.673 \log \text{FQp} + 1.244 \log \text{MAq} + 0.401 \log \text{S} + 0.0486 \text{ SC} \\ + 0.482 \text{ S/A} + 0.0280 \text{ BC} - 0.0036 \text{ OC} + 0.942 \text{ R} + 0.0086 \text{ RC} \dots \dots \dots \dots (5)$$

$$\text{SS-SSf} = -1.639 + 0.240 \text{ C} + 0.00514 \text{ EB} \dots \dots \dots \dots \dots \dots \dots \dots \dots \dots (6)$$

An indication of the overall efficiency of the equations is given by comparing the observed sediment discharge with that computed by using the equations. Such a comparison for all 29 watersheds of the study, using (5) to obtain computed values, is shown in Figure 13. (Mean values of 0.30 and 9.1 were used for R and S in the equation when specific values were missing.) For those 17 watersheds of the Willamette Basin for which data on eroding banks were available, a comparison of the observed sediment discharge with that computed by (6) is shown in Figure 14. The comparisons indicate that a satisfactory separation of the high from the low sediment producing watersheds can be made by using the equations. Of equal importance, however, are the regression coefficients of the individual variables in the equations, for they are measures of the importance of various "causes" of sediment discharge from watersheds.

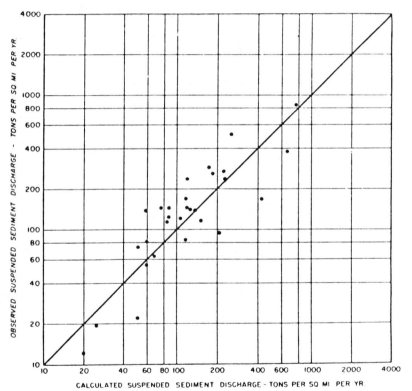

Fig. 13--Comparison of observed suspended-sediment discharge with that computed by (5), for 29 watersheds of western Oregon

Streamflow characteristics influencing suspended-sediment discharge--Two variables, the mean annual runoff and the peakedness of runoff, were used to characterize the streamflow in the watersheds. Both variables were highly significant in relation to sediment discharge.

The effects of mean annual streamflow and discharge peakedness on sediment discharge are given by the partial regression coefficients of 1.24 for MAq and 1.67 for FQp in (5). An index of the quantitative variation of sediment discharge which was associated with streamflow differences between watersheds may be obtained by substituting the extreme values of MAq and FQp from Table 6 in (5). For example, substituting for MAq the values 7.48 cfs/sq mi for Wilson watershed and

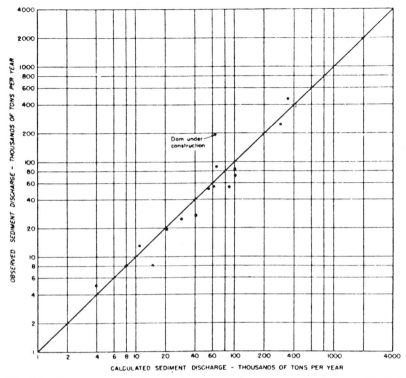

Fig. 14--Comparison of observed suspended-sediment discharge with that
computed by (6), for 17 watersheds of the Willamette Basin, Oregon

1.04 cfs/sq mi for Big Butte watershed indicates that if all other variables were the same for the
two watersheds the sediment discharge of the Wilson watershed would be 11.6 times that of the
Big Butte watershed because of differences in mean annual streamflow alone. Similarly, between
watersheds with the extreme values of peakedness of streamflow FQp, the sediment discharge of
the watershed whose flows are most peaked would be 3.7 times that of the watershed with the least
peaked flows. The regression results permit rating watersheds as to their suspended-sediment
discharge on the basis of streamflow characteristics.

Topographic characteristics influencing suspended-sediment discharge--Slope, area, and elevation were the three topographic characteristics tested for their effects on sediment discharge.
Slope of streams one mile in length was taken as an index of the steepness of the land areas. Difficulties in determining a value of the variable for a watershed because of differences in topographic mapping scales and standards between watersheds were overcome by using some relations of
stream physiography found by HORTON [1945]. For each watershed the streams of each order
number were traced on the best available topographic map. Average length and slope of each order
number were determined. Average stream length was plotted on log scale against order number
on arithmetic scale [see HORTON, 1945, Fig. 10], a trend drawn, and the order number of the
streams of one-mile length read off to the nearest tenth of an order number. Then the average
stream slope was plotted against order number [see HORTON, 1945, Fig. 6] and the slope corresponding to the order number of streams one mile long was read off. In the regression analysis
this slope variable, the slope of the streams of one-mile length, was significant at about eight to
one odds against a chance effect.

The effect of differences in slope on sediment discharge is given by the regression coefficient
of 0.401 for S in (5). The magnitude of the effect may be indicated by substituting in the equation
the greatest and least values of S for the watersheds given in Table 6. Such a substitution indicates
that the watershed with the greatest slope would have sediment discharge 2.2 times that of the
watershed with the least slope because of the slope difference alone. The regression results permit rating watersheds as to their suspended-sediment discharge based on a slope characteristic.

Sediment production per unit area varied very little with area of watershed when other variables of the watershed were taken into account. (The regression coefficient for the area variable A in (1) is nearly unity.) Elevation was not indexed separately but was a factor included in the rain and snow-melt frequencies in the streamflow peakedness variable FQp.

Soil characteristics influencing suspended-sediment discharge--The problem in evaluating soils was, how could the characteristics of the soils which influence sediment production be determined and used? Four steps were involved: sampling the soil, measuring the physical characteristics of the soil, relating the physical characteristics to measured sediment production, and applying the results in rating of soil as to erosibility.

Sampling the soils of a watershed--Determining the soil characteristics of a watershed by sampling was largely a problem of how to minimize the number of samples without undue loss of information. It was not feasible to sample all of the infinite variation in soils. Soil variations may be associated with every combination and level of each of the soil-forming factors: parent material, topography, climate, organisms, and time [JENNY, 1941, p. 15]. In soil sampling some sort of selection and averaging must obviously be resorted to. Therefore, some standard conditions of certain soil-forming factors were set up under which all soil samples were taken.

What variability in soil-forming factors should be sampled and which excluded by selection of the standard conditions depend on the variability of soils and the dominant erosion processes in the study area. In this study of sediment discharge, the soil of the forested areas of the watersheds was youthful and profile development was limited; therefore, it was thought possible to restrict soil sampling to differences in the single soil-forming factor--parent material. JENNY [1946] has called this a determination of a parent material sequence or lithosequence, obtained by arranging in order of magnitude the quantitative values of some physical characteristics of the soils. If these physical characteristics are found to be related to measured sediment production, the lithosequence becomes an erosibility sequence, and within the errors of determination, the relative erosibility of the types of parent material may be given numerical values. When this has been done an evaluation of the soil erosibility in terms of production of suspended sediment has been obtained.

A geologic map furnished a basis for locating and segregating the different parent materials in the 29 watersheds. Soil samples of the zero- to six-inch depth of soil were taken in each of the geologic types. The samples were all taken under a single set of conditions defined as standard: at slopes of about 30 pct, on west exposures, and under full natural cover, with the attempt being made to take samples at elevation of 2000 feet and where mean annual precipitation was about 60 inches. (The last two conditions could not be met for all geologic types.)

Differences in soils attributable to geology were measured from these soil samples, taken at the defined standard conditions in each geologic type. These soil differences may be evaluated for their effect on sediment discharge. The effects on sediment discharge of differences in soil in areas where runoff, topography, and land use were different from the standard condition were evaluated by the watershed variables. This was how the number of soil samples needed was reduced to feasible limits. In the forest-land areas, 42 soil samples were taken, three samples in each of the 14 major geologic types.

Measuring physical characteristics of the soils--The soil samples were analyzed in the laboratory. The analyses measured physical characteristics which MIDDLETON [1930] found to be qualitatively related to erosion and which had been found quantitatively related to sediment production in a recent study in Southern California [ANDERSON, 1951]. These characteristics are: particle size distribution from gravel through clay, and the suspension per cent. Sieving was used to determine soil fractions with diameters greater than two mm. Particle size distribution smaller than two mm was determined by the hydrometer method [BOUYOUCOS, 1936]. The dispersion procedure for the ultimate silt plus clay analysis was: two grams of sodium hexametaphosphate were added to about 35 grams of soil in 200 cc of water, and the mixture was stirred in a milk-shake machine for ten minutes. The hydrometer method was also used in analyzing particle size distribution in the undispersed soil in obtaining Middleton's suspension per cent.

Relating soil characteristics to measured suspended sediment--Once the laboratory measures were obtained, it was necessary to test which of these measured characteristics of soils were different for different soil types. It was reasoned that unless a soil characteristic was different between soil types, it could not possibly be diagnostic in explaining differences in sediment discharge. Therefore, an analysis of variance was used to test the significance of differences between soil types in some physical characteristics expected to be related to erosibility. The characteristics, in decreasing order of significance, were: surface area of particles coarser than 0.05 mm diameter, aggregated silt plus clay, Middleton's dispersion ratio, ultimate silt plus clay, Middleton's suspension per cent.

The next problem was to choose from among the soil characteristics, those which should be related to physical processes in suspended-sediment discharge. Two processes were considered: (1) the supply process, represented by the component of the soil which produces the fraction of erosion which is caught and measured as suspended sediment; ultimate silt plus clay was taken as an index of the supply factor; (2) the binding process, represented by the fraction of the soil which tends to bind the soil together versus the amount of soil surface in the non-binding fraction which requires binding. The surface area of particles coarser than 0.05 mm diameter divided by the aggregated silt plus clay was taken as an index of how effectively the soil was bound. It was called the surface-aggregation ratio.

The soil characteristics for the 14 major soil-geologic types of western Oregon are given in Table 8, Col. 3 - 10. Weighting each of these by the per cent of area of each type in a watershed gave average values of the soil variables for the watersheds, Table 6, Col. 4 - 6. Including the two soil variables in regression analyses (3) gave a measure of the significance of soil differences in affecting sediment discharge. The test of significance indicated 19 to one odds against a chance effect. The quantitative effects of soil variations were obtained by using the regression coefficients for the silt plus clay SC, and the surface-aggregation ratio S/A from (5) to calculate the relative erosibility of the soil developed on the various geologic types. The resulting relative soil erosibility for the 14 geologic types is given in Table 8, Col. 11. It may be seen that the most erosible soil, that developed on the Eocene volcanic, is 11 times as erosible as the soil developed on alluvium. The geologic map, together with these ratings of erosibility, locate and evaluate the differences in suspended-sediment discharge attributable to soil.

Table 8-- Physical characteristics of surface soil and relative erosibility of major soil-geologic types of western Oregon

Geologic types	Geological symbol[a]	Particle size distribution diameters-mm				Suspension[b]	Aggregated si+cl[c]	S/A[d]	Se[d]	Relative erosibility[e]
		<0.05	0.05-2	2-5	>5					
		pct	pct	pct	pct	pct	pct	$\frac{cm^2}{gm\ pct}$	$\frac{pct}{100}$	
Alluvium	Qa	25.1	10.6	27.6	36.7	5.7	19.4	0.37	143	0.7
Recent volcanic	Qv	16.9	73.5	5.3	4.3	10.2	6.7	2.75	2450	7.7
Young volcanic	QPv	19.6	54.9	12.0	13.5	11.0	8.6	2.04	2290	4.7
Columbia basalt	PMv	20.1	20.2	26.2	33.5	6.8	13.3	0.59	420	1.0
Miocene marine	Mm	32.0	10.4	31.9	25.7	8.6	23.4	0.26	227	2.6
Oligocene marine	Øm	24.9	16.4	34.6	24.1	7.1	17.8	0.41	329	1.8
Eocene marine	Em	24.2	19.1	28.2	28.5	7.2	17.0	0.62	427	1.6
Eocene volcanic	Ev	40.5	12.6	26.8	20.1	10.0	30.5	0.17	182	6.2
Cretaceous marine	K	31.0	27.0	23.8	18.2	15.4	15.6	0.57	858	3.3
Jurassic Triassic	JTR	19.4	14.1	28.7	37.8	6.1	13.3	0.53	331	0.9
Old intrusive	Ji	20.2	50.6	22.8	6.4	10.8	9.4	1.64	1660	2.8
Carboniferous volcanic	Cv	30.9	34.9	20.9	13.3	13.6	17.3	0.68	891	3.7
Carboniferous	C	31.1	22.4	23.6	22.9	17.8	13.3	0.60	1070	1.8
Devonian	D	25.2	6.7	22.2	45.9	5.9	19.3	0.23	189	1.2
Agricultural[f]	..	19.5	3.8	57.4	19.3	4.6	14.9	0.35	101	...

[a]Geologic Map of United States, U. S. Geological Survey, 1932.
[b]MIDDLETON [1930]; all percentages given are per cent of whole soil.
[c]Ultimate silt plus clay (particles < 0.05 mm in diameter) minus suspension.
[d]Defined in text.
[e]From (5) using particle size < 0.05 mm for SC and values of S/A from this table; relative values assigned by taking PMv geology as 1.0.
[f]Based on two samples.

Land-use characteristics influencing suspended-sediment discharge--Four land-use characteristics were analyzed for their effects on sediment discharge. These were (1) the recent cut-over area, measured from detailed county maps prepared by the Forest Survey of the Forest Service; (2) the cultivated area, measured from the Land and Water Inventory Maps of the Soil Conservation Service; (3) the length of eroding main channels, measured by the CORPS OF ENGINEERS [1950]; and (4) the area of roads, measured from maps used for first two characteristics.

Recent cut-over and roads together had a highly significant effect on sediment, as did eroding banks. Cultivation was significant at about ten to one odds. Thus, sediment discharge was found to increase as the area of recent cut-over, bare cultivation, and roads increased and with the increases in the length of eroding main channel banks. The regression coefficients of (5) and (6) permit estimation of the quantitative effects of land-use and eroding channel banks on sediment discharge.

If timber cutting were to increase from the present annual cut of 0.6 pct of the watersheds to 1.5 pct, which would result from full utilization of the forest on a 70-year rotation of cutting, then the regression results indicate that suspended sediment would increase by 18 pct due to the cutting alone. However, the associated probable increase in road development might have a much greater effect. If future roads, built to the same standards and hence having the same effects as present roads, were to occupy 0.55 pct of the forested watersheds instead of the present 0.1 pct, then, the regression results indicate that the suspended-sediment discharge would be 2.6 times as great. Together logging and road development in the future could increase sediment discharge by three times if no preventive measures are taken.

Control of erosion from cultivated lands and eroding main channel banks would markedly affect sediment discharge. The regression results indicate that conversion of bare cultivated land to other types of cultivation would nearly eliminate the converted land as a source of suspended sediment. Control of eroding banks of the main channels would reduce sediment discharge by five tons per year per running foot of eroding bank.

In all, the study indicates that a large part of the present or potential sediment discharge is attributable to land-use and channel bank condition. To some degree these are subject to control. Obviously a reduction in the area occupied by the several kinds of present land-use or in the length of eroding channel bank could reduce sediment discharge. But modification of present land-use methods might also lead to significant reductions in sediment discharge, even though the area subject to use were extended. Discussion of such modifications, and their effectiveness, is beyond the scope of this paper.

Application of the results

Map of erosion potential--It was reasoned that sediment production from parts of watersheds with certain characteristics will be the same or proportional to that from whole watersheds with those characteristics. To the degree that this is true, the regression results may be applied to segregate and rate the major source areas of sediment production.

Forest-land erosion potential--The forest land was segregated in 39 complexes with different values of streamflow, topography, or soil variables. Eq. (5) was then solved for each complex, inserting the individual values of the streamflow, topography, and land-use variables and mean (1950) values for the land-use variables. The means used for the land-use variables were 6.1 pct for recent cut-over and 0.1 pct for roads. The calculated suspended sediment production was called the "erosion potential." The individual results for the 39 complexes were grouped to form ten erosion-potential zones for the forest lands. The erosion-potential zones are shown on a map (Fig. 12), with the potential being expressed as average annual suspended sediment production in tons per square mile.

The present sediment production from large forest areas may be expected to be approximately the same as the potential. For small areas where land-use conditions may be far from average, the sediment production expected is the potential times a factor which depends on the deviation of the land use from average. The partial regression coefficients for roads R and recent cut-over RC in (5) permit calculation of this land-use factor, and hence calculation of sediment production.

Agricultural land erosion potential--The regression results, Eq. (6), indicate an average rate of sediment production from cultivated lands of 240 tons per square mile per year. Time was not available to analyze the soil samples which had been taken in the agricultural land soil types to permit determination of their relative erosibility. Therefore another basis of erosibility was used, the Soil Conservation Service Erosion Classes. The erosion classes, 0, 1, 2, and 3, were assigned relative erosion production rates of 0, 1, 4, and 7, respectively, based on the definitions of the respective classes as having top-soil losses of 0, 0-25, 25-75, and greater than 75 pct. The location of these erosion classes are given in Soil Conservation Service Soil and Water Inventory Maps; the relative areas of these four classes in western Oregon were: Class 0, 0.5 pct; Class 1, 67.4 pct; Class 2, 31.7 pct; and Class 3, 0.4 pct. The breakdown of sediment production rates was then: for Class 0, zero; for Class 1, 120 tons/sq mi yr; for Class 2, 480 tons/sq mi yr; and for Class 3, 840 tons/sq mi yr. The location of these agricultural land erosion zones is shown on the erosion potential map (Fig. 12).

The location of eroding banks is not shown on the map, but is given for the Willamette Basin by CORPS OF ENGINEERS [1950].

Major sources of sediment production--The regression results permit segregation of the three major sources of sediment production in the Willamette Basin above Salem, Oregon. Using (6) and the values from Table 6 for sediment production from forest land SSf, for the cultivated land area C and for the length of eroding banks EB, the present average annual sediment production of 1,955,000 tons for the Willamette watershed above Salem may be segregated as follows: from the 5460 sq mi of forest land, 24 pct; from the 1820 sq mi of agricultural land, 22 pct; and from the 205,000 ft of eroding main channel, 54 pct.

Areal limitations of results--The results here reportedly may not always apply to all small local areas. Deviations in sediment production from the estimated potentials will occur where local soil differences occur because of local variation in geology or fault zones, where local slopes are widely different from average for a zone; and where unusual disturbances occur, such as road fills paralleling a stream for long distances, local logging disturbances that are unusually severe, or dam building in process. The map values apply only to the broad erosion potential zones. Differences in the erosion potential between zones indicate the relative need for precautions in land development.

Needs and priorities in further research--Further extensive research of the type of this study is needed to obtain relations between specific criteria of land condition and sediment discharge. For example, the area of bare soil exposed at stream margins and the area of other bare land might be set up as criteria and their values determined for the individual watersheds. If these were included in regression analyses together with the "erosion potential" of the watersheds, an evaluation might result which would permit more effective land management for reduction of sediment discharge from watersheds.

Another need is for estimation of other types or measures of erosion such as surface soil loss and sediment deposition. In studies leading toward such estimations, the methods and some of the variables used in this study should be applicable. For example, in a study of erosion measured as deposition, it is a worthwhile hypothesis that the soil binding variable S/A would be important in its present form, but the supply variable SC should be modified.

Intensive research into economic methods of offsetting adverse effects of land use on sediment production may be justified, especially in those zones where the erosion potential is high. Research priorities suggested by this study are (1) eroding-bank control, (2) road-development methods and precautions, (3) bare-cultivation control, and (4) logging operation methods and precautions.

Summary and conclusions

(1) Suspended sediment sampling results were used to obtain average annual suspended-sediment discharge from 29 watersheds of western Oregon by relation of sediment to streamflow and by use of streamflow frequencies.

(2) Average annual suspended-sediment discharge of watersheds varied from as little as 12 to as much as 840 ton/sq mi.

(3) The variation in sediment discharge was associated with differences in streamflow, topography, soils, bank conditions, and land use, which were numerically evaluated by multiple regression to permit estimation of their independent effects on sediment discharge. The effects may be illustrated by these results: (a) The watershed with the greatest annual streamflow would be expected to have a sediment discharge 11.6 times that of the watershed with the least flow. Similarly, the watershed whose flows were most peaked would have a sediment discharge 3.7 times that of the watershed with the least peaked flows. (b) The watershed with the steepest slopes, as characterized by the slope of the streams of one mile length, would have sediment discharge 2.2 times that

of the watershed with the least slope. (c) Sediment-discharge differences associated with differences in the soil developed on the 14 major geologic types of the mountain areas were significant. The most erosible soil, that developed on Eocene volcanic, was 11 times as erosible as the soil developed on alluvium. (d) If under future development of forest lands, using present methods, the recent cut-over area should occupy 1.5 pct of the watershed area, and roads 0.55 pct, suspended-sediment discharge would be expected to be three times as great as at present. Most of this increase would be associated with the extension of the road area. Conversion of bare cultivated land to other types of cultivation would largely eliminate these lands as sources of suspended sediment. Prevention of bank cutting in the main channels of the Willamette River and tributaries would eliminate 54 pct of the present sediment discharge.

(4) As an aid to land and water managers in planning resource development, a map of the erosion potential of western Oregon was prepared, with erosion potential being defined as the expected annual suspended-sediment discharge if all lands, both forest and agricultural, were in the average 1950 land-use condition.

(5) It is concluded that in the areas of high erosion potential extra precautions should be taken in land-use development, that erosion potential zones should be considered in assigning priorities of research in methods of offsetting adverse effects of land use, and that intensive investigation should be made into methods of controlling sediment production from eroding channel banks, roads, bare cultivated, and logged areas.

Acknowledgment--Grateful acknowledgment is made to G. L. Hayes and E. A. Core for suggestions regarding land use variables, to T. G. Storey for help in obtaining values of these variables for the watersheds, and to M. S. Barton for collecting most of the soil samples. Appreciation is expressed to D. M. Ilch, under whose administrative guidance the study was conducted, for his encouragement and suggestions.

References

ANDERSON, H. W., Flood frequencies and sedimentation from forest watersheds (with discussions), Trans. Amer. Geophys. Union, v. 30, pp. 567-586, and v. 31, pp. 621-623, 1949 and 1950.

ANDERSON, H. W., Physical characteristics of soils related to erosion, J. Soil Water Conserv., v. 6, pp. 129-133, 1951.

BOUYOUCOS, C. J., Directions for making mechanical analyses of soils by the hydrometer method, Soil Sci., v. 42, pp. 225-229, 1936.

CORPS OF ENGINEERS, Methods and analysis of sediment loads in streams, Rep. 8, Measurement of sediment discharge of streams, St. Paul District Sub-Office, Iowa City, 92 pp., 1948.

CORPS OF ENGINEERS, Columbia River and tributaries, Northwestern United States, v. 5, app. J, pt. 1, pl. 64 and 65; pt. 2, Table VI-45, House Doc. 531, 81st Cong. 2nd Ses., 1950.

GOTTSCHALK, L. C., A method of estimating sediment accumulation in stock ponds, Trans. Amer. Geophys. Union, v. 28, pp. 621-625, 1947.

GOTTSCHALK, L. C., and G. M. BRUNE, Sediment design criteria for the Missouri Basin loess hills, U. S. Soil Cons., Ser. Tech. Paper 97, pp. 1-21, 1950.

HORTON, R. E., Erosional development of streams and their drainage basins, hydrophysical approach to quantitative morphology, Geol. Soc. Amer. Bull., v. 56, pp. 275-370, 1945.

JENNY, H., Factors of soil formation, McGraw Hill, 281 pp., 1941.

JENNY, H., Arrangement of soil series and types according to functions of soil-forming factors, Soil Sci., v. 61, pp. 375-391, 1946.

MIDDLETON, H. E. Properties of soils which influence soil erosion, U. S. Dept. Agric. Tech. Bull. 178, pp. 1-16, 1930.

SNEDECOR, G. W. Statistical methods, 4th ed., Iowa State College Press, 485 pp., 1946.

California Forest and Range Experiment Station,
     Forest Service, U. S. Department of Agriculture,
          Berkeley, California

(Manuscript received August 24, 1953, presented at the Thirty-Fourth Annual Meeting, Washington, D. C., May 5, 1953; open for formal discussion until September 1, 1954.)

14

CLIMATIC FACTORS IN SOIL EROSION

F. Fournier

This article was translated expressly for this Benchmark volume by M. P. Mosley of the Ministry of Works and Development, New Zealand, from "Les facteurs climatiques de l'érosion du sol" in Assoc. Géographes Français Bull. **203**:97–103 (1949), *by permission of the publisher, Association de Géographes Français.*

It is always valuable to use quantitative data to evaluate the evolution of the natural environment. In general, when one tries to study the problem closely--that is, to evaluate the evolution of a given phenomenon on a small area--the influence of local factors disturbs to such a degree the effect of more general factors that one can obtain significant relationships only with a very large number of measurements.

The objective of this study is to evaluate the influence of climatic factors upon soil erosion. In this field, the number of measurements upon small areas appears to be insufficient to permit the establishment of significant correlations. We therefore decided that one could obtain quantitative data on erosion intensity by using data for sediment load in streams, expressed as mean quantities of solid material lost each year by a unit area of a drainage basin. In this study, we have used only data for suspended sediment transport. This sediment has been eroded from the soil surface by running water following rainfall.

Data for the transport of coarse bed load is of little use because it is difficult to estimate, and the number of particles rolling along the bed depends upon current velocity.

Because the climatic factor influencing soil erosion appears to be rainfall, we have used planimetry on rainfall maps to calculate mean monthly and mean annual rainfall upon the drainage basins.

RELATION BETWEEN MEAN ANNUAL PRECIPITATION OVER A DRAINAGE BASIN AND MEAN ANNUAL SEDIMENT LOAD OF THE WATERCOURSE

Initially we attempted to establish a correlation between mean annual precipitation over and mean annual sediment yield from a watershed (Figure 1). The graph does not immediately illuminate the relationship, but it permits certain observations:

1. Disregarding mountainous watersheds, the data points are distributed, with some scatter, along a parabolic curve.

2. With reference to this curve, the data points
   for mountainous watersheds, scattered over the
   whole graph, are aberrant.
3. Examination of the points that are found in
   moving from the upper to the lower branch of
   the parabola shows that the curve reflects a
   climatic classification.

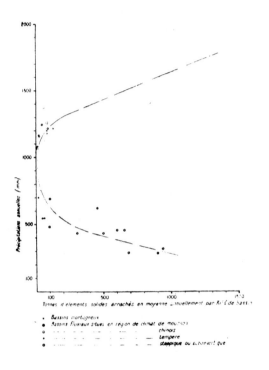

Figure 1   Relationship between annual precipitation (mm)
and mean annual sediment yield (t km^{-2}).
[*Editors' Note:* The data points should have
been plotted using mean annual sediment yield,
the dependent variable, on the ordinate.]

One first encounters points for the Ganges and the Red River,
situated in a monsoonal "central Hindu" climate. The Ganges received
1932 mm and the Huang Ho 1686 mm, four-fifths of the precipitation
falling in June to September.

Then follow the Si-Kiang basin, with a similar climate, although
tending to have some characteristics of the Chinese climate; the
Yang-tse-Kiang basin, with a typical Chinese climate; and basins of

the southeast United States with a characteristic climate similar to the Chinese climate--a heavy mean annual precipitation (over 1300 mm) distributed over all the seasons but with a maximum in summer and a minimum in autumn.

The points that follow are those of temperate climate rivers, mainly in North America: first, those for which monthly rainfall distribution is analogous to that in the southeast United States; and then those tending to have a uniform distribution of rainfall throughout the year. For this group, one finds a minimum erosion rate. Only 2 to 3 t km^{-2} yr^{-1} of solid matter is lost by the rivers flowing to the Atlantic in the northeast United States, whose watersheds receive monthly precipitation amounts that are never less than 80 mm and never greater than 98 mm.

When one descends toward the lower branch of the curve, one finds points for drainage basins whose climates, although temperate, become increasingly continental and sediment loads increase. Then along the lower part of the parabola come, from right to left, a succession of points representative of basins situated mainly in increasingly steppe-like climates before becoming semiarid (Vaar, Afrine, Indies, Rio-Grande, Huang Ho, Tigris, Colorado, and its tributaries). The rainfall received by these rivers is concentrated in a short part of the year.

This first attempt at correlation is not satisfactory, but it demonstrates that regions with greatest erosion rates are those whose rainfall distribution through the year is decidedly variable. We are therefore forced to find a coefficient that will describe this variability and to establish a new correlation.

ESTABLISHMENT OF A CLIMATIC EROSION FACTOR

The ratio (mean annual precipitation)/(mean number of rain days) could be the required factor, but the quantity of rain required for a day to be defined as a rain-day is not the same for all meteorological services (for example, in the United States it is 0.01 inch and in India it is 0.1 inch). This ratio is therefore difficult to use. After much research, the ratio (maximum monthly rainfall)/(annual rainfall) finally gave a coefficient usable for defining the effect of climate. If all the year's rainfall falls in one month, the ratio is equal to unity. But if rainfall tends to be uniformly distributed over the year, the ratio tends to be 1/12, or 0.08.

The relationship between this coefficient (on the ordinate) and values of erosion (on the abscissa) suggests the following conclusions:

1. The graph is no longer a parabolic curve but has two straight segments resulting from the grouping of points in two distinct series: rivers with mountainous watersheds and rivers having the overall character of lowland channels.
2. Comparison of the distribution of the points in this graph with that in Figure 1 (which shows the influence of mean annual rainfall) suggested the use of a ratio that combines the coefficients used in the two cases as a coefficient accounting for the effect of climate.

This led us finally to use the following coefficient:

$$\frac{(\text{maximum monthly rainfall})^2}{(\text{annual rainfall})} \times \text{annual rainfall},$$

which is mathematically equivalent to

$$\frac{(\text{maximum monthly rainfall})}{(\text{annual rainfall})} \times \text{maximum monthly rainfall}.$$

This turns out to be an association of an index of nonuniformity of precipitation over the year with an index of the absolute value of precipitation.

The correlation between the coefficient thus calculated and the value taken as the measure of erosion (suspended sediment load per km^2 of drainage basin) permits us to obtain Figure 2. The data points are grouped into three series.

In the first, they are situated around two straight lines that intersect at a value of annual erosion of 100 t km^{-2}. None of the rivers considered here has the character of mountainous watercourses.

Around the first line are grouped the temperate climate rivers. Moving from the origin, where those rivers are located whose precipitation is distributed uniformly through the year, a succession of points representing drainage basins with a temperate climate that has a progressively more pronounced continental character are found along the line. At an erosion rate of 80 to 90 t km^{-2} are found basins with a Chinese-type climate in the United States.

Around the second straight line are grouped points representing rivers with steppe, Chinese, and monsoonal or tropical climates.

The second series is located around a single line whose origin closely coincides with that of the first straight line of the preceding series. Around this are grouped in homogeneous fashion the points for mountainous drainage basins. The slope of this line is less than that of the two lines of the preceding series, which suggests that for the given climatic conditions, the effect is much more marked in mountainous than in lowland regions.

Finally, the third group of points, few in number, corresponds to basins with high relief but with a subarid climate: the Colorado and its tributaries and the Rio Grande. The quantity of sediment transported appears to be independent of climate. Is this a consequence of such a strong influence of environment that the climatic factor is unable to account for the effect of environment? The cause of the aberrance of this series arises rather from the fact that the notion of the mean has no sense in the subarid zone. If one studies these rivers annually, establishing for each year values of the climatic factor and of erosion, the points obtained on the graph would come to be found around the line representing rivers in mountainous regions.

One may thus conclude that precipitation influences soil erosion through its intensity and through its distribution over the year. By using a factor that includes these two influences, one obtains the most logical and satisfactory grouping of data.

Examination of the correlations established by using the factor nicely demonstrates its value, because values of erosion increase with it, but it also provides evidence for the influence of land slope.

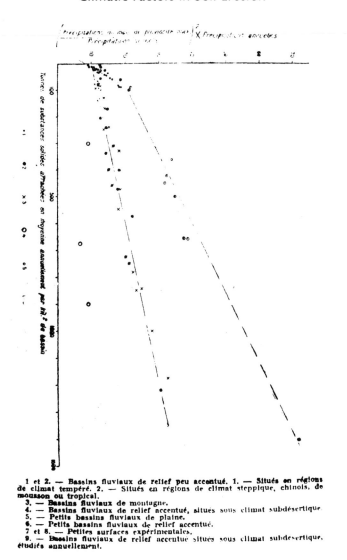

1 et 2. — Bassins fluviaux de relief peu accentué. 1. — Situés en régions de climat tempéré. 2. — Situés en régions de climat steppique, chinois, de mousson ou tropical.
3. — Bassins fluviaux de montagne.
4. — Bassins fluviaux de relief accentué, situés sous climat subdésertique.
5. — Petits bassins fluviaux de plaine.
6. — Petits bassins fluviaux de relief accentué.
7 et 8. — Petites surfaces expérimentales.
9. — Bassins fluviaux de relief accentué situés sous climat subdésertique, étudiés annuellement.

Figure 2 Relationship between the climate coefficient

$$\left(\frac{\text{precipitation in the month of maximum rainfall}}{\text{annual precipitation}}\right)^2$$

times annual precipitation and mean annual sediment yield (t km^{-2}).
[*Editors' Note:* The data points should have been plotted using mean annual sediment yield, the dependent variable, on the ordinate.]

EVALUATION OF THE RESULTS

From the outset we have signaled the difficulties encountered in using data obtained from experimental plot studies. In the course of this work, we have also established that use of data for solid load for torrents of the Appenines with drainage areas of several hundred km^2 gives aberrant results. One cannot consider erosion as a homogeneous phenomenon. Generally, in the torrent channels where masses of water move rapidly, huge masses of gravel particles may be rolled by the force of the water. On the other hand, landslides have been reported during intense rains. These mechanisms may finally produce erosion, but it is clear that the material thus entrained can only be on very steep slopes, and the phenomena can only be localized; the most obvious example is the alluvial cone. These localized phenomena do not appear in the data used here because it is difficult to measure coarse sediment moving along the riverbed and it is less important in large rivers, the velocity of which is relatively low.

It is no less true that these phenomena would have a tendency to occur on steeper slopes or where the value of the climatic factor is greater.

With the aim of verifying this point and also the results obtained, we included on the graph points representative of the phenomenon of erosion, calculated (1) for small nontorrent course drainage basins (up to 700 km^2 in area) for which were available measures of suspended sediment transport and (2) for experimental plots at the U.S. agricultural experiment stations.

One can make the following conclusions:

1. In small drainage basins, values for erosion, consequent upon the action of climate, confirmed the results already obtained. The data points lie around the lines for plain lands or mountain lands in accord with the degree of topographic relief of the basins.
2. On the experimental plots, areas covered with their natural vegetation have, for a given value of the climatic coefficient, values of erosion that deviate little from their theoretical values.

In contrast, for cultivated soils, erosion can attain extremely high levels and, for a given climatic factor, the points are greatly displaced toward the right of the graph. It is possible that this effect would be all the more marked as the slope and the climatic factor themselves increase, but the amount of the experimental data does not permit this question to be considered further.

15

Reprinted from *Am. Geophys. Union Trans.* 39:1076–1084 (1958)

Yield of Sediment in Relation to Mean Annual Precipitation

W. B. Langbein and S. A. Schumm

Introduction—The yield of sediment from a drainage basin is a complex process responding to all the variations that exist in precipitation, soils, vegetation, runoff, and land use. This study is aimed only toward a discernment of the gross variations in sediment yield that are associated with climate as defined by the annual precipitation. Such a study may contribute to an understanding of the effects of climatic change on erosion and of the regional variations in sediment yield. Data on sediment yields are now available in sufficient number for this kind of study, although still quite deficient in geographic coverage. Two major sources of sediment data exist. Records collected at about 170 gaging stations of the U. S. Geological Survey, where sediment transported by streams is measured, is one source of data; whereas, the other source of data is provided by the surveys of sediment trapped by reservoirs. Both kinds of data are used in this study.

Precipitation data—Precipitation is used as the dominant climatic factor in the study of sediment yield, because it affects vegetation and runoff. However, the effectiveness of a given amount of annual precipitation is not everywhere the same. Variations in temperature, rainfall intensity, number of storms, and seasonal and areal distribution of precipitation can also affect the yield of sediment. For example, *Leopold* [1951] in an analysis of rainfall variation in New Mexico, found that despite the absence of any trend in annual rainfall, changes in the number of storms produced a significant influence upon erosion. Although analyses of these effects are beyond the scope of this study, the effect of temperature, which controls the loss of water by evapotranspiration, can be readily taken into account. As is well known, the greater the temperature, the greater are the evapotranspiration demands upon soil moisture; hence, less moisture remains for runoff. More precipitation is required for a given amount of runoff in a warm climate than in a cool climate. Therefore, instead of using actual figures of annual precipitation, it is preferable to use figures of precipitation adjusted for the effect of annual temperature. However, in lieu of carrying out these extended computations, it appears possible to use the data on annual runoff which already reflect the influence of temperature. Annual runoff data are available for all the gaging station records and for most of the reservoir records. Because of the well-established relationships between annual precipitation and runoff, it is readily possible to estimate precipitation from the runoff figures.

We shall define effective precipitation as the amount of precipitation required to produce the known amount of runoff. Figure 1 shows a relationship between precipitation and runoff based on data given in Geological Survey Circular 52 [*Langbein*, 1949]. This graph has been used to convert known values of annual runoff to effective precipitation, based on a reference temperature of 50°F. In a warm climate, with temperature greater than 50°, the precipitation so estimated would be less than the actual amount of precipitation; in a cool climate, the effective precipitation so estimated would be more than the actual amount. This is the desired relationship.

Sediment-station data—In recent years a number of records of sediment yield, as measured at sediment-gaging stations, have become available. Annual loads were computed for about 100 stations giving preference to the smaller drainage areas in any region. All parts of the country,

181

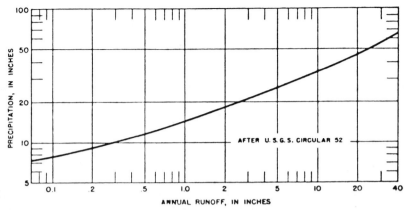

FIG. 1 – Relation between annual precipitation and runoff for a mean temperature of 50° F

where sediment records are collected, are represented.

The annual sediment loads were first arranged according to effective precipitation. They were next assembled into the class groups shown in Table 1, and the arithmetic averages were then computed for each group. Within any group the loads may vary tenfold, reflecting geologic and topographic factors not considered in this study. Each group mean is subject to a standard deviation of about 30 pct. The group averages are plotted in Figure 2. The curve shown was fitted to the data, subject to the condition (1) that it did not depart more than one standard deviation (30 pct) from any of the plotted group means, and (2) that it show zero yield for zero precipitation.

There is considerable opportunity for bias in the figures for load, because the relatively few records prohibit any high degree of selectivity. Few of the rivers drain areas in their primeval environment and moreover, land use can greatly affect the sediment yield. Farming, grazing, road construction, and channelization tend to increase sediment yield; reservoirs impound and, therefore, delay the movement of sediment. If these effects are uniform countrywide, then the overall results might be free of bias in the statistical sense, even though the absolute magnitudes may not be representative of primeval conditions. However, there is considerable variation, particularly with respect to intensity of agricultural operations, which perhaps are most intensive in the midcontinent region. The effects of various kinds of land use upon erosion vary with climate, physiography, soil type, and original vegetation. One surmises that the effect of cultivation is greater in the humid region, where effective pre-

TABLE 1 – *Group averages for data at sediment stations*

Range in effective precipitation	Number of records in each group	Average effective precipitation	Average yield
inch		inch	tons/sq mi
Less than 10	9	8	670
10 to 15	17	12.5	780
15 to 20	18	17.5	550
20 to 30	20	24	550
30 to 40	15	35	400
40 to 60	15	50	220

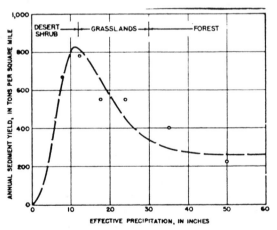

FIG. 2 – Climatic variation of yield of sediment as determined from records at sediment stations

cipitation is more than about 30 inches, because of the great contrast between original forest cover and tillage. The erosive reaction of some soil types to cultivation is evident for some small drainage basins in the humid region, which have sediment yields that approach or exceed those usual even in the arid country and are far above those to be expected within their particular range of annual precipitation. For example, sediment yield from small drainage basins (0.1 to 1.0 sq mi) in the loess hills of Iowa and Nebraska is very high, largely because of poor conservation practices on wind deposited soils. These rates "are among the highest found anywhere in the country" [*Gottschalk* and *Brune*, 1950, p. 5].

Another source of bias is the relatively non-uniform distribution of sediment-gaging stations. Most are in the central part of the country, whereas virtually none is available in the Pacific Coast Region, in New England, or in the Gulf Region.

Reservoir sediment data—Although preference was given to the smaller drainage areas in using gaging-station records of suspended sediment, opportunities for choice in this regard were severely limited. Fortunately, surveys of sedimentation in reservoirs are more numerous, so that there was opportunity to be more selective in choosing reservoirs below small drainage areas, which on that account were presumed to be more indicative of sediment yield nearer the source. Data on reservoir sedimentation were compiled by the *Federal Inter-Agency River Basin Committee* [1953]. Rates of sedimentation were obtained from surveys of sediment accumulation, expressed as an annual rate in acre feet or tons per square mile of drainage area. For those reservoirs where the bulk density of the deposits was determined, the annual rates per square mile are given in terms of tons, otherwise the rates are given in terms of acre feet. In these cases, volumes in acre feet were converted into tons by assuming a density of deposit of 60 lb/cu ft, an average of reported densities. In selecting reservoirs, preference was given to those with capacities exceeding 50 ac ft/sq mi of drainage area, in order to select those which trap a large portion of the sediment that enters the reservoir.

Reservoirs with less than five square miles of drainage area appear to have highly variable rates of sedimentation. For very small areas, rates of sediment yield are greatly influenced by details of land use and local features of the terrain [*Brown*, 1950]. For this reason, reservoirs having drainage areas between 10 and 50 sq mi were used. Because no reservoirs in desert areas were listed in the Inter-Agency compilation, data for desert reservoirs were obtained from unpublished records collected by the U. S. Geological Survey. However, because these reservoirs were on drainage areas of ten square miles or less, rates of sediment yield for these desert reservoirs were adjusted downward to obtain equivalent rates from drainage areas of 30 sq mi, according to the 0.15 power

rule explained below. The sediment data were arranged according to effective precipitation and grouped as shown in Table 2.

Group averages of the reservoir data are plotted in Figure 3. The general shape of the resulting curve is quite similar to the one obtained from the records of suspended sediment measured at river stations. The most evident difference is that the yields are about twice those indicated by the sediment-station records. There are significant differences between the two kinds of records. The sediment-station records do not include bed

TABLE 2. – *Group averages for reservoir data*

Range in effective precipitation	Number of reservoirs in each group	Average effective precipitation	Average yield	Remarks
inch		inch	tons/ sq mi	
8–9	31	8.5	1400	15 reservoirs in San Rafael Swell, Utah, and 16 in Badger Wash, Colo.
10	38	10	1180	26 reservoirs in Twenty-mile Creek basin, Wyo., 7 in Corn-field Wash, N. Mex., and 5 general.
11	12	11	1500	General
14–25	18	19	1130	General
25–30	10	27.5	1430	General, including debris basins in Southern Calif. considered as one observation.
30–38	20	35.5	790	General
38–40	11	39	560	General
40–55	18	45	470	General
55–100	5	73	440	General

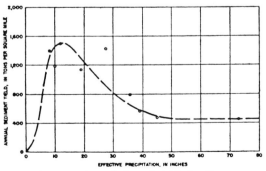

FIG. 3 – Climatic variation of yield of sediment as determined from reservoir surveys

load, which, being the coarser fraction of the load, is trapped by a reservoir. The effects are variable depending on relative amounts of bed and suspended loads at gaging stations and on the trap efficiency of the reservoirs. Moreover, reservoirs are generally built in terrain that offers favorable sites, which means drainage basins with steep slopes and hence higher rates of net erosion. However, in the comparison made here most of the difference is probably due to the effect of size of drainage area. Several studies have shown that sediment yields decrease with increased drainage area, reflecting the flatter gradients and the lesser probability that an intense storm will cover the entire drainage basin. Assuming that the graphs shown by *Brune* [1948] are correct for this effect, rates of sediment yield are inversely proportional to the 0.15 power of the drainage area. Noting that the drainage areas used for Figure 2 average about 1500 sq mi and those for Figure 3 about 30 sq mi, the sediment yields for reservoir data should average about $(1500/30)^{0.15}$, or 1.8 times that for the sediment-station data. This correction applied to the reservoir data would very nearly account for most of the difference between the curves of Figures 2 and 3.

Figures 2 and 3 appear to show a maximum sediment yield at about 12 inches annual effective precipitation, receding to a uniform yield from areas with more than 40 inches effective precipitation. The lack of data for climates with less than 5 inches of annual precipitation makes it difficult to determine the point of maximum yield with accuracy. Available data indicate, however, that it is at about 12 inches or less.

In a similar study of erosion rates and annual precipitation for large rivers of the world (Fig. 4), *Fournier* [1949] notes that the drainage basins are located on a parabolic curve in relation to their climatic character. For example, the upper limb of the parabola (greater than 43 inches) is formed by rivers typical of a monsoon climate: Ganges, Fleuve Rouge (Hung Ho), Yangtze, and some basins of southeastern United States; the middle segment of the curve between 24 and 43 inches of rainfall is formed by drainage systems located in regions with essentially equally distributed annual rainfall, as the Atlantic coastal rivers of the northeastern United States; the lower limb of the curve below 24 inches is formed by rivers draining regions of the more continental steppe or semiarid climates: Vaal, Indus, Rio Grande, Hwang-Ho, Tigris, and Colorado. Fournier con-

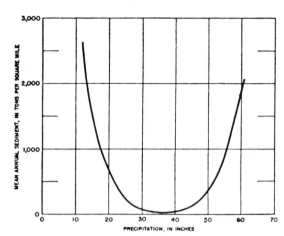

Fig. 4 – Relation of sediment yield to precipitation [after *Fournier*, 1949]

cluded that the regions of maximum erosion are those in which monthly rainfall varies greatly, the monsoon and steppe climates.

The midpart of his curve shows an annual yield of only five to seven tons per square mile which Fournier attributes to basins in which rainfall is uniform throughout the year such as those in the northeastern United States. These figures are much below that indicated by the few available records in that region which range from a minimum of about 40 tons/sq mi for Scantic River in Connecticut to 370 tons/sq mi for Lehigh River in Pennsylvania.

The lower limb of Fournier's curve terminates at a precipitation of about 12 inches, reaching an annual yield of about 2500 tons/sq mi. Although rates as high and even higher occur in many areas in the arid country, the figure of 2500 tons/year seems somewhat high as an average. The upward trend cannot continue if there is zero sediment yield (in rivers) for zero precipitation; the curve seemingly must reverse its upward trend and swing downward towards the origin.

The upper limb of Fournier's curve (above 43 inches precipitation) shows sharply increasing sediment yield with increasing precipitation, a trend that is not evident in Figures 2 and 3. However, it is possible that additional information in such areas of great rainfall as northern California and the Pacific Northwest may introduce an increasing trend in this part of those graphs.

Analysis of the climatic variation in sediment yield—The variation in sediment yield with climate can be explained by the operation of two factors each related to precipitation. The erosive influence of precipitation increases with its amount, through its direct impact in eroding soil and in generating runoff with further capacity for erosion and for transportation. Opposing this influence is the effect of vegetation, which increases in bulk with effective annual precipitation. In view of these precepts, it should be possible to analyze the curves shown in Figures 2 and 3 into their two components, the erosive effect of rainfall and the counteracting protective effect of vegetation associated with the rainfall.

These opposing actions can be represented by mathematical expressions of the following form

$$S = aP^m \frac{1}{1 + bP^n} \qquad (1)$$

in which S is annual load in tons per square mile, P is effective annual precipitation, m and n are exponents, and a and b are coefficients. The factor aP^m in the above equation describes the erosive action of rainfall in the absence of vegetation. The die-away factor $1/(1 + bP^n)$ represents the protective action of vegetation. The factor aP^m increases continuously with increase in precipitation, P, whereas the factor $1/(1 + bP^n)$ is unity for zero precipitation, and decreases with increases in precipitation.

Eq. (1) can not be evaluated by the usual least-squares method. Hence it was evaluated by trial and error, graphical methods yielding the following approximate results.

$$S = \frac{10P^{2.3}}{1 + 0.0007P^{3.33}}$$

for sediment station data and

$$S = \frac{20P^{2.3}}{1 + 0.0007P^{3.33}}$$

for reservoir sediment data.

The factor $P^{2.3}$ describes the variation in sediment yield with constant cover. Analyses of measurements of rainfall, runoff, and soil loss made on small experimental plots operated by the Department of Agriculture [Musgrave, 1947], indicate that, other factors the same, erosion is proportional to the 1.75 power of the 30-minute rainfall intensity. However, it is rather difficult to draw a connection between the intensity of 30-minute precipitation and annual precipitation. Inspection of Yarnell's [1935] charts indicates one relationship exists in the eastern part of the country and another in the western areas. However, in both areas 30-minute intensities vary

with annual rainfall to some power greater than unity. Hence, one can conclude that erosion will vary regionally to some power of the annual precipitation greater than 1.75.

The second factor, $1/(1 + 0.0007 P^{3.33})$, equals $S/aP^{2.3}$. This function, as graphed in Figure 5, purports to isolate the variation in sediment yield caused by differing degrees of vegetative cover. This function varies as shown in Table 3.

There is a good deal of information on the relation between different vegetal covers and rates of erosion in a given climate. However, most of this information deals with cultivated lands and

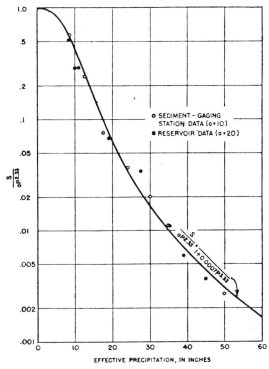

FIG. 5 – Decrease in relative sediment yield with increasing precipitation

TABLE 3 – *Variation in sediment yield associated with vegetative cover*

Effective precipitation	Vegetative cover[a]	$\frac{1}{1 + 0.0007\,P^{3.33}}$
inch		
7	Desert shrub	0.69
13	Desert shrub	0.23
20	Grassland	0.06
30	Grassland	0.017
40	Forest	0.006
50	Forest	0.003

[a] Associated with effective precipitation.

few of the vegetal data are in quantitative terms. *Musgrave* [1947] attempted a quantitative evaluation of relative erosion based on data collected at experimental watersheds in the Pacific Northwest. The results agree quite well with the results given in Table 3.

Cover	Relative erosion
Row crops or fallow	1.0 to 0.60
Small grains, grass hayland, crested wheat grass	0.05
Pasture, excellent condition, and forests	0.01 to 0.001

Formula (1) may be generalized as

$$S \propto R/V \qquad (2)$$

where R is annual runoff, and V is mass density of vegetation. In any given region the two factors operate separately; thus, sediment loads may vary with runoff depending on land use and vegetal conditions. For example, *Brune* [1948] shows that, for a given land condition in the Midwest, sediment yield increases with runoff, and that, for a given runoff, sediment yield varies enormously with percent of tilled land. The present study, however, treats of the broad climatic variation in which both runoff and vegetation are each uniquely related to effective precipitation.

With increasing precipitation, sediment yield varies as shown on Figures 2 and 3, but runoff increases as shown on Figure 1. The ratio between sediment yield and runoff is a measure of the concentration. This quantity is generally reported in parts per million (ppm) by weight and may be computed by dividing sediment yield in tons per square mile by runoff computed in tons per square mile. Figure 6 shows results of this computation for the data in Table 2. The concentration decreases sharply with increasing precipitation.

Annual precipitation, as indicated by the annual runoff, is used as the sole climatic measure. We have considered differences in precipitation intensity and its seasonal distribution only so far as these influences are reflected in the amount of annual runoff. For example, low precipitation regimes are characteristically more variable than those of humid regions [*Conrad*, 1946], and, indeed, the short-period excesses in intensity show up in the runoff. However, we repeat, as we wrote in our introduction, that although climatic influences on sediment are more complex, a good deal can be learned from consideration of the annual precipitation.

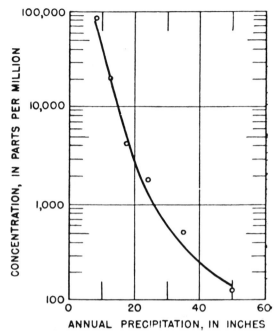

Fig. 6 – Variation of concentration with annual precipitation

Precipitation and vegetation—There can be no question of the highly significant effect of vegetation on erosion. For this reason, we have assembled information on climatic variation and vegetal bulk. The information on vegetal bulk contained in Table 4 was obtained mainly from published sources, ranging in reliability from carefully weighed quadrats to forest statistics and two estimates based on examination of photographs (for references, see Table 4). However, considering the more than 1000-fold variation in vegetal weight, as between desert shrubs to forests, great precision does not seem to be needed for the rough kind of study that seems possible at this time. Some of the data on vegetal weights were given directly in pounds per acre or equivalent. The forest data were obtained by dividing the reported cubic-foot volumes of saw-timber and pole-timber trees, given in millions of cubic feet for each state, by the respective forest area in acres. Unit weights of 45 lb/cu ft were used for hardwoods, 35 for soft woods, and 40 for mixed forests.

Table 4 also includes data on mean annual precipitation and temperature applicable to each case. The climatologic data were not usually given in the references cited and were obtained from U.S. Weather Bureau reports.

Figure 7 shows a plot of precipitation against

TABLE 4 – *Climatologic data and data on weight of vegetation*

Location	Type of vegetation	Mean annual precip.	Mean annual temp.	Weight of vegetation	Reference for vegetal bulk
		inch	°F	lb/ac	
Las Vegas, Nev.	Desert shrub	5	65	100	*McDougal* [1908, pl. 28]
Salt Lake Desert, Lakeside, Utah	Desert shrub	8	50	400	*McDougal* [1908, pl. 24]
Clark Co., Idaho	Sagebrush	12	40	891	*Blaisdell* [1953]
Fremont Co., Idaho	Sagebrush	12	40	1,273	*Blaisdell* [1953]
Coconino Wash, Ariz.	Grass	15	45	1,886	*Clements* [1922]
Burlington, Colo.	Grasses	17	52	2,251	*Weaver* [1923]
Phillipsburg, Kans.	Grasses	22	52	3,230	*Weaver* [1923]
Lincoln, Nebr.	Grasses	27	51	4,467	*Weaver* [1923]
Sandhills, Nebr.	Wheat grass	18	49	4,000	*Smith* [1895]
Lincoln, Nebr.	Grasses	27	51	6,224	*Kramer* and *Weaver* [1936]
Fraser forest, Colo.	Lodgepole pine	25	32	43,000	*Wilm* and *Dunford* [1948]
Rocky Mt. States	Conifers	28	38	54,000	*U. S. Forest Service* [1950]
Northeast Central States	Mixed forest	30	43	64,000	*U. S. Forest Service* [1950]
Northeast States	Hardwood forest	42	45	55,000	*U. S. Forest Service* [1950]
Southeast States	Mixed forest	51	60	48,000	*U. S. Forest Service* [1950]
Pacific Coast States	Conifers	64	47	150,000	*U. S. Forest Service* [1950]
Serro do Navio, Amapa Terr., Brazil	Hardwood forest	120		870,000	Field estimate by M. G. Wolman, 1956

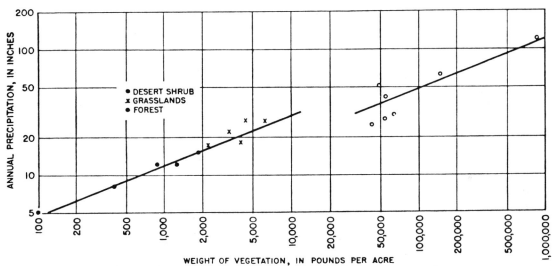

FIG. 7 – Relation between precipitation and weight of vegetation per unit area

vegetation weight. For the data available, the correlation seems quite high with a decided break between forest and nonforest types. The graph indicates that for equivalent rainfall, forests have about five times as much weight as grasses. With a longer life span, trees should understandably show greater total weight in place, although perhaps the annual growth (= annual decay for equilibrium) would be less than for grasses.

The seeming fact that desert shrubs are on the lower continuation of the line defined by the grasses seems anomalous. In arid and semi-arid regions the increase in vegetal bulk with rainfall rather simply reflects increasing opportunity for a greater number of plants, greater opportunity for each plant to reach maximum development for the species environment, and opportunity for growth of larger species. However, this does not explain the variation of forest bulk with rainfall greater than needed to satisfy optimum evapotranspiration demands for the climate. Forests are areas of water surplus in the climatic sense, yet vegetal density seems to vary with precipitation and temperature. Among the eastern states, for example, the vegetal bulk per unit area in Maine and North Carolina are about the same. The lesser annual precipitation in Maine appears to be compensated by a lesser temperature; whereas, in North Carolina higher annual precipitation is compensated by a higher temperature. The forested areas of Washington, Oregon, and California have about the same temperature; the forest densities seem to follow precipitation as follows:

State	Precipitation inch	Vegetation lb/ac
Washington	80	177,000
Oregon	60	158,000
California	53	120,000

The variation in unit weight shown in Figure 7 is made up of two components, one due to variation in weight among different communities of the same vegetation type, and that due to variation in weight among different types. The latter is very likely the dominant factor in the relationship on Figure 5. Beyond a certain limiting precipitation, say that for which precipitation is adequate to meet all evapotranspiration requirements, differences in vegetal bulk may reflect not so much growth factors as differences in plant types or associations. The heavy vegetal bulk in the Pacific Northwest, for example, may be the reflection of a difference in plant type rather than a direct effect of precipitation on growth.

We are considering here only gross relations, ignoring rather important variations that might be due to differences in species, topographic setting, or moisture conditions that might favor or discourage growth. For example, there are patches of timber in the valley bottoms, in the Great Plains, with weight densities far exceeding that of the grasslands. The data used to define this relationship are admittedly crude and subject to bias. The forest statistics, for example, generally exclude bark, leaves, flowers, fruit, and most branches. The ratio of these parts to the whole tree decreases with age. The existing data are for stands in various degrees of maturity, and most existing data exclude roots. The ratio of roots to aboveground growth is variable among different kinds of plants and may be a large source of error.

Then again, although one might conceive that the maximum amount of plant material should theoretically be correlative with climate, other factors such as aspect, depth, and nature of soil are of major influence. Ideally, vegetal den-

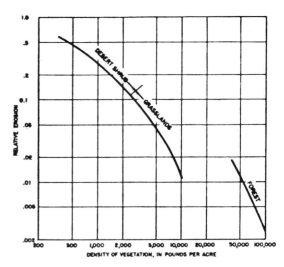

FIG. 8 – Relative erosion compared with density of vegetation

sities should be studied locally to arrive at a normal density for the regional climate. However, this kind of study would be beyond the scope of this discussion. Only the evident fact that vegetal densities are so variable over the range of climates experienced in this country makes it at all possible to use the existing data.

Interpreting the graph in Figure 5 as an indication of relative erosion associated with vegetation, as shown on Figure 7, the relationship shown on Figure 8 can be drawn. According to this graph, a relative change in vegetal density is effective on erosion throughout the climatic range, although the break between trees and grass suggests that per pound, grass is more effective in retarding erosion than trees.

Erosion and climate change—Examination of Figures 2 and 3 may be useful in visualizing not only variations in rates of net erosion between climatic zones in the United States but also the probable change in rates of erosion and stream activity during a climatic change.

Within the 0- to 12-inch precipitation zone an increase in annual rainfall would apparently be followed by an increase in erosion and vice versa; whereas, between about 12 to 45 inches of rainfall, erosion should decrease with increased precipitation. Above 45 inches of precipitation, erosion should remain about constant with increased precipitation, although Fournier's curve (Fig. 4) shows a marked increase in sediment yield above 43 inches of precipitation.

The direction of a change in sediment yield

with changing rainfall appears to be dependent on the amount of precipitation before the change. For example, in a drainage basin located in a region with mean precipitation ranging from about 10 to 15 inches, a change either to a wetter or drier climate might result in a decrease in erosion, in the one case owing to increased density of vegetation and in the other case owing to a decrease in runoff. The above discussion assumes unchanged temperature, but perhaps a change in mean annual precipitation would be accompanied by an inverse change in mean annual temperature, further enhancing its effects.

A change in stream character and activity, with climate change, can probably be understood best in relation to the changes in the ratio of sediment load to discharge as precipitation increases or decreases. Referring to Figure 6, it is apparent that as annual precipitation decreases, the concentration of sediment per unit of runoff increases. This suggests quite strongly that, other factors being the same, the increasing sediment loads associated with increasing dryness will cause aggradation, in an amount depending on the magnitude of the climatic change. *Mackin* [1948, pp. 493–495] has summarized changes to be expected in stream activity with changes in load and discharge. In every case, an increase in load or decrease in discharge with constant load results in aggradation and vice versa.

The decrease in annual runoff with decreased precipitation will necessitate an adjustment of stream gradient and shape according to established principles [*Leopold* and *Maddock*, 1953], such that the width and depth of the channel should decrease and gradient increase. These changes are consistent with aggradation. Of course, an increase in precipitation might be expected to result in degradation as sediment concentration decreases. The increased discharge will result in an increase in channel width and depth and a decrease in gradient. Numerous exceptions to the above generalizations can be cited, especially when glaciation, deforestation, cultivation, or a change in base level become important.

REFERENCES

BLAISDELL, J. P., Ecological effects of planned burning of sagebrush-grass range on the Upper Snake River Plains, *U. S. Dept. Agr. Tech. Bul. 1075*, 1953.

BROWN, C. B., Effects of soil conservation, *Applied sedimentation*, P. D. Trask (ed), pp. 380–406, John Wiley and Sons, 707 pp., 1950.

BRUNE, GUNNAR, Rates of sediment production in midwestern United States, *Soil Cons. Serv. TP-65*, 40 pp., 1948.

CLEMENTS, F. E., Destruction of range by prairie dogs, *Carnegie Inst. Washington Yearbook 21*, 1922.

CONRAD, V. A., *Methods in climatology*, Harvard Univ. Press, 1946.

FEDERAL INTER-AGENCY RIVER BASIN COMM., *Summary of reservoir sedimentation surveys for the United States through 1950*, Subcommittee on Sedimentation, Sedimentation Bul., 31 pp., 1953.

FOURNIER, M. F., Les facteurs climatiques de l'erosion du sol, *Bul. Assn. Geogr. Francais 203* (Seance du 11 Juin 1949), 97–103, 1949.

GOTTSCHALK, L. C., AND G. M. BRUNE, Sediment design criteria for the Missouri Basin loess hills, *Soil Cons. Serv. Tech. Pub. 97*, 21 pp., 1950.

KRAMER, J., AND J. E. WEAVER, Relative efficiency of roots and tops of plants in protecting the soil from erosion, *Nebr. Univ. Cons. & Surv. Div. Bul. 12*, p. 94, 1936.

LANGBEIN, W. B., Annual runoff in the United States, *U. S. Geol. Surv. Circ. 52*, 11 pp., 1949.

LEOPOLD, L. B., Rainfall frequency: An aspect of climatic variation, *Trans. Amer. Geophy. Union*, **32**, 347–357, 1951.

LEOPOLD, L. B., AND T. MADDOCK, JR., The hydraulic geometry of stream channels and some physiographic implications, *U. S. Geol. Surv. Prof. Paper 252*, 57 pp., 1953.

MACKIN, J. H., Concept of the graded river, *Bul. Geol. Soc. Amer.*, **59**, 463–512, 1948.

McDOUGAL, D. T., Botanical features of North American deserts, *Carnegie Inst. Washington Pub. 99*, 1908.

MUSGRAVE, G. W., The quantitative evaluation of water erosion—A first approximation, *J. Soil Water Cons.*, **2**, 133–138, 1947.

SMITH, T. G., Forage conditions of the prairie region, *U. S. Dept. of Agr., Yearbook*, 1895.

WEAVER, J. E., Plant production as a measure of environment, *Carnegie Inst. Washington Yearbook 22*, 1923.

WILM, H. G., AND E. G. DUNFORD, Effect of timber cutting on water available for streamflow, *U. S. Dept. of Agr. Tech. Bull. 968*, 12–14, Nov. 1948.

YARNELL, D. L., Rainfall-intensity frequency data, *U. S. Dept. Agr. Misc. Pub. 204*, 67 pp., 1935.

U. S. Forest Service, Basic forest statistics of the U. S. as of Jan. 1945, Sept. 1950.

U. S. Geological Survey, Washington, D. C. (W.B.L.) and U. S. Geological Survey, Denver Federal Center, Denver, Colorado (S.A.S.)

(Communicated manuscript received April 25, 1958.)

16

Reprinted from pages F-1–F-9, F-16, and F-17–F-23 of *U.S. Geol. Survey Water-Supply Paper 1535F*, 1961, pp. 1–23

SOLUTES IN SMALL STREAMS DRAINING SINGLE ROCK TYPES, SANGRE DE CRISTO RANGE, NEW MEXICO

By John P. Miller

ABSTRACT

The solute content of stream waters draining a single rock type was found to be essentially uniform regardless of the size of area drained. This implies a steady-state relation for the various weathering mechanisms through a considerable range in conditions of slope, soil, vegetation, and hydrology.

Average solute concentrations of waters draining quartzite, granite, and sandstone are in the proportion 1:2.5:10. Most of the solute content of waters draining sandstone is derived from carbonate cement and thin limestones, which together constitute less than 1 percent of the rock. Relative mobilities of the various elements (ratio of percentage in water to percentage in rock) agree generally with previous estimates based on less satisfactory data.

Calculated values for rates of chemical denudation suggest considerably greater importance of this factor in landscape sculpture than is commonly believed.

INTRODUCTION

Weathering of rocks results in three kinds of products: (a) residual minerals, (b) authigenic minerals, and (c) solutes. The first two categories of materials are relatively immobile, except in the presence of vigorous transporting agencies, and tend to accumulate at the site of weathering. Soluble products, on the other hand, move readily with percolating ground waters and ultimately into the streams.

Investigations of weathering processes have consisted largely of chemical and mineralogical comparisons between unweathered rock and the overlying weathered mantle. Studies of weathering processes based primarily on composition of the soluble products are very few in number, despite the existence of a voluminous literature on water chemistry. A principal reason is that practically all regular sampling stations are at sites where stream waters draining several different rock types are mingled. This limitation of the available data led to the generalized conclusions of Smyth (1913) and Polynov (1937) about relative mobility of the common elements during weathering. In both cases, average compositions of diverse rocks were compared with average compositions of stream waters. Smyth's estimate applies to the entire surface of the earth, whereas Polynov considered only the surface underlain by igneous rocks. Recently, Anderson and

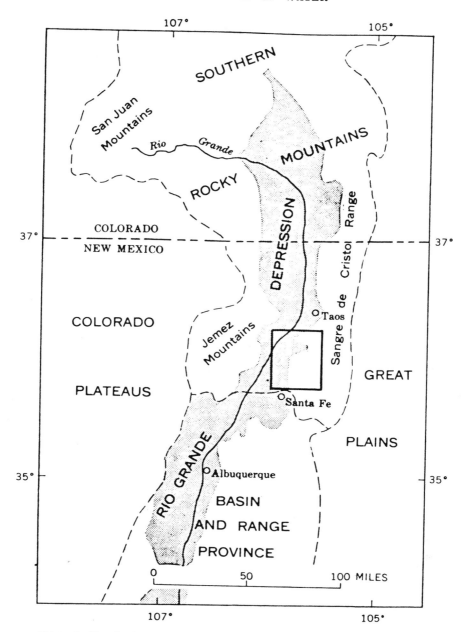

FIGURE 1.—Map showing location of study area in relation to regional physiographic features.

Hawkes (1958) have published estimates of relative mobility which apply to specific rock types in New England.

This paper describes the chemical character of stream waters derived from several small drainage basins, each of which is underlain

191

by one of three "uniform" rock types (granite, metaquartzite, and sandstone). Estimation of relative mobilities of the common elements and differences in chemical denudation rates between lithologic types were the primary objectives of this study.

ACKNOWLEDGMENTS

Cheerful assistance in the somewhat arduous task of sample collection was rendered by Dean Gerber, Santa Fe, N. Mex. Arthur Montgomery, Lafayette College, Pennsylvania, and P. K. Sutherland, University of Oklahoma, contributed unpublished information on mineralogical and chemical properties of rocks in the area studied and criticized the manuscript. The principal work of this study was done by the chemical analysts. To them, and also to other U.S. Geological Survey colleagues who aided in this project, special thanks are due. The suggestions of R. M. Garrels and R. Siever, Harvard University, at various stages of this investigation, are greatly appreciated.

PHYSICAL SETTING

The area studied in this report is located at the southern end of the Sangre de Cristo Range (fig. 1). The mountains rise 5,000 to 6,000 feet above the Rio Grande depression to the west and the Great Plains to the east, and reach a maximum elevation of 13,100 feet. Climate varies markedly with altitude. Average annual precipitation ranges from about 12 inches at the lower elevations to 35 inches or more on the highest peaks. Differences in precipitation and temperature are clearly indicated by zonation of vegetation. In order of their occurrence from low altitude to high—piñon, yellow pine, aspen, spruce, and fir dominate the luxuriant forest which covers the mountains up to the timberline at 11,000 to 11,500 feet. Above timberline, there are clumps of dwarf trees and several kinds of sedges and grasses similar to varieties found in arctic regions.

STREAM CHARACTERISTICS

Streams in the southern Sangre de Cristo Range drain either to the Pecos River or to the Rio Grande. They flow in deep, narrow canyons, and the channels are covered with coarse gravels derived largely from adjacent valley walls. Runoff is characterized by a period of high discharges during the late-spring snowmelt season and by sporadic floods resulting from intense rainfall during summer and early fall. However, the streams never carry much suspended sediment. Even after heavy rains, they are merely cloudy rather than truly turbid.

Irrigation is widely practiced in this section of New Mexico, but all sampling stations discussed in this paper are more than a mile upstream from diversions. Most of the area considered is wilderness which, except for grazing of cattle and sheep and the presence of a few prospect pits, is unaffected by the activities of man.

A detailed account of stream characteristics in this part of the Sangre de Cristo Range has been published by Miller (1958).

SUMMARY OF GEOLOGY

A generalized geologic map and structure section, showing the major lithological and structural features, is given in figure 2. The oldest rocks of the Sangre de Cristo Range are Precambrian quartzites, schists, and granites, which underlie most of the mountain crest and are also exposed in several deep canyons of the Pecos drainage basin. A thick section of sedimentary rocks overlies this basement complex. Except for a thin (<100 feet) limestone of Mississippian(?) age, the entire sedimentary column is of Pennsylvanian age consisting in the areas where water samples were taken of several thousand feet of thick nonmarine sandstones alternating with shales and thin marine limestones.

Structurally, this part of the range is a broad north-trending anticline broken at several places by major faults that also trend roughly north. As can be seen in figure 2, the western flank of the anticline is greatly eroded and is buried by Tertiary sediments. Like the adjacent Picuris area (Montgomery, 1953), the Precambrian metamorphic rocks are complexly folded with axes trending roughly from east to west.

Pleistocene glaciation of the southern Sangre de Cristo Range consisted of numerous small valley glaciers, which produced rugged peaks and broad, deep cirques but left relatively little glacial and glaciofluvial debris in most valleys. Patterned ground, block fields, and other surficial features produced by Pleistocene and Recent frost action are abundant above 10,000 feet.

Chemical and mineralogical composition of rock types drained by the stream waters of the southern Sangre de Cristo Range are discussed under appropriate headings.

GRANITE

The granitic rocks of this area, referred to by Montgomery (1953) as the Embudo granite, include coarse-grained quartz monzonite, dark-biotite granite and light-colored, well-foliated granite, and also abundant pegmatites and quartz veins. Table 1 gives a chemical analysis and estimated mode of Embudo granite in the Picuris Range, which is a few miles north of the principal area considered here. According to Montgomery (written communication, November 1959), the Embudo granite is identical in both areas.

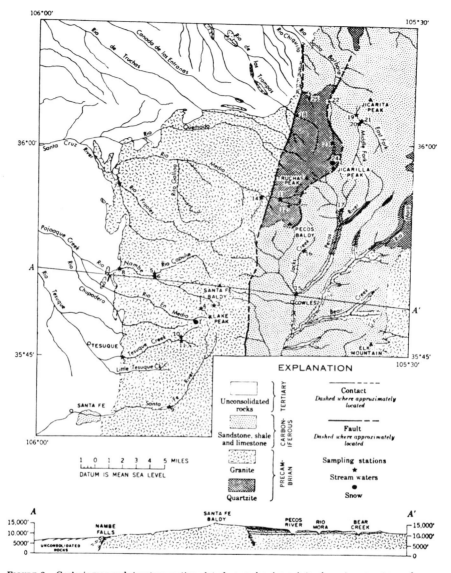

FIGURE 2.—Geologic map and structure section of study area showing points where stream waters and snow were sampled.

QUARTZITE

The metaquartzite considered in this study is the lower part of the Precambrian Ortega formation of Montgomery (1953). Although quartzite is the principal lithologic type in this formation, there are many thin interbedded layers of sillimanite-kyanite gneiss and also distinctive beds of several different schists, most of which, however,

194

TABLE 1. *Mode and chemical composition, in percent, of Embudo granite*

[Average of three analyses reported by Montgomery, 1953]

Mode		Chemical composition	
Quartz	30	SiO_2	72. 12
Microcline	30	TiO_2	. 42
Albite and oligoclase	31	Al_2O_3	14. 24
Biotite	6	Fe_2O_3	1. 48
Muscovite	1	FeO	. 99
Epidote	1	MnO	. 09
Sericite	1	MgO	. 52
Chlorite	Tr.	CaO	1. 56
Magnetite and ilmenite	Tr.	Na_2O	3. 17
Sphene	Tr.	K_2O	4. 03
Apatite	Tr.	P_2O_5	. 12
Zircon	Tr.	S	. 01
Allanite	Tr.	BaO	. 07
Pyrite	Tr.	ZrO_2	Tr.
		CO_2	. 03
		H_2O	. 77
			99. 62

do not occupy areas that drain to the streams considered in this report. The quartzite is many thousands of feet thick and dips steeply throughout the area.

Montgomery, who is making a detailed petrologic study of these rocks, prepared the mineralogical data given in table 2. No chemical analyses are available.

SANDSTONE

The rocks designated here as "sandstone" actually include also such diverse lithologic types as conglomerates, shales, mudstones, and thin limestones—all of Pennsylvanian age and up to 7,000 feet thick. In the area where water samples were taken, units called "limestone" in measured sections constitute 1 to 2 percent of the total thickness. However, the limestones are impure, and the $CaCO_3$ equivalent of the entire section, including calcareous cement in some of the terrigenous rocks, is 0.5 to 1.0 percent (P. K. Sutherland, written communication, November 1959). Roughly 25 percent of the total section is shale and mudstone. The remainder is a highly varied sequence of coarse terrigenous rocks (conglomerates, estimated 15 percent; sandstone, estimated 60 percent). Quartz and quartzite are the predominant detrital constituents, but feldspar and igneous and metamorphic rock fragments are present in small amounts throughout the sequence. In some layers, feldspathic constituents form a major proportion of the rock. Silica is the predominant cement, with carbonate minor in most beds.

TABLE 2.—*Estimated modes in percent of quartzite (Ortega formation) in areas draining to various sampling stations*

[Data from A. Montgomery, written communication]

	Station 12	Station 13	Stations 18, 23, and 25
Quartz	97	89	97
Muscovite	.5	[1] 2	1
Biotite	.5	2	
Feldspar		2	
Hornblende		2	
Staurolite	.5		
Sillimanite	Tr.	1	.5
Ilmenite and hematite	.5	1	.5
Epidote and clinozoisite	.1	Tr.	.1
Tourmaline		.3	
Garnet	Tr.	.2	
Kyanite	Tr.	.2	
Andalusite	Tr.	Tr.	
Piedmontite		Tr.	
Apatite		Tr.	
Beryl		Tr.	
Idocrase		Tr.	

[1] Biotite and chlorite.

Detailed chemical and mineralogical analyses of these rocks have not been made, and sampling them for the purpose of making detailed comparisons with water samples would be a major undertaking.

Throughout the area where water samples were collected, the sandstone sequence is flat lying or very gently dipping.

SAMPLING PROCEDURE

Altogether, 23 samples of stream waters were collected. The samples were distributed according to lithologic terrain of basin as follows: granite, 11; quartzite, 5; and sandstone, 7. In addition, two samples of snow were obtained, one each from granite and quartzite terrains. The total number of samples, and to some extent the location of sampling stations, was determined primarily by available time and manpower. In most cases, it was necessary to walk 5 to 15 miles and backpack the necessary equipment and the samples.

All water samples were collected between June 26 and July 8, 1958, except for sample 25 (taken July 26). This sampling period was chosen because it follows the peak snowmelt runoff and precedes the summer rains, which ordinarily begin during the first half of July. No rain fell during the month of June, but a few very small rains had occurred by July 26. At the time of sampling, all streams were at slightly less than half-bankfull stage.

The location of sampling stations is shown on figure 2. No special statistical sampling design was used. Rather, insofar as possible, the stations were arranged for each rock type so that samples from drainage areas of various sizes would be obtained. Although altitudes of sampling stations range from 6,700 feet to 10,400 feet, three-fourths of them are above 9,000 feet. Location and extent of rock exposure account for the considerable differences in altitude ranges of sampling stations on the three rock types:

	Altitude (feet)
Granite	6, 700–10, 400
Quartzite	9, 400–10, 100
Sandstone	8, 300–10, 200

The procedure at each stream sampling station was as follows:

1. The pH of the water was determined by means of a portable pH meter. First, the meter was standardized with buffer solution, which previously had been brought to stream temperature. Next, water was dipped from the stream in a polyethylene beaker, the electrodes were immersed, and the pH was read. Two or more determinations were made at each station.

All field pH measurements are listed in table 3, to indicate degree of reproducibility of results. The field pH values listed in table 4 are averages of the data reported in table 3.

2. Two half-gallon samples were collected in polyethylene bottles at each station. Foreign matter (mostly spruce needles) was removed by filtration through a clean cotton handkerchief, which was thoroughly rinsed in the stream at each station. Cloth was used rather than filter paper because of the greater speed of filtration. The sample collected for spectrographic trace-element analysis was acidified to pH 6.0–6.5 with HCl; the other sample was not acidified or treated in any way prior to analysis.

The samples of snow were collected from snowfields above timberline. The surficial dust-covered snow was scraped away, and clean snow was ladled directly into polyethylene bottles with a silver spoon.

ANALYTICAL METHODS

All samples were analyzed by methods regularly used by the U.S. Geological Survey. Haffty (1960) has given a brief description of the residue method for spectrographic determination of trace elements in water.

It is exceedingly difficult to evaluate precision of the methods and accuracy of the results. From the discussion by Hem (1959), it appears that the precision for the common elements is ordinarily in the range 2 to 10 percent. However, in these dilute waters, concentrations of some elements are near the limits of sensitivity and the

[*Editors' Note:* Table 3 has been omitted.]

results are probably somewhat less accurate. Precision of spectrographic methods is in the range 10 to 20 percent.

[*Editors' Note:* The detailed results of the analyses have been omitted.]

RELATIVE MOBILITY OF THE ELEMENTS

The stream waters described in this report were derived entirely from snowmelt that had percolated through thin surficial deposits and underlying bedrock. Traveltime of meltwater from snowbank to stream did not exceed 3 months and probably was much less. Despite great differences in altitude, slope, vegetation, drainage area, and runoff conditions, waters draining a specific rock type are remarkably uniform in composition. This relation, shown in figure 3, implies that solute concentration is the result of weathering processes operating at uniform intensity and presumably under steady-state conditions over broad areas. Thus, different basins draining the rock of the same type yield waters of similar composition provided there is no dilution of ground waters by water flowing overland directly into channels. Also, because stream flow is several times faster than subsurface flow, the solute content must be acquired by the water before it reaches the channels. Therefore, the concentrations of various elements in these waters may be considered the result of differences in mobility under the prevailing conditions of weathering.

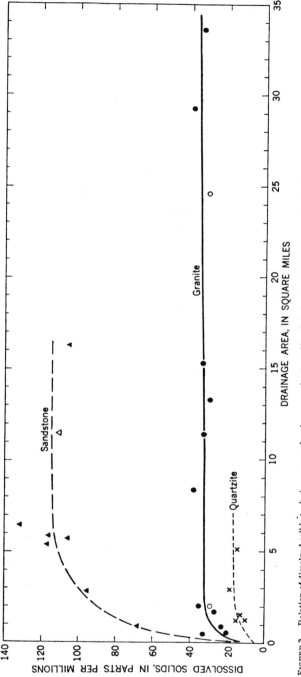

FIGURE 3.—Relation of dissolved solids to drainage area, showing approximate uniformity of waters derived from rock of each type. Open circles and triangles represent weighted-mean values for locations below tributary junctions; that is, data for station pairs (5 and 6, 10 and 11, 20 and 21) were combined by weighting according to prevailing discharge of each tributary at time of sampling. Each curve intersects the y-axis at the average value of dissolved solids in snow ($\approx$ 5 ppm).

Previous workers (Smyth, 1913; Polynov, 1937; Anderson and Hawkes, 1958) have defined the ratio

$$\frac{\text{Percent of element in water}}{\text{Percent of element in rock}}=\text{relative mobility},$$

and this designation is used here also. Data on chemical composition of the Embudo granite of Montgomery (1953) of the Sangre de Cristo Range are adequate for reasonably accurate calculation of relative mobilities. In the case of quartzite and sandstone, estimated chemical compositions, computed from modes, permit only comparison of order of mobility.

Calculated values of relative mobilities are given in table 7. The order of mobility for the various rocks can be summarized as follows:

Granite_____Ca>Mg>Na>Ba>K>Si>Fe=Mn>Ti>Al
Quartzite____Ca>>Na>>K>Mg>Fe>Si>Al
Sandstone___Ca>>>Na>K>Si=Al

For granite, the order and also specific values agree reasonably well with previous work, except for the finding of Anderson and Hawkes (1958) that magnesium is the most mobile element of certain New England granites. Comparative data for the other rocks are not available.

The exceedingly large values for relative mobility of calcium in quartzite and sandstone serve to emphasize the variety of factors affecting the ratio. In particular, the proportion of relatively inert minerals present in a rock may tend to obscure significant relations between highly reactive constituents present in small quantities. Further consideration of relative mobility should include analyses of soil waters and also laboratory determination of solubilities both for bulk rock and for its individual constituent minerals.

RATE OF CHEMICAL DENUDATION

The yield of dissolved products from a basin is commonly expressed as a denudation rate, meaning the average rate of lowering of the land surface. Accurate determinations require more detailed hydrologic and chemical information than is available for most areas. Estimation of denudation rates from the Sangre de Cristo data involves the following assumptions.

1. Average concentrations of dissolved solids reported for waters derived from each rock will be considered as mean annual values. Because of low-flow conditions at the time of sampling, they may be closer to maximum annual concentrations; but specific data on this point are lacking. However, study by C. H. Hembree and F. H. Rainwater (written communication, Jan. 22, 1960) of waters derived

TABLE 7.—*Composition of dissolved constituents in waters compared with composition in source rocks, and calculated relative mobilities of various constituents*

[Values are recalculated to 100 percent]

Constituent	Source	Granite Percent	Granite Relative mobility	Quartzite Percent	Quartzite Relative mobility	Sandstone Percent	Sandstone Relative mobility
Si	Water	38.24	0.57	37.09	≈0.4	5.95	≈0.06
	Rock	67.30		≈95		≈96	
Al	Water	.14	.01	.26	≈.1	.08	≈.1
	Rock	15.07		≈2		≈1	
Mg	Water	4.98	8.03	1.32	≈4	7.08	
	Rock	.62		≈.3		_____	
Ca	Water	32.01	14.42	42.38	≈200	81.48	≈200
	Rock	2.22		≈.2		≈.4	
Ba [1]	Water	.16	1.33	.24		.23	
	Rock	.12		_____		_____	
Na	Water	18.41	3.92	11.49	≈60	3.47	≈4
	Rock	4.70		≈.2		≈.8	
K	Water	5.51	.96	6.62	≈7	1.54	≈2
	Rock	5.70		≈1		≈.9	
Fe	Water	.52	.14	.55	≈.5	.16	
	Rock	3.62		≈1		_____	
Ti	Water	.01	.02	.01	≈.05	.01	
	Rock	.50		≈.2		_____	
Mn	Water	.02	.14	.04	≈2	.003	
	Rock	.14		≈.02		_____	

[1] Includes Sr.

from granite in the Wind River Range, Wyo., indicates that concentration at any given station changes very slightly with variation in discharge. Furthermore, as in the present study, their data show that concentration is independent of drainage area.

TABLE 8.—*Runoff data for gaging stations in the Sangre de Cristo Range*

Stream and location	Altitude (feet)	Drainage area (sq mi)	Average annual discharge (cfs per sq mi)	Average annual runoff (inches)	Years of record	Length of record (years)
Rio Tesuque near Santa Fe	7,100	11	0.29	3.9	1936–50	15
Rio Nambe near Nambe	6,200	37	.29	3.9	1933–40, 1942–50	17
Santa Cruz River at Cundiyo	6,460	86	.37	5.0	1931–57	27
Pecos River near Pecos	7,505	189	.51	6.9	1931–57	27
Santa Fe River near Santa Fe	7,718	18	.46	6.2	1913–18, 1919–27, 1928–57	42
Rio Santa Barbara near Llano	8,300	38	.77	10.4	1953–57	5

2. The linear increase of precipitation and runoff with altitude, indicated by station data, is extrapolated to the crest of the range. Available data on runoff in the southern Sangre de Cristo Range, summarized in table 8, are from stations fairly well distributed through the altitude range 6,200 to 8,300 feet. Precipitation data given in the Climatic Summary of the United States to 1930 (U.S. Department of Agriculture) consist of records of more than 10 years duration for 11 stations ranging in altitude from 6,000 to 9,000 feet.

3. The data for snow (table 4) and the results of Junge and Werby (1958) indicate that the approximate average concentration of precipitation in this area is 5 ppm of dissolved solids. In order to correct values of dissolved solids in stream waters, the average atmospheric contribution must first be weighted according to the ratio of precipitation to runoff. Altitudinal variation in this correction, assuming 5 ppm dissolved solids in precipitation, is given below.

Altitude (feet)	Precipitation (inches)	Runoff (inches)	Ratio	Correction (ppm)
7,000	13. 3	5. 0	2. 7	−13
8,000	17. 5	8. 6	2. 0	−10
9,000	21. 7	12. 0	1. 8	−9
10,000	25. 9	15. 5	1. 7	−8
11,000	30. 1	19. 0	1. 6	−8
12,000	34. 3	22. 4	1. 5	−7

Quartzite terrains in this area are restricted to high altitudes, but granite and sandstone extend from the base of the range to the crest.

4. Specific gravity of all three rock types is considered to be 2.6. Calculated values of denudation rates are plotted in figure 4. These results emphasize that denudation depends both on the chemical susceptibility of the rock to weathering and on the total quantity of solvent, which in this case is closely related to altitude. In general, these streams carry very little suspended load, and it may be that downwearing by chemical denudation is a major process of landscape sculpture in mountainous regions.

Assuming that these estimates of denudation rates are reasonably accurate, it seems very unlikely that erosion surfaces can be preserved unaltered for millions of years. For example, the curves of figure 4 indicate that 1 million years of weathering at 10,000 feet could cause sandstone terrains to be lowered 48 feet, granite terrains by 11 feet, and meta-quartzite terrains by 3 feet. Thus, solution weathering alone would considerably lower any surface and in lithologically contrasting areas would produce rugged relief within a few million years.

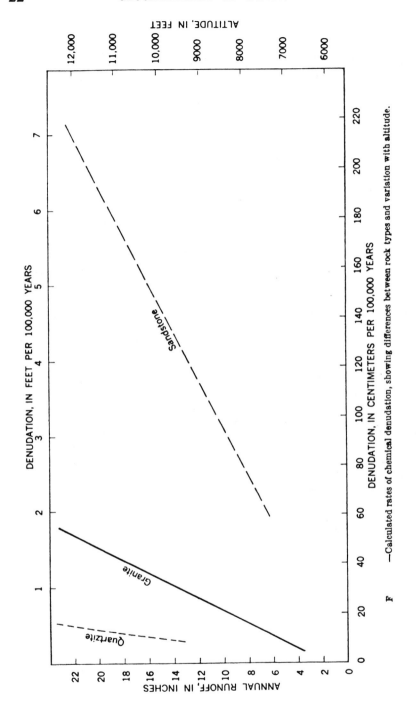

F —Calculated rates of chemical denudation, showing differences between rock types and variation with altitude.

CONCLUSIONS

The approximately uniform concentration of solutes in waters draining a single rock type indicate a steady-state relation for various weathering mechanisms affecting the system rock–water. This generalization applies for a considerable range in conditions of slope, soil, vegetation, and hydrology. Such a relation appears to be potentially useful for predicting the chemical character of composite waters derived from several rock types.

Estimated values of denudation rates indicate that solution weathering deserves greater consideration in several kinds of geomorphic problems.

Future investigations of the kind described here should be expanded to include experimental determinations of rock and mineral solubilities.

REFERENCES CITED

Anderson, D. H., and Hawkes, H. E., 1958, Relative mobility of the common elements in weathering of some schist and granite areas: Geochim. et Cosmochim. Acta, v. 14, p. 204–210.

Haffty, Joseph, 1960, Residue method for common minor elements: U.S. Geol. Survey Water-Supply Paper 1540–A, 9 p.

Hem, J. D., 1959, Study and interpretation of the chemical characteristics of natural water: U.S. Geol. Survey Water-Supply Paper 1473, 269 p.

Junge, C. E., and Werby, R. T., 1958, The concentration of chloride, sodium, potassium, calcium, and sulfate in rain water over the United States: Jour. Meteorology, v. 15, p. 417–425.

Miller, J. P., 1958, High mountain streams; effects of geology on channel characteristics and bed material: New Mexico Bur. Mines Mem. 4, 53 p.

Montgomery, A., 1953, Precambrian geology of the Picuris Range, north-central New Mexico: New Mexico Bur. Mines Bull. 30, 89 p.

Polynov, B. B., 1937, Cycle of weathering: London, Murby. (Translation by A. Muir)

Rainwater, F. H., and Thatcher, L. L., 1960, Methods for collection and analysis of water samples: U.S. Geol. Survey Water-Supply Paper 1454, 301 p.

Smyth, C. H., 1913, The relative solubilities of the chemical constituents of rocks: Jour. Geology, v. 21, p. 105–120.

17

Reprinted from *Science* **156**:1734–1737 (1967)

AMAZON RIVER: ENVIRONMENTAL FACTORS THAT CONTROL ITS DISSOLVED AND SUSPENDED LOAD

Ronald J. Gibbs

Department of Geology, University of California, Los Angeles

Abstract. *Analytical results of sampling during both wet and dry seasons along the Amazon River, at its mouth, and from 16 tributaries reveal that the physical weathering dominant in the Andean mountainous environment controls both the overall composition of the suspended solids discharged by the Amazon and the amount of dissolved salts and suspended solids discharged.*

The purpose of this study was to determine the factors that control erosion in the Amazon River system. The amount and composition of suspended solids and the amount of dissolved salts carried by this system are evaluated. Analysis of the composition of the dissolved salts will be reported upon completion.

The Amazon River basin was selected as the area for the present study because (i) the influence of man on its properties is negligible; (ii) the effects of local natural anomalies are minimized by the large area of the drainage basin; and (iii) the series of 16 large tributaries that drain the Amazon's wide range of geologic and climatic source areas make possible the sampling of the material before it is mixed with the materials from other tributaries, thereby permitting the testing of the hypothesized controlling factors.

Periodic (monthly to semiannual) sampling of the suspended solids, bottom sediments, and water, and the measurement of salinity, temperature, and pH were accomplished throughout the seasonal cycle in these 16 tributaries (accounting for more than 90 percent of tributary discharge into the Amazon River) and along the Amazon. The *in situ* conductivities were taken as a measure of the concentration of dissolved salts.

Suspended solids were removed from each of 74 samples of water (20 liters each) in the field, mainly by pressure molecular filtration (size of pores, 0.45 μ), and the material was stored in a small amount of river water with Hutner's (1) volatile organic preservative (a mixture of o-cholorobenzene, n-butyl chloride, and 1,2-dichloroethane) to prevent decay by microbes and alteration of the distribution of the particles according to size.

Fifty samples of the suspended material were analyzed by x-ray diffraction for mineral composition after separation, by size, into fractions ($< 2\ \mu$, 2 to 20 μ, and $> 20\ \mu$) and removal of organic material and iron oxide coatings, according to procedures published elsewhere (2). Analytical results were considered in connection with the environmental factors of geology, elevation, climate, and vegetation. As the possible controlling factors, nine parameters related to these four environmental factors were measured for each tributary basin from appropriate maps and data: (i) areal percentage of "calcic" rocks (limestone, dolomite, and volcanic rocks other than rhyolite); (ii) areal percentage of igneous and metamorphic rocks (mainly Precambrian shield areas of acid- to intermediate-type rocks); (iii) areal percentage of continental sedimentary rocks; (iv) areal percentage of marine sedimentary rocks, excluding "calcic" rocks; (v) areal percentage of "calcic" rocks in the upper third of each tributary basin; (vi) mean elevation (using 13 elevation intervals) above the mouth (base level) of the tributary; (vii) mean temperature for the months of minimum and of maximum temperature and the mean annual temperature; (viii) mean precipitation for the months of minimum and of maximum precipitation and the mean annual precipitation; and (ix) areal percentage of broadleaf evergreen vegetation (high values representing dense jungle and low values representing grasslands or shrubs, or both). Analytical results and measured values of the environmental parameter data are compiled separately (3), as are the results of the multiple regression and correlation analyses of the data (3).

The salinity distribution of the Amazon River and its tributaries (Fig. 1) shows the dilution of the contribution of higher-salinity (ranging from about 250 to 85 parts per million) tributaries from mountainous environments by the tributaries with lower salinity (ranging from about 50 to 4 ppm) from tropical environments as they join the Amazon's eastward flow to the ocean.

A plot of the salinity versus the elevation of each tributary shows a strong overall covariance for the wet and dry seasons. Of the nine parameters studied, elevation was the overall dominant factor controlling salinity (accounting for 85 percent of the variability of salinity), the other parameters being of far less importance.

Calculations were made of the total weight of dissolved salts eroded from each tributary per year, on the basis of both measured conductivities and calculated discharges (3). The calculation of tributary discharges was made by subdivision of measured discharges of both the upper Amazon (above the Rio Negro) and the lower Amazon (excluding the Negro and Tapajós for which separate discharge estimates are available) (4). The basis for subdivision of the discharges was the mean value for annual precipitation obtained for each tributary basin (3) and the assumption of equal runoff factors with-

in each portion, upper and lower, of the Amazon basin. These calculations of the load of salts clearly indicate the dominant contributions of the Andean environment. Total dissolved salts eroded from the mountainous environment were computed by combining the estimated amounts of dissolved salts eroded from the mountainous headwaters of the Marañon, Ucayali, Napo, Içá, Japurá, and Madeira rivers. On this basis about 85 percent of the total dissolved salts discharged by the Amazon is supplied from the 12 percent of the total area of the basin, comprising the mountainous-type environment.

A discharge-weighted mean salinity of 36 ppm was calculated for the Amazon from the salinity data I obtained for the mouth of the Amazon (3) and from discharge data for this river (4). On this basis, a provisional discharge-weighted mean salinity of 114 ppm was calculated for the river water of the world. This value is lower

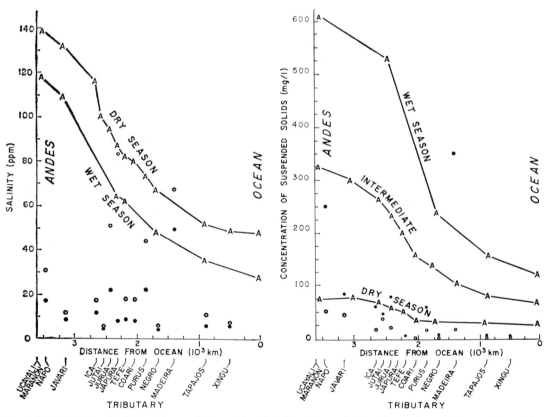

Fig. 1 (left). Variation of salinity of Amazon River and its tributaries. *A*, Concentrations of main stream of Amazon. Open circles, concentrations in dry seasons; closed circles, concentrations in wet seasons of indicated tributaries.   Fig. 2 (right). Variation of suspended solids of Amazon River and its tributaries. *A*, Concentrations of main stream of Amazon. Open circles, concentrations in dry seasons; closed circles, concentrations in wet seasons of indicated tributaries.

Table 1. Sediment load-weighted mean mineralogic composition by size of fractions of solid material transported by six tributaries from tropical environments, five tributaries from mountainous environments, and from the Amazon River at its mouth in weight percent (see 6).

Fraction	Minerals (%)						
	Quartz	Plagio-clase	K-feldspar	Kao-linite	Mica	Montmo-rillonite	Chlorite
Tributaries from tropical environment							
<2 μ	6	0	0	88	5	<1	<1
2 to 20 μ	48	5	9	26	13	0	<1
Tributaries from mountainous environment							
<2 μ	8	0	0	29	30	29	3
2 to 20 μ	42	8	6	13	26	0	5
Amazon at its mouth							
<2 μ	7	0	0	31	33	27	2
2 to 20 μ	40	7	7	14	27	0	5

than the 120 ppm mean salinity calculated by Livingstone (5).

Concentrations of the suspended solids in the Amazon River and its tributaries (Fig. 2) show the dilution of the waters with high concentrations from mountainous environments by the dilute rivers from tropical environments. The bed load, or that material carried within 50 cm of bottom, contributes only a small portion of the total solid material carried, with bed-load transportation for the mixed and mountainous tributaries estimated at 2 to 3 percent and 5 to 10 percent, respectively, of total sediment transported. Concentration of these solids was controlled mainly by elevation and, to a lesser extent, by precipitation. The range of the concentration of suspended solids of each tributary was controlled by elevation; variations within each range, however, were due to seasonal variations in precipitation.

Annual rates of erosion of suspended solids of the tributaries, based on both the above calculated discharges and the measured concentrations (3), indicate the dominant contributions of the Andean environment. The total suspended solids eroded from the mountainous environment were estimated by combining the estimated amounts of suspended solids eroded from the mountainous headwaters of the Marañon, Ucayali, Napo, Içá, Japurá, and Madeira rivers. On this basis about 80 percent of the total suspended solids discharged by the Amazon is supplied from the 12 percent of the total area of the basin comprising the mountainous-type environment.

Mean diameter (ranging from 1 to 13 μ) of the particles of the material transported showed a positive covariance with elevation. Distribution of particles according to size also varied with the elevation of each tributary basin; these particles ranged from coarse sand (1 mm) to fine clay (< 0.1 μ) for mountainous tributaries and from coarse clay (2 μ) to fine clay (< 0.1 μ) for low-elevation tributaries.

Composition of the fractions (< 2 μ and those from 2 to 20 μ) of the solid material discharged from mountainous and tropical tributaries and by the Amazon at its mouth is given in Table 1. Composition of the clay mineral suite discharged by the Amazon into the ocean strongly resembles that of the clay mineral suite from mountainous tributaries and, it will be noted, is radically different from the composition of the clay mineral suite from tropical tributaries. Among the minor constituents identified, gibbsite, though barely detectable in the samples of suspended solids of the discharge of the Amazon, was identified by its characteristics 4.85 Å peak in some of the samples from 12 of the 16 tributaries.

Composition of the 2- to 20-μ fraction of the solid material discharged by tributaries from the mountainous environments and that discharged by the Amazon into the ocean is also remarkably similar. Compositional averages for the fraction > 20 μ are not included in Table 1 because analytical results indicated that many of the rivers either did not transport material of this size or transported so little that results of mineralogic analysis are not applicable to this discussion.

Rivers from the tropical environments contain the highest percentage of clay minerals and are far richer in kaolinite (up to 94 percent of the < 2-μ fraction) than tributaries from the mountainous environments. However, the facts that mountainous rivers have transport rates for minerals more than 100 times those of tropical rivers, and that the overall composition distribution of minerals discharged into the ocean differs only slightly from that of material eroded by tributaries from the mountainous environments are accounted for by the much higher concentrations of suspended solids in the waters from these mountainous environments.

Of the nine environmental parameters, elevation was the most significant in controlling the greatest number of concentrational and compositional parameters of the Amazon River system. Elevation determined, to a great extent, the concentrations of dissolved salts and suspended solids, the particle size, and concentrations of many of the various minerals of the suspended solids. The other factors proved dominant for only a few of the minerals of the suspended solids.

Control of the dissolved and suspended loads of the Amazon River system by physical weathering dominant in the Andean environment is indicated by the following observations: (i) the overall composition of the suspended solids discharged by the Amazon differs only slightly from that of the suspended solids eroded from the mountainous environment; and (ii) about 85 percent of the total amount of dissolved salts and suspended solids discharged is eroded from the 12 percent of the total area of the Amazon basin comprising the Andean-type environment.

Since the sediments carried by the Amazon River obviously do not indicate the existence of the immense area of the basin's tropical rain forest (probably one of the largest in geologic history), it is important to use caution in paleoenvironmental studies of sediments and sedimentary rocks in order not to overlook similar circumstances that may have occurred in the geologic past.

References and Notes

1. S. H. Hutner and C. A. Bjerknes, *Soc. Exp. Biol. Med.* **67**, 393 (1948).
2. R. J. Gibbs, *Amer. Mineral.* **50**, 741 (1965); *Clay Minerals*, in press.
3. ———, thesis, Univ. of California, San Diego (1965); *Bull. Geol. Soc. Amer.*, in press.
4. R. E. Oltman, paper presented at Associação de Biologia Tropical, Simpósio sôbre a Biota

Amazônica, Belém, Brazil, June 1966; R. E. Oltman, H. O'R. Sternberg, F. C. Ames, L. C. David, *U.S. Geol. Surv. Circ. No. 486* (1964).
5. D. A. Livingstone, *U.S. Geol. Surv. Prof. Pap. 440* (1963), chap. G.
6. Table 1 is based on data for the following six tropical environment tributaries: Tefé, Coari, Negro, Tapajós, Xingu, and Araguari; the following five mountainous environment tributaries: Marañon, Ucayali, Napo, Içá, and Japurá; and, for the Amazon at its mouth, sampling locations in the main channel off Macapá, Brazil, above the influence of sea water.
7. This research was accomplished at Scripps Institution of Oceanography, La Jolla, California. Field work was performed with the co-operation of Instituto Nacional de Pesquisas da Amazônia, Manaus-Amazonas, Brazil.

14 March 1967

208

18

Reprinted from *Geol. Soc. America Bull.* **81**:3015–3032 (1970), courtesy of the Geological Society of America

Geochemical Balance of a Small Watershed and Its Geomorphic Implications

EMERY T. CLEAVES *Maryland Geological Survey, Johns Hopkins University, Baltimore, Maryland 21218*

ANDREW E. GODFREY *Department of Geology, Vanderbilt University, Nashville, Tennessee 37203*

OWEN P. BRICKER *Department of Earth and Planetary Sciences, Johns Hopkins University, Baltimore, Maryland 21218*

ABSTRACT

A detailed input-output study of a small forested watershed draining the Wissahickon Formation in the Piedmont of Maryland revealed that chemical solution is five times as effective in removing material as is mechanical erosion. Solution weathering removes 16.9 tons/sq mi/yr of material compared with 3.2 tons/sq mi/yr by mechanical erosion.

Plant activity during the growing season increased the concentration of silica, bicarbonate, calcium, and potassium, thus increasing total dissolved solids by one-third. Autumn leaf fall also caused a short-term increase of these ions.

Rainfall does not simply dilute floodwaters as the concentration of sulfate, potassium, and calcium increases whereas silica and bicarbonate decrease in concentration during a flood cycle. Our data suggest that during the first half of a flood cycle, both the flood water and the dissolved solids in it come from an area in and immediately adjacent to the flood plain.

The weathering model derived from our study suggests that on a long-term basis approximately one-half of the erosion of the Pond Branch watershed is caused by chemical solution of the silicate minerals kaolinite, vermiculite, biotite, and oligoclase. This contrasts to short-term ratio of solutional to mechanical weathering of five to one.

INTRODUCTION

Fluvial, or mechanical erosion, has dominated theories of landscape development and the interpretation of landforms. Denudation by solution or chemical weathering has generally been considered significant only in terrains underlain by limestone or marble. In the Piedmont of the eastern United States, Hack (1960, p. 95) has speculated that as relief of the landscape is reduced, the manner in which waste is removed may shift from mechanical toward chemical weathering. In order to study the relative importance of chemical weathering in a typical ridge and ravine landscape of the Piedmont, we selected a small forested watershed underlain by a single rock type and made a detailed input-output analysis, particularly emphasizing the geochemistry of the system. This paper expands upon an earlier report based upon the first year's period of study (Bricker and others, 1968).

Pond Branch is in the Piedmont Province of Maryland, 8 mi north of the Baltimore City line (Fig. 1). The stream is dammed just above its junction with Baisman Run. The 95-acre watershed is forested, except for 3 acres along a pipeline. Pond Branch was selected for study because: (1) it is a perennial stream; (2) the pond behind the dam had been surveyed for sediment accumulation 9 yrs before this study; (3) the watershed is almost entirely forested; and (4) the watershed is underlain by one rock type, the Lower Peletic Schist Member of the Wissahickon Formation (Southwick and Fisher, 1967).

Two assumptions were made in this study; first, that the biomass is in dynamic equilibrium; and second, that all waters draining from the watershed originate as precipitation on the watershed. The stability of the biomass is suggested by the following evidence: (1) Trees of all size classes from first-year saplings to trees with crowns 100 ft high occur in the watershed; (2) the watershed contains little evidence of a first growth forest, such as greenbrier, poison ivy, Virginia creeper, and so on; (3) dead falls

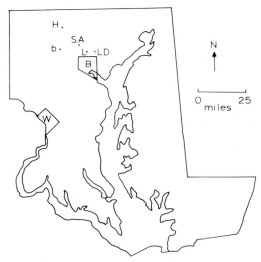

Figure 1A. Location of study area. Letters mean as follows: B, Baltimore; b, Baltimore No. 1 Watershed; H, Hampstead; L, Lutherville; LD, Loch Raven Dam; SA, study area; W, Washington, D. C.

TABLE 1. AVERAGE MODAL ANALYSIS OF SIX WISSAHICKON FORMATION SAMPLES

Mineral	Average Percent	Range Percent
quartz	42.4	50.5 − 36.5
plagioclase (An_{22} Ab_{78})	10.6	16.8 − 1.3
muscovite	23.9	30.8 − 16.1
biotite	9.8	14.3 − 4.5
staurolite	9.1	24.6 − 1.4
garnet	1.2	3.7 − 0
kyanite	T	T − 0
tourmaline	0.1	0.2 − T
zircon	0.1	0.5 − T
apatite	0.2	0.7 − 0.1
iron oxide	T	0.3 − 0
chlorite	T	0.1 − 0
opaques	2.2	2.5 − 1.2

Quartz occurs as thin lenses, pods, and irregular knots.

Stream and rainwater samples were collected periodically from 1 March 1966 through 31 May 1968. A small number of ground-water, interflow, and stemflow samples were also obtained. The samples were collected and stored in polyethylene bottles to minimize contamination. Rain samples were collected at the site of the stream gauge and at nearby Lutherville (the home of one of the authors). No significant difference was observed in chemical composition of the rainwater between the two sites. The rain samples were collected by inserting a polyethylene funnel into a bottle and mounting the apparatus 1 to 2 ft above ground level. Stream water samples were obtained by immersing the bottles in the stream immediately above the gauge. Ground-water samples came from a seep in Trench No. 1 (Fig. 2), some 15 ft below ground level. Interflow water was collected during flood events at a rock ledge where water percolating through the O and A soil horizons flowed out over the rock. Stemflow was collected from a tree adjacent to the gauge site.

The waters were analyzed for silica (aqueous), aluminum, iron, magnesium, calcium, sodium, potassium, bicarbonate, sulfate, and chloride by methods cited in Bricker and others (1968); pH was measured in the field and laboratory using a glass electrode. Total dissolved solids reported are the sum of the constituents analyzed. Total dissolved solids for 7 samples were determined gravimetrically after evaporation. The results were essentially the same as from a summation of constituents.

The minerals in the unweathered Wissahickon Formation were identified with a petrographic microscope, and modal analyses

are present in the watershed; and (4) the last logging in the area was in 1958, and was a selective rather than clear-cutting operation.

Precipitation upon the watershed appears to be the only source of the water eventually discharged in stream flow. Rock structure and lithology are favorable for local recharge, and there is no indication of any ground-water contribution to the stream from outside the basin. Apparently sufficient ground water can be stored in the thick saprolite which underlies the interfluves to maintain base flow in the stream. Calculations of water stored in the saprolite in the basin indicate that base-flow discharge should be sustained for a minimum of 25 days and a maximum of 349 days. Water from ground-water storage sustained base flow for 187 days during a period when no significant recharge occurred in the Pond Branch watershed.

The rock underlying Pond Branch watershed is the lower Peletic Schist Member of the Wissahickon Formation (Southwick and Fisher, 1967). The schist consists of quartz, plagioclase, muscovite, biotite, and staurolite with garnet, kyanite, apatite, tourmaline, zircon, and chlorite as minor accessories (Table 1). Pyrite was not observed in the thin sections examined but was present in outcrop in the watershed. The rock is strongly contorted, medium- to coarse-grained schist with well-developed schistosity.

210

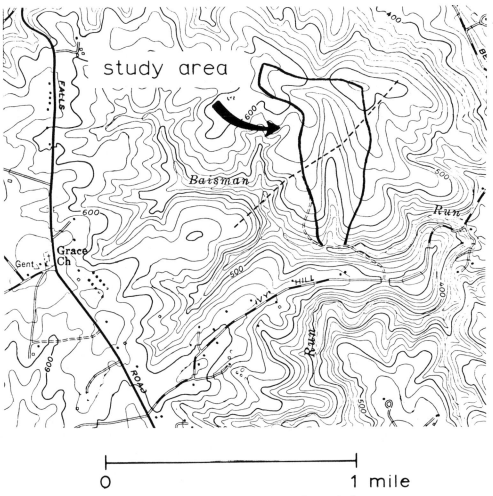

Figure 1B. Study area. Dashed line indicates pipeline.

were calculated. Clay mineral weathering products were identified by X-ray diffraction (Brown, 1961). Halloysite was differentiated from kaolinite on the basis of data cited by Brindley and others (1963).

The rock samples were collected from exposures of bedrock; saprolite samples were secured from auger holes and trenches (Fig. 2). Weathered muscovite and pseudomorphs after oligoclase and biotite were hand-picked from samples to help establish mineral weathering sequences.

Sediment accumulation in the pond was measured in 1957, 1966, and 1968. In the 1957 and 1966 surveys, a probe was used to fix upper and lower surfaces of sediment. To determine sediment accumulation caused by construction

of a pipeline across the watershed in the fall of 1967, we placed stakes in the pond, and the accumulated sediment was subsequently measured in May 1968.

GEOMORPHOLOGY

Chemical weathering presently is the dominant agent of erosion in the Pond Branch watershed. Acid-charged water percolating through the saprolite and rock reacts with many of the minerals. Some of the reactants are removed in solution in ground water and discharged into the surface water. Other reactants are reconstituted as clay minerals and oxides. However, sculpturing of the watershed by this process is not obvious in the landforms. Rather, the physical features of the watershed would suggest mechanical erosion. Evidence of mechanical erosion in the watershed includes the narrow valley at the lower end of the drainage basin (Fig. 3), escarpments along the flood plain, headcuts near the start of perennial flow, and apparent lateral shifting of the valley.

Lateral shifting of the stream from east to west is suggested by the asymmetry of the narrow valley side slopes (Fig. 3) as the lower portion of the western slope is considerably steeper than the eastern one. Small hanging valleys on the west side and fanlike forms on the east side (Fig. 2) also suggest lateral movement toward the west. A scarp between the flood plain and the valley sides suggests a westward

shifting of the stream (Fig. 2). The scarp is almost continuous from the lower end to the upper end of perennial flow along the west side but is present only intermittently along the east side.

Despite indications of a westward shift of the stream, no cause is discernible. Foliation in the schist strikes essentially east-west (Fig. 2); consequently the stream is not migrating down the dip of foliation. No other structural or textural features of the rock are apparent which could cause the stream shift.

Evidence of chemical weathering includes the thick saprolite mantling the interfluves and shallow closed depressions in the upper end of

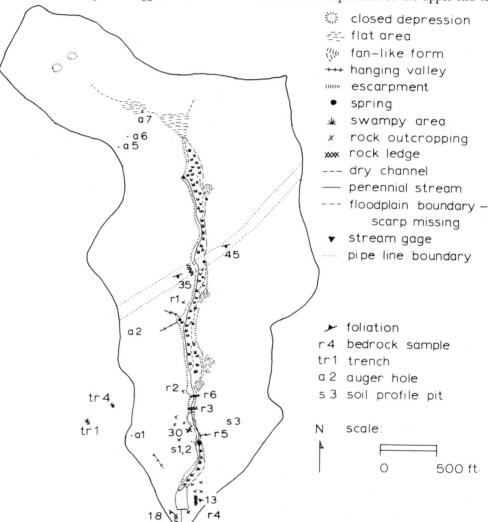

Figure 2. Map showing sample localities and landforms.

the stream basin. The interfluve and swale parts of the watershed are underlain by saprolite 40 to 80 ft deep. Several factors show this: distribution of rock exposures; springs, which commonly occur at the bedrock-saprolite interface; auger holes; trenches; and the pipeline excavation.

Two shallow closed depressions are located at the head of the watershed (Fig. 2). Each is circular, about 200 ft in diameter and 1 ft deep. The floors are flat. Similar features developed on granite in the Piedmont of North Carolina have been described by LeGrande (1952). Apparently these features are solution sinks similar to those found in limestone terrains. At Pond Branch, the location of the depressions directly upvalley from the head of the stream channel suggests that they are formed by solution.

PARTICULATE MATTER

The rate of mechanical erosion in the watershed was estimated from sediment accumulation measurements in the pond and from concentration of suspended matter in the stream water. Prior to construction of a pipeline, both base flow and flood flow carried similar amounts of suspended matter. The average concentration was 6.7 ppm, ranging from a low of 0.4 ppm to a high of 11.8 ppm. At these concentrations, 3.2 tons of suspended matter per sq mi/yr is transported from the watershed.

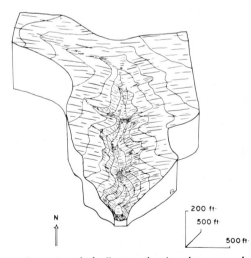

N

200 ft

500 ft

500 ft

Figure 3. Block diagram showing the topography of the watershed. The watershed is comprised of two areas, an upper, broad, gently sloping swale and a lower, asymmetric, narrow, steep-sided valley.

After construction of the pipeline began in October 1967, concentrations of suspended matter in flood flow as high as 96.5 ppm were measured. Concentration of suspended matter did not change in base flow. Sediment settled out in the pond, forming a layer .25 to 5 in. thick. Measurement of the accumulated sediment (organic and inorganic) showed that erosion from the construction area added 15.4 tons of sediment to the Pond in an 8-month period, a rate of 145.4 tons/sq mi/yr. The sediment was derived from only 3 acres of the watershed. These data demonstrate that only when the leaf litter and root mat is effectively breached, for example by construction or farming, are large volumes of sediment eroded from previously forested land and transported away in streams (*see also* Ursic, 1965; Wolman and Schick, 1967). When forest cover is intact, the quantity of suspended matter is exceedingly small.

HYDROLOGIC BALANCE

To formulate data on rate of chemical weathering, it is necessary first to determine the hydrologic balance for the watershed. The hydrologic balance (Table 2) was computed from precipitation, stream flow, and groundwater measurements using the following equation:

$$(1) \qquad P = R + ET + S_{GW}$$

where P equals precipitation, R equals run-off, ET equals evapotranspiration, and S_{GW} equals change in ground-water storage. Changes in soil moisture are assumed to be insignificant compared with the other components of equation (1). Evapotranspiration was calculated as the arithmetic sum of precipitation minus runoff and change in ground-water storage.

Precipitation data for the Pond Branch watershed were obtained from the U.S. Weather Bureau station at Loch Raven Dam (U.S. Weather Bur. Monthly Climatological Summaries) and supplemented by rain gauge data from the watershed for June-September 1967, and for January-December 1967 by data from an unofficial Weather Bureau station at Lutherville, Maryland.

A general period of drought affected the Northeastern United States from 1961 to 1966. The drought broke in the fall of 1966. The Pond Branch hydrologic budget reflects this change in climatic conditions, as precipitation and runoff in 1967 significantly increased

TABLE 2. POND BRANCH HYDROLOGIC BUDGET (INCHES)

	Precipitation	Total Runoff	Change in Storage	Evapotranspiration	Flood Flow	Base Flow
1964 (Aug.–Dec.)	13.3	3.6	−0.3	10.0	0.3	3.3
1965	34.3	6.4	−0.7	28.6	0.4	6.0
1966	38.5	5.2	−0.6	33.9	0.2	5.0
1967	44.9	7.2	+4.9	32.8	1.1	6.1
1968 (Jan.–May)	16.0	4.3	+1.3	10.4	0.6	3.7
Total	147.0	26.7	+4.6	115.7	2.6	24.1

Note 1. Although the geochemical balance was determined for 1 March 1966 through 31 May 1968, we calculated the hydrologic budget for the longer period to better compare the Pond Branch hydrologic data to other watersheds.

Note 2. For the period 1 March 1966 through 31 May 1968, precipitation totalled 32,136,000 cu ft, base flow 5,709,000 cu ft (88.5 percent of total runoff), flood flow 740,000 cu ft, and ground-water storage increased by 2,111,000 cu ft. For the period, the water-volume data for runoff have been increased by 10 percent (see text for explanation). The runoff data shown in the table have not been adjusted by this factor.

over 1966 and 1965, and the ground-water reservoir was replenished (Table 2).

Runoff data were compiled from stream-flow measurements of Pond Branch from 1 March 1966 to 31 May 1968, and for 1 August 1964 to 28 February 1966 by correlation with the hydrographic record of Baisman Run to which Pond Branch is tributary. The drainage area of Baisman Run at the point of measurement is 1.5 sq mi.

A small dam was constructed upstream from the pond and the stream gauge installed in the slack water behind the dam. We estimate that leakage through and around the dam was 10 percent of the measured runoff, and that this percentage loss was the same for both base flow and flood flow.

Base flow at Pond Branch from March 1966 through May 1968 averaged 2 l/sec and varied from a low of 0.5 l/sec in August 1966 to a high of 4.8 l/sec in May 1968. The base flow comprised 87 percent of the runoff. Flow rate was measured with a calibrated plastic container and stop watch. In general, base flow exhibited a seasonal fluctuation, being lower from late spring through summer, coincident with the growing season, and higher from November into March. For the period prior to March 1966, the base-flow contribution to runoff was assumed to average 2 l/sec.

In general, we believe that the base-flow portion of runoff represents a conservative volume. The actual base-flow discharge probably exceeded the calculated volume. This is

particularly true for the growing season, as base flow can fluctuate diurnally by a factor of at least 2 (see Fig. 5).

Storm runoff from March 1966 through May 1968 comprised 13 percent of the total runoff (Table 2). The greatest pulse of runoff took place on August 26 and 27, 1967, when 114,500 cu ft of water was discharged. This discharge resulted from a series of thunderstorms passing over the watershed. The second largest volume of storm runoff (99,000 cu ft) was discharged on May 28, 1968 as a consequence of a general area storm. Altogether, 64 flood events were measured in Pond Branch.

Ground-water storage increased appreciably during the period of study, as estimated from a U.S. Geological Survey observation well. The well, Car. Bf-1, at Hampstead, Maryland, in Wissahickon Formation has a similar topographic setting as the ridges bordering Pond Branch. From water-level records of this well, saprolite thickness at Pond Branch, and gravity yield of the saprolite, we estimate that water stored in the saprolite increased by 2,100,000 cu ft from March 1966 through May 1968. For the entire period, August 1964 through May 1968, water in storage increased by 4.6 in. (Table 2).

WEATHERING OF THE SILICATE MINERALS

To evaluate the input of dissolved solids from the weathering of the silicate minerals in the Wissahickon Formation, it is necessary to determine the minerals being altered, and the secondary minerals and dissolved solids being formed. From these data it is then possible to estimate the amount of dissolved solids discharged into stream runoff.

The weathering of two minerals in the Wissahickon Formation, oligoclase and biotite, contributes the bulk of the dissolved solids to the stream water (Bricker and others, 1968), even though the major constituents of the rock are quartz and muscovite (Table 1). These two minerals alter to clay minerals and oxides, and release ions to solution. Plagioclase weathers to kaolinite, halloysite, and gibbsite in the weathering environment at Pond Branch (Table 3). Feldspar is the sole mineral source of calcium and sodium. Analyses of feldspar from trenches 1 and 4 (Table 3) indicate that plagioclase alters to kaolinite, and kaolinite to gibbsite where feldspar-quartz veins occur in the schist. Apparently the feldspar-quartz veins serve as conduits in the schist, collecting the water percolating through the saprolite and channeling its flow. As a result of movement of water along the veins, the silica is leached to such an extent that gibbsite is formed.

Saprolite and weathered rock in well-drained areas in the watershed (hilltops and steep valley sides) also contain gibbsite (Table 4). Note that gibbsite is absent from the relatively poorly drained swale. Because the occurrence of the gibbsite is related to drainage conditions in the watershed, we believe that gibbsite is a product of the modern weathering environment, and not an inheritance of a previous weathering regimen.

Based upon X-ray, petrographic, and chemical data, the weathering reactions may be formulated:

(2) plagioclase → kaolinite

$$Na_{0.78}Ca_{0.22}Al_{1.22}Si_{2.78}O_{8.0} + 0.61\ H_2O$$
$$+ 1.22\ H_2CO_3 \rightarrow 0.61\ [Al_2Si_2O_5(OH)_4]$$
$$+ 1.56\ SiO_2 + 0.78\ Na^+ + 0.22\ Ca^{++}$$
$$+ 1.22\ HCO_3^-$$

(3) kaolinite → gibbsite

$$Al_2Si_2O_5(OH)_4 + H_2O \rightarrow 2\ Al(OH)_3$$
$$+ 2\ SiO_2$$

Biotite alters to vermiculite and kaolinite in the weathering environment at Pond Branch

TABLE 3. X-RAY ANALYSES OF WEATHERED MINERALS: OLIGOCLASE, BIOTITE, AND MUSCOVITE

	Sample Number	Vermiculite	10 Å	Kaolinite	Halloysite	Gibbsite	Mix-layer 10Å–14Å	Mix-layer 7Å–10Å	Other	Feet below Surface	Location and Comments
Feldspar in feldspar-rich layers in schist	R4-fsp	—	x	x	x	—	—	x	plagioclase	4	trench 1
	R7-fsp	—	x	x	—	—	—	—	plagioclase	7	trench 1
	R21-fsp	—	x	x	?	—	—	—	plagioclase	4	trench 4
	R25-fsp	—	x	?	x	—	—	x		8	trench 4
	R29-fsp	—	x	?	x	—	—	x	plagioclase	12	trench 4
Feldspar in feldspar-quartz veins	R12-fq	—	x	—	—	x	—	—	plagioclase	9	trench 1
	R14-fq	—	x	—	—	x	—	—	plagioclase	8	trench 1
	R15-fq	—	x	x	x	x	—	x	plagioclase	7	trench 1
	R17-fq	—	x	—	—	x	—	—	plagioclase	13.5	trench 1
	R32-fq	—	x	x	x	—	—	—		5	pipeline
	R34-fq	—	—	x	x	—	—	—	plagioclase	5	pipeline
Biotite	R39-bio	x	x	x	—	x	x	x	biotite	2.5	weathered rock at base of S 3; interleaved with muscovite
	R42-bio	—	—	x	—	—	x	x		6	kaolinite psuedomorph after biotite A 1
Muscovite	R1-mus	—	x	x	—	?	—	—		1	trench 1
	R5-mus	—	x	x	—	?	x	—		5	trench 1
	R9-mus	—	x	x	—	—	x	x		9	trench 1
	R19-mus	—	x	x	—	—	—	—		2	trench 4
	R29-mus	—	x	x	—	—	x	x		10	trench 4

Symbols: x means present; — means absent; ? means possibly present

TABLE 4. X-RAY ANALYSES OF SILT-CLAY SIZE FRACTION OF SAPROLITE FROM AUGER HOLES
AND SOIL PROFILES

Sample No.	Topographic Position	Depth (in feet)	Quartz	10Å	10–14Å	7–10Å	Vermiculite	Kaolinite	Gibbsite	Comments
R40	hilltop	3.5–4	x	x	x	—	x	x	x	auger hole 1
R41	hilltop	5.0–5.3	—	x	x	—	—	x	x	A 1
R42	hilltop	5.8–6.0	x	x	x	x	—	x	x	A 1
R43	hilltop	5.0–5.2	—	x	—	—	—	x	—	A 2
R44	hilltop	5.8–6.0	x	x	x	—	—	x	x	A 2
R45	hilltop	7.2–7.5	x	x	x	x	—	x	x	A 2
R49	hilltop	0.0–1.5	x	x	x	—	x	x	—	A 5; upland area adjacent to swale
R50	hilltop	2.0–3.5	x	x	—	—	x	x	—	A 5
R51	hilltop	3.5–4.0	x	x	—	—	x	x	—	A 5
R53	slope	0.0–1.5	x	x	—	—	x	x	—	A 6; A soil horizon; slope leading down to swale
R54	slope	1.5–2.5	x	x	—	—	x	x	—	A 6; B soil horizon
R55	slope	2.5–4.0	x	x	—	—	—	x	—	A 6; B–C transition zone
R56	slope	4.0–5.0	x	x	—	—	—	x	—	A 6; C soil horizon
R57	swale	0.0–0.2	x	x	—	—	x	x	—	A 7; A soil horizon
R58	swale	0.2–1.5	x	x	—	—	x	x	—	A 7; A soil horizon
R59	swale	1.5–3.0	x	x	x	—	x	x	—	A 7; B soil horizon
R60	swale	3.0–4.0	x	x	—	—	x	x	—	A 7; B soil horizon
R61	swale	4.0–4.8	x	x	—	x	—	x	—	A 7; C soil horizon
R35	valley side	0.3	x	x	—	—	—	—	—	S 1; A soil horizon
R36	valley side	2.0	x	x	—	—	x	x	?	S 1; B soil horizon
R37	valley side	1.3	x	x	x	—	x	x	?	S 2; A soil horizon
R38	valley side	2.0	x	x	—	—	x	x	x	S 2; B soil horizon

Symbols: x means present; — means absent; ? means possibly present

(Table 3) with mixed-layer clays, 7Å–10Å and 10Å–14Å, forming intermediary products. Biotite is the major mineral source of magnesium, and, with muscovite, is a source of potassium. Vermiculite and 10Å–14Å mixed-layer clay, distinctive of biotite weathering at Pond Branch, are evident in samples from the trenches, soil profiles in the narrow valley, and auger holes (Table 4). The weathering reactions can be generalized as:

biotite ⟶ vermiculite

⟶ kaolinite

⟶ gibbsite

The solid arrows indicate reactions that are supported by X-ray analyses; the dashed arrows indicate reactions that probably occur but cannot be confirmed. Equations for the weathering of biotite to vermiculite and to

kaolinite may be formulated (biotite composition from Hopson, 1964, p. 83):

(4) biotite → vermiculite

$$3[K_2(Mg_3Fe_3)Al_2Si_6O_{20}(OH)_4] + 8 H_2O + 12 H_2CO_3 + 2[(Mg_3Fe_3)Al_3Si_5O_{20}(OH)_4 \times 8 H_2O] + 6 K^+ + 3 Mg^{2+} + 3 Fe^{2+} + 8 SiO_2 + 12 HCO_3^-$$

(5) vermiculite → kaolinite

$$2[(Mg_3Fe_3)Al_3Si_5O_{20}(OH)_4] \cdot 8 H_2O + 30 H_2CO_3 \rightarrow 3 Al_2Si_2O_5(OH)_4 + 6 Mg^{2+} + 6 Fe^{2+} + 4 SiO_2 + 29 H_2O + 30 HCO_3^-$$

Muscovite alters to illite and kaolinite in the Pond Branch watershed (Table 3). Unlike biotite, muscovite is not rapidly destroyed by the weathering processes; mica persists throughout the profiles in the trenches and auger holes. The 10Å–14Å mixed-layer clay may be an

216

TABLE 5. AVERAGE DISSOLVED SOLIDS IN POND BRANCH WATERS (IN PPM)

	SiO_2	Na	K	Ca	Mg	HCO_3	Cl	SO_4	pH	Total Solids	No. of Samples
Precipitation	0.1	0.2	0.2	0.3	0.1	0.3	0.6	1.7	4.6	3.5	24
Base Flow	9.3	1.7	0.9	1.4	0.8	7.7	2.1	1.3	6.7	25.1	38
Flood Flow*	6.9	1.4	1.3	1.9	1.3	5.0	2.2	6.3	6.2	26.3	6
Ground Water	8.1	1.8	0.8	0.5	0.5	5.4	2.9	1.0	6.1	21.0	5
Interflow	3.0	0.7	2.1	1.2	1.3	0.0	1.7	18.3	4.5	28.4	6
Stem Flow	2.3	0.4	18.6	18.3	3.6	40.7	8.1	37.8	6.1	129.8	1

* Samples are of flood crests only and do not include those collected on rising and declining stages. The average value for pH was determined from laboratory measurements.

intermediate weathering product of muscovite, but more probably is an intermediate stage in biotite alteration to kaolinite. Gibbsite may be a weathering product of muscovite, but this cannot be confirmed.

From the evidence at hand, we cannot establish how much muscovite is being altered to illite and kaolinite. Although muscovite-like flakes persist in the A-soil horizon, the flakes in both saprolite and soil commonly have lost their luster, are whitish rather than pearly, and are punky. The probable weathering reactions are:

(6) muscovite → illite (composition adapted from Reesman and Keller, 1967)
$$KAl_2AlSi_3O_{10}(OH)_2$$
$$\rightarrow K_{0.6}Al_2AlSi_3O_{10}(OH)_2 + 0.4\ K^+$$

(7) muscovite → kaolinite
$$2[KAl_2AlSi_3O_{10}(OH)_2] + 3\ H_2O$$
$$+ H_2CO_3 \rightarrow 3\ Al_2Si_2O_5(OH)_4 + 2K^+$$
$$+ 2\ HCO_3^-$$

WATER CHEMISTRY

The chemistry of the precipitation, ground water, and runoff is the last element needed to calculate a geochemical balance for the watershed. Precipitation is believed to be the sole source of water introduced into the watershed. That process adds about 38 percent of the dissolved solids eventually discharged in runoff. Dissolved solids in precipitation average 3.5 ppm (Table 5), and vary from 2.2 to 7.5 ppm. Sulfate is the major anion, and hydrogen and calcium are the principle cations.

Of the dissolved solids considered in this study, chloride is the only species likely to be contributed entirely by precipitation because it is not present in the rock in the watershed. If this assumption is valid, then chloride concentrations calculated from runoff and storage figures should equal actual concentrations de-

termined from the water analyses (Table 6), thereby providing a check on the hydrologic balance. The equations in Table 6 assume that the chloride comes directly from precipitation, that the ground-water basin is water tight, and that steady state conditions exist for chloride in the hydrologic cycle. Calculated chloride concentrations in runoff and storage agree, within experimental error, with measured values for 1967 (Table 6).

Base flow comprised about 90 percent of the runoff and carried approximately 90 percent of the dissolved solids out of the watershed from 1 March 1966 through 31 May 1968. The major dissolved solid is silica, the primary cation is sodium, and the primary anion is bicarbonate (Table 5).

Concentrations of the dissolved solids in base flow vary with time (Fig. 4). Total dissolved solids are lowest in concentration in winter and early spring (21 to 24 ppm) and highest in summer and early fall (26 to 30 ppm). Average concentration is 25.1 ppm.

TABLE 6. CHLORIDE CONCENTRATIONS FOR 1967: WATER ANALYSES AND CALCULATED VALUES

Water analyses		calculated
Cr	2.2 ppm	2.1 ppm
C_s	2.9 ppm	3.1 ppm

Equations*: $Cr = \dfrac{Cp\ (P)}{R}$, $C_s = \dfrac{Cp\ (P)}{S}$

R = runoff = 7.2 in.
S = storage = 4.9 in.
P = precipitation = 44.9 in.
Cp = average chloride concentration in 1967 precipitation = 0.34 ppm
Cr = average chloride concentration in 1967 runoff
C_s = average chloride concentration in 1967 storage water

* Equations *from* Juang and Johnson, 1967

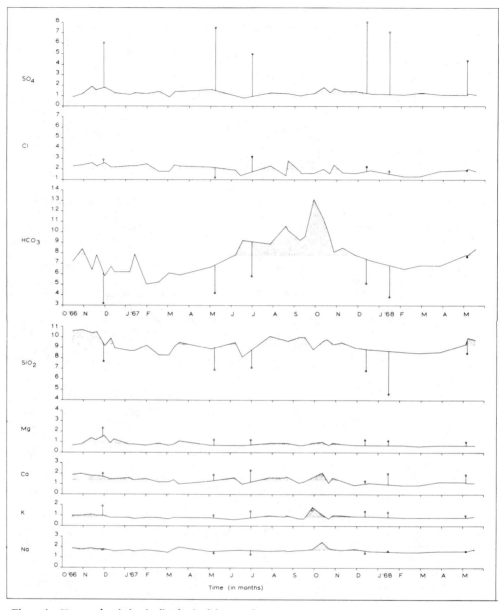

Figure 4.   Temporal variation in dissolved solids. Base-flow variation shown by solid line. Flood-flow concentration shown by circle on stilt. Stippled areas indicate concentration of a particular ion which exceeds average value. Concentration of ion is in parts per million.

Bicarbonate and silica best illustrate the seasonal variation. The July to October 1967 hump is quite obvious (Fig. 4). Calcium and potassium show a similar trend. Chloride and sulfate do not follow this pattern. This suggests that sulfate, like chloride, enters the watershed mainly in precipitation.

Seasonal variation of the dissolved solid species apparently results from the biologic activity of the deciduous forest. During the warmer months, the decay of organic matter and the generation of CO_2 by plant roots increase the carbonic acid content of waters in contact with mineral matter. The result is

218

more vigorous chemical weathering. This relationship is indicated by the silica and bicarbonate curves in Figure 4.

A second biologic effect is related to autumn leaf fall. Note (Fig. 4) the increase in the concentration in October 1967 of calcium, potassium, and bicarbonate. Apparently, leaf decay in the stream produces a short-term increase in the concentrations of these constituents. A similar effect was observed in Quantico Creek, Virginia, by Slack and Feltz (1968).

The volume and composition of base flow of the stream was monitored for a 24-hr period beginning May 17, 1968 at 10:00 a.m. The maximum discharge observed, 4.8 l/sec, occurred at 12:00 midnight; the minimum discharge, 2.4 l/sec, occurred at 5:00 p.m. (Fig. 5). We presume this is primarily due to the diurnal differences in biomass water demand. The concentrations of Mg, Ca, Na, K, SO_4, and SiO_2 remained constant over the 24-hr period (Fig. 5). Chloride appears to show a slight increase from the beginning to the end of the 24-hr period. This was unexpected, and we have no explanation for this observation at the present time. Titration alkalinity displayed a sharp maximum at 10:00 p.m., but neither pH nor any of the other dissolved constituents showed a corresponding shift. We attribute this increase in titration alkalinity to biologic release of an organic proton acceptor during the late afternoon and evening.

Flood events or sharp rises in stream flow comprised approximately 10 percent of runoff, and removed about 10 percent of the dissolved solids. Chemical composition of the flood water differs significantly from base flow (Fig. 4); silica and bicarbonate were markedly lower in flood events, chloride higher in some events and lower in others, and potassium and calcium higher. These differences indicate that precipitation and storm runoff do not simply dilute the flood water. Except for direct fall on the stream channel, precipitation falling on the watershed in the area of the flood plain rapidly picks up soluble salts from the trees, leaf litter, and the A horizon of the soil and moves them into the stream during a flood event.

Flood runoff, and the dissolved solids in it, are apparently generated in the relatively restricted area of the watershed in and adjacent to the flood plain. All of the water discharged by the stream during the rising stage of a flood can be accounted for by direct fall of rain upon the stream and flood plain (Table 7). However, during the falling stage, sources adjacent to the flood plain add substantial quantities of water to the flow (Table 7; Dunne, 1969, p. 216–218).

The chemical composition of the interflow water suggests that it is a major factor influencing the chemical variation of flood flow (Table 5). Those ions that are higher in flood flow are generally high in the interflow: sulfate, potassium, and magnesium. Ions that are lower in flood flow, silica, bicarbonate, and sodium, are also low in the interflow.

The volume and composition of one flood flow was monitored on May 5, 1968. Total dis-

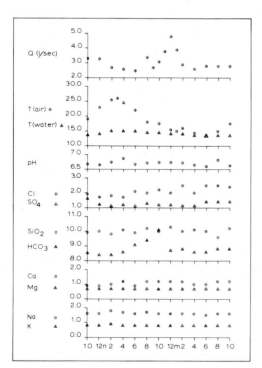

Figure 5. Twenty-four-hour base-flow cycle. Time, in hours, shown along horizontal scale. Vertical scale shows: liters per second, temperature in degrees centigrade, and dissolved solids in parts per million.

TABLE 7. PRECIPITATION AND DISCHARGE FROM TWO THUNDERSTORMS (CU FT)

	Rising Stage		Entire Storm	
	Precip. on		Precip. on	
Storm	Flood plain	Q	Flood plain	Q
1	5,800	1,100	25,900	33,200
2	10,100	3,000	30,200	130,000

solved solids and various species show an orderly change with change in discharge (Fig. 6).

During the flood event, silica concentration decreased by only 10 percent (9.4 ppm to 8.4 ppm), although stream discharge increased by 284 percent (3.2 to 9.1 l/sec). These figures suggest that a substantial portion of the precipitation comprising the flood event passes through the near-surface soil adjacent to the stream and picks up silica on its way to the stream channel. Otherwise, the silica concentration would be lower than it is. In part, precipitation displaces water in the soil, water being discharged into the stream by translatory

flow as a result of pressure exerted on a saturated system, and in part, precipitation passes through the soil into the stream. The rain water, moving through the soil, reacts with the silicate minerals and biologic material, picking up silica (and other ions) before discharging into the stream.

Unlike silica, cation concentrations increased with increased discharge. Two processes appear to be responsible for the increase. Rain water penetrates the upper part of the soil and flushes soluble salts out of organic debris. Second, hydrogen ions from the rain displace cations from exchange sites on clay minerals, amorphous material, and organic material. Note

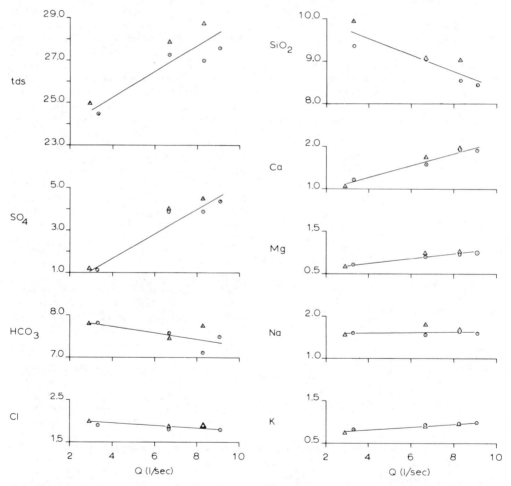

Figure 6.   Changes in water composition during a flood event. Concentration of dissolved solids is in parts per million. Abbreviations and symbols: Tds, total dissolved solids; Q, discharge, in liters per second; circle-rising stage (including flood crest); triangle-falling stage.

that the average pH of the rain is 4.66 compared to flood flow pH of 6.15 (Table 5).

Of the cations, calcium shows the most marked increase. That the calcium source is not primarily mineralogic in flood flow is indicated by the contrast in the silica:calcium ratio, 6.6:1 in base flow compared to 2.7:1 in flood flow. Rather, analysis of stem flow (Table 5) suggests that the source is biologic.

Magnesium and potassium also increase in concentration in flood flow. These ions are more concentrated in interflow and stem flow than in flood flow, suggesting a biologic source. When the leaf litter and upper soil horizon are flushed by rain, magnesium and potassium are released and transported into the stream. Under base flow conditions, these ions are stored as soluble salts in the upper soil profile and do not contribute to base flow. Sodium concentration remained essentially constant during this event, but in other flood events, it showed a decrease in concentration at peak instantaneous discharge. This suggests that the primary source of sodium, like silica, is rock weathering.

The anions, like the cations, do not show a simple dilution or concentration with increased flood flow. The increase in the concentration of sulfate is probably due to its high concentration in both the precipitation and water flushing through the biomass. During the sampled flood cycle, the chloride concentration remained constant; however, during other storms, it increased in some and decreased in others. This suggests that chloride is reflecting its concentration in rain, and is independent of any mineralogic source. The concentration of bicarbonate decreases, following the trend of silica and sodium. This suggests that the primary source of bicarbonate, like silica and sodium, is the weathering of silicate minerals (Bricker and others, 1968, p. 132).

Total dissolved solids, in general, increased during a flood event (Fig. 6), and concentrations in discharge in the falling stage were greater than concentrations in discharge in the rising stage. The increased concentration of dissolved solids on the falling stage compared to the rising stage suggests changing source areas for the flood water. We believe the primary source area for the rising stage is the flood plain, as translatory flow (Hewlett and Hibbert, 1966), stemflow, interflow, and direct rainfall on the stream channel contribute various quantities of water. During the falling stage, water comes from a larger area than

during the rising stage, and is in contact with mineral and plant matter for a longer time period.

GEOCHEMICAL BALANCE

Introduction

The geochemical budget of a watershed may be formulated as: *input* (dissolved materials in precipitation + dissolved materials contributed by mineral weathering) = *output* (dissolved materials in water leaving the watershed + dissolved materials in water temporarily stored in the watershed + material temporarily taken up by the biomass). For the Pond Branch watershed (Fig. 7), the balance (Table 8) was calculated from the hydrologic, chemical, and mineral weathering data developed in the previous sections of this paper. The biomass column is the algebraic sum of the other four quantities. If all sources of input, reactions within the watershed, and pathways for materials leaving the watershed are considered, the residual column in Table 8 will be zero. Deviations from zero indicate disequilibrium of the biomass and/or sources or sinks of materials other than those considered. Bias in sampling and analytical error may also result in deviations from zero, but we believe that errors from these two factors are minor.

Precipitation

Precipitation contributes about 38 percent of the dissolved material input into the watershed, largely in the form of sulfate, chloride, bicarbonate, and calcium ions (Table 8). Significant quantities of Na, K, and Mg are also added to the watershed in precipitation. Chloride ion is contributed primarily by rain, and since this ion does not take part in any weathering reactions in the watershed, it should behave as a conservative species in the over-all water balance. The fact that the chloride input is reasonably well balanced by output (Tables 6 and 8) suggests that the hydrologic balance is essentially correct.

Sulfate is also primarily atmospheric in source. Graphical portrayal of base-flow data (Fig. 4) shows that both chloride and sulfate ions vary independently of silica, which comes mainly from rock weathering. Consequently, we believe that sulfate also is primarily of atmospheric origin, even though small amounts of sulfate probably are generated in the watershed by weathering of pyrite.

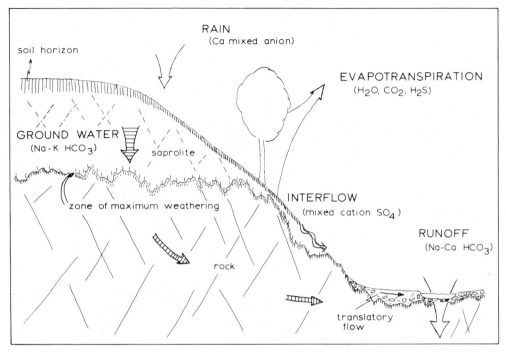

Figure 7. Pond Branch weathering model.

Hydrogen ion introduced by precipitation rapidly reacts with weatherable minerals in the watershed releasing cations into solution. Approximately 25 kg of hydrogen ion came into the watershed in precipitation, but only 0.03 kg left in runoff.

The principal sources of the dissolved solids in the precipitation are probably industrial pollutants and sea salt. Industrial pollution is indicated by the large amount of sulfate (7.22 kg/acre/yr; Table 8, precipitation column) the high chloride relative to sodium (the ratio of Na to Cl in sea water is 0.55; in Pond Branch precipitation, the ratio is 0.34), and the low *p*H. Industrial pollution may also be a major source of the Ca, K, and Mg. A second, and apparently minor source of cations and silica, may be atmospheric dust.

Weathering

Weathering of silicate minerals in the watershed accounts for the silica, alkali, and alkaline earth elements in excess of the amount introduced in rain. From the weathering reactions (equations 2–5), the water chemistry (Table 5), and the hydrologic budget (Table 2), the amount of dissolved solids and clay minerals formed by weathering has been calculated

(Table 9). Silica is assumed to be in dynamic equilibrium with the biomass, although it can be utilized by plants during growth (Lovering, 1959; Miller, 1963). Sodium and magnesium apparently are not essential plant nutrients in the sense that calcium and potassium are essential; therefore, for the calculations in Table 9, we have assumed that magnesium and sodium are, like silica, in dynamic equilibrium with the biomass. Each of these 3 dissolved solids, consequently, equal zero in the biomass column of Table 8.

To calculate the amount of silica and calcium generated by the weathering of oligoclase to kaolinite (eq. 2), sodium is used as the limiting factor. This is because oligoclase is the only significant source of sodium, and sodium is minimally affected by the biomass. Applying this constraint to equation 2, we calculate that 1195 kg of silica are released by the weathering of oligoclase and that 844 kg of silica are generated by other weathering reactions. Also released, as oligoclase weathers to kaolinite, are 112 kg of calcium.

Both the weathering of biotite to vermiculite (eq. 4) and vermiculite to kaolinite (eq. 5) are sources of magnesium. Because vermiculite is present in the watershed (Table 4), more

TABLE 8. GEOCHEMICAL BALANCE (1 MARCH 1966 THROUGH 31 MAY 1968)

(IN KILOGRAMS PER ACRE PER YEAR)

	INPUT		=	OUTPUT		
	Precipitation	Weathering		Runoff	Storage	Biomass
SiO₂	0.42 +	9.53	=	7.69 +	2.26	+0
Na	0.86 +	1.06	=	1.42 +	0.50	+0
K	0.68 +	0.93	=	0.79 +	0.22	+0.60
Ca	1.28 +	0.52	=	1.21 +	0.14	+0.45
Mg	0.42 +	0.46	=	0.74 +	0.14	+0
HCO₃	1.28 +	11.60	=	6.77 +	1.51	+4.60
Cl	2.55 +	..	=	1.79 +	0.81	−0.05
SO₄	7.22 +	..	=	1.59 +	0.28	+5.35
TDS	14.71 +	24.10	=	22.00 +	5.86	+10.95

Note: hydrogen ion, 0.11 kg/acre/yr in precipitation and less than 0.0002 kg/acre/yr in runoff

No alumina or iron was found in precipitation samples, and only trace amounts in runoff.

vermiculite has to be generated in the weathering of biotite than is consumed in the production of kaolinite. Subject to this constraint, magnesium is partitioned between the two reactions so that the amount of potassium evolved from the process described in equation 4 is minimized. Consequently, we allocated approximately 63 percent of the available magnesium to equation 4 and 37 percent to equa-

tion 5. Subject to these limits, we calculate that 199 kg of potassium and 530 kg of silica are released to solution from these two reactions. Ferrous iron is also released to solution, but is oxidized to hydrous ferric oxides and remains in the watershed.

The last weathering reaction to be considered, kaolinite to gibbsite (eq. 3), is limited by the amount of silica remaining in solution (314 kg). This amount of silica, applied to equation 3, indicates that 5.2 kilomoles of gibbsite form by weathering of kaolinite (Table 9, part B).

To recapitulate briefly, data presented in Table 9, part B, indicate that 3 clay minerals, vermiculite, kaolinite, and gibbsite, are being formed in the weathering environment at Pond Branch. Five moles of oligoclase weather for every mole of biotite that weathers; this compares to a mole ratio of 3 oligoclase to 1 biotite in a unit volume of fresh rock (calculated from Table 1). The comparison indicates that oligoclase weathers at a greater rate than biotite. The bulk of the silica input comes from weathering of oligoclase (Table 9, part A; eq. 2), but significant amounts also are generated by alteration of clay minerals (eq. 5 and eq. 3). Sodium and calcium are derived from oligoclase, potassium primarily from biotite, and magnesium from biotite and vermiculite.

Runoff

The dissolved load of the stream is measured as the dissolved material per unit volume times

TABLE 9. DISSOLVED SOLIDS AND CLAY MINERALS FROM WEATHERING OF SILICATE MINERALS

A. Dissolved solids (in kilograms)*

Reaction	SiO₂	Na	Ca	Mg	K	HCO₃
oligoclase to kaolinite (eq. 2)	1195	229	112	..	..	948
biotite to vermiculite (eq. 4)	408	..	..	61	199	620
vermiculite to kaolinite (eq. 5)	122	..	..	37	..	915
kaolinite to gibbsite (eq. 3)	314	..	..	..	..	..
Total	2039	229	112	98	199	2483

B. Silicate minerals consumed (minus sign) or formed (plus sign) (in kilomoles)

Reaction	olig	bio	vermic	ka	gb
oligoclase to kaolinite (eq. 2)	−12.8	..	..	+7.8	..
biotite to vermiculite (eq. 4)	..	−2.8	+1.7	..	..
vermiculite to kaolinite (eq. 5)	..	..	−1.0	+1.5	..
kaolinite to gibbsite (eq. 3)	..	..	..	−2.6	+5.2
Total	−12.8	−2.8	+0.7	+6.7	+5.2

*Calculated from Table 8: Kg/acre/yr times 95 acres times 2.25 yrs equals dissolved solids

the volume of water passing the gauge, and takes into consideration variations due to flood flow and annual cycles.

Storage

Substantial amounts of dissolved solids have accumulated in the watershed in ground water (Table 8, storage). This reservoir of dissolved constituents fluctuates considerably with fluctuation in volume of stored water (Table 2), and if unaccounted for, could cause a significant error in the calculation of the geochemical budget. The fluctuations in storage result from short-term climatic variations.

Biomass

The net change of ions in the watershed has been derived indirectly by computing the algebraic sum of precipitation, weathering, runoff, and storage (Table 8, biomass). Chloride does not appear to be immobilized in the watershed. Sodium, magnesium, and silica are assumed to be in dynamic equilibrium with the biomass. Potassium, calcium, and sulfate apparently are being taken up in greater amounts by the biomass than are returned by decay of organic matter. In addition to the geochemical budget calculations, biomass uptake of K and Ca is suggested by the increase of these elements in runoff during flood events, as rain water moving through the upper soil horizons and decaying organic matter flushed out these ions. Calcium and potassium also increase in concentration at the time of autumn leaf fall. Work by Miller (1963, p. 409, Table 17) in a beech forest in New Zealand supports our data on immobilization of K and Ca by the biomass.

Sulfate appears to be accumulating in large amounts in the watershed (5.34 kg/acres/yr; Table 8). The biomass utilizes sulfur, which, in part, is returned to stream water by the decay of organic material. The occurrence of soluble sulfate in the biomass is indicated by the increase of sulfate during flood flow. Other pathways certainly exist by which sulfur is removed from the watershed, as there is no indication of the formation of secondary sulfur minerals. One such pathway is bacterial reduction of sulfur to hydrogen sulfide. Both the bogs along the flood plain and the swampy upper end of the pond were noted to have a strong hydrogen sulfide odor, indicating bacterial reduction of sulfur. When the water level in the pond was lowered, large gas bubbles with a hydrogen sulfide odor surfaced through the swampy ooze. However, the quantitative importance of this pathway and the presence of other pathways cannot be determined from the data at hand.

WEATHERING MODEL: A SUMMING UP

The landforms in the Pond Branch watershed suggest major mechanical erosion, but our study indicates that chemical denudation is the dominant erosional process at present. For the period of study, the average rate of removal of dissolved solids derived from rock weathering was 16.9 tons/sq mi/yr compared to 3.2 tons/sq mi/yr of particulate matter. Combining chemical denudation and particulate matter erosion rates (3.0 tons/yr/95 acres), and dividing by weight of rock removed from the valley (7,695,000 tons), we calculate that 2,565,000 yrs would be required to erode the present valley at Pond Branch. We consider this period to be a maximum estimated age.

However, the present 5 to 1 ratio of chemical to mechanical weathering probably is not typical, as 42 percent (by volume) of the rock is quartz. Quartz is not readily soluble, yet it has to be removed from the watershed, otherwise a thick deposit of sand and gravel would accumulate in the area of the flood plain. Field evidence indicates that the flood plain is thin, less than 5 ft thick, and composed of boulders in a sand-silt-clay matrix. Bedrock is at or near the surface along much of the perennial stream. Consequently, mechanical (fluvial) erosion must occasionally purge the flood plain area. High intensity events of low frequency presumably are responsible, as no evidence exists in the watershed to indicate movement of significant amounts of sand and boulders since the dam was built in 1957. These events apparently are responsible for shaping the valley.

Consequently, the long-term rate of denudation probably is closer to 1:1 than 5:1. Assuming that chemical weathering removes 16.9 tons/sq mi/yr, and that mechanical weathering removes a similar amount, we calculate that 1,540,000 yrs would be required to erode the present valley at Pond Branch.

Presumably, the topography of much of the eastern Piedmont in Maryland has developed in late Pliocene and Pleistocene time. The present erosional rate at Pond Branch apparently represents a lull between catastrophic events. However, chemical denudation appears to be just as important an erosional process as mechanical erosion in the forested Piedmont region in the mid-Atlantic states.

A simple solution weathering model of the Pond Branch watershed can be constructed by dividing the basin into three areas: channel, flood plain and valley sides–upland. Precipitation falling on the watershed is essentially devoid of silica and low in pH. Rain that falls in the channel dilutes water contributed to the stream from the flood plain and adjacent areas but does not interact with biomass or mineral matter. Rain that falls on the flood plain penetrates into the shallow subsurface, displacing existing water. In the process of passing through the flood plain, the water of a calcium–mixed anion type changes to mixed cation–bicarbonate sulfate type through interaction with minerals and flushing out of soluble salts from the biomass (Table 5).

Precipitation that falls outside of the flood plain area, and that is not immediately evaporated, penetrates into the soil and underlying saprolite. Much of the water in the ground is transpired by the biomass, and large quantities of CO_2 and possible sulfur compounds are lost. The remaining water eventually passes through the saprolite and rock into the stream. In passing through the soil, saprolite, and rock, the water reacts with the minerals; some of the products of weathering are carried away as dissolved solids, and some are reconstituted as new minerals. In its passage, the water changes from a calcium-mixed anion composition in rain to sodium-potassium bicarbonate in ground water and finally to a sodium-calcium bicarbonate composition in base flow. The transition in water composition reflects the weathering of vermiculite to kaolinite, kaolinite to gibbsite, and oligoclase and biotite to clay minerals, sesquioxides, and aqueous silica.

ACKNOWLEDGMENTS

We wish to thank M. Gordon Wolman and Charles B. Hunt of Johns Hopkins University for critically reviewing the manuscript. Rainfall data for an unofficial Weather Bureau station at Lutherville, Maryland, was provided by W. J. Moyer of the Earth Sciences Services Administration, Weather Bureau State Climatologist for Maryland.

This study was supported, in part, by grants from the Petroleum Research Fund of the American Chemical Society (665-G2) and the National Science Foundation (GP 2660).

REFERENCES CITED

Bricker, O. P., Godfrey, A. E., and Cleaves, E. T., 1968, Mineral-water interaction during the chemical weathering of silicates: Am. Chem. Soc. Advances Chemistry Ser. 73, p. 128–142.

Brindley, G. W., Santos, P. S., and Santos, H. S., 1963, Mineralogical studies of kaolinite-halloysite clays; Part I. Identification problems: Am. Mineralogist, v. 48, p. 897–910.

Brown, G., Editor, 1961, The X-ray identification and crystal structures of clay minerals: London, Mineralogical Society, 544 p.

Dunne, Thomas, 1969, Runoff production in a humid area: Ph.D. dissert., The Johns Hopkins Univ., Baltimore, Maryland, 249 p.

Hack, J. T., 1960, Interpretation of erosional topography in humid temperate regions: Am. Jour. Sci., v. 258-A, p. 80–87.

Hewlett, J. D., and Hibbert, A. R., 1966, Factors affecting the response of small watersheds to precipitation in humid areas: Internat. Symposium Forest Hydrology: Oxford and New York, Pergamon Press, p. 275–290.

Hopson, C. A., 1964, The crystalline rocks of Howard and Montgomery Counties, in The Geology of Howard and Montgomery Counties: Baltimore, Maryland Geol. Survey, p. 27–215.

Juang, F.H.T., and Johnson, N. M., 1967, Cycling of chlorine through a forested watershed in New England: Jour. Geophys. Research, v. 72, no. 22, p. 5641–5647.

LeGrande, H. E., 1952, Solution depressions in diorite in North Carolina: Am. Jour. Sci., v. 250, p. 566–585.

Lovering, T. S., 1959, Significance of accumulator plants in rock weathering: Geol. Soc. America Bull., v. 70, p. 781–800.

Miller, R. B., 1963, Plant nutrients in hard beech, Part III, The cycle of nutrients: New Zealand Jour. Sci., v. 6, p. 388–413.

Reesman, A. L., and Keller, W. D., 1967, Chemical composition of illite: Jour. Sed. Petrology, v. 37, no. 2, p. 592–596.

Slack, K. V., and Feltz, H. R., 1968, Tree leaf control on low flow water quality in a small Virginia stream: Environmental Sci. and Technology, v. 2, p. 126–131.

Southwick, D. L., and Fisher, G. W., 1967, Revision of stratigraphic nomenclature of the Glenarm Series in Maryland: Maryland Geol. Survey Rept. Inv. no. 6, 19 p.

U.S. Weather Bureau Monthly Climatological

Summaries, August 1964–May 1968.

Ursic, S. J., 1965, Sediment yields from small watersheds under various land uses and forest covers, *in* Proceedings of Federal Interagency Sedimentation Conference 1963: U.S. Dept.

Agriculture Misc. Pub. 970, p. 49.

Wolman, M. G., and Schick, A. P., 1967, Effects of construction on fluvial sediment, urban and suburban areas of Maryland: Water Resources Research, v. 3, no. 2, p. 451–464.

MANUSCRIPT RECEIVED BY THE SOCIETY JANUARY 23, 1970

REVISED MANUSCRIPT RECEIVED MAY 22, 1970

19

Copyright © 1968 by the Geological Society of America

Reprinted from *Geol. Soc. America Bull.* **79**:1573–1588 (1968), courtesy of the
Geological Society of America

Speculations Concerning Paleohydrologic Controls of Terrestrial Sedimentation

S. A. SCHUMM   *Department of Geology, Colorado State University, Fort Collins, Colorado*

Abstract: The relations that have been recorded among modern climatic, phytologic, and hydrologic data are used to speculate about the effects of evolving vegetation on the hydrologic cycle. At present the peak of erosion rates occurs in semiarid regions, whereas during prevegetation time erosion rates rose to a plateau, the magnitude of which depended upon the erodibility and weathering characteristics of the rocks. With the appearance of terrestrial vegetation and its colonization of the earth's surface, erosion rates decreased, as did runoff and flood peaks.

A review of the relations existing between the morphologic and hydrologic characteristics of river channels demonstrates that fluvial sedimentary deposits are significantly different depending upon the nature of the sediment load moved through the channel.

Combining the conclusions obtained from an analysis of hydrologic relations with conclusions concerning effects of type of sediment load upon river morphology, it is possible to speculate on the changing nature of the land phase of the hydrologic cycle before and during the colonization of the landscape by vegetation.

During prevegetation time, bed-load channels moved coarse sediments from their sources and spread them as sheets on piedmont areas. With increased plant cover, alluvial deposits were stabilized, but large floods caused periodic flushing of sediment from the system, thereby creating cyclic sedimentary deposits.

The influence of climate change on the volume and type of sediment moved from an erosional system became more pronounced as the effect of vegetation on the hydrologic cycle increased. Finally, with the appearance of grasses during the Cenozoic Era, the relations between climate, vegetation, erosion, and runoff became much as today except for the influence of man.

CONTENTS

Introduction 1574
Acknowledgments 1574
Hydrology 1574
Paleohydrology 1576
River morphology and sedimentation 1578
  Channel morphology 1578
  River adjustment 1581
  Nature of channel sediments 1581
Paleochannels and fluvial sedimentation . . . 1582
  Prevegetation 1583
  Primitive vegetation 1583
  Modern vegetation 1584
Conclusions 1585
References cited 1586

Figure
1. Hypothetical series of curves illustrating relations between precipitation and relative sediment yield during geologic time . . 1575
2. Hypothetical series of curves illustrating the relations between precipitation and runoff during geologic time 1575

3. Generalized diagrams illustrating types of single and multiple channel systems and their morphologic and sediment characteristics 1580
4. Randomly spaced flood events 1585

Plate
1. Bed-load and suspended-load channels. . . ⎫
2. Braided bed-load channel of Rakaia River, Canterbury Plain, South Island, New Zealand. . . . . ⎬ See
3. Part of anastomosing suspended-load channel pattern north of Murray River near Yallakool, New South Wales, Australia ⎪ Plate Section
4. Riverine Plain near Darlington Point, New South Wales, Australia . . . . ⎭

Table
1. Classification of stable alluvial channels . . . 1579

INTRODUCTION

Geologists have long been aware that the hydrologic cycle is an integral part of the cycle of denudation and reconstruction of continents. For example, Hadding (1929) has suggested that the Lower Archean leptites of Sweden are pyroclastics that were eroded, transported, and redeposited by the initial functioning of the hydrologic cycle during and following the first rains, and Holmes (1965, p. 193) has incorporated the concept of the hydrologic cycle into his scheme of the geologic cycle of rock change.

Throughout most of geologic time, rivers have moved the denudation products of the lands toward the sea or to interior basins of deposition. Fluvial sediments have been identified in sedimentary deposits of all ages, and their association with coal, oil, uranium, and gold has caused considerable interest in these deposits and in paleochannels in particular. Recently geomorphic and engineering studies of modern rivers have been used to provide a basis for the interpretation of paleochannels and their associated fluvial sediments (Allen, 1965b; Moody-Stuart, 1966). In addition, there is an increasing interest in the effect of climate as a control of cyclic sedimentary deposits of fluvial origin (Swann, 1964; Hollingworth, 1962; Beerbower, 1961). In such studies it is tempting to make the uniformitarian assumption that modern hydrologic relations can be applied directly to stratigraphic problems of the ancient past. However, the very important influence of vegetation on the hydrologic cycle requires consideration, and Kaiser (1931), Tricart and Cailleux (1952), and Russell (1956), among others, have warned that rates of erosion and landform evolution have been altered by progressive colonization of the continents by vegetation. Hence, to understand better the environment under which paleochannels formed and transported fluvial sediments to a terrestrial or littoral site of deposition, it is mandatory that the hydrologic as well as the climatic characteristics of the system be evaluated.

The objectives of this paper are to consider the effects of the advent and evolution of terrestrial vegetation on runoff and sediment yield and to speculate on the manner in which these paleohydrologic variables influenced the morphology of river systems and their sediments.

Much of the presentation will necessarily be a review of the results of recent pertinent geomorphic and hydrologic research. For example,

only after a discussion of modern climatic and hydrologic relations can deductions be made concerning the hydrologic characteristics of the past. Further, in order to apply the paleohydrologic deductions to an explanation of paleochannel characteristics, river behavior in response to changing hydrologic regimen will be considered. It may seem presumptuous to attempt to use modern hydrologic data as bases for paleohydrologic deductions; nevertheless, relations have been developed which, if expressed in the most general terms, may be useful in directing our thinking concerning the interpretation of ancient fluvial deposits.

The discussion that follows should not be construed as an attack on the principle of uniformity because all the deductions concerning paleohydrology are based upon modern hydrologic data. In fact, an effort to apply these data to an understanding of Quaternary paleohydrology has already been made (Schumm, 1965), and an appeal has been issued to the hydrologic profession for assistance in the application of hydrologic data to geologic problems (Schumm, 1967).

ACKNOWLEDGMENTS

I thank E. D. McKee and R. F. Hadley of the U.S. Geological Survey and R. J. Weimer of the Colorado School of Mines for their criticisms and encouragement. An opportunity was provided during the 1967 National Science Foundation Institute of Fluvial Geomorphology, Colorado State University, to discuss the material presented herein with several geologists; among them were K. F. Bik, Harvey Blatt, and V. E. Gwinn, who made valuable suggestions.

HYDROLOGY

Some general relations that exist between mean annual temperature, precipitation, runoff, and sediment yield for the United States have been developed previously, and these are shown as curve 4 on both Figures 1 and 2. The effects of geology and drainage-basin morphology on these data have not been identified, but, for Figure 1, weak rocks and high-relief drainage basins would produce sediment in excess of that suggested by curve 4, whereas resistant rocks or low-relief drainage basins would produce less (Corbel, 1964). For Figure 2, impermeable soils or steep gradients would produce more runoff than indicated by curve 4, whereas permeable soils and gentle gradients should produce less.

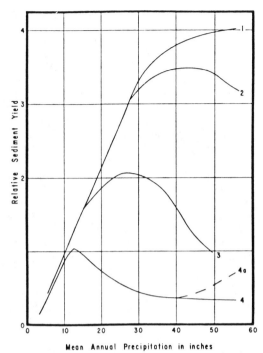

Figure 1. Hypothetical series of curves illustrating the relation between precipitation and relative sediment yield during geologic time. Curve 1, before the appearance of land vegetation; curve 2, after the appearance of primitive vegetation on the earth's surface; curve 3, after the appearance of flowering plants and conifers; curve 4, after the appearance of grasses. Curve 4a shows the increase in sediment yield rates for tropical monsoonal climates (Fournier, 1949). The peak of modern sediment yield (curve 4) is shown at a relative sediment yield of about 1. The relation is presented in this manner in order to discourage attempts to extrapolate specific sediment-yield rates to the geologic past. The curves show only the relative differences to be expected for a given rock type.

In spite of these qualifications, the two curves show in a general way the hydrologic differences that can be expected among the climatic regions of the United States, although the data have been adjusted to a mean annual temperature of 50° F (Langbein and Schumm, 1958).

Using a somewhat different line of reasoning, the curves can also be used to demonstrate the hydrologic effects of a climate change, and additional curves have been developed to evaluate the effects of temperature changes of the Quaternary Period on land forms of nonglaciated areas (Schumm, 1965). For a higher average temperature, sediment-yield curve 4

of Figure 1 and runoff curve 4 of Figure 2 will shift to the right.

Curve 4 of Figure 1 shows that the sediment yield from a drainage basin will increase from a minimum in a region of no precipitation to a peak at about 13 inches of precipitation (about 800 tons per square mile per year for the United States). The change in slope of curve 4 between 10 and 15 inches of precipitation is the result of increased effectiveness of vegetation in protecting the soil and retarding erosion, that is, a transition from desert shrubs to grassland. Sediment-yield rates decrease further to a very low value (about 300 tons per square mile per year for the United States) in forested regions of high rainfall. However, Fournier (1949, 1960) has suggested that in tropical monsoon climates sediment-yield rates will increase as indicated by curve 4a. Curve 4 was developed for uniform annual precipitation (Langbein and Schumm, 1958), but highly seasonal precipitation should increase sediment yields (Douglas, 1967). Fournier's conclusion is supported by the mass-movement investigations of the Indian Geological Survey in the eastern Himalayas, where heavy precipitation, fires, and land use have caused high sediment yields due to the mass movement of soil into the river channels (Dutt, 1966). However, the maximum sediment yield (curve 4) would also

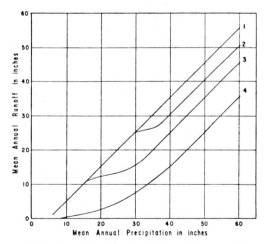

Figure 2. Hypothetical series of curves illustrating the relations between precipitation and runoff during geologic time. Curve 1, before the appearance of land vegetation; curve 2, after the appearance of primitive vegetation on the earth's surface; curve 3, after the appearance of flowering plants and conifers; curve 4, after the appearance of grasses.

increase considerably if sediment-yield data were available for semiarid drainage basins that are geologically and geomorphically similar to the very rugged Himalayan drainage basins.

Curve 4 of Figure 2 shows how runoff will increase with increased precipitation for uniform monthly precipitation at a mean annual temperature of 50° F (Langbein and others, 1949). Between about 5 and 20 inches of precipitation, runoff increases slowly with increased precipitation because the quantity of water lost through transpiration and interception by vegetation increases as precipitation increases. However, at about 40 inches of precipitation the water requirements of vegetation and evaporation have been met, and above 40 inches of precipitation runoff increases directly with precipitation.

It has been pointed out that, depending upon the climate before a climate change, increased precipitation will increase runoff but sediment-yield rates may increase or decrease (Schumm, 1965). In very arid regions an increase in precipitation will increase sediment yields because the vegetational cover will not improve sufficiently to retard erosion effectively, whereas under an initially semiarid climate an increase of precipitation will improve the vegetational cover sufficiently to cause a decrease in sediment-yield rates (Fig. 1, curve 4). Therefore, changes in the quantity of sediment moving out of a drainage system cannot be considered as a function of precipitation and runoff alone. The vegetational influence must also be evaluated, but this becomes increasingly difficult as the more remote geologic periods are considered.

PALEOHYDROLOGY

Paleohydrology can be defined as the science of the waters of the earth, their composition, distribution, and movement on ancient landscapes from the occurrence of the first rainfall to the beginning of historic hydrologic records. It is impossible to consider water movement on a landscape without becoming involved with the various problems of sediment transport, and therefore the inclusion of the word "composition" in the definition stipulates that the quantity and type of sediments moved through paleochannels must be considered a part of paleohydrology.

Certain types of modern hydrologic data may be applicable to the remote geologic past. For example, the effect of total destruction of vegetation on the hydrologic regimen of a drainage basin, hopefully, provides a tantalizing glimpse into the hydrology of the distant past.

An excellent example is provided by the Ducktown area of Tennessee, which was completely denuded by the poisonous effects of smelter fumes. The average annual precipitation in this region is between 55 and 60 inches; the characteristic vegetation is an oak-chestnut type of forest, and thin soils are underlain by shale and sandstone (Brater, 1939). The drainage basins from which data have been obtained are small, ranging from 6 to 89 acres in area. For two basins of comparable size (forested, 19 acres; barren, 16 acres) the records show that peak discharges were 9 times greater from the barren basin. No quantitative information is available for runoff volumes or sediment yields, but total runoff occurred in about 135 minutes for the forested drainage basin and in about 35 minutes, or one-fourth of the time, for the bare basin, and sediment yields were increased greatly.

Another source of information is provided by drainage basins from which vegetation has been removed by burning or where small areas have been treated to destroy the vegetation for experimental purposes. The sediment yield from these basins increases on the order of a hundred-fold (Storey and others, 1964). In the steep and rugged watersheds of the San Gabriel and San Bernardino Mountains, partial burning of the brush cover in this area of Mediterranean climate has caused major increases in both peak discharge and sediment yield (Anderson and Trobitz, 1949). As an example, a fire in 1936 reduced plant cover density from about 60 to 30 percent on the 4.8-square-mile Day Canyon watershed, thereby increasing the peak discharge by a factor of 2.4. In another case, the burning off of 20 percent of the plant cover caused a 90-percent increase in sediment yield.

All the drainage basins from which these data have been obtained are small, but, as vegetational cover becomes less effective, the sediment yields, runoff, and peak discharge of even the largest drainage system will increase significantly. However, some caution must be exercised before the erosion rates that have been measured in denuded areas can be applied to the erosion rates of prevegetation times. The modern rates are the result of erosion of soil and regolith, whereas, if vegetation had never been present, the rates would be slower, being dependent upon the speed of rock weathering and upon the rate of erosion of weakly cemented rocks.

For our purposes it will suffice to state with confidence that the presence of vegetation will reduce total runoff from a region and that up to a limit, depending upon intensity and duration of precipitation, the presence of vegetation will reduce flood peaks and increase the length of time of flooding. Kittredge (1948, p. 253) concludes, after a review of the evidence, that

Surface runoff from areas of well-established and undisturbed vegetation, if any, occurs only in heavy rains and is less than 3 percent of the precipitation, whereas, it starts after lighter rains and may be over 60 percent from denuded and cultivated lands.

He concludes with regard to flood peaks (1948, p. 271) that

In floods resulting from surface runoff or melting snow the heights of the crests are reduced and the times of occurrence are retarded by reforestation. Peak flows from forested areas rarely exceed 60 sec-ft per square mile, whereas from eroded or denuded land they may be 500 to 1,000 or more sec-ft per square mile.

Although the influence of forest vegetation in reducing the runoff and peak discharge of floods has been determined to be appreciable, its influence may be negligible during those major storms when the rainfall exceeds the total storage capacity of the drainage basin.

Using what is known about vegetational influences on sediment yield and runoff, an attempt has been made to show how the sediment-yield and runoff curves could have evolved through geologic time (Figs. 1, 2). As suggested by Russell (1956), four divisions of geologic time are of interest: (1) prevegetation time, (2) time during colonization of alluvial areas by primitive vegetation, (3) time during colonization of interfluves by modern types of flowering plants, and (4) time following the appearance of grasses.

It is difficult to establish the limits of these three phases of vegetational evolution. However, it is generally agreed that, although vascular land plants have been identified in the Cambrian Period (Andrews, 1961, p. 49), significant colonization of the land did not occur before the Devonian Period (Dunbar, 1960, p. 171) about 400 m.y. ago.

During the vast span of time between the Devonian Period and the end of the Paleozoic Era, the important forms of vegetation were probably confined to nearshore and coastal-plain areas. Primitive flowering plants have been identified in sediments of Permian and

Triassic age, but these could not have survived outside of tropical rain forests (Andrews, 1961, p. 183). Conifers were abundant in the Jurassic, and by Middle Cretaceous modern deciduous forests had appeared. These plants were capable of surviving seasonal changes of climate and climate fluctuations (Dunbar, 1960, p. 304, 337). Therefore, between the Devonian and Cretaceous Periods (400 m.y. to 135 m.y. ago) vegetation was progressively expanding its range as new forms evolved.

With the appearance of vegetation capable of surviving abrupt and possibly severe climatic fluctuations, the less favorable habitats were colonized, and between the Cretaceous Period and the Miocene Epoch (between 135 and 25 m.y. ago) the interfluve or upland areas were colonized to the extent that plant cover was effective in holding weathered material on the hillslopes and lowering the denudation rates. With the appearance of grasses in Miocene time (Dunbar, 1960, p. 432, 435) the effects of vegetation on the hydrologic cycle must have been as they are today.

During Precambrian and early Paleozoic time (prior to 400 m.y. ago) the lack of an effective vegetational cover would have permitted sediment-yield rates to increase with increased precipitation to a maximum rate, dependent only upon the erodibility or weathering rates of the rocks (Fig. 1, curve 1). On Figure 1 this maximum rate is arbitrarily shown as four times greater than the average rates of sediment yield of the present. It is assumed that the severe microclimatic environment, characterized by extremes of temperature and rapid changes of temperature at the ground surface, would have caused rapid weathering. The uninhibited transport of the weathered material would have produced rapid rates of denudation. The prevegetation denudation rate for weakly cemented sedimentary rocks would have been much greater than that shown, but it would have been much less for massive igneous rocks.

On Figure 2, curve 1 indicates that when transpiration and interception losses do not occur, runoff increases directly with precipitation after some loss due to infiltration and evaporation. This loss is assumed to have been about 5 inches, so runoff is zero from regions which received less than 5 inches of precipitation annually. At present, runoff from regions receiving less than 5 inches of annual precipitation is minimal (Langbein and others, 1949). Although some concavity may exist at the lower end of curve 1, and although curve 1 may indicate a quantity of runoff too great for even the prevegetation situation, the significant factor is that runoff would have increased directly with precipitation. Further, flood peaks would have been much greater in the absence of vegetation.

Curve 1 indicates that runoff was about 50 percent of precipitation at 10 inches of annual precipitation, but that runoff increased to 75 percent of precipitation at 20 inches and to 80 percent at 30 inches. Above 40 inches of precipitation, curve 4 parallels curve 1 but is lower and indicates a 20-inch water loss due to vegetational influences. The position of curve 1 on Figure 2 was established by assuming that it must be parallel to the upper portion of curve 4 and that forest vegetation can consume 20 inches of water annually in a humid region (Kittredge, 1938, 1948, Figs. 19, 20).

To summarize, during prevegetation time, runoff occurred more frequently as floods, and both sediment yield and runoff increased with increased precipitation. The sediment-yield rates increased with precipitation to a plateau, depending upon the erodibility of the rocks forming the drainage basin, and runoff for a given amount of precipitation was greater than at present.

As colonization of the land surface by primitive plants progressed during late Paleozoic and early Mesozoic time (between 400 and 135 m.y. ago), the presence of vegetation decreased erosion in certain parts of the landscape, and the peak of the sediment-yield curve would have decreased somewhat (Fig. 1, curve 2). Also, a discontinuity would have appeared in the runoff-precipitation curve for those areas of greater rainfall where vegetation was present and where water was utilized by the primitive vegetation (Fig. 2, curve 2). Primitive vegetation probably occurred only on coastal plains and in humid valleys, and therefore its effects initially may have been local. Sediment was shed continually from the barren hillslopes and temporarily stored in the valleys as in prevegetation times, but, at least in the downstream areas, the vegetation would have tended to stabilize the sedimentary deposits and the stream channels.

As more modern and hardier plants evolved during the Mesozoic and early Cenozoic (between about 125 and 25 m.y. ago), less favorable habitats were colonized and the vegetation moved up onto the interfluve areas, further decreasing sediment yields and compressing the peak of the curve (Fig. 1, curve 3). The discontinuity in the runoff-precipitation relation would have shifted into drier regions (Fig. 2, curve 3), and, with the appearance of grasses during late Cenozoic time (about 25 m.y. ago), erosion and runoff conditions similar to those of the present but without man's influence would have prevailed (Figs. 1 and 2, curve 4).

Gilluly (1964) concluded from an estimate of the volume of Triassic and younger sediments of the Atlantic coast of the United States that the average rate of Mesozoic and Cenozoic erosion was about equal to the present rates of erosion. However, present erosion rates may be on the average twice that of the natural rates (Douglas, 1967) because of man's influence on the land. For this reason curve 3 (Fig. 1) shows Mesozoic and early Cenozoic erosion as being about double that of the present (Fig. 1, curve 4).

The curves of Figures 1 and 2 provide some basis for paleohydrologic interpretation, and they illustrate the manner in which sediment yield and runoff will vary with climate. However, the curves of Figure 1, although providing information on the relative volumes of sediment delivered from a system during a given climate, provide no information on the type of sediment to be expected. It is known that the material held in solution in rivers increases as the climate becomes more humid (Durum and others, 1961) and that the presence of vegetation in humid regions retards the movement of sediment from the system and promotes weathering to yield a finer-grained product (Garner, 1959). Therefore, not only will sediment yield decrease with increased precipitation but the material will be finer for sediment-yield curves 3 and 4 of Figure 1, whereas fluvial sediments during prevegetation time (curve 1) and during colonization of the land by primitive vegetation (curve 2) were coarse-grained because weathering of the material on the hillslopes was precluded by rapid erosion.

In conclusion, the very different hydrologic characteristics of the geologic past, as outlined above, should be reflected in the characteristics of paleochannels and their sedimentary deposits.

RIVER MORPHOLOGY AND SEDIMENTATION

Channel Morphology

The morphology of stable alluvial river channels is determined by the water and sediment that move through the channel. The greater the flow of water (Qw) the larger will be channel width (w), depth (d), and meander wavelength (l) and the smaller will be channel gradient (s), as follows:

$$Qw \propto \frac{w, d, l}{s}. \tag{1}$$

However, for a given discharge channel dimensions are significantly influenced by the type of sediment load that is transported through the channel and that determines the character of its bed and banks. A significant increase in the ratio of bed load to total sediment load (Qs) will cause an increase in channel width, meander wavelength, and slope but a decrease in depth and sinuosity (P; ratio of channel length to valley length).

$$Qs \propto \frac{w, l, s}{d, P}. \tag{2}$$

As width appears in the numerator of equation 2 and depth in the denominator, a change in the type of sediment load will significantly influence channel shape (width-depth ratio).

On the basis of the relations of equation 2, alluvial river channels have been classified (Table 1; Fig. 3A; Pl. 1) according to the type of sediment load as bed-load, mixed-load, and suspended-load channels (Schumm, 1963a). Actually, a continuum of channels exists, and within this classification braided, meandering, and straight channels can be recognized (Fig. 3), depending upon the type of sediment forming the perimeter of the channel (Table 1). The percentage of silt and clay (sediment finer than 0.074 mm) exposed in the channel perimeter is inversely proportional to the percentage of the total sediment load that is sand, and it is an index of the ratio of bed load to total sediment load or type of sediment load passing through an alluvial channel (Schumm, in press).

In addition to the three classes of river channels there are two major types of river systems. These may be referred to as single-channel and multiple-channel rivers (Fig. 3; Table 1).

Within the single-channel category, straight channels are bed-load channels whereas the meandering channels are mixed- or suspended-load channels (Table 1; Fig. 3). Both types have been recognized in sedimentary deposits (Moody-Stuart, 1966). Exceptions to this generalization do occur, however, because where valley gradient is very low, even a suspended-load channel can be straight (Schumm, 1963a). Braided channels (Fig. 3; Pl. 1, fig. 1; Pl. 2) are single-channel bed-load rivers which at low water have islands of sediment or relatively permanent vegetated islands exposed in the channels (Doeglas, 1962). Any single channel can be part of a multiple-channel system, and, in fact, most river channels are part of a dendritic drainage pattern; however, the multiple-channel rivers considered here are those that are flowing on an alluvial surface and are distributary systems. Of the several types of multiple-channel systems perhaps only alluvial-fan distributaries may be difficult to find in the stratigraphic record. Certainly delta distributary systems have been recognized (Busch, 1959), as have alluvial-plain

TABLE 1. CLASSIFICATION OF STABLE ALLUVIAL CHANNELS

Type of sediment load	Channel sediment (percentage of silt and clay in channel perimeter M)	Bedload (percent of total load)	Type of river	
			Single channel	Multiple channel
Suspended load and dissolved load	>20	<3	Suspended-load channel. Width-depth ratio < 10; sinuosity > 2.0; gradient relatively gentle.	Anastomosing system.
Mixed load	5 to 20	3 to 11	Mixed-load channel. Width-depth ratio > 10, < 40; sinuosity < 2.0, > 1.3; gradient moderate. Can be braided.	Delta distributaries. Alluvial plain distributaries.
Bed load	<5	>11	Bedload channel, width-depth ratio > 40; sinuosity, < 1.3; gradient relatively steep. Can be braided.	Alluvial fan distributaries.

233

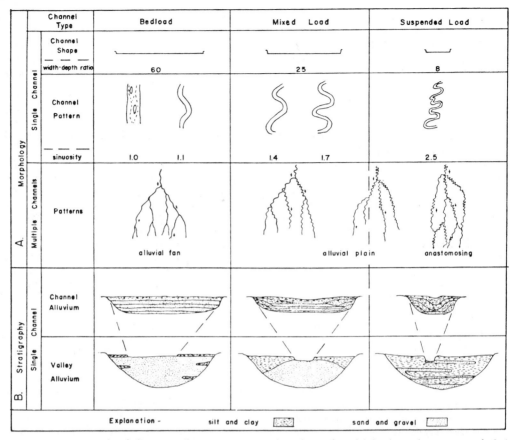

Figure 3.   Generalized diagrams illustrating types of single- and multiple-channel systems and their morphologic and sediment characteristics. (A) Channel morphology. (B) Alluvial stratigraphy. Diagrammatic sketches of channel and valley-fill deposits of bed-load, mixed-load, and suspended-load channels. Vertical exaggeration is about fivefold; for example, compare channel shape as shown in Figure 3A with that of Figure 3B.

systems of channels (Craig and others, 1955). Alluvial-fan distributaries will probably always be of the bed-load type (Fig. 3A), whereas alluvial-plain distributaries and delta distributaries, depending upon the type of sediment moved through the channels, can be composed of any of the three channel types. The anastomosing-channel systems, however, will probably be composed primarily of suspended-load channels (Pl. 1, fig. 2; Pl. 3).

Clarification is needed regarding multiple-channel systems that branch and rejoin in contrast to the permanently separated distributary systems of alluvial fans and plains and deltas. The terms "braiding" and "anastomosing" have been used synonymously for braided-river channels in this country, but elsewhere, particularly in Australia, "anas-

tomosing" is a common term applied to multiple-channel systems on alluvial plains (Hills, 1960, p. 61; Whitehouse, 1944). The channels transport flood waters and, because of the small sediment load moved through them, aggradation, if it is occurring, is a slow process. As a result, these low-gradient suspended-load channels are quite stable (David and Browne, 1950, v. 2, p. 11). An anastomosing system of channels appears on a map of the Holocene deposits of the Netherlands (Zonneveld, 1956; see Potter and Pettijohn, 1963, Figs. 7 to 9). The differences between braided and anastomosing patterns are shown on Figure 3A and Plates 1, 2, and 3.

Although each of the multiple-channel types has been identified in the geologic record, only the single-channel streams will be discussed here because each channel of a multiple-channel system can be classified according to its morphology and sediment characteristics without regard to the rest of the system. In addition, it is only during floods that all channels of a multiple-channel system are functioning, for at low discharge a single-channel can accommodate the total discharge.

River Adjustment

One of the most intriguing problems of geology and civil engineering is to determine the nature and extent of river adjustment to a change of hydrologic regimen. The empirical relations expressed by equations 1 and 2 indicate that a significant alteration of hydrologic regime can cause an adjustment involving many aspects of a river channel, and, in fact, a complete transformation of channel morphology or river metamorphosis can occur.

A brief review of some modern examples of river adjustment will be presented here to provide a basis for discussion of the paleo-hydrologic control of paleochannels and of the nature of fluvial sediments that reach a depositional site. These examples demonstrate that sediment type, annual discharge, and flood discharge do influence river channels as suggested by the equations. For example, a reduction in both peak discharge and annual discharge has caused a major reduction in the width of the North and South Platte rivers, the Arkansas River, and the Red River. In addition, each of these channels has become more sinuous, suggesting that the quantity of bed load moved through the channels has been reduced. On the other hand, the highly sinuous, narrow, deep Cimarron River channel of the nineteenth century was transformed by a major flood to an extremely wide, straight channel through which considerable quantities of sand were moved (Schumm and Lichty, 1963). Similarly, an increase in runoff, flood peaks, and major influx of coarse sediments to meandering channels of New Zealand has transformed them to straight, wide, braided channels (Campbell, 1945).

A revealing example of river metamorphosis resulting from a climate change is illustrated by the channels shown on Plate 4. The modern Murrumbidgee River has a suspended-load channel of high sinuosity and low width-depth ratio. The youngest paleochannel (ancestral river, Pels, 1964) is larger, but its shape and sinuosity are similar to those of the modern river, indicating that it carried a greater dis-

charge but that the type of sediment load was similar to that of the modern river. In fact, borings into this aggraded paleochannel reveal that a large proportion of the channel fill is silt and clay, which is the predominant sediment load of the Murrumbidgee River. The oldest paleochannel (prior stream, Butler, 1958) is of a very different type. It is a wide and relatively shallow channel of low sinuosity and high width-depth ratio (Schumm, in press). The prior streams are, in fact, bed-load and mixed-load channels. The older paleochannels were functioning when the vegetational cover on the catchment was not effective in preventing erosion and transportation of large quantities of sand to the alluvial plain. Subsequently a more humid climate prevailed, and the sand load of the stream decreased because the vegetation became denser and more effective in the prevention of erosion.

Equations 1 and 2 indicate that, with an increase in mean annual discharge and a decrease in the proportion of bed load, a channel can become more sinuous, shorten its meander wavelength, become narrower and deeper, and decrease its gradient. All of these changes appear to have occurred as the bed-load and mixed-load paleochannels (prior streams) became suspended-load paleochannels (ancestral rivers).

A renewed trend toward dryness has decreased runoff, but the existing vegetation is capable of protecting the soils of the drainage basin, and, although the channel dimensions have decreased, the pattern, form, and gradient of the modern channel is very like that of the youngest paleochannel.

The gradient of the modern river is about half that of the older paleochannel, however, and this decrease in gradient was accomplished by an increase in channel length through development of a highly sinuous course rather than by incision.

The equations and the examples demonstrate that the type of sediment load moved through a channel exerts a major control on the channel morphology, and therefore one may conclude that each type of channel must be associated with a characteristic sedimentary deposit.

Nature of Channel Sediments

Just as channel morphology differs among various types of channels, so the mechanics of erosional and depositional processes differ (Schumm, 1960, 1961). It follows that the nature of channel-fill deposits will vary markedly, depending upon the type of sediment load moved through the channels (Fig. 3B). Not only will the width-depth ratios of the channels differ, but the stratification and the distribution of sediments in the channels will be dissimilar. The channel fill of the bed-load type of stream is a sheet deposit with a flat upper surface which is covered by finer flood-plain sediment (Fig. 3B). The mixed-load channel fill (Fig. 3B) is a combination of sandy vertical-accretion deposits and lateral-accretion deposits of fine sediment which yield a convex-up sand deposit. In the aggraded suspended-load channel, sand is found at the very base of the deposit. The construction of berms within the channel forms a sigmoid pattern of cross-stratification near each bank, and further filling of the channel yields a broadly concave stratification. All of the sand units of the channel fill deposits will show planar and trough cross-stratification (McKee and Weir, 1953).

The end members of this triad of channel types are recognized by Moody-Stuart (1966) in the fluvial deposits of Devonian age in Spitsbergen. Generally, however, what a stratigrapher sees in the field is not a single channel fill but rather a valley fill which may be composed of channel fills of several types of rivers. It is certainly unlikely that a fluvial deposit 100 feet deep and several miles wide represents a single-channel deposit.

As the fluvial models of Allen (1965b) and Moody-Stuart (1966) indicate, valley-fill deposits differ depending upon channel characteristics which in turn reflect the hydrologic regimen of the system. The diagrammatic cross sections of Figure 3B suggest how valley-fill deposits of the three river types might appear in section.

The valley-fill stratigraphy of a bed-load channel (Fig. 3B) is based upon changes that have occurred along the bed-load streams of Kansas and Texas (Schumm and Lichty, 1963), where during years of low peak discharge a veneer of fine sediments is deposited over the deep deposit of cross-bedded sand and gravel to form a flood plain. The result is a flat-topped sand body capped by a thin deposit of fine alluvium. This flood plain may be destroyed periodically by major floods and the fine sediment swept downstream. In the valley of the Canadian River of northern Texas borings by the Bureau of Reclamation at the site of the Sanford Dam show a veneer of fine sediment over 150 feet of sand and fine gravel.

The valley-fill stratigraphy of a mixed-load channel (Fig. 3B) is based partly upon borings into the prior-stream channels adjacent to the Murrumbidgee River. The mixed-load valley fill of Figure 3B is essentially a lens-shaped deposit of cross-bedded sand. The base of the deposit is composed of sand and some gravel, whereas the upper part is sand flanked by silt, sand, and clay. As aggradation progressed more of the fine suspended-load sediments were deposited along the flanks of the channel to form a well-developed flood plain.

The valley-fill stratigraphy associated with a suspended-load channel (Fig. 3B) is based upon information obtained from bores into the Murrumbidgee River. Lateral migration of this suspended-load channel leaves a sheet of channel sand which is overlain by point-bar deposits of lateral accretion and overbank deposits. Only about one-third or less of this deposit will be composed of sand or coarser sediments.

The results of aggradation by bed-load and suspended-load streams on an alluvial plain would not be unlike the braided- and meandering-stream depositional models of Allen (1965a, Figs. 35c and 35d). Further, according to the classification of sand bodies proposed by Potter (1963), some pods and ribbon deposits could be discontinuous sand bodies deposited by suspended-load rivers, whereas the dendroids and belt deposits could have been deposited by mixed- and bed-load channels. Sheet deposits could have been formed by all three of the river types. Thin sheet deposits could have been formed by a meandering suspended-load or mixed-load channel, whereas a thick fluvial sheet deposit could have been the result of bed-load channel deposition over an alluvial plain.

PALEOCHANNELS AND FLUVIAL SEDIMENTATION

Because the quantity and nature of both sediment and water discharge from a drainage system varies between climatic regions, it is evident that the changing effects of evolving vegetation on the hydrologic cycle (Figs. 1, 2) would have had a significant role in the establishment of paleochannel morphology and on the character of the deposits related to these ancient channels.

The evidence will be found in the rocks, and, although the effects of phytological deserts (Urwüsten) on sedimentary deposits have been considered by several authors (Schwarzbach,

1963, p. 68; Twenhofel, 1932, p. 793–795), Pettijohn (1957, p. 683) states that "It is difficult to demonstrate any secular changes in the character of the sediments" although there may be "small or second-order differences in the average composition of sediments of various ages." The thesis presented in this paper is that, although the mineralogy and chemical composition of sediments may be relatively unchanged throughout geologic time (for a different opinion, *see* Cayeux, 1941), the distribution of sediments, the form of the deposits, and the occurrence of abrupt changes of lithology may all be attributed in part to the vastly different erosion conditions and hydrology of ancient times.

Considering what is known about the hydrologic and sedimentologic controls of river morphology, let us speculate on the behavior of ancient rivers and on the types of fluvial deposits that might be expected in the geologic record during prevegetation time, during colonization of alluvial areas by primitive vegetation, and during colonization of interfluve areas by modern types of vegetation.

Prevegetation

During the time before significant terrestrial vegetation appeared, the hydrologic regimen of arid regions would have been much as today, and even in humid regions the products of weathering were swept from hillslopes and divides into well-developed networks of stream channels. Unlike the humid regions of today where weathering products are stored on hillslopes as well as in alluvial valleys, the bare prevegetation hillslopes should have lost material as rapidly as it weathered to a size that could be transported. These immature sediments accumulated in valleys and on pediment surfaces. Sediment production proceeded at a rapid rate in hot humid climates, but an arid-appearing landscape existed everywhere. Bedload channels were ubiquitous and wide braided streams occupied the entire floor of alluvial valleys. Upon leaving the sediment source areas the channels would not have been confined, and vast alluvial-plain piedmont deposits formed. Depending upon the environment of deposition and tectonics, either great thicknesses of clastics accumulated or sheetlike deposits formed, as floods spread across piedmont areas, reworking and sorting the sediments. Gravels could have been spread as sheets over erosional surfaces (Stokes, 1950), and some

questionable Paleozoic tillites (Schwarzbach, 1963) may have formed in this manner.

No time lag between climatic fluctuations and river response existed. The runoff of wet years flushed vast quantities of sediment from the source areas, whereas dry years permitted accumulation of sediment in the system as mass wasting and local runoff continued to strip weathered material from the interfluves.

Under such conditions varvelike flood deposits could have formed. Each flood event deposited a layer of sand or relatively coarse sediment which was covered by fine sediments deposited during the waning of the flood waters. Such deposits are recognized on modern flood plains (Schumm and Lichty, 1963), and some Precambrian rocks, which cannot be related to glaciation (Eskola, 1932), and which are composed of large varves may have formed in this manner. Of course, any isostatic or tectonic uplift in the sediment source areas caused a pulse of sediment to be moved out of the drainage basins into the depositional areas, thereby inundating either piedmont deposits or an erosional landscape with a flood of coarse sediment (Hamblin, 1958; Schumm, 1963b).

The water table in these fluvial deposits fluctuated greatly in response to wet and dry seasons, and under these circumstances red beds could have formed under humid, subhumid (Krynine, 1950), and arid conditions (Walker, 1967).

Primitive Vegetation

With the appearance of terrestrial vegetation, initially confined to the nearshore environment but eventually colonizing the alluvial valleys of sediment source areas, the delivery of sediment from interfluve areas was not inhibited and sheetlike alluvial deposits continued to form (Meckel, 1967). The presence of vegetation on coastal plains undoubtedly stabilized channels in that part of their courses, but large floods could have caused periodic shifts in the channel position and an influx of coarse sediment. The floods could have been responsible for some cyclic sedimentary deposits, the cutouts which occur in coal measures (Robertson, 1952; Niyogi, 1966, p. 970), the ventifact reported by Mantle in an English coal swamp deposit (Schwarzbach, 1963, p. 68), and the juxtaposition of meandering and straight river channels in coal measures (Thiadens and Haites, 1944, facing page 22). Hence, the appearance of vegetation provided the opportunity for additional discontinuities in the stratigraphic record, and these effects were intensified as vegetation moved into valleys of the sediment source areas. In addition, suspended-load and mixed-load channel deposits became more common.

A sequence of events which has been postulated to explain valley filling and subsequent trenching of alluvium in the western United States (Schumm and Hadley, 1957) may be pertinent here. Sediment moving off poorly vegetated hillslopes and from tributary channels can be stored in a main valley through stabilizing effects of vegetation. Trenching of these deposits and removal of stored sediment will take place episodically when the valley gradient is locally steepened by deposition to the extent that erosion will occur during floods. The removal of sediment can be promoted by any factor that weakens vegetation cover on the valley floor, such as a series of drought years or overgrazing or both.

Before modern vegetation colonized interfluve areas, the situation in many ancient valleys resembled that of modern semiarid regions. Valley slopes and small tributaries provided an abundant supply of sediment to the main channel. During years of normal precipitation much of the sediment remained within the valley system, and these deposits then supported vegetation with its stabilizing influence. The reworking of stored sediment by a meandering river moved fine sediment out of the valley and across the well-vegetated alluvial plain to the sea. The vegetation of the alluvial plain stabilized the channel and inhibited the movement of coarse sediment at normal flows (Robertson, 1952). However, continued accumulation of sediment in upstream valleys caused these alluvial deposits to become less stable, and eventually a major flood could have triggered a phase of channel metamorphosis and sediment movement from the source area. Perhaps in humid regions runoff was always adequate to cope with the supply of sediment, but where erosion rates were very high and where the quantity of water percolating into coarse alluvium was large, such a sequence of events was possible.

If we think in terms of a threshold flood, which when exceeded causes export of stored sediment from a drainage system, a climatic or hydrologic pattern can be imposed on the cycle of erosion and deposition such that floods of a certain recurrence interval or frequency trigger pulses of sediment movement from the system. That is, over a long period of years a net accumulation of sediment occurs due to the high yield from barren interfluves and the stabilization of valley alluvium by vegetation. Steepening of the valley slope must occur during this period of sediment accumulation, and the stability of the valley or its ability to withstand erosion must decrease with time. Eventually a flood of a given recurrence interval, perhaps 1000 years, will modify a sinuous channel to the extent that sediment export in excess of accumulation will occur over a short span of time, and sediment that accumulated over a period of years will move through a straight steep-gradient, bed-load channel to the sea.

This can be illustrated by Figure 4. It is assumed that with climate unchanging, deposition in the drainage system will proceed at a constant rate, thereby steepening the gradient of valley floors and causing a decrease in stability of these alluvial deposits. If during 10,000 years ten major floods (1000-year recurrence interval) occur randomly, then the coincidence of major floods and reduced valley stability will produce periods of high sediment yield when erosion removes accumulated sediment, reduces the valley gradient, and restores the valley system to stability. The movement of coarse sediments transforms channels of the piedmont area from suspended-load to mixed- or bed-load channels, and they deliver essentially equally spaced pulses of coarse sediment to the depositional basins. Between these periods of high and coarse sediment yield, lower rates of sediment yield would prevail and finer-grained sediments would leave the system. The result should be a cyclic sedimentary deposit which could resemble the "fining upward" fluvial sequences described by Allen (1965a). This hypothesis adds a hydrologic explanation of their origin in addition to the previously suggested controls of channel wandering, variation in base level, and periodic diastrophic activity (Allen, 1965a, p. 243).

Modern Vegetation

As vegetation developed on interfluves and became much more effective in retarding erosion, effects of a climate change on sediment-yield rates became more pronounced. The narrowing of the peak of the sediment-yield curve, as vegetation spread over the surface of the drainage basin, established a situation such that a change of climate at or below 25 inches of precipitation (Fig. 1) could produce considerable variation in the amount and character of sediment yield.

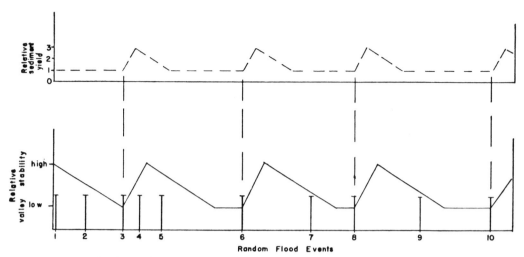

Figure 4. Randomly spaced flood events (vertical lines) of large magnitude, when they coincide with a period of low stability of the valley alluvium (solid line), cause incision and removal of accumulated sediment. High sediment yields (upper dashed line) are characteristic of short periods when the accumulated alluvium is flushed from the system.

With the appearance of modern vegetation the rivers of humid regions became relatively stable. Rivers draining humid regions are competent to transport sediment yielded to them from vegetated valley sides. Storage of sediment as soil on interfluves provides an opportunity for additional weathering, and the resulting fluvial sediments are mature and fine grained. Under these circumstances only a major climate change, diastrophism, or man's activities can cause changes in the type of sediment moved from the system or river-channel metamorphosis. Further, it is only in semiarid regions at present that pulses of sediment move out of tectonically undisturbed drainage basins. The discharge of sediment from basins in humid regions will be relatively constant except when a catastrophic hydrologic event or a eustatic or tectonic base-level change interferes with the normal progress of landscape denudation and movement of sediment through the system.

CONCLUSIONS

A variety of hydrologic data and geomorphic observations has been reviewed in order to permit speculation on the changes in the hydrologic cycle, river morphology, and fluvial sedimentary deposits during geologic time. Before the appearance of significant plant cover, denudation and runoff rates were high and floods were large, and coarse sediments

were spread as sheets over piedmont areas even in humid climates. Under these conditions bed-load channels and channel deposits were common, and in many fluvial sedimentary deposits individual flood events should be recorded.

During colonization of the earth by vegetation, a hydrologic mechanism was operative which could have formed cyclic sedimentary deposits. The stratigraphic record of this hydrologically unstable time contains an abundance of interruptions and cyclic deposits (Beerbower, 1961). In fact, in the coal measures of the Netherlands, Carboniferous coals show numerous cycles of river activity, whereas Tertiary lignites formed with only minor interference (Thiadens and Haites, 1944). Other causes were operative to cause major breaks in the stratigraphic record of earth history, but hydrologic factors should no longer be ignored in the interpretation of ancient fluvial or near-shore sedimentary deposits.

Obviously, the evidence concerning the paleohydrology of the remote past is as nebulous as that concerning its paleoclimatology, and the generalizations presented in this paper probably will be of little value in offering an explanation of a specific fluvial deposit. Nevertheless, the relations presented will open certain areas to further inquiry and speculation while at the same time provide limits to the extrapolation of modern hydrologic data for explanation of past events.

239

REFERENCES CITED

Allen, J. R. L., 1965a, Fining upwards cycles in alluvial successions: Geol. Jour. (Liverpool and Manchester), v. 4, p. 229–246.

—— 1965b, A review of the origin and characteristics of recent alluvial sediments: Sedimentology, v. 5, p. 89–191.

Anderson, H. W., and Trobitz, H. K., 1949, Influence of some watershed variables on a major flood: Jour. Forestry, v. 47, p. 347–356.

Andrews, H. N., Jr., 1961, Studies in paleobotany: New York, Wiley, 487 p.

Beerbower, J. R., 1961, Origin of cyclothems of the Dunkard Group (Upper Pennsylvanian-Lower Permian) in Pennsylvania, West Virginia, and Ohio: Geol. Soc. America Bull., v. 72, p. 1029–1050.

Brater, E. F., 1939, The unit hydrograph principle applied to small watersheds: Am. Soc. Civil Engineers Proc., v. 65, p. 1191–1215.

Busch, D. A., 1959, Prospecting for stratigraphic traps: Am. Assoc. Petroleum Geologists Bull., v. 43, p. 2829–2843.

Butler, B. E., 1958, Depositional systems of the Riverine Plain in relation to soils: Commonwealth Scientific and Industrial Research Organization Soil Pub. 10, 35 p.

Campbell, D. A., 1945, Soil conservation studies applied to farming in Hawke's Bay: Part 2, Investigations into soil erosion and flooding: New Zealand Jour. Sci. Tech., sec. A, v. 27, p. 147–172.

Cayeux, Lucien, 1941, Causes anciennes et causes actuelles en geologie: Paris, Masson, 79 p.

Corbel, Jean, 1964, L'erosion terrestre, étude quantitative: Annales Géographie, v. 73, p. 385–412.

Craig, L. C., and others, 1955, Stratigraphy of the Morrison and related formations, Colorado Plateau region, a preliminary report: U. S. Geol. Survey Bull., 1009-E, p. 125–166.

David, T. W. E., and Browne, W. R., 1950, The geology of the Commonwealth of Australia: 3 vols., London, Arnold.

Doeglas, D. J., 1962, The structure of sedimentary deposits of braided streams: Sedimentology, v. 1, p. 167–190.

Douglas, Ian, 1967, Man, vegetation and the sediment yields of rivers: Nature, v. 215, p. 925–928.

Dunbar, C. O., 1960, Historical geology: (2nd ed.) New York, Wiley, 500 p.

Durum, W. H., Heidel, S. G., and Tison, L. J., 1961, Worldwide runoff of dissolved solids: U. S. Geol. Survey Prof. Paper 424-C, p. 326–329.

Dutt, G. N., 1966, Landslides and soil erosion in the Kalimpong Subdivision, Darjeeling District, and their bearing on North Bengal floods: Geol. Survey India Bull., ser. B, no. 15, pt. 1, p. 61–77.

Eskola, Pentti, 1932, Conditions during the earliest geological times as indicated by the Archean rocks: Suomalaisen Tedeakatemian Toimituk sia, Annales Acad. Sci., Fennicae, ser. A, v. 36, 74 p.

Fournier, M. F., 1949, Les facteurs climatiques de l'erosion du sol: Assoc. Géographer Français Bull., v. 203, p. 97–103.

—— 1960, Climat et erosion: Paris, Presses Univ. France, 201 p.

Garner, H. F., 1959, Stratigraphic-sedimentary significance of contemporary climate and relief in four regions of the Andes Mountains, Geol. Soc. America Bull., v. 70, 1327–1368.

Gilluly, James, 1964, Atlantic sediments, erosion rates, and the evolution of the Continental Shelf: some speculations: Geol. Soc. America Bull., v. 75, p. 483–492.

Hadding, Assar, 1929, The first rains and their geological significance: Geol. Fören. Förh. and Lingar, v. 51, p. 19–29.

Hamblin, W. K., 1958, Cambrian sandstones of northern Michigan: Michigan Dept. Conserv., Geol. Survey Div., Pub. 51, 146 p.

Hills, E. S., 1960, The physiography of Victoria: Melbourne, Whitcombe and Tombs, 292 p.

Hollingworth, S. E., 1962, The climate factor in the geologic record: Geol. Soc. London Quart. Jour., v. 118, p. 1–21.

Holmes, Arthur, 1965, Principles of physical geology: London, Nelson, 1288 p.

Kaiser, Erich, 1931, Der Grundsatz des Aktualismus in der Geologie: Zeitschr. Deut. Geol. Gessellschaft: v. 83, p. 389–407.

Kittredge, Joseph, 1938, The magnitude and regional distribution of water losses influenced by vegetation: Jour. Forestry, v. 36, p. 775–778.

—— 1948, Forest influences: New York, McGraw-Hill, 394 p.

Krynine, P. D., 1950, Petrology, stratigraphy, and origin of Triassic sedimentary rocks of Connecticut: Connecticut Geol. and Nat. History Survey Bull. 73, 247 p.

Langbein, W. B., and others, 1949, Annual runoff in the United States: U. S. Geol. Survey Circ. 52, 14 p.

Langbein, W. B., and Schumm, S. A., 1958, Yield of sediment in relation to mean annual precipitation: Am. Geophys. Union Trans., v. 39, p. 1076–1084.

McKee, E. D., and Weir, G. W., 1953, Terminology for stratification and cross-stratification in sedimentary rocks: Geol. Soc. America Bull., v. 64, p. 381–390.

Meckel, L. D., 1967, Origin of Pottsville conglomerates (Pennsylvanian) in the central Appalachians: Geol. Soc. America Bull., v. 78, p. 223–258.

Moody-Stuart, M., 1966, High and low sinuosity stream deposits with examples from the Devonian of Spitsbergen: Jour. Sed. Petrology, v. 36, p. 1101–1117.

Niyogi, Dipankar, 1966, Lower Gondwana sedimentation in Saharjuri Coalfield, Bihar, India: Jour. Sed. Petrology, v. 36, p. 960–972.

Pels, Simon, 1964, The present and ancestral Murray River system: Australian Geog. Studies, v. 2, p. 111–119.

Pettijohn, F. J., 1957, Sedimentary rocks: New York, Harper, 718 p.

Potter, P. E., 1963, Late Paleozoic sandstones of the Illinois basin: Illinois Geol. Survey Rept. Inv. 217, 92 p.

Potter, P. E., and Pettijohn, F. J., 1963, Paleocurrents and basin, analysis: Berlin, Springer-Verlag, 296 p.

Robertson, Thomas, 1952, Plant control in rhythmic sedimentation: Compte Rendu 3rd Cong. pour l'avancement des études de stratigraphie et de geol. du Carbonifere: v. 2, p. 515–521.

Russell, R. J., 1956, Environmental changes through forces independent of man, p. 453–470 in Thomas, W. L., Jr., Editor, Man's role in changing the face of the earth: Chicago, Univ. Chicago Press.

Schumm, S. A., 1960, The effect of sediment type on the shape and stratification of some modern fluvial deposits: Am. Jour. Sci., v. 258, p. 177–184.

—— 1961, Effect of sediment characteristics on erosion and deposition in ephemeral-stream channels: U. S. Geol. Survey Prof. Paper 352-C, p. 31–70.

—— 1963a, A tentative classification of alluvial river channels: U. S. Geol. Survey Circ. 477, 10 p.

—— 1963b, The disparity between present rates of denudation and orogeny: U. S. Geol. Survey Prof. Paper 454-H, 13 p.

—— 1965, Quaternary paleohydrology, p. 783–794 in Quaternary of the United States, Wright, H. E., Jr., and Frey, D. G., Editors: Princeton, New Jersey, Princeton Univ. Press.

—— 1967, Paleohydrology: Application of modern hydrologic data to problems of the ancient past: Internat. Hydrology Symposium (Fort Collins) Proc., v. 1, p. 185–193.

—— in press, River adjustment to altered hydrologic regimen, Murrumbidgee River and paleochannels, Australia: U. S. Geol. Survey Prof. Paper 598.

Schumm, S. A., and Hadley, R. F., 1957, Arroyos and the semiarid cycle of erosion: Am. Jour. Sci., v. 255, p. 161–174.

Schumm, S. A., and Lichty, R. W., 1963, Channel widening and flood-plain construction along Cimarron River in southwestern Kansas: U. S. Geol. Survey Prof. Paper 352-D, p. 71–88.

Schwarzbach, Martin, 1963, Climates of the past (translated by Muir, R. O.): London, Van Nostrand, 328 p.

Stokes, W. L., 1950, Pediment concept applied to Shinarump and similar conglomerates: Geol. Soc. America Bull., v. 61, p. 91–98.

Storey, H. C., Hobba, R. L., and Rosa, J. M., 1964, Hydrology of forest lands and range lands, sec. 22, 52 p., in Chow, Ven-Te, Editor, Handbook of applied hydrology: New York, McGraw-Hill.

Swann, D. H., 1964, Late Mississippian rhythmic sediments of Mississippi valley: Am. Assoc. Petroleum Geologists Bull., v. 48, p. 637–658.

Thiadens, A. A., and Haites, T. B., 1944, Splits and washouts in the Netherlands coal measures: Mededeelingen Gol. Stichting ser. C-11-1, no. 1, 51 p.

Tricart, J., and Cailleux, A., 1952, Conditions anciennes et actuelles de la genese de peneplains: Internat. Geog. Union Proc. (Washington), p. 396–399.

Twenhofel, W. H., and others, 1932, Treatise on sedimentation: Baltimore, William and Wilkins, 926 p.

Walker, T. R., 1967, Formation of red beds in modern and ancient deserts: Geol. Soc. America Bull., v. 78, p. 353–368.

241

Whitehouse, F. W., 1944, The natural drainage of some very flat monsoonal lands: Australian Geographer, v. 4, p. 183–196.
Zonneveld, J. I. S., 1956, Fluvial deposits, p. 103–106 *in* Pannekoek, A. J., *Editor*, Geological history of the Netherlands: S-Gravenhage, Staatsdrukkerig en vitgeuerijbedrigf, 147 p.

MANUSCRIPT RECEIVED BY THE SOCIETY FEBRUARY 14, 1968
REVISED MANUSCRIPT RECEIVED MAY 24, 1968
PUBLICATION AUTHORIZED BY THE DIRECTOR, U. S. GEOLOGICAL SURVEY

Figure 1. Bed-load channel of the Wairau River, South Island, New Zealand.

Figure 2. Suspended-load channel of the Saline River near Vesper, Nebraska.

BED-LOAD AND SUSPENDED-LOAD CHANNELS

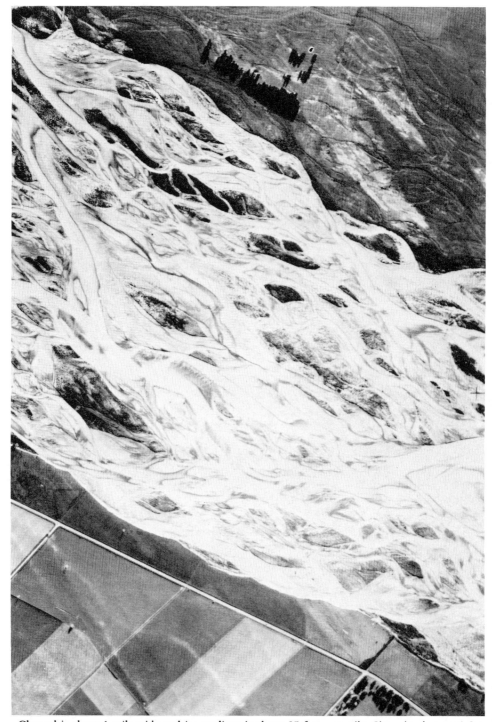

Channel is about 1 mile wide and its gradient is about 25 feet per mile. Sinuosity is near 1.0 and width-depth ratio is in excess of 100. Flow is from left (west) to right (east). Reproduced with permission of the Surveyor General, Lands and Survey Department, New Zealand.

BRAIDED BED-LOAD CHANNEL OF THE RAKAIA RIVER,
CANTERBURY PLAIN, SOUTH ISLAND, NEW ZEALAND

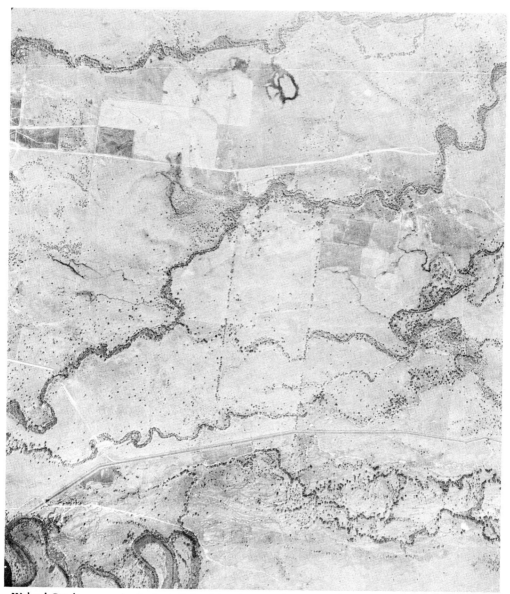

Wakool Creek appears in the lower left of the photograph. Wakool Creek is about 100 feet wide and its gradient is about 1 foot per mile. Sinuosity is greater than 2.0 and width-depth ratio is below 10. Flow is from right (east) to left (west). All of the channels visible on the photograph are anabranches of the Wakool Creek multiple-channel system. Reproduced with permission of New South Wales Department of Lands.

PART OF ANASTOMOSING SUSPENDED-LOAD CHANNEL PATTERN NORTH OF
MURRAY RIVER NEAR YALLAKOOL, NEW SOUTH WALES, AUSTRALIA

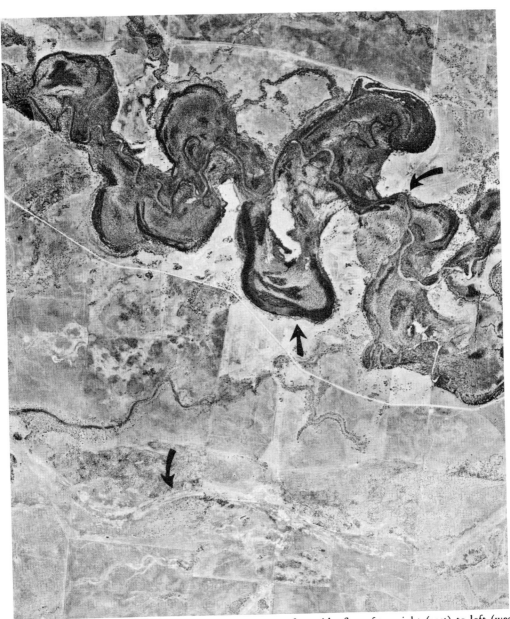

The sinuous Murrumbidgee River, which is about 200 feet wide, flows from right (east) to left (west) across the photograph (upper arrow). It is confined to an irregular flood plain on which the youngest paleochannel is preserved as large oxbow lakes (middle arrow), which are evidence of a past time of high discharge (ancestral stream). The trace of a low-sinuosity older paleochannel (prior stream) crosses the lower quarter of the photograph (lower arrow). Reproduced with permission of New South Wales Department of Lands.

RIVERINE PLAIN NEAR DARLINGTON POINT, NEW SOUTH WALES, AUSTRALIA

Part IV

MAN'S INFLUENCE ON EROSION

Editors' Comments
on Papers 20, 21, and 22

20   WOLMAN
A Cycle of Sedimentation and Erosion in Urban River Channels

21   DUNNE
Sediment Yield and Land Use in Tropical Catchments

22   LIKENS et al.
Excerpts from *Effects of Forest Cutting and Herbicide Treatment on Nutrient Budgets in the Hubbard Brook Watershed-Ecosystem*

That various forms of human activity affect particulate and solute yields and erosion processes in general has been documented in many publications and symposia proceedings (for example, see Boughton, 1970; Slaymaker and McPherson, 1973). The papers included in this part emphasize the first two of the following three principles of environmental impact: (1) substantial human interference with nature causes a disruption of existing dynamic equilibria, bringing forth significant increases in erosion; (2) the changes in yield are complex, manifold, and interrelated; and (3) the high yields usually abate with time. These responses also occur due to changes triggered by nonanthropogenic sources such as fires, hurricanes, or earthquakes.

The papers included in this part cannot cover the whole field of anthropogenic effects on erosion and sediment yields, although they include a wide range of activities, including deforestation, agricultural development, overgrazing, and urbanization. Other activities that may influence erosion and sediment or solute yields are hydraulic mining and strip-mining (Gilbert, 1917; Collier, Pickering, and Musser, 1970), dam construction and river control works (Kellerhals and Gill, 1973), application of agricultural chemicals such as fertilizers and pesticides, and drainage and irrigation. Not all of these activities cause an increase in sediment yields. Dam construction, for instance, may result in upstream storage of much of the sediment being transported by a river, and diversion of sediment-laden water on to irrigated fields may significantly reduce suspended sediment concentrations downstream, although solute concentration may be increased conversely.

The impact of human activity upon net rates of denudation and sediment yield and its modification of geologic erosion rates has been somewhat neglected until recently. For example, Fournier (Paper 14) ignored human activity as a factor controlling erosion rates, and Langbein and Schumm (Paper 15) and Corbel (1964) alluded to the significance of anthropogenic erosion without attempting to correct their data accordingly. Since the mid-1960s, however, the impact of human activity has been increasingly considered by, among others, Douglas (1967), Judson (1968), Meade (1969), and Janda (1971). Judson estimated, for example, that the mass of material carried to the oceans by rivers was 9.3×10^9 t y^{-1} before human activity had begun to have a significant effect, rising to 24×10^9 t y^{-1} at the present time, and Meade estimated that present sediment loads in rivers of the U.S. Atlantic Coast states are four to five times greater than in pre-European times.

Although they may be appreciable, human impacts upon erosion rates are not necessarily simple, and their quantitative determination may be difficult. Natural and anthropogenic erosion must be distinguished, and Tricart (Paper 2) has pointed out that erosion on a watershed and the net sediment yield from that watershed may be quite different because of the effect of sediment storage as colluvium, flood-plain deposits, and valley fills (Happ, 1945). For example, Trimble used data for a number of U.S. southern piedmont rivers to demonstrate that upland erosion has been equivalent to ground lowering of 95 mm/100 yrs but that sediment yields were equivalent to only 5.3 mm/100 yrs (a sediment delivery ratio of 6 percent). Much of the eroded sediment has been stored as valley fill deposits, although as sediment loads from the headwaters decline in response to conservation measures, the valley fills themselves may become sediment sources for contemporary erosion (Costa, 1975).

Paper 20 presents M. G. Wolman's model of the response of sediment yields to different stages of land use and economic development, based on data for the Maryland piedmont (Wolman and Schick, 1967). Wolman demonstrated that human impact may be considerable—up to four orders of magnitude increase in sediment yield during the urban construction period. It is not due simply to the removal of the protective cover of vegetation from a watershed but may be indirect, by increased flood peaks and stream flow, which in turn cause increased bank erosion (Anderson, Paper 13; Hammer, 1972). A noteworthy feature of Wolman's model is the eventual decrease in sediment yields to levels comparable to those from forested areas, comparable to the declining sediment yields in Megahan's (1974) model for areas disturbed by forest development and road construction. In general, erosion rates and sediment yields decline rapidly after the initial disturbance of a geomorphic system, whether by natural catastrophe or human activity; Graf (1977) has suggested a rate law in fluvial geomorphology to formalize this.

Because of the complex combination of factors that control erosion rates and sediment yields, it is difficult to identify the precise effects and relative importance of each. This is particularly the case with human activity, which Wolman's model suggests can be highly variable and which can be intentionally modified by appropriate conservation measures. In his discussion of the Langbein-Schumm rule, Wilson (1973) stressed the multivariate nature of the problem but concluded that land use is the most important factor, on both a local and a global scale. Paper 21, by Thomas Dunne, is a specific test of Wilson's conclusion and hence of the validity of simple relationships like the Langbein-Schumm rule. This paper is also of significance because it presents comprehensive sediment yield data for part of Africa, about which relatively little information had been available until major research programs in Kenya (Dunne, Dietrich, and Brunengo, 1978), Tanzania (Rapp, Berry, and Temple, 1972), and Tunisia (IAHS, 1977) began to produce results.

In an extension of the approach used by Fournier (Paper 14), Langbein and Schumm (Paper 15), and Jansen and Painter (1974), Dunne attempted to predict sediment yields for forty-six Kenyan watersheds in terms of topography, precipitation or runoff, and land use, using multiple regression analysis. Only 11 percent of the variability in sediment yield was statistically explained by an equation including relief ratio and runoff, but 77 percent was explained by an equation that also included land use type. A plot of sediment yield as a function of runoff and land use (Paper 21, figure 3) shows no sign of the Langbein-Schumm rule (Paper 15, figure 2) except for forested and lightly grazed grassland watersheds, which are not affected by human activity. Instead sediment yields progressively increase with runoff for each land use, the increase being steepest for pasture, intermediate for agriculture, and lowest for forest. The overwhelming impact of land use in

Dunne's study area is also manifested by a comparison with Tertiary, Pliocene/Quaternary, and drier glacial Quaternary erosion rates (22, 77, and up to 200 t km^{-2} y^{-1}, respectively), estimated from the elevations of dated erosion surfaces (Dunne, Dietrich, and Brunengo, 1978). Present-day erosion rates in forested or lightly grazed areas are similar to the estimated geologic erosion rates, but in heavily grazed areas they range up to 20,000 t km^{-2} y^{-1}. This disparity between natural and anthropogenic erosion rates indicates that the soil profile may continue to erode at greatly increased rates for only limited periods before being completely removed. Dunne, Dietrich, and Brunengo (1978) consider that this period is only two hundred to five hundred years in Kenya. Erosion rates must then decline to balance geologic weathering rates, although the disparity between natural and accelerated rates usually decreases with time.

Much of the research into human impacts upon erosion and sediment yields has concentrated upon particulate matter because of its high visibility and clear economic and practical importance. Solute loads are receiving increasing interest, however, as water resource and land managers realize their broader implications for eutrophication, quality of domestic and industrial water supply, and nutrient losses from ecosystems. Foresters in particular have attempted to draw up nutrient budgets for forest ecosystems under different management regimes as an aid to their management on a sustained yield basis. A number of integrated, interdisciplinary studies of nutrient cycling in forest ecosystems have been undertaken. They include the H. J. Andrews Experimental Forest, Oregon (Fredriksen, 1971), the Priest River Experimental Forest, Idaho (Snyder, Haupt, and Belt, 1975), the Taita watersheds, New Zealand (Claridge, 1970), and Hubbard Brook, New Hampshire (Likens et al., 1977).

All of these studies have considered the effects of such forest management practices as clear-cut logging and slash burning upon nutrient losses. We reprint here an extract from a much-quoted paper reporting results from the study at Hubbard Brook (Paper 22). The experimental procedure involved clear felling of the forest and inhibition of vegetation regrowth for two years by application of herbicides; the fallen trees were left in place and were not burned. Hence the experiment did not simulate the effects of any standard forest management practice or of any natural catastrophe such as wildfire or tree mortality due to disease. The results are for the worse-case situation in which the normal functioning of the ecosystem is completely disrupted; a subsequent study in which normal clear-cutting procedures were followed produced increases in solute loads, but of a much smaller magnitude (Pierce et al., 1972).

The Hubbard Brook study once again emphasizes the importance of the organic subsystem to the behavior of the whole watershed ecosystem (Paper 22, figure 15). The authors indicated that nutrient losses are related to alteration of the nitrogen cycle; nitrate nitrogen (an anion) no longer taken up by living plants moves in large quantities through the soil, facilitating the loss of the cations (Ca, Mg, K, Na, Al) by leaching. They pointed out that the observed nutrient losses represent a "mining" of the nutrient capital of the ecosystem rather than an increase in weathering rates and so cannot be sustained indefinitely. In fact, studies such as those by Pierce et al. (1972) and Fredriksen, Moore, and Norris (1975) indicate that nutrient losses decrease rapidly with revegetation. Precise trends are a complex function of such factors as soil porosity and texture (which control soil erodibility, pathways and rates of water movement to the streams, degree of adsorption of nutrients) and the chemical and other characteristics of the precipitation (which control rates of nutrient uptake). Brown (1974) considers that the complexity of the interactions among these many factors precludes generalization of management effects in different forest types.

REFERENCES

Boughton, W. C., 1970, Effect of Land Management on Quantity and Quality of Available Water, *Australian Water Resources Council, Research Program Report 179.*

Brown, G. W., 1974, *Forestry and Water Quality*, Oregon State University Bookstores, Corvallis, Oregon.

Brune, G. M., 1953, Trap Efficiency of Reservoirs, *Am. Geophys. Union Trans.* **34**:407–419.

Claridge, G. G. C., 1970, Studies in Element Balances in a Small Catchment at Taita, New Zealand, *Internat. Assoc. Sci. Hydrology Pub.* **96**:523–540.

Collier, C. R., R. J. Pickering, and J. J. Musser, 1970, Influences of Strip Mining on the Hydrological Environment of Parts of Beaver Creek Basin, Kentucky, 1955–1966, *U.S. Geol. Survey Prof. Paper 427-C*, 80p.

Corbel, J., 1964, L'érosion terrestre, étude quantitative, *Annales Géographie* **73**:385–412.

Costa, J. E., 1975, Effects of Agriculture on Erosion and Sedimentation in the Piedmont Province, Maryland, *Geol. Soc. America Bull.* **86**: 1281–1286.

Douglas, I., 1967, Man, Vegetation and the Sediment Yield of Rivers, *Nature* **215**:925–928.

Dunne, T., W. E. Dietrich, and M. J. Brunengo, 1978, Recent and Past Rates of Erosion in Semi-Arid Kenya, *Zeitschr. Geomorphologie* **29**:91–100.

Fredriksen, R. L., 1971, Comparative Chemical Quality of Natural and Disturbed Streams Following Logging and Slash Burning, in *Symposium on Forest Land Use and the Stream Environment*, ed. J. T. Krygier and J. D. Hall, Oregon State University, Corvallis, Oregon, pp. 125–137.

Fredriksen, R. L., D. G. Moore, and L. A. Norris, 1975, The Impact of Timber Harvest, Fertilisation and Herbicide Treatment on Streamwater Quality in Western Oregon and Washington, in *Forest Soils and Forest Land Management, North American Forest Soils Conf., 4th Proc.*, ed. B. Bernier and C. H. Winget, Laval University, Quebec, pp. 283–313.

Gilbert, G. K., 1917, Hydraulic Mining Debris in the Sierra Nevada, *U.S. Geol. Survey Prof. Paper 105*, 105p.

Graf, W. L., 1977, The Rate Law in Fluvial Geomorphology, *Am. Jour. Sci.* **277**:178–191.

Hammer, T. R., 1972, Stream Channel Enlargement Due to Urbanisation, *Water Resources Research* **8**:1530–1540.

Happ, S. C., 1945, Sedimentation in South Carolina Piedmont Valleys, *Am. Jour. Sci.* **243**:113–126.

International Association of Scientific Hydrology, 1977, Proceedings, Paris Symposium on Erosion and Solid Matter Transport in Inland Waters, *Internat. Assoc. Sci. Hydrology Pub.* **122**:211–232, 260–268, 278–291.

Janda, R. J., 1971, An Evaluation of Procedures Used in Computing Chemical Denudation Rates, *Geol. Soc. America Bull.* **82**:67–80.

Jansen, J. M. L., and R. B. Painter, 1974, Predicting Sediment Yield from Climate and Topography, *Jour. Hydrology* **21**:371–380.

Judson, S., 1968, Erosion of the Land, *Am. Scientist* **56**:356–374.

Kellerhals, R., and D. Gill, 1973, Observed and Potential Downstream Effects of Large Storage Projects in Northern Canada, *Com. Internat. Grands Barrages, Proc. 11th Conf.*, Madrid, pp. 731–754.

Likens, G. E., F. H. Bormann, R. S. Pierce, J. S. Eaton, and N. M. Johnson, 1977, *Biogeochemistry of a Forested Ecosystem*, Springer-Verlag, New York.

Meade, R. H., 1969, Errors in Using Modern Stream-Load Data to Estimate Natural Rates of Denudation, *Geol. Soc. America Bull.* **80**:1265–1274.

Megahan, W. F., 1974, Erosion over Time on Severely Disturbed Granitic Soils: A Model, *U.S. Forest Service Research Paper INT-156*, 14p.

Pierce, R. S., C. W. Martin, C. C. Reeves, G. E. Likens, and F. H. Bormann, 1972, Nutrient Loss from Clearcutting in New Hampshire, in *Proceedings of a Symposium on Watersheds in Transition*, ed. S. C. Gallany. T. G. McLaughlin, and W. D. Striffler, American Water Resources Association and Colorado State University, Fort Collins, pp. 285–295.

Rapp, A., L. Berry, and P. Temple, eds., 1972, Studies of Soil Erosion and Sedimentation in Tanzania, *Geog. Annaler* **54A**:3–4, 105–379.

Slaymaker, H. O., and H. J. McPherson, 1973, Effects of Land Use on Sediment Production, in *Fluvial Processes and Sedimentation, Proceedings of the Hydrology Symposium*, National Research Council of Canada, Department of Environment, Canada, pp. 158–183.

Snyder, G. G., H. F. Haupt, and G. H. Belt, 1975, Clearcutting and Burning

Slash Alter Quality of Streamwater in Northern Idaho, *U.S. Forest Service Research Paper INT-168*, 34p.

Trimble, S. W., 1977, The Fallacy of Stream Equilibrium in Contemporary Denudation Studies, *Am. Jour. Sci.* **277**:876–887.

Wilson, L., 1973, Variations in Mean Annual Sediment Yield as a Function of Mean Annual Precipitation, *Am. Jour. Sci.* **273**:335–349.

Wolman, M. G., and A. P. Schick, 1967, Effects of Construction on Fluvial Sediment, Urban and Suburban Areas of Maryland, *Water Resources Research* **3**:451–464.

20

Copyright © 1967 by the Editor, Gunnar Østrem

Reprinted from *Geog. Annaler* **49A**:385–395 (1967)

A CYCLE OF
SEDIMENTATION AND EROSION
IN URBAN RIVER CHANNELS

BY M. GORDON WOLMAN

Department of Geography, Johns Hopkins University

ABSTRACT. Historical evidence and contemporary measurements indicate in the Piedmont of Maryland that successive changes in land use have been accompanied by changes in sediment yield and in the behavior of river channels. Sediment yields from forested areas in the pre-farming era appear to have been less than 100 tons/sq.mi/year. Yields from agricultural lands in the same region at a later time range from 300 to 800 t/sq.mi. on large drainage areas. Subsequently, on lands marginal to expanding urban centers, a decline in active farming may be accompanied by a decline in sediment yield. In marked contrast, areas exposed during construction can produce sediment loads in excess of 100,000 t/sq.mi./year. Small channel systems become clogged with sand during this construction period. While sediment deposited in channels during construction is gradually removed by subsequent clearer flows, rates of removal are slow and hampered by deposition of debris. Increased runoff from urban areas coupled with a decline in sediment yields to values on the order of 50 to 100 t/sq.mi. promote continued bank erosion and channel widening. Maximum observed rates of bank erosion were on the order of 1.0 foot per year. Raw banks adjacent to coarse cobble bars and widespread deposits of flotsam and debris attest to the flood regimen of urban rivers. Canalization in concrete does not eliminate such debris nor does it eliminate deposition of sediment as local changes in gradient, excessive channel width, and debris accumulation foster deposition even in canalized reaches.

Equilibrium and a cycle of change

Students of geomorphology have long debated the meaning and value of concepts such as grade and equilibrium applied to the behavior of stream channels. These and similar phrases generally denote a condition of balance, stability, or both in the characteristics and behavior of a river channel. However, logical and semantic difficulties demand that phrases associated with the concept of equilibrium must be used with care and circumscribed by qualifications. Thus, progressive degradation over geologic time is inconsistent with a too rigid definition of equilibrium which implies stability in elevation, gradient, and channel form. At

the same time, over somewhat shorter periods of time, slow but progressive degradation may yet be associated with near constancy or stability of channel form. With only small changes in inflow of water and sediment, channel form and even channel gradients may remain relatively stable.

While a universally applicable concept of equilibrium may be difficult to formulate because of the problems posed by varying time scales and rates of change of channel gradients and channel forms, the concept of equilibrium can be useful in dealing with the response of channel systems to significant changes in the values of the independent variables such as discharge and sediment load over shorter intervals of time. Under these conditions a reasonable working hypothesis, perhaps paraphrasing Mackin (1948), might be that over a period of years channel slope and form are adjusted to the quantity of water and to the quantity and characteristics of the sediment load provided by the drainage basin. Each of the independent variables, water and sediment, are in turn related to the soil, lithology, vegetation, and climate of the region. This statement of adjustment allows for momentary scour and fill and for short-term trends, measured in years, in channel behavior associated with high water or drought.

If a set of river channels are in equilibrium with prevailing conditions in a drainage basin, it follows that major disturbances on the drainage basin will result in changes in channel form and behavior. The process of urbanization of the landscape constitutes a major interruption of "prevailing" conditions on a watershed. If prior to urban development, a kind of equilibrium prevailed in which channel gradient and form were related to water

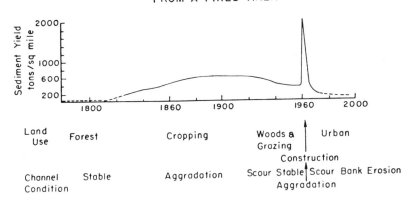

Figure 1. The cycle of land use changes, sediment yield, and channel behavior in a Piedmont region beginning prior to the advent of extensive farming and continuing through a period of construction and subsequent urban landscape.

and sediment derived from the watershed, then the sequential changes which occur as urban development takes place on the watershed can be expected to alter markedly the equilibrium forms and may result in the eventual establishment of new conditions of equilibrium. At the present time the disturbance of equilibrium can be documented, but as the data here will show, it is currently difficult to determine whether a new equilibrium will be established or whether instead a condition of disequilibrium will persist.

The process or cycle of urbanization on the watershed that is reflected in the river channels of a region consists of three stages; 1) an initial stable or equilibrium condition in which the landscape may either be primarily agricultural or dominated by forests, 2) a period of construction during which bare land is exposed to erosion, and 3) a final stage consisting of a new urban landscape dominated by streets, rooftops, gutters, and sewers. This theoretical cycle is sketched in Figure 1 along with estimates of the quantity of sediment derived from the watershed, the presumed channel behavior, and the sequence of land use. The data are based upon experience in the Middle Atlantic region of the United States. The conditions outlined in Figure 1 are described in the following paragraphs. These serve both as an outline and as a summary of the data presented and evaluated in the body of the paper.

In accord with the historical evidence, Figure 1 shows a modest yield of sediment prior to the farming era given here as beginning around 1700 A.D., and a significant increase in sediment yield during the farming period to an average value of about 600 tons/square mile. A decline in yields to a value of perhaps 300 tons/ sq.mi. is shown for the period immediately preceding construction based on the observation that in the environs of some of the major urban centers much farmland may be put in grass or allowed to return to brush and forest while awaiting development. With the onset of clearing for construction sediment yields rise to perhaps several thousand or more tons/sq. mi. during a short interval of perhaps one to three years. The interval is short, of course, only where a single or isolated unit of land and channel is considered since progressive development of a large drainage area will affect downstream reaches of channel for a longer period. Following construction, if the entire area has been developed, sediment yields should be expected to decline to values as low as or lower than those experienced prior to the farming era. This condition is shown by a dashed line on Figure 1.

Changes in the watershed are accompanied by changes in channel characteristics. During the farming era historical evidence indicates considerable accumulation of sediment within channels (Gottschalk, 1945). Subsequently, in areas returned to brush and forest, much of the fine-grained sediment appears to have been removed returning the channels to a condition in which the channel bed was composed of gravel with lesser amounts of silts and sands. With the onset of construction large quantities of sand are delivered to channel systems and new sandbars and dunes may blanket the bed of the channel. If vegetation becomes established on the bars, channels are constricted, or locally banks may erode, accompanied by an increase in flooding at the channel constrictions. Upon completion of streets and sewerage systems sediment derived from the watershed decreases while the rapidity of runoff is increased. Channel bars and vegetation may be removed by flows of clear water. At the same time the absence of a fresh supply of sediment may result in progressive channel erosion without concomitant deposition.

While this general configuration of the process of urbanization and its effect on channel systems is reasonable, it is the purpose of this paper to review, add to, and evaluate the evidence for successive stages in the cycle. Because the economic consequences of these changes in the natural landscape are significant, it is hoped that a better understanding of the processes may be helpful in evaluating alternative methods of managing the land surface as well as the channel and riverine bottom lands.

Changes in sediment production

Sediment yields from agricultural and forested regions are well documented (U.S. Dept. of Agriculture, 1964) and are not repeated here. The upper part of Table 1 shows sediment yields for wholly forested regions as well as for areas of mixed farming and forests. Data from large areas wholly in forest are quite limited but current estimates suggest that yields may be less than 100 tons per square mile. In contrast the figure for the Gunpowder Falls at two successive intervals is of particular interest. From 1914 to 1943, a period of intense farming on the watershed, sediment yield was approximately 800 tons/sq. mi. During a later period when much land was returned to grazing and to forest in the immediate vicinity of the city of Baltimore, sediment yield declined to approximately 200 tons/sq. mi. or one-quarter of the earlier figure. The dip in the curve in Figure 1 is based upon this observation.

Sediment yields from areas undergoing construction may exceed by several hundred-fold the yields from lands in forests and grazing, or by several fold areas in agriculture. As the selected values in Table 1 show, yields may exceed 100,000 tons/sq. mi. over very small areas. On larger drainage basins in which the

Table 1. Sediment yield from drainage basins under diverse conditions.

RIVER AND LOCATION	DRAINAGE AREA SQUARE MILES	SEDIMENT YIELD TONS/SQ. MILE/YR.	LAND USE
BROAD FORD RUN, MD.	7.4	11	FORESTED: ENTIRE AREA
HELTON BRANCH, KY.	0.85	15	SAME
FISHING CREEK, MD.	7.3	5	SAME
GUNPOWDER FALLS, MD.	303	808	RURAL - AGRICULTURAL, 1914-1943, FARMLAND IN COUNTY 325,000 TO 240,000 AC.
SAME		233	RURAL - AGRICULTURAL, 1943-1961, FARMLAND IN COUNTY 240,000 TO 150,000 AC.
SENECA CREEK, MD.	101	320	SAME
BUILDING SITE, BALTO., MD.	0.0025	140,000	CONSTRUCTION: ENTIRE AREA EXPOSED
LITTLE FALLS BRANCH, MD.	4.1	2,320	CONSTRUCTION: SMALL PART OF AREA EXPOSED
STONY RUN, MD.	2.47	54	URBAN: ENTIRE AREA

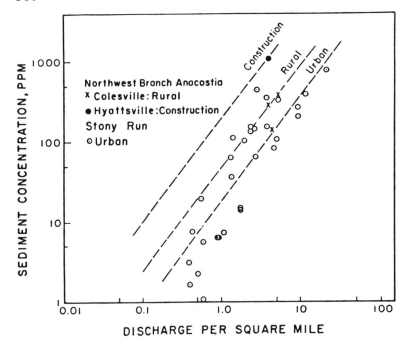

Figure 2. Curves relating sediment concentration and discharge in streams from three drainage areas differing in land use. The drainage area of the Northwest Branch of the Anacostia River above Colesville (21.3 sq.mi.) is rural, between Colesville and Hyattsville (45.2 sq. mi.) considerable land is exposed to construction, while Stony Run (2.5 sq.mi.) lies within the city of Baltimore. Curves suggest highest concentrations from areas undergoing construction with successively lower values for rural and urban watersheds.

entire area is not undergoing construction, yields may still exceed several thousand tons per square mile (Wolman and Schick, 1967).

Comparison of sediment rating curves also indicates that for a given discharge or frequency of flow, sediment concentrations may be twice or more than those from similar areas not subject to construction (Figure 2). Keller (1962) reports sediment loads 3 to 5 times as high. As one would expect, the quantity of sediment derived from the areas undergoing construction is a function of gradient, quantity and intensity of precipitation, characteristics of the soil, and topographic discontinuities at the construction site. However, even in the absence of precipitation, large quantities of suspended sediment may result from construction activities where heavy machinery operates directly in the stream channels. Thus on a clear day without precipitation concentrations in a local channel reached 3300 ppm and sediment load followed a diurnal cycle in accord with the variation in construction activity (Figure 3). While the total load derived from this source may be small, the turbidity created in the flow is significant.

The yield of sediment from urban areas following completion of construction is less

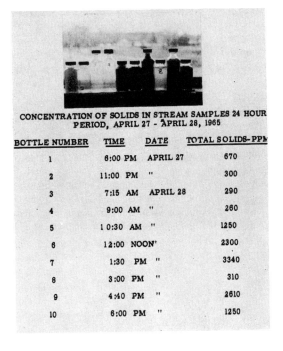

CONCENTRATION OF SOLIDS IN STREAM SAMPLES 24 HOUR PERIOD, APRIL 27 - APRIL 28, 1965

BOTTLE NUMBER	TIME	DATE	TOTAL SOLIDS-PPM
1	6:00 PM	APRIL 27	670
2	11:00 PM	"	300
3	7:15 AM	APRIL 28	290
4	9:00 AM	"	260
5	10:30 AM	"	1250
6	12:00 NOON	"	2300
7	1:30 PM	"	3340
8	3:00 PM	"	310
9	4:40 PM	"	2610
10	6:00 PM	"	1250

Figure 3. Diurnal variation in sediment concentration observed in Herring Run on a clear day as a result of construction activity within the channel. (Data from Whitman, 1965).

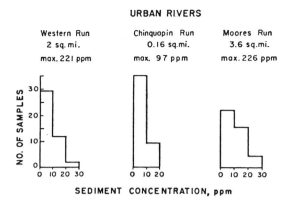

Figure 4. Histograms showing concentrations of suspended solids in three streams in the Baltimore Metropolitan Area. Location of these streams is shown on the map in figure 5.

well documented. In most large cities where measurements have been made construction has continued at successive locations on the drainage area above the measuring site. However, some data are available for streams in the Baltimore area.

Periodic spot observations in three streams in the Baltimore area suggest the order of magnitude of average sediment yields and concentrations in urban areas (Figure 4). The drainage basins of both Western Run and Chinquapin Run are underlain by crystalline rocks of the Piedmont while Moores Run is primarily in

Figure 5. Map showing streams in the Baltimore Metropolitan region referred to in the text.

the Coastal Plain (Figure 5). (The channel of Chinquapin Run is pictured below in Figure 8). None of these samples were collected during storms, and hence concentrations are probably somewhat too low. Nevertheless the values are low and average sediment yields from these urban areas appear to be small.

More detailed observations on both low and storm flow in streams in the Baltimore area show similarly low concentrations of suspended sediment (Brosky, 1966). During a summer storm with a peak flow of 17 cfs from a drainage area of 2.5 square miles (a flow equaled or exceeded about one percent of the time) the peak concentration of suspended solids was only 439 ppm (Figure 6). A maximum concentration of 793 ppm was observed for a flood flow with a recurrence interval of approximately 1.5 years. Brosky noted that in the absence of construction sites on the watershed, suspended solids were almost exclusively granular and without clays characteristic of samples of suspended solids from areas undergoing construction.

A rough estimate indicated that the average fallout of dry solids on the watershed, measured by the city nearby, of 58.7 tons/month in summer and 117 tons/month in winter exceeded the quantity of material removed in solution and suspension. For the storm shown in Figure 6 suspended solids amounted to about 77 percent of the dissolved load. Preliminary estimates indicate that the amount of sediment removed from the basin may be considerably less than the dry fallout but available data is not sufficiently accurate to warrant detailed comparison. However, concentrations are low even in storm periods and a crude computation suggests that average annual clastic load is on the order of 50 t/sq.mi.

Evidence of the reduction of sediment supply from urban areas is provided by a survey of sediment in large culverts draining new developments. Of 14 drains surveyed in the suburban region, only 3 showed 20 per cent or more of the end or cross-sectional area of the culvert occupied by sediment. Furthermore, at two of these three, the surrounding suburban development had only been completed within the preceding year. Where development had been completed five or more years, sediment covered 10 per cent or less of the culvert cross-section.

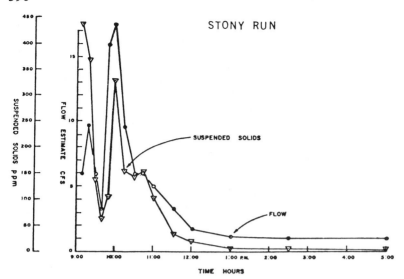

Figure 6. Hydrograph showing variations in flow and in concentration of suspended solids in Stony Run during a storm on July 29, 1964. Peak discharge represents a flow equaled or exceeded on the order of 1% of the time or about three days per year.

To permit rapid removal of runoff, storm drainage culverts are often placed on high gradients. In addition, a number of estimates indicate that peak runoff from impervious areas (Carter, 1961) may exceed by 2 to 6 times the peak runoff from the same area prior to urban development. Thus it is not surprising that sediment is progressively removed from culverts following completion of construction. The rate of removal should depend upon the timing and magnitude of the runoff, the capacity and gradient of the drain. These observations all indicate that the inflow or renewal of sediment is reduced allowing the flow to remove from the drains sediment accumulated during the period of construction.

Data representing the full transition from "natural" or agricultural conditions, through construction, to a completely urbanized watershed at a single location were not available to the author. Therefore curves relating suspended sediment concentration to discharge for three locations on two different streams are compared. The watershed of Stony Run lies within Baltimore City (drainage area 2.5 sq. mi.) and contained no area exposed to construction. In contrast most of the area on the Northwest Branch of the Anacostia above Colesville (drainage area 21.3 sq.mi.) is rural and agricultural while extensive areas are undergoing construction in the intervening region between Colesville and Hyattsville (drainage area 45.2 sq.mi.). A straight line was fitted

by eye to the data for Stony Run and curves parallel to it were drawn for the Northwest Branch of the Anacostia at Colesville and at Hyattsville. Thus the slopes of the curves differ from those originally drawn by Keller (1962). Three points including an especially low concentration are shown for the Colesville station for comparison. As Keller (1962) pointed out initially, the curves indicate that for a given flow, concentrations from the areas undergoing construction may be 5 times greater than from the rural areas. In addition, the curve added here for the completely urban watershed appears to lie below that of the rural area. Eighty-five per cent of the points for the "urban river" fall below the Colesville or "rural" curve (see Fig. 2).

In summary, the data appear adequate to support the contention that sediment yields during construction exceed yields not only from forests but from agricultural lands as well. Less well documented but suggestive is the evidence that sediment yield several years after completion of urban development is very low, perhaps as low as or lower than sediment yields from completely forested areas.

Channel behavior

In forested regions where sediment yields are low, stream channels in the crystalline Piedmont flow on beds of cobble gravel interspersed with finer grained deposits and occasional bedrock outcrops and within banks generally

1963

1966

Figure 7. Successive photographs of sand bars developed in the channel of Jones Falls 100 yards downstream from major highway construction completed in 1963. Vegetation on bars appears during summer months but has not become established as shown by the winter picture in 1966.

composed of silty or sandy loam. An influx of sediment laden water derived from construction on the watershed can be expected to result in extensive deposition of sand bars and dune sand generally coarser than the finer sediments carried in suspension prior to the advent of construction. Such a generalization would not be true, of course, if the inflow of sediment was small relative to the transporting capacity of the receiving channel, i.e., 100 acres of exposed land contributing sediment directly to the Mississippi River is unlikely to create new channel forms within the Mississippi. Virtually every large metropolitan center in the region, however, contains a myriad of small streams which may be affected. In metropolitan Baltimore the formation of deltas at the confluence of two channels, of sand bars, and banks of

sand dunes over pre-existing gravel beds have been observed (Wolman and Schick, 1967). Less clear is the progression of channel changes following completion of construction.

Comparison of photographs of the channel of Jones Falls at the time of completion of a super-highway, which involved massive earth-moving, with the same reach three years later (Figure 7) shows little or no change in the size of bars. The reach shown is 100 to 200 yards downstream from the highway construction. Upstream from the reach the area is primarily rural. Further downstream, however, within the urban area, comparison of photographs taken four years apart indicates that some bars may have been removed from mid-channel and from the outside of bends, but in general the channel continues to contain extensive sand deposits particularly where piles of debris and bridge piers encourage deposition. Photographs of nearby Roland Run, taken $3^{1}/_{2}$ years apart, indicated that with lessening of suburban development on the watershed, some sand and gravel deposits have been scoured from an upstream reach. In a reach 300 yards downstream deposition appears to have increased upstream from a small bridge opening. The intervening steeper channel contains little or no fresh deposits of gravel and sand.

Colby (1964) has shown that at a depth of one foot and a velocity of 2 feet per second the rate of sand transport will be about 5 tons per day per foot of width. In an urban channel at a drainage area of 2 square miles these conditions are reached on the order of one percent of the time. Assuming a deposit of sand 0.5 foot thick over a reach $^{1}/_{4}$ mile long and a unit weight of 100 pounds per cubic foot, without additional inflow of sand, removal of the deposit would require about 7 days or two years. This figure is of course hypothetical but observation of streams in the Baltimore area suggest that the period will be considerably longer. Some observers (Guy, 1963, Wolman and Schick, 1967), noted that channels may be cleared of sediment in 5 to 7 years. However, continuing observations indicate that channel curvature, local flattening of slope, establishment of vegetation, and particularly trash and debris may inhibit removal of sediment for even longer periods.

Both the expected increase in runoff from urban areas and the absence of sediment should contribute to an increase in channel erosion and to an increase in channel width. An increase in the number of peak flows particularly would tend to increase the amount of bank erosion. With a decrease in the available sediment, deposition would not keep pace with erosion, as it might under "normal" conditions of flood-plain formation, thus promoting progressive widening of the channel.

Exposure of raw banks in miles of urban river channels suggests bank erosion. As Hadley and Schumm (1961) have observed, however, raw banks are not prima facie evidence of high rates of bank erosion. Detailed observations of 7200 feet of the channel of Western Run in northwest Baltimore indicate that active bank erosion is occurring along a distance of about 580 feet of the total length

of 14,400 feet of channel bank. The channel was straightened and deepened beginning about ten years ago to an average gradient of one per cent. Maximum observed erosion was about 2.2 feet on the outside of an aligned curve constructed 3½ years ago. With the exception of irregularities at tree roots, at points adjacent to gravel bars, and at junctions of concrete culverts, average erosion is probably less than one foot per year. Some slumping can be seen near the top of the higher banks but the result appears to be a gentler side slope not yet attacked by the shallow flow at the base.

As the photographs of Chinquapin Run (Figure 8) show, little or no vertical buildup of point bars is taking place in the urban channels. Similar conditions were observed on Western

1960

1967

Figure 8. Photographs of Chinquapin Run north of Northern Parkway in Baltimore City. Photographs in 1967 show development of vegetation, particularly establishment and growth of locust trees. Comparison of photographs B and D reveals some deposition along the left bank with erosion of 2 to 3 feet along the near vertical right bank. Comparison of photographs A and C suggest scouring of sand on the point bar as well as some channel erosion just downstream from the point bar. Some channel widening is in evidence in both photographs.

Run. Locust trees and grass have established adjacent to and on the low bar along the left margin of the channel in Figure 8D. Lateral deposition to a height of 1.5 feet or less is also evident. The right bank is steeper in 1967 than in 1960 and has receded about 3 feet in the 7-year period. Comparison of Figures 8A and 8C over the same time interval indicates that the channel has widened particularly in the bend by removal of the sands on the point bar and by erosion downstream from the point bar along the left bank (left foreground of the photograph 8C). A locust tree about 3 years old has become established adjacent to the left or concave bank suggesting some stability at that point. Nearby some slump is also evident along the concave bank at the right edge of the photograph. One hundred yards downstream from the reach in Figure 8 on the outside of a bend, highwater (estimated frequency once or twice per year) on January 26, 1967, undermined a tree 18 inches in diameter resulting in local lateral erosion of about one foot. These observations establish the fact that erosion is taking place albeit at a relatively slow pace. However, because such erosion is progressive and unaccompanied by comparable accretion, a net widening appears to be taking place above the elevation of the coarse cobble bed. Because continued widening will reduce

the depth of flow for a given discharge, the rate of lateral erosion should be expected to decline. Nevertheless, in the absence of an equivalent inflow of sediment, a new equilibrium in transport will not be established.

Because of the great variability of natural channels, it is difficult to make statistically adequate comparison of channel shape and size before and after urbanization. Figure 9 is an attempt at such a comparison and the data suggest that the width of channels in urban areas may be somewhat larger than in comparable channels in "natural" or agricultural areas.

The erosion and flood characteristics of the urban river may be better demonstrated by the visual, subjective, impression of the channel than by any current objective or measurable parameters. The combination of raw banks, exposed cobble bars, and debris including flotsam strewn about the floodplain and channel margins all convey the impression of frequent flooding and the transient character of the alluvial features. Thus, the comparative photographs in Figure 10 show two aspects of the results of flooding in two completely different environments. The top photographs (Figures 10A and 10B), were taken immediately after the record hurricane flood of August, 1955, in Connecticut. They show the abrupt channel widening which commonly results from the deposition of coarse cobble bars and the characteristic flotsam and debris deposited on the channel margins by the floodwaters. The lower photographs show precisely the same erosion features and ubiquitous debris in several urban rivers in the Baltimore metropolitan region. Similar exposed banks coupled with tangles of debris on the margins of the channel can be seen over many miles of urban river channel. While attempts have been made to quantify this evidence, no readily mappable parameters have described these conditions as well as the overall visual impact. The existence of market carts and tricycles with densities up to 14 per linear mile is not an uncommon measurable parameter but perhaps one of dubious comparative significance.

Flood debris, eroding banks, scour holes, and exposed bars all appear to suggest the development of an erosive regimen in the urban river channel following completion of development on the watershed. These effects

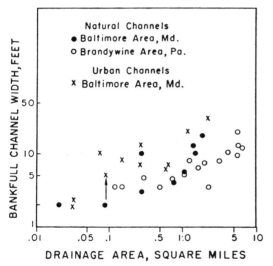

Figure 9. Comparisons of drainage area and channel width for streams in the Piedmont region under rural and urban conditions. Data indicate that at least some of the urban channels show an expected increase in width.

Figure 10. Comparative photographs illustrating the apparent flood regimen of urban channels. Upper photographs were taken following the hurricane flood of August, 1955, in Connecticut. Photograph A at Lewis Atwood Brook near Waterbury, Connecticut, shows characteristic widening associated with deposition of coarse cobbles following the flood. Comparable deposition and channel widening is shown on Western Run below Simmonds Avenue in Baltimore, Maryland. Photograph B, Farmington River near Unionville, Connecticut, shows characteristic debris deposited by flood waters comparable to debris illustrated by the photograph (D) of Herring Run at Pioneer Drive in Baltimore, Maryland.

may be attributed to the combined action of an increase in the magnitude and number of peak flows as well as to a decrease in the availability of sediment derived from the watershed.

Social response to changing channel behavior

Recognizing the potential value of river bottom lands for recreational use, a number of metropolitan areas in the United States have been moving toward reservation of floodplain lands for parks and open spaces. Where such use is contemplated an effort has also been made to avoid canalizing stream channels with concrete or other materials in order to preserve a more natural environment.

Accompanying this trend toward reservation of open spaces in a natural condition, however, has been a public demand for maintenance of urban steram channels against the ravages of erosion and the accumulation of rat-infested debris. One rather common response to this demand, and in the eyes of some a response completely at odds with preservation of the natural scene is the canalization of extensive reaches of channel in concrete. Several assets of such concrete channels are assumed to be rapid dispersal of storm drainage, an increased potential for self-cleaning, and low maintenance costs. Aside from aesthetic considerations it is important to recognize that deposition and erosion may be subject to the same controls in floodways as in the preexisting alluvial channels. Abrupt flattening of gradients in broad floodways and the accumulation of debris may induce deposition at precisely those locations where such deposition previously occurred. Careful and sometimes expensive designs may mitigate such problems, but in many cases it is likely that removal of sediment and debris and continuous channel maintenance will be required regardless of design.

Because the urban river poses both opportunities for recreational land use as well as problems in control and maintenance, it is important that alternative plans for control and use of these rivers be developed in accord with some understanding of the principles of their behavior. The evidence suggests that even in the relatively restricted field of erosion and sedimentation in alluvial channels there are significant physical consequences resulting from urban development of entire watersheds. Recognition of these consequences, while it solves no problems, can perhaps serve a purpose in demonstrating the need for forethought in planning for the appropriate use of the riverine environment. As always, the appropriate combination of aesthetics, economics, and physical limitations is not a constant but must vary from city to city and from river to river.

Acknowledgement

The author is indebted to Messrs. D. L. Brosky and I. L. Whitman for their kindness in permitting him to use hitherto unpublished data on sediment in streams in the metropolitan region, and to J. Prussing of the Department of Public Works of the City of Baltimore for providing maps of channel works on Western Run.

References

Brosky, D. L., 1966: Solids in a small urban Watershed at extreme flows. 29 pp., unpublished.

Carter, R. W., 1961: Magnitude and frequency of floods in suburban areas. *U. S. Geol. Survey Prof. Pap.* 424-B, pp. B9 – B11.

Colby, B. R., 1964: Practical computations of bed-material discharge. *J. Hydr. Div., Am. Soc. Civ. Eng.*, v. 90, pp. 217–246.

Gottschalk, L. C., 1945: Effects of soil erosion on navigation in Chesapeake Bay. *Geogr. Rev.* v. 35, pp. 219—237.

Guy, H. P., 1963: Residential construction and sedimentation at Kensington, Md. Paper presented at *Federal Inter-Agency Sedimentation Conf.*, Jackson, Miss., Jan., 1963, 16 pp.

Hadley, R. F. and *Schumm, S. A.*, 1961: Sediment sources and drainage basin characteristics in Upper Cheyenne River basin. *U. S. Geol. Survey Water Supply Paper* 1531B, pp. 137–196.

Keller, F. J., 1962: Effect of urban growth on sediment discharge, Northwest Branch Anacostia River basin, Maryland. *U. S. Geol. Survey Prof. Pap.* 450-C, pp. C129–C131.

Mackin, J. H., 1948: Concept of the graded river. *Geol. Soc. Amer. Bull.*, v. 59, pp. 463—512.

U. S. Dept. of Agriculture, 1964: Summary of reservoir sediment deposition surveys made in the United States through 1960. *Misc. Publ.* 964.

Whitman, I. L., 1965: Erosion and sediment transport on Herring Run due to construction activities. 25 pp., unpublished.

Wolman, M. G. and *Schick, A. P.*, 1967: Effects of construction on fluvial sediment; urban and suburban areas of Maryland. *Water Resources Res.*, v. 3, No. 2.

Reprinted from *Jour. Hydrology* **42**:281–300 (1979)

SEDIMENT YIELD AND LAND USE IN TROPICAL CATCHMENTS

THOMAS DUNNE

Department of Geological Sciences and Quaternary Research Center, University of Washington, Seattle, WA 98195 (U.S.A.)

(Received July 20, 1978; accepted for publication January 15, 1979)

ABSTRACT

Dunne, T., 1979. Sediment yield and land use in tropical catchments. J. Hydrol., 42: 281—300.

Analysis of sediment yields from 61 Kenyan catchments allows the refinement of regional relationships between the yields and their major controls. Land use, which has been ignored in earlier regional analyses, is the dominant control, but within each land-use category it is possible to recognize the effects of the climatic and topographic variables that other writers have stressed. The long-term geologic rate of erosion in these tropical environments is estimated to vary between 20 and 200 t km^{-2} yr.$^{-1}$, depending mainly upon climate. These values agree closely with Douglas' estimate for undisturbed catchments. An analysis of the major sources of sediment in disturbed catchments suggest that rural roads contribute a large, and commonly ignored, fraction of the sediment leaving agricultural areas. The temporal pattern of sediment transport is also affected by land use, and emphasizes the significance of rare wet periods in the removal of soil from tropical catchments.

STATEMENT OF THE PROBLEM

Various attempts have been made to relate the sediment yields of catchments to simple climatological indices, such as annual rainfall (Langbein and Schumm, 1958; Wilson, 1973), seasonality of precipitation (Fournier, 1960), and annual runoff (Douglas, 1967; Dendy and Bolton, 1976). Wilson (1972, 1973) reviewed these studies and concluded that no single relationship is valid on a worldwide basis, and that even within a relatively uniform area, the most important single control is land use. He did not have data on land use to support this conclusion, but nevertheless was able to demonstrate his point from circumstantial evidence. Jansen and Painter (1974) incorporated an index of natural vegetation into their multiple-regression analysis of sediment yields from catchments throughout the world, but could not allow for the extensive alterations of the cover that had occurred due to land use. Their results did not show a strong influence of cover on sediment yields; climate and topography were more important controls. The influence of land use was stressed at a recent symposium (I.A.H.S., 1974), but no attempt was made to examine its effect on the regional and global relationships referred to above.

In his global assessment, Wilson (1973) pointed out, as many other authors have done, the paucity of measured sediment yields from tropical catchments. Fournier (1960) and Jansen and Painter (1974) incorporated some tropical data into their analyses. In Africa, there have been few studies of sediment yields beyond broad estimates for the largest rivers of the continent (Holeman, 1968). In Tanzania, Rapp et al. (1972a, b) have carried out a broad program of soil erosion research including studies of sediment yields from small basins (1.5—640 km^2) with mixed land use and vegetation cover. Temple and Sundborg (1972) measured the sediment yield of the 156,600-km^2 Rufiji basin in Tanzania.

I have compiled all the available sediment yields from Kenyan catchments to examine Wilson's contention that land use is the most important factor affecting regional and global variations in sediment yield. The data show clearly the overwhelming influence of land use on any attempt to construct general curves. Under a single land use, however, it is possible to recognize the effects of runoff and topography. The effect of catchment size on sediment yield does not seem to make any important difference to the results that are being stressed here.

After demonstrating the magnitude of the land-use effect, I estimate the long-term geologic rate of erosion in Kenya, and discuss the major sources contributing to accelerated soil loss. The report also considers briefly the frequency of river discharges that transport sediment. The study supplements the small body of data on African sediment yields.

PHYSICAL GEOGRAPHY OF KENYA

Geology

A generalized geologic map of Kenya is given in Fig. 1. The following account is based upon the review by Baker (1970). Precambrian rocks outcrop over large areas of the country. In western Kenya, these rocks consist mainly of folded volcanics and sediments intruded by numerous granitic bosses and batholiths. In eastern Kenya and in the area immediately east of the Rift Valley the Precambrian rocks consist of folded metasediments and granitic intrusions. After considerable tectonic activity throughout the Precambrian, the region has since behaved as a cratonic block characterized by broad warping and epeirogenic movements of the crust during the late Cainozoic. In the eastern quarter of the country, marine sediments of Mesozoic and Cainozoic ages cover the Precambrian basement.

Extensive faulting and volcanism began in the Miocene, and continued through the Pleistocene as the Rift Valley was formed. Faults with displacements between 1700 and 3000 m formed single and multiple escarpments bounding the Rift Valley. Volcanism during the middle and late Cainozoic generated great thicknesses of basic and intermediate lavas and pyroclastic rocks which form extensive plateaux and mountains in and around the margins of the Rift Valley.

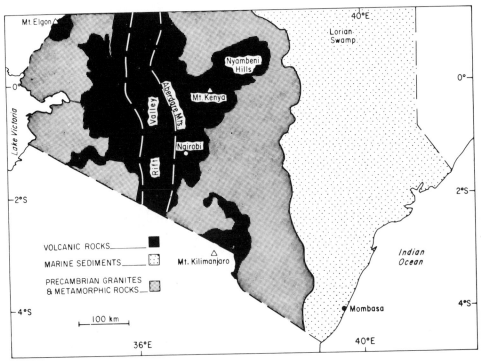

Fig. 1. Geologic map of central and southern Kenya. Source: Government of Kenya (1970).

The rocks produced by the history outlined above can be classified into the three groups shown in Fig. 1 on the basis of their general elevation, the manner in which they weather and give rise to landforms, and on their general pedologic and hydrologic properties.

Topography and drainage

The topography and drainage pattern of Kenya strongly reflect the geologic conditions outlined above. Most of eastern Kenya is covered by rolling topography, which rises from about 250—300 m at a range of coastal hills to 900 m high mountain ranges inland. This metamorphic region is interrupted by occasional volcanic ridges and hills. In the centre of the country volcanic plateaux range between 1200 and 2700 m on either side of the Rift Valley. Volcanic mountains, such as Mount Kenya (5200 m), the Aberdares (4250 m), Mount Elgon (4250 m) and the Nyambeni Range (2150 m) rise above these plateaux and form the major drainage divides of the country. The Rift Valley cuts through these volcanic highlands as a 50 km wide trough about 1000 m deep. To the west of the Rift Valley, the volcanic plateau slopes down to a lowland draining to Lake Victoria at an elevation of 1150 m. Steep, dissected terrain developed on Precambrian rocks lies north and south of this lowland.

All perennial streams originate in the volcanic highlands of the country or in the Precambrian mountains in the wet lake region. West of the Rift Valley, all streams drain to Lake Victoria. East of the Rift Valley, two major stream systems, the Tana and Athi-Galana, flow from the volcanic highlands across semi-arid terrain to the Indian Ocean. The lower areas of these drainage basins contribute runoff only during short wet seasons. A third major drainage system, the Ewaso (Uaso) Nyiro, drains from Mount Kenya, the Nyambeni Hills, and the northeast part of the Aberdare Mountains, then flows across an arid lowland to the inland Lorian Swamp. The Rift Valley is also an area of internal drainage. Details of the drainage network are provided by the national atlas (Government of Kenya, 1970).

Climate

The pattern of mean annual rainfall (Fig. 2) is closely correlated with topography. Values range from 250 mm in the northeastern region to more than 3500 mm high on the southeastern flank of Mt. Kenya. Most rain occurs in two seasons: March—May and October—December.

Mean daily air temperatures also vary with altitude. In the coolest month (usually July or August) temperatures average about 24°C at sea level and in the arid lowlands, and less than 6°C high on the volcanic mountains. In the hottest season (February—March) the variation is from 35°C down to about 12°C.

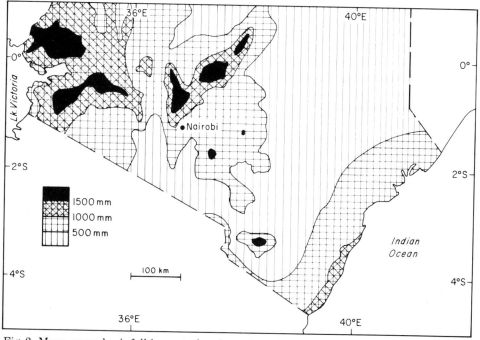

Fig. 2. Mean annual rainfall in central and southern Kenya. Source: Government of Kenya (1970).

The distribution of runoff strongly reflects the rainfall pattern. Mean annual runoff varies from over 2000 mm/yr. on the upper slopes of Mt. Kenya to virtually zero in the dry lowlands in the northeastern part of the study area. Mean annual runoff (Q) is approximately related to mean annual rainfall (P) by the expression:

$$Q = 0.000033P^{2.27}$$

where both quantities are expressed in millimetres and P ranges from 500 mm to at least 3000 mm.

Soils

The distribution of soil types reflects the patterns of geology, topography and climate. The volcanic highlands are covered with strongly leached dark-red and dark-brown loamy soils. Under their original forest cover these soils have moderate to high amounts of organic material (2—10% carbon) in the A-horizon, well-aggregated structure, and high infiltration capacities; their resistance to erosion is high. When these soils are cultivated their organic content, infiltration capacities, and resistance to erosion are substantially decreased.

Around the lower slopes of the volcanic and Precambrian highlands dark-brown and black loams, clays and clay loams occur on extensive footslopes with impeded drainage. Planosols cover the interfluves, and rendzinas, phaeozems and vertisols cover the middle and lower footslopes. The last two soil types have deep cracks and absorb large amounts of water during the early part of each rainy season. As the clays swell, however, the volume of storm runoff increases sharply and significant erosion occurs even on gradients of 0.01—0.02 (Dunne, 1977; Dunne et al., 1978b).

The gently sloping topography on the Precambrian lowlands of eastern Kenya is mantled with sandy loams and sandy clay loams, ranging in depth from 0.1 to more than 1.0 m. The topsoils contain little organic material (<1% carbon) and are poorly aggregated. The subsoils are blocky and some contain caliche.

Vegetation and land use

Much of the natural vegetation of Kenya has been removed to promote cultivation, pasture growth, or the manufacture of charcoal. Above an altitude of 2000—2100 m on the volcanic highlands dense forests survive with a thick ground cover of ferns, shrubs and dead vegetation. Below this altitude only remnants of the original woodland survive, mainly on the steepest slopes and on rocky areas. The woodland has a canopy cover of about 90% and a ground cover of shrubs and herbs. Most of the woodland has been replaced by hand-cultivated smallholdings covered by annual crops such as maize, pyrethrum and vegetables, or by perennial crops such as tea, coffee and bananas. The gradients of these small holdings vary from horizontal on narrow ridgetops and valley bottoms to about 70% on hillsides. Very little soil conservation is practised, though there has been a small, recent increase in the use of terracing and strip cropping. This zone also contains some large, commercial estates which grow tea, coffee, wheat, sugar and sisal. The estates tend to be on the best agricultural land, on the gentlest slopes, and have the greatest amount of capital available for soil conservation. Although some soil conservation works such as terracing, strip cropping and mulching are used on these lands, they also contribute significantly to the sediment yields of their regions.

In the areas with less than 750 mm of rainfall, agriculture is marginal and is mixed with pastoralism. Crops such as maize, millet and vegetables are grown, often on the steepest, highest slopes of an area, where rainfall is a little greater than the regional average. The areas which are not cultivated tend to be heavily grazed and browsed grassland and bushland. The large regions of Kenya with rainfall of less than 500 mm are generally occupied by nomadic pastoralists who rear cattle, sheep, goats and camels on grasslands and bushed grassland. Wild herbivores are present in large numbers in some of these subhumid areas.

Further details of the vegetation and land use of the country are given by Pratt et al. (1966) and Government of Kenya (1970).

THE STUDY

During the period 1948—1968, sediment concentrations were measured by the depth-integrating method at a large number of stations throughout the southern half of the country (the only portion containing perennial streams). At 61 stations the data were sufficient to allow construction of sediment rating curves (Dunne, 1974). Daily discharge records for the same stations could then be used to calculate sediment yields for basins over a wide range of geologic, topographic and climatic conditions. The discharge records are of variable but generally high quality.

For most stations, flow data are available for the period 1960—1970 or for 1956—1970. For a few stations only 3—5 yr. of discharge records are available. The flow records do not always coincide with the period of sediment sampling. This lack of comparability causes some uncertainties in the interpretation of results, but the overall picture is a coherent one and sheds new light on sediment yields from both disturbed and undisturbed areas of a tropical country.

During the period 1948—1968 there was no consistent change in the sediment rating curve from any station except for one large sand-bedded stream in a grazing region (Dunne, 1976, fig. 2). New sediment measurements were made by G.S. Ongweny of the University of Nairobi during 1974—1977 for one of the rivers included in this study. The measurements show that during the period since 1968 there has been a shift in the sediment rating curve for the Thiba River, which drains the southern slopes of Mt. Kenya. At another station where the original sediment rating curve was known to be inadequate, Dunne and Ongweny (1975) made an estimate of the sediment yield which is borne out by Ongweny's (1977) measurements.

The bedload transport rate could not be measured, but field observations suggest that this contribution is relatively small. In the volcanic highlands the river channels are steep and are floored by bedrock or by a thin layer of lava cobbles and boulders which can be moved only in exceptional floods. The soils of the region consist mainly of sandy clay loams which travel in suspension after entering a river channel. In the semi-arid regions where pelitic metamorphic rocks are exposed, sand constitutes a larger fraction of the load and in the larger lowland rivers I have observed bedload transport. At present, there is no way of assessing its relative importance in these streams, although other geomorphologists have concluded that in most low-land rivers the bedload contribution to total load is less than 10% [see Judson and Ritter (1964) and Gregory and Walling (1973) for summaries]. This value is less than other uncertainties in the data due to the variability of run-off and sediment concentrations. I have therefore ignored bedload transport.

RESULTS

Mean annual sediment yields were calculated for each basin and plotted against mean annual runoff in Fig. 3. In a few catchments most of the runoff is supplied by the higher portion of a catchment while the intermediate or lower portions contribute most of the sediment yield. My field observations indicate, however, that this complication does not distort the general pattern of the data to be discussed here.

In each basin the dominant land use was classified into one of the following categories: (1) completely forested; (2) forest covers 51—100% of the basin and the remainder is cultivated; (3) agricultural land occupies more than half of the catchment while the remainder is forested; and (4) rangeland. This crude classification was made by measuring the extent of forests shown on 1:250,000-scale maps made during the sediment sampling period, by examining aerial photographs, and by field inspection of each catchment to assess the most widespread land use. In a few catchments in north central Kenya where cultivation and grazing had expanded somewhat since the period of sediment sampling, the earlier land use was used. A more quantitative treatment of this variable is not possible with the agricultural census statistics that are currently available.

The lines in Fig. 3 separate basins with different vegetation covers. The separation is not complete, of course, because of variations in hillslope gradient, degree of dominance of the characteristic land use, record duration, quality of the sediment rating curves and other characteristics. Even with such a coarse grouping of catchments according to their dominant use, however, a pattern is evident. Forested catchments lose 20—30 t km^{-2} yr.$^{-1}$. The

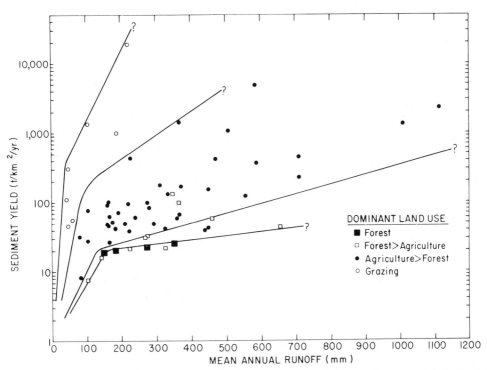

Fig. 3. Mean annual sediment yield and mean annual runoff for catchments with indicated dominant land uses.

sediment yields of agricultural lands vary enormously with runoff. At one end of the scale, a basin with flat topography developed on permeable lavas and with sparse rainfall loses approximately 10 t km^{-2} yr.$^{-1}$. In the wettest, steepest cultivated catchments the soil loss exceeds 4000 t km^{-2} yr.$^{-1}$. The great variability among agricultural basins results from differences in the type of agriculture and the proportion of the catchment that is cultivated, as well as differences in climate, soil erodibility and other physical properties. The sediment yields from rangeland catchments are also variable; the driest catchments lose less than 100 t km^{-2} yr.$^{-1}$, while $2 \cdot 10^4$ t km^{-2} yr.$^{-1}$ are exported from the wettest, steepest grazed catchment.

For a single land-use type, there is a general increase of sediment yield with runoff. Higher runoff is associated with heavier rainfall and therefore with greater kinetic energy for hillslope erosion and stream transport of the eroded sediment. There are striking differences, however, in the rates at which sediment yield increases with runoff. As the average density of cover decreases from forest to rangelands, the sediment loss becomes more sensitive to variations in runoff. The data also suggest that catchment steepness is an important control, although it is correlated to some degree with rainfall and runoff.

Multiple regression analysis was used to define the effects of the three variables thought to control sediment yields. The literature on erosion (e.g.,

Wischmeier and Smith, 1965; Kirkby, 1969) at both the hillslope and catchment scales suggests that sediment yields should vary with the multiplicative effects of: rainfall or runoff, topography, and cover density and other aspects of land use. Differences in soil erodibility are important in some regions. In this survey of Kenyan data, it was possible to obtain indices of land use, runoff and topography. For the last variable the relief ratio (basin relief/mainstream length), defined by Schumm (1954), was used. The analysis was confined to independent catchments. If sediment yields were available for a downstream station and for a station on a tributary, only the value from the latter was used. This requirement excluded 15 stations from the 61 shown in Fig. 3. The remaining 46 were distributed among the land-use categories as follows: forest 4; forest dominant 9; agriculture dominant 28; and rangeland 5.

The data were combined into the following form:

$$SY = UQ^a S^b \tag{1}$$

where SY = mean annual sediment yield (t km^{-2} yr.$^{-1}$), Q = mean annual runoff (mm), S = relief ratio (dimensionless), and U indicates land use expressed in terms of the four categories in Fig. 3. In the logarithmic form of eq. 1 used in the regression analysis, $\log U$ is a dummy variable, which takes on values of zero and one according to the land use (Rao and Miller, 1971). The complete form of the equation used is, therefore:

$$\log(SY) = \delta_0 + \delta_1 U_1 + \delta_2 U_2 + \delta_3 U_3 + a\log Q + \alpha_1(U_1 \log Q) + \alpha_2(U_2 \log Q)$$
$$+ \alpha_3(U_3 \log Q) + b\log S + \beta_1(U_1 \log S) + \beta_2(U_2 \log S) + \beta_3(U_3 \log S)$$

When the catchment is completely forested, $U_1 = U_2 = U_3 = 0$. Where forest covers 51–100% of the catchment, $U_1 = 1$; $U_2 = U_3 = 0$. If forest covers 0–50% of a basin that is otherwise cultivated, $U_2 = 1$; $U_1 = U_3 = 0$. For rangelands, $U_3 = 1$, and the other values are zero. The symbols a, b, α_i, β_i and δ_i are regression parameters.

When this regression model was used on the field data, the following equation resulted:

$$\log(SY) = 0.19 - 1.19 U_1 - 1.04 U_2 + 0.44 U_3 + 0.46\log Q + 0.82(U_1 \log Q)$$
$$+ 1.02(U_2 \log Q) + 1.98(U_3 \log Q) - 0.03\log S + 0.50(U_1 \log S)$$
$$+ 0.54(U_2 \log S) + 1.15(U_3 \log S) \qquad (R^2 = 0.77) \tag{2}$$

Use of the same model without the land use variable produced the equation:

$$\log(SY) = -0.06 + 0.69\log Q - 0.23\log S \qquad (R^2 = 0.11) \tag{3}$$

An F-test using the residual sums of squares from the two equations indicated that the reduction in explained variance due to the inclusion of the land-use variable is significant at the 0.01 level. Land use, therefore, is an extremely important variable that cannot be ignored in the construction of global or regional analyses of erosion rates.

The regression analysis also revealed several other important details about the controls of erosion and their interaction. If the four completely forested catchments are examined alone (i.e., if the dummy variables are all set to zero in eq. 2, the result is:

$$SY = 1.56 \, Q^{0.46} S^{-0.03} \qquad\qquad (R^2 = 0.98) \quad (4)$$

The inclusion of the topographic variable does not significantly reduce the explained variance and the equation simplifies to:

$$SY = 2.67 \, Q^{0.38} \qquad\qquad (R^2 = 0.98) \quad (5)$$

Although it is based upon only four small catchments with less than 400 mm of runoff, this equation is similar to one calculated by Douglas (1967) for subtropical forested catchments in Australia (see Fig. 4).

Eq. 5 indicates that sediment yields of forested catchments increase only slowly with rainfall and runoff and not significantly with topographic steepness. This latter statement presumably cannot be extended to some of the highest and steepest catchments in the world which lie at active plate margins (Ahnert, 1970; Janda, 1972; Li, 1975), but through the range of relief ratio encountered in the forest lands of Kenya ($0.01 \leqslant S \leqslant 0.09$) the effect of topography is not apparent. This conclusion is in general agreement with a number of plot and field studies which suggest that if the vegetation cover is sufficiently dense, sediment yield will be low even on steep hillsides.

A similar analysis of data from the other land use categories in eq. 2 produces the following relationships:

For forest > agriculture ($n = 9$):

$$SY = 0.10 Q^{1.28} S^{0.47} \qquad\qquad (R^2 = 0.76) \quad (6)$$

For agriculture > forest ($n = 28$):

$$SY = 0.14 Q^{1.48} S^{0.51} \qquad\qquad (R^2 = 0.74) \quad (7)$$

For rangeland ($n = 5$):

$$SY = 4.26 Q^{2.17} S^{1.12} \qquad\qquad (R^2 = 0.87) \quad (8)$$

Although the sample sizes for forests and rangelands approach the number of independent variables and therefore produce misleadingly high correlation coefficients, the results indicate that as the density of cover decreases the effects of increasing runoff and topographic steepness become more important. Whereas under forest the sediment yield is proportional only to the 0.38 power of runoff and the zeroth power of relief ratio, the corresponding exponents in the case of rangeland are 2.17 and 1.12. The interactions between land use (or cover density) and the other two variables are represented by the positive coefficients of the $(U_i \log R)$ and $(U_i \log S)$ terms in eq. 2.

THE EFFECT OF CATCHMENT SIZE

Brune (1948) showed that sediment yield per unit area generally decreased with catchment size for basins in the upper Mississippi Valley. The same effect has been recognized in several other studies (e.g., Maner, 1958; Roehl, 1962; Dendy and Bolton, 1976). The ratio of basin sediment yield to the amount of soil eroded from hillsides is called the sediment delivery ratio. For fixed conditions of climate and land use this fraction commonly decreases with catchment size because small upstream basins are generally steep and have limited floodplains in which sediment can be stored. It is questionable whether this concept applies to the evolution of catchments over geologic time. However, there is considerable evidence (e.g., Happ et al., 1940; Haggett, 1961; Trimble, 1975; Dietrich and Dunne, 1978) that when some disturbance, such as land use or catastrophic landsliding, increases sediment yields over the long-term background value, a portion of the sediment is stored temporarily on foot-slopes and in valley-floor deposits.

Brune estimated that the sediment yield per unit area decreased in proportion to the −0.15 power of basin area. Langbein and Schumm (1958) used this factor to correct the yields of small upland catchments to the equivalent yield from basins of 3885 km^2 (1500 mi.^{2}). Wilson (1973) suggested that this exponent should be closer to −0.30, but used the lower value for comparability with the earlier results. He corrected all basin yields to their equivalents at 259 km^2 (100 mi.^{2}).

Most of the Kenyan catchments referred to in this paper are deeply incised into volcanic highlands. The hillslopes are steep and straight; the valley floors are steep and narrow with limited floodplain development, and most river channels are floored by bedrock. Extensive storage of sediment in the manner described by Haggett (1961) and Trimble (1975) is not possible in catchments with areas less than 100 km^2 and in most of the catchments smaller than 2000 km^2. In these basins (which constitute 87% of the sample studied) the sediment yield is an accurate reflection of average hillslope erosion. It is unlikely, therefore, that catchment size has any important effect on the results of the earlier statistical analysis, except in the sense that relief ratio is weakly, but significantly, correlated with drainage area:

$$S = 0.11A^{-0.23}, \qquad r^2 = 0.31, \qquad p < 0.005$$

Following the example of earlier authors for purposes of comparison, I corrected all sediment yields to the equivalents at 259 km^2 and repeated the statistical analysis described above. The results are essentially the same as those of the earlier analysis except that the exponents of the relief ratio are lower because the correction for drainage area removes a portion of the effect of steepness. The regression equations are:

For forest ($n = 4$, i.e. almost equal to the number of variables):

$$SY_{259} = 4.23Q^{0.27} \qquad\qquad (R^2 = 0.86) \qquad (5a)$$

For forest > agriculture ($n = 9$):

$$SY_{259} = 0.15Q^{1.14}S^{0.34} \qquad\qquad (R^2 = 0.71) \qquad (6a)$$

For agriculture > forest ($n = 28$):

$$SY_{259} = 0.06Q^{1.52}S^{0.33} \qquad\qquad (R^2 = 0.74) \qquad (7a)$$

For rangeland ($n = 5$):

$$SY_{259} = 1.33Q^{2.15}S^{0.71} \qquad\qquad (R^2 = 0.85) \qquad (8a)$$

The summary regression equation including the dummy variables is:

$$\log(SY_{259}) = 0.11 - 0.93U_1 - 1.33U_2 + 0.01U_3 + 0.44\log Q + 0.70(U_1\log Q)$$
$$+ 1.08(U_2\log Q) + 1.71(U_3\log Q) - 0.10\log S + 0.44(U_1\log S)$$
$$+ 0.43(U_2\log S) + 0.81(U_3\log S) \qquad (R^2 = 0.76) \qquad (2a)$$

The analysis that includes corrections for catchment size does not yield more precise results than the original equation. On the basis of the earlier discussion of Kenyan physiography it seems better not to include the correction until data become available on the effects of catchment size.

GEOLOGIC RATE OF EROSION

Because of the overwhelming influence of land use on the sediment yields of these tropical catchments it is difficult to find areas in which the geologic rate of erosion can be evaluated. The same problem has complicated the efforts of others (e.g., Langbein and Schumm, 1958; Fournier, 1960; Schumm, 1963; Wilson, 1973).

Douglas (1967) reported sediment yields from catchments in eastern Australia that were "selected to avoid as much human disturbance as possible". The regression line which Douglas derived from these data is plotted in Fig. 4. It fits the data from the four forested catchments in Kenya very closely. The range of yields from the small Kenyan sample ($\sim 20{-}30$ t km^{-2} yr.$^{-1}$) therefore, seems to be representative of undisturbed, humid catchments under tropical forest in tectonically stable areas. Dunne et al. (1978a) have also shown that such a yield is compatible with the evidence from erosion surfaces dating from the early and middle Cainozoic when most of Kenya was covered with forest or woodland.

It is more difficult to estimate the background rate in the drier areas with a cover of grass and bush. The evidence from late Cainozoic erosion surfaces in southern Kenya suggests that as the climate of the region became more arid during the Pleistocene the long-term erosion rate increased, and was roughly equivalent to sediment yields of $75{-}200$ t km^{-2} yr.$^{-1}$ (Dunne et al., 1978a). The curve of Douglas (1967) for undisturbed lands (Fig. 4) shows a peak sediment yield of 115 t km^{-2} yr.$^{-1}$ in semi-arid regions. Five sediment-yield values (three measured from whole basins and two calculated by subtraction of

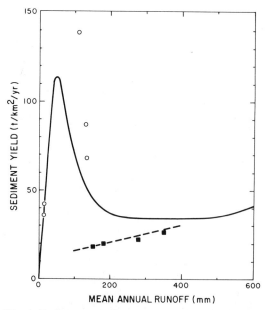

Fig. 4. Sediment yield and mean annual runoff from nine Kenyan catchments with little or no human disturbance during the period of sediment sampling. The *open circles* indicate catchments with a cover of dry woodland, bush and grass; the *solid squares* represent catchments that are completely forested. The *dashed line* is Douglas' (1967) regression line for forested catchments in eastern Australia. The *solid curve* is Douglas' (1967) summary of sediment yields for undisturbed catchments.

yields measured at two stations along a river) are available for Kenyan basins that have a cover of dry woodland, bush and grass and that were grazed only lightly at the time of the sediment measurements. These bush-covered catchments are different from the rangeland catchments included in the earlier statistical analysis. The latter are heavily grazed by large nomadic herds of domestic livestock and wild herbivores, while the five bush-covered catchments are managed on a conservative commercial basis and the entry of wild game is discouraged. The grazing pressure is probably not much greater than that due to wild herbivores alone in earlier times. The sediment yields of these lands vary from 50 to 140 t km^{-2} yr.$^{-1}$, and agree quite closely with the Douglas curve for undisturbed lands in Fig. 4.

SOURCES OF SEDIMENT

In the forested regions landsliding is rare in spite of the heavy rainfall and the steepness of the hillsides. Soil creep appears to be the only significant mechanism that contributes sediment to stream channels. Sediment yields in the range 20—30 t km^{-2} yr.$^{-1}$ are equivalent to average creep velocities of roughly 2 mm yr.$^{-1}$ with the soil thickness ($\sim$1 m) and drainage density ($\leqslant$ 5 km km^{-2}) that characterize the montane zones of Kenya. Such velocities seem reasonable in the light of field measurements in moist, tropical environments by other workers. Eyles and Ho (1970) measured a velocity of 3.2 mm yr.$^{-1}$ in a 0.4 m deep soil on a gradient of 0.18, and Lewis (1976) documented velocities of 1.8 and 2.3 mm yr.$^{-1}$ in 0.4 m deep soils on gradients of 0.32—0.35.

In cultivated catchments the sediment sources are more diverse. It is important to identify them separately and to indicate their relative importance in order that soil conservation measures can be planned accordingly. The sediment yield of the agricultural or grazed portion of each catchment was separated from that of the forested portion by the following method. The area of forest was measured on 1:50,000- or 1:250,000-scale maps. In catchments west of the Rift Valley, an average sediment yield of 20 t km^{-2} yr.$^{-1}$ (close to the only measured value of 18.3 t km^{-2} yr.$^{-1}$ from that region) was assigned to the forest. East of the Rift Valley, where rainfall intensities are generally higher (Taylor and Lawes, 1971, fig. 1), an average of 30 t km^{-2} yr.$^{-1}$ was used. The statistical analysis presented above and comparison with other published results suggest that sediment yields from Kenyan forests are unlikely to vary greatly from these values.

When the contribution of the forest has been subtracted from the total sediment yield, soil loss from the agricultural and grazed subcatchments can be calculated, and plotted in the form shown in Fig. 3. The results are similar except that there are only three fields on the graph: forest, agriculture and rangelands. The upper envelopes for cultivated and grazed areas are raised somewhat, as several agricultural areas, for example, have sediment yields that range from 1000 to 5000 t km^{-2} yr.$^{-1}$. Because the runoff values had to be calculated for each subcatchment by subtraction, and because it was not possible to define a relief ratio for the lower portion of a catchment, I did not subject the data to a regression analysis.

The frequency distributions of sediment yields from the subcatchments with various land uses are shown in Fig. 5. The median yield from cultivated regions is not particularly high (90 t km^{-2} yr.$^{-1}$), but the largest decile of the distribution includes yields from 1000 to 5000 t km^{-2} yr.$^{-1}$.

A great deal of research (e.g., Roose, 1975) has focused attention on soil erosion from cultivated fields, and the problem is obvious from field inspection during tropical rainstorms. Signs of intense sheetwash and rilling are visible on some cultivated areas. It is not yet clear, however, how much of the sediment from an agricultural catchment originates from the fields. The 5,000—20,000-m^2 smallholdings which characterize the Kenyan highlands have a cover of annual and perennial crops, weeds and grasses. The plots are bare at the beginning of each rainy season, but their rough surfaces inhibit runoff and soil erosion.

Runoff and erosion in the agricultural areas are most intense on dirt roads and tracks. There are no direct measurements of the sediment contribution from this source, but it is probably high. The network of roads and tracks is dense and is connected to the stream network. The roadbed and adjoining

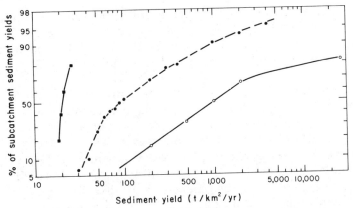

Fig. 5. Cumulative frequency of sediment yields from catchments and subcatchments under three land uses. *Squares* represent forest; *solid circles* indicate agricultural lands, and *open circles* represent rangelands. The yields for cultivated and grazed subcatchments were calculated by subtracting the soil loss from the forested portion of a basin from the total catchment yield.

banks expose dense, compacted subsoil with a low infiltration capacity. These thoroughfares are heavily travelled and constantly disturbed by vehicles, animals and pedestrians. Some of the sediment eroded from the roads is washed onto adjoining fields, but the larger part is conveyed to stream channels by roadside ditches.

A few measurements from rural roads and construction sites in North America indicate the probable magnitude of soil loss on Kenyan roads (Table I). Although they are highly variable, these data suggest a lower limit of about 10,000 t km^{-2} yr.$^{-1}$.

TABLE I

Measured rates of soil loss from rural roads and construction sites in North America

Type of surface	Location	Rainfall (mm yr.$^{-1}$)	Rainfall erosivity index	Soil loss (10^3 t km^{-2} yr.$^{-1}$)	Source
Bulldozed skid trails; gradients < 0.10	N Georgia	1200—1500	300	9.9	Hornbeck and Reinhart (1964)
Logging roads	W Oregon	2400	—	23	Swanson and Dyrness (1975)
Building sites	Maryland	1100	200	20—50	Wolman and Schick (1967)
Road cuts; gradients 0.47—1.0	N Georgia	1200—1500	300	20—60	Diseker and Richardson (1961, 1962)

The values of the rainfall erosivity index are taken from U.S.S.C.S. (1975a).

   In an attempt to assess the approximate magnitude of this sediment contribution in Kenyan catchments the areas of a sample of roads were measured
in the Chania River basin, located on the deeply dissected backslope of the
Rift Valley escarpment 40 km north of Nairobi. The agricultural zone has a
density of at least 0.80 km (4800 m^2) of main rural roads and 1.4 km (7000 m^2)
of secondary road per square kilometer, as well as a dense network of footpaths that are not shown on the 1:50,000 maps used for the length measurements. Road gradients ranged from 0.02 to 0.10. The rainfall erosivity (RE)
value was estimated from a regression between this factor and the median
annual rainfall constructed with tropical data from the Hawaiian Islands
(U.S.S.C.S., 1975b) and Tanzania (Rapp et al., 1972b). The resulting equation
is:

$$RE = 2.0\,P(mm)^{0.71}, \qquad n = 41; \qquad r^2 = 0.88; \qquad p < 0.005$$

   The RE value estimated for the agricultural zone of the Chania basin is 300,
which is similar to the figures for the eastern U.S.A. in Table I. A soil loss rate
of 10^4—$2 \cdot 10^4$ t yr.$^{-1}$ per square kilometer of road seems to be appropriate,
therefore. Such a rate would generate 118—236 t yr.$^{-1}$ per square kilometer
of catchment. Since the sediment yield of this agricultural zone is 720 t km^{-2}
yr.$^{-1}$, the roads probably contribute between 15 and 35% of the total basin
sediment yield. Footpaths cover at least as large an area as the roads and they
cross steeper land (gradients $\leqslant 0.25$). It is likely, therefore, that their sediment contribution is of the same order as that from roads, although many of
them drain onto fields.
   A similar calculation for a region of large commercial grain farms on flat
topography in western Kenya, 250 km NW of Nairobi, predicts an annual
sediment contribution from roads of approximately 50 t km^{-2} of catchment
in a region where the sediment yields from agricultural lands range from 23
to 80 t km^{-2} yr.$^{-1}$. The contribution from roads through rangelands is insignificant. Dunne et al. (1978b) have used measurements of tree-root exposure
to document high rates of soil loss from heavily grazed hillsides in Kenya.
   The tentative conclusion from these very rough calculations is that erosion
from rural roads and footpaths may generate a large part of the sediment
yield of catchments and may cause an over-estimation of the degradation of
cultivated lands. The magnitude of this contribution increases with population density as the network of roads and footpaths becomes denser and is
disturbed more intensely. The solutions to this erosion problem, of course,
are quite different from those required on the cultivated fields.

SEDIMENT TRANSPORT AND FLOW FREQUENCY

   Wolman and Miller (1960) examined the relative amounts of sediment
transported by flows of different frequencies. They concluded that events of
moderate frequency, occurring at least once or twice per year, carry the
largest portion of the sediment load, and that 90% of the sediment is trans-

ported by discharges which occur at least once every five years.

The Kenyan catchments lose an average of 41% of their sediment in the highest one percent of daily flows, but this average varies between land use types, as indicated in Table II. The highest ten percent of flows carry an average of 80% of the mean annual yield. These indices of the relative transporting efficacy of various flows do not show any country-wide correlation with basin characteristics such as area, annual runoff, flow variability, or sediment yield. Within a single large basin, however, the relative importance of the highest flows decreases with increasing catchment size (Fig. 6). As drainage area increases downstream the river retains the competence and capacity to transport large amounts of sediment on a greater number of days per year. The rapid discharge fluctuations in the upland catchments are also evened

TABLE II

Percentage of the long-term mean annual sediment yield carried by the highest one percent and highest ten percent of flows from catchments with various land uses

Land use	Highest one percent of flows		Highest ten percent of flows	
	range	median	range	median
Forest	10—70	35	40—90	60
Forest > agriculture	5—60	35	40—90	60
Agriculture > forest	10—80	43	40—90	81
Grazing	20—80	75	60—92	95

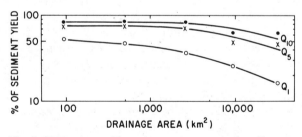

Fig. 6. Percentage of long-term mean annual sediment yield carried by flows greater than certain values in catchments of various size in the Tana River basin. The subscripts on the Q-values represent percentiles on the long-term flow duration curve.

out by channel storage, so that the effects of the most catastrophic sediment-transporting flows are damped out.

The highest one percent or ten percent of flows mentioned above do not refer to a regularly occurring short period in each year. During late 1961 and early 1962, Kenya suffered catastrophic floods from a period of heavy rainfall following a severe drought. The recurrence interval of the resulting peak discharge is unknown, but in most catchments it was the largest in records which vary in length from 15 to 33 yr.

To estimate the amount of sediment carried by the highest daily discharge, most sediment rating curves have to be extrapolated, although the most important rivers were sampled near the peak of the flood. In the semi-arid grazing lands of northern Kenya, the effects of this catastrophe were particularly severe, and in some areas gullies were initiated where none had been present before. In the 15,300-km^2 Uaso Nyiro basin above Archer's Post, 100 km north of Mt. Kenya, the equivalent of 4.75 yr. of sediment transport was accomplished during the month of highest flows. This yield amounted to 22% of all sediment lost during the 21-yr. record.

SUMMARY

Sediment yields in Kenyan catchments that have suffered little or no disturbance are low throughout a range of climate. Although few catchments remain undisturbed, their sediment yields are closer to levels proposed by Douglas (1967) than to the values of Langbein and Schumm (1958) and Fournier (1960), confirming Douglas' assertion that frequently quoted yields do not represent long-term geologic conditions.

The most dramatic differences in basin sediment yield result from differences in land use, as Wilson (1973) hypothesized. For the major land-use types in Kenya, it is possible to construct general relations between sediment yield, runoff and topographic steepness of the kind that have been sought by several hydrologists. The yields become more sensitive to variations in runoff and topography as the vegetation thins from forest through agricultural crops to rangeland. The resulting regression equations are still not useful for prediction elsewhere, as the residuals are large and suggest the influence of other variables such as the duration of record, type of agriculture, and regional variations of lithology and soil. It seems unlikely that any relationship will give a precise prediction of yields for a large, variable region. However, the equations extend the attempts of Langbein and Schumm, Fournier, and others to define the interactions between the major controls of soil loss on a regional scale.

ACKNOWLEDGEMENTS

This study was supported by the McGill—Rockefeller Program, the American Philosophical Society, and the Explorers' Club. Records of runoff and sediment sampling were made available by the Ministry of Water Development, Government of Kenya. George S. Ongweny and Vincent Dunne assisted with the fieldwork. David Dethier and Lee Wilson helped me to improve an earlier draft of the manuscript, while Potluri Rao gave advice on statistical analysis.

REFERENCES

Ahnert, F., 1970. Functional relationships between denudation, relief, and uplift in large mid-latitude drainage basins. Am. J. Sci., 268: 243—263.

Baker, B.H., 1970. Tectonics of Kenya Rift Valley. Ph.D. Thesis, University of East Africa, Nairobi.

Brune, G., 1948. Rates of sediment product᾽ ı in midwestern United States. U.S. Soil Conserv. Serv., Tech. Publ., 65, 40 pp.

Dendy, F.E. and Bolton, G.C., 1976. Sediment yield—runoff—drainage area relationships in the United States. J. Soil Water Conserv., 31: 264—266.

Dietrich, W.E. and Dunne, T., 1978. Sediment budget for a small catchment in mountainous terrain. Z. Geomorphol., Suppl., 29: 135—149.

Diseker, E.G. and Richardson, E.C., 1961. Roadside sediment production and control. Trans. Am. Soc. Agric. Eng., 4: 62—68.

Diseker, E.G. and Richardson, E.C., 1962. Erosion rates and control measures on highway cuts. Trans. Am. Soc. Agric. Eng., 5: 153—155.

Douglas, I., 1967. Man, vegetation, and the sediment yields of rivers. Nature (London), 215: 925—928.

Dunne, T., 1974. Suspended sediment data for the rivers of Kenya. Rep. to Ministry of Water Development, Nairobi, 108 pp.

Dunne, T., 1976. Evaluation of erosion conditions and trends. In: S.H. Kunkle and J.L. Thames (Editors), Guidelines for Watershed Management, F.A.O. Conservation Guide 1, U.N. Food and Agriculture Organization, Rome, pp. 53—83.

Dunne, T., 1977. Intensity and controls of soil erosion in Kajiado District, Kenya. Consult. Rep., Kenya Wildlife Management Project, Food and Agriculture Organization, Nairobi, 167 pp.

Dunne, T. and Ongweny, G.S.O., 1976. A new estimate of sediment yields in the Upper Tana catchment. Kenyan Geogr., 2: 20—38.

Dunne, T., Brunengo, M.J. and Dietrich, W.E., 1978a. Cainozoic erosion rates in Kenya estimated from erosion surfaces. (Unpublished.)

Dunne, T., Dietrich, W.E. and Brunengo, M.J., 1978b. Recent and past rates of erosion in semi-arid Kenya. Z. Geomorphol., Suppl., 29: 91—100.

Eyles, R.J. and Ho, R., 1970. Soil creep on a humid tropical slope. J. Trop. Geogr., 31: 40—42.

Fournier, F., 1960. Climat et érosion. Presses Universitaires de France, Paris, 201 pp.

Government of Kenya, 1970. Atlas of Kenya. Government Printer, Nairobi, 103 pp.

Gregory, K.J. and Walling, D.E., 1973. Drainage Basin Form and Process. Halsted, N.Y., 456 pp.

Haggett, P., 1961. Land use and sediment yield in an old plantation tract of the Sierra do Mar, Brazil. Geogr. J., 127: 50—62.

Happ, S.C., Rittenhouse, G. and Dobson, G.C., 1940. Some principles of accelerated stream and valley sedimentation. U.S. Dep. Agric., Tech. Bull. 695, 133 pp.

Holeman, J.N., 1968. The sediment yield of major rivers of the world. Water Resour. Res., 4: 737—748.

Hornbeck, J.W. and Reinhart, K.G., 1964. Water quality and soil erosion as affected by logging in steep terrain. J. Soil Water Conserv., 19: 23—27.

I.A.H.S. (International Association of Hydrological Sciences), 1974. Effects of man on the interface of the hydrological cycle with the physical environment. I.A.H.S. Publ. No. 113, 157 pp.

Janda, R.S., 1972. Testimony concerning stream sediment loads in northern California and southern Oregon presented in public hearings before the California Regional Water Quality Control Board, North Coast Region, Eureka, Calif., 25 pp.

Jansen, J.M.L. and Painter, R.B., 1974. Predicting sediment yields from climate and topography. J. Hydrol., 21: 371—380.

Judson, S. and Ritter, D.F., 1964. Rates of regional denudation in the United States. J. Geophys. Res., 69: 3395—3401.

Kirkby, M.J., 1969. Infiltration, throughflow and overland flow. In: R.J. Chorley (Editor), Water, Earth and Man, Methuen, London, pp. 215—228.

Langbein, W.B. and Schumm, S.A., 1958. Yield of sediment in relation to mean annual precipitation. Am. Geophys. Union Trans., 39: 1076—1084.

Lewis, L.A., 1976. Soil movement in the tropics: a general model. Z. Geomorphol., Suppl., 25: 132—144.

Li, Y., 1975. Denudation of Taiwan Island since the Pliocene epoch. Geology, 4: 105—107.

Maner, S.B., 1958. Factors affecting sediment delivery rates in the Red Hills physiographic area. Am. Geophys. Union Trans., 39: 669—675.

Ongweny, G.S., 1977. Problems of soil erosion and sedimentation in selected water catchment areas in Kenya with special reference to the Tana River. Aqua, 1: 85.

Pratt, D.J., Greenway, P.J. and Gwynne, M.D., 1966. A classification of East African rangeland, with an appendix on terminology. J. Appl. Ecol., 3: 369—382.

Rao, P. and Miller, R.L., 1971. Applied Econometrics. Wadsworth, Belmont, Calif., 235 pp.

Rapp, A., Murray-Rust, D.H., Christiansson, C. and Berry, L., 1972a. Soil erosion and sedimentation in four catchments near Dodoma, Tanzania. Geogr. Ann. Ser. A, 54: 255—318.

Rapp, A., Axelsson, V., Berry, L. and Murray-Rust, D.H., 1972b. Soil erosion and sediment transport in the Morogoro River catchment, Tanzania. Geogr. Ann. Ser. A, 54: 125—155.

Roehl, J.W., 1962. Sediment source areas, delivery ratios, and influencing morphological factors. Symp. on Continental Erosion, Bari, I.A.H.S., Publ. No. 59, pp. 202—213.

Roose, E.J., 1975. Erosion et ruissellement en Afrique de l'Ouest. ORSTROM, Abidjan, 72 pp.

Schumm, S.A., 1954. The relation of drainage basin relief to sediment loss. Symp. on Continental Erosion, Rome, I.A.H.S., Publ. No. 36(1), pp. 216—219.

Schumm, S.A., 1963. The disparity between present rates of denudation and orogeny. U.S. Geol. Surv., Prof. Pap. 454-H.

Swanson, F.J. and Dyrness, C.T., 1975. Impact of clearcutting and road construction on soil erosion by landslides in the West Cascade Range, Oregon. Geology, 3: 393—396.

Taylor, C.M. and Lawes, E.F., 1971. Rainfall intensity—duration—frequency data for stations in East Africa. East Afr. Meteorol. Dep., Tech. Mem. 17, 29 pp.

Temple, P.H. and Sundborg, A., 1972. The Rufiji River, Tanzania: hydrology and sediment transport. Geogr. Ann., Ser. A, 54: 345—368.

Trimble, S.W., 1974. Man-induced soil erosion on the Southern Piedmont, 1700—1970. Soil Conserv. Soc. Am., Spec. Publ., Ankeny, Iowa.

Trimble, S.W., 1975. Denudation studies: can we assume steady state? Science, 188: 1207—1208.

U.S.S.C.S. (U.S. Soil Conservation Service), 1975a. Procedure for computing sheet and rill erosion on project areas. U.S. Dep. Agric., Portland, Oreg., Tech. Release 51, 16 pp.

U.S.S.C.S. (U.S. Soil Conservation Service), 1975b. Guidelines for use of the universal soil-loss equation in Hawaii. U.S. Dep. Agric., Honolulu, Hawaii, Tech. Note 1, 40 pp.

Wilson, L., 1972. Seasonal sediment yield patterns of United States rivers. Water Resour. Res., 8: 1470—1479.

Wilson, L., 1973. Variations in mean annual sediment yield as a function of mean annual precipitation. Am. J. Sci., 273: 335—349.

Wischmeier, W.H. and Smith, D.D., 1965. Predicting rainfall—erosion losses from cropland east of the Rocky Mountains. U.S. Dep. Agric., Washington, D.C., Agric. Handb. 282, 47 pp.

Wolman, M.G. and Miller, J.P., 1960. Magnitude and frequency of forces in geomorphic processes. J. Geol., 68: 54—74.

Wolman, M.G. and Schick, A.P., 1967. Effects of construction on fluvial sediment, urban and suburban areas of Maryland. Water Resour. Res., 3: 451—464.

22

Reprinted from pages 23–25 and 41–47 of *Ecol. Monogr.* **40**:23–47 (1970), by permission of the publisher, Duke University Press

EFFECTS OF FOREST CUTTING AND HERBICIDE TREATMENT ON NUTRIENT BUDGETS IN THE HUBBARD BROOK WATERSHED-ECOSYSTEM[1]

Gene E. Likens[2], F. Herbert Bormann, Noye M. Johnson, D. W. Fisher, and Robert S. Pierce

(Accepted for publication July 31, 1969)

Abstract. All vegetation on Watershed 2 of the Hubbard Brook Experimental Forest was cut during November and December of 1965, and vegetation regrowth was inhibited for two years by periodic application of herbicides. Annual stream-flow was increased 33 cm or 39% the first year and 27 cm or 28% the second year above the values expected if the watershed were not deforested.

Large increases in streamwater concentration were observed for all major ions, except NH_4^+, $SO_4^=$ and HCO_3^-, approximately five months after the deforestation. Nitrate concentrations were 41-fold higher than the undisturbed condition the first year and 56-fold higher the second. The nitrate concentration in stream water has exceeded, almost continuously, the health levels recommended for drinking water. Sulfate was the only major ion in stream water that decreased in concentration after deforestation. An inverse relationship between sulfate and nitrate concentrations in stream water was observed in both undisturbed and deforested situations. Average streamwater concentrations increased by 417% for Ca^{++}, 408% for Mg^{++}, 1558% for K^+ and 177% for Na^+ during the two years subsequent to deforestation. Budgetary net losses from Watershed 2 in kg/ha-yr were about 142 for NO_3-N, 90 for Ca^{++}, 36 for K^+, 32 for SiO_2-Si, 24 for Al^{+++}, 18 for Mg^{++}, 17 for Na^+, 4 for Cl^-, and 0 for SO_4-S during 1967–68; whereas for an adjacent, undisturbed watershed (W6) net losses were 9.2 for Ca^{++}, 1.6 for K^+, 17 for SiO_2-Si, 3.1 for Al^{+++}, 2.6 for Mg^{++}, 7.0 for Na^+, 0.1 for Cl^-, and 3.3 for SO_4-S. Input of nitrate-nitrogen in precipitation normally exceeds the output in drainage water in the undisturbed ecosystems, and ammonium-nitrogen likewise accumulates in both the undisturbed and deforested ecosystems. Total gross export of dissolved solids, exclusive of organic matter, was about 75 metric tons/km[2] in 1966–67, and 97 metric tons/km[2] in 1967–68, or about 6 to 8 times greater than would be expected for an undisturbed watershed.

The greatly increased export of dissolved nutrients from the deforested ecosystem was due to an alteration of the nitrogen cycle within the ecosystem.

The drainage streams tributary to Hubbard Brook are normally acid, and as a result of deforestation the hydrogen ion content increased by 5–fold (from pH 5.1 to 4.3).

Streamwater temperatures after deforestation were higher than the undisturbed condition during both summer and winter. Also in contrast to the relatively constant temperature in the undisturbed streams, streamwater temperature after deforestation fluctuated 3–4°C during the day in summer.

Electrical conductivity increased about 6–fold in the stream water after deforestation and was much more variable.

Increased streamwater turbidity as a result of the deforestation was negligible, however the particulate matter output was increased about 4–fold. Whereas the particulate matter is normally 50% inorganic materials, after deforestation preliminary estimates indicate that the proportion of inorganic materials increased to 76% of the total particulates.

Supersaturation of dissolved oxygen in stream water from the experimental watersheds is common in all seasons except summer when stream discharge is low. The percent saturation is dependent upon flow rate in the streams.

Sulfate, hydrogen ion and nitrate are major constituents in the precipitation. It is suggested that the increase in average nitrate concentration in precipitation compared to data from 1955–56, as well as the consistent annual increase observed from 1964 to 1968, may be some measure of a general increase in air pollution.

TABLE OF CONTENTS

Introduction 24	Ions 32
The Hubbard Brook Ecosystem 25	*Ammonium and Nitrate* 32
Methods and Procedures 25	*Sulfate* 34
Hydrologic Parameters 25	*Chloride* 35
Precipitation Chemistry 26	*Calcium, Magnesium, Potassium, and Sodium* ... 36
Chemical Input Through Herbicide Application.. 27	*Aluminum* 38
Streamwater Parameters 28	*Dissolved Silica* 38
Temperature 28	*Bicarbonate* 38
Dissolved Oxygen 29	Effect of Nitrification on Cation Losses 39
Turbidity 30	Nutrient Budgets 41
Particulate Matter 31	General Discussion and Significance 43
pH 31	Conclusions 45
Electrical Conductivity 31	Literature Cited 46

INTRODUCTION

Management of forest resources is a worldwide consideration. Approximately one-third of the surface of the earth is forested and much of this is managed or deforested by one means or another.

Forests may be temporarily or permanently reduced by wind, insects, fire, and disease or by human activities such as harvesting or management utilizing physical or chemical techniques. Management goals range from simple harvest of wood and wood products, to increased water yields, to military stratagems involving defoliation of extensive forested areas.

Despite the importance of the forest resource, there is very little quantitative information at the ecosystem level of understanding on the biogeochemical interactions and implications resulting from large-scale changes in habitat or vegeta-

[1] This is Contribution No. 14 of the Hubbard Brook Ecosystem Study. Financial support for this work was provided by NSF Grants GB 1144, GB 4169, GB 6757, and GB 6742. The senior author acknowledges the use of excellent facilities and resources at the Brookhaven National Laboratory during the preparation of part of this manuscript. Also, we thank J. S. Eaton for special technical assistance, and W. A. Reiners and R. C. Reynolds for critical comments and suggestions. Published as a contribution to the U. S. Program of the International Hydrological Decade, and the International Biological Program.

[2] Present address: Division of Biological Sciences, Cornell University, Ithaca, New York 14850.

tion. This gap in our understanding results because it is particularly difficult to get quantitative ecological information that allows predictions about the entire ecosystem. The goal of the Hubbard Brook Ecosystem study is to understand the energy and biogeochemical relationships of northern hardwood forest watershed-ecosystems as completely as possible in order to propose sound land management procedures.

The small watershed approach to the study of hydrologic-nutrient cycle interaction used in our investigations of the Hubbard Brook Experimental Forest (Bormann and Likens 1967) provides an opportunity to deal with complex problems of the ecosystem on an experimental basis. The Hubbard Brook Experimental Forest, maintained and operated by the U.S. Forest Service, is especially well-suited to this approach since ecosystems can be defined as discrete watersheds with similar northern hardwood forest vegetation and a homogeneous bedrock, which forms an impermeable base (Bormann and Likens, 1967; Likens, *et al.*, 1967). Thus, the six small watersheds we have used at Hubbard Brook provide a replicated experimental design for manipulations at the ecosystem level of organization.

All vegetation on Watershed 2 (W2) was cut during the late fall and winter of 1965, and subsequently treated with herbicides in an experiment designed to determine the effect on 1) the quantity of stream water flowing out of the watershed, and 2) fundamental chemical relationships within the forest ecosystem, including nutrient relationships and eutrophication of stream water. In effect this experiment was designed to test the homeostatic capacity of the ecosystem to adjust to cutting of the vegetation and herbicide treatment. This paper will discuss the results of this experimental manipulation in comparison to adjacent, undisturbed watershed-ecosystems.

THE HUBBARD BROOK ECOSYSTEM

The hydrology, climate, geology, and topography of the Hubbard Brook Experimental Forest have been reported in detail elsewhere (Likens, *et al.*, 1967).

The climate of this region is dominantly continental. Annual precipitation is about 123 cm (Table 1), of which about one-third to one-fourth is snow. Although precipitation is evenly dis-

TABLE 1. Average annual water budgets for Watersheds 1 through 6 of the Hubbard Brook Experimental Forest. Watershed 2 has been excluded from the averages for 1965–68; 1967–68 is based on Watersheds 1, 3, and 6 only

Water Year (1 June–31 May)	Precipitation (P) (cm)	Runoff (R) (cm)	P-R (Evaporation and Transpiration) (cm)
1963–64	117.1	67.7	49.4
1964–65	94.9	48.8	46.1
1965–66	124.5	72.7	51.8
1966–67	132.5	80.6	51.9
1967–68	141.8	89.4	52.4
1963–68	122.2	71.8	50.4
1955–68	122.8	71.9	50.9

tributed throughout the year, stream flow is not. Summer and early autumn stream flow is usually low; whereas the peak flows occur in April and November. Loss of water due to deep seepage appears to be minimal in the Hubbard Brook area (Likens, *et al.*, 1967). The bedrock of the area is a medium to coarse-grained sillimanite-zone gneiss of the Littleton Formation and consists of quartz, plagioclase and biotite with lesser amounts of sillimanite. The mantle of till is relatively shallow and has a similar mineral and chemical composition to the bedrock. The soils are podzolic with a pH less than 7. Despite extremely cold winter air temperatures, soil frost seldom forms since insulation is provided by several centimeters of humus and a continuous winter snow cover (Hart *et al.*, 1962).

METHODS AND PROCEDURES

Precipitation is measured in the experimental watershed with a network of precipitation gauges,

approximately 1 for every 12.9 hectares of watershed. Streamflow is measured continuously at stream-gauging stations, which include a V-notch weir or a combination of V-notch weir and San Dimas flume anchored to the bedrock at the base of each watershed.

Weekly samples of precipitation and stream water were obtained from the experimental areas for chemical analysis. Rain and snow were collected in two types of plastic containers, 1) those continuously uncovered or 2) those uncovered only during periods of rain or snow. One-liter samples of stream water were collected in clean polyethylene bottles approximately 10 m above the weir in both the deforested and undisturbed watersheds. Chemical concentrations characterizing a period of time are reported as weighted averages, computed from the total amount of precipitation or streamflow and the total calculated chemical content during the period. Details concerning the methods used in collecting samples of precipitation and stream water, analytical procedures, and measurement of various physical characteristics have been given by Bormann and Likens (1967), Likens, *et al.* (1967), and Fisher, *et al.* (1968).

During November and December of 1965 all trees, saplings and shrubs of W2 (15.6 ha) were cut, dropped in place, and limbed so that no slash was more than 1.5 m above the ground. No roads were made on the watershed and great care was taken to minimize erosion. No timber or other vegetation was removed from the watershed. Regrowth of vegetation was inhibited by aerial application of the herbicide, Bromacil ($C_9H_{13}BrN_2O_2$), at 28 kg/ha on 23 June 1966. Approximately 80% of the mixture applied was Bromacil and 20% was largely inert carrier (H. J. Thorne, personal communication). Also, during the summer of 1967, approximately 87 liters of an ester of 2, 4, 5-trichlorophenoxyacetic acid (2, 4, 5-T) was individually applied to scattered regrowths of stump sprouts.

The results reported cover the period immediately following the cutting of the vegetation on W2, 1 January 1966 through 1 June 1968.

[*Editors' Note:* Detailed information on hydrology, precipitation, and streamwater parameters has been omitted.]

NUTRIENT BUDGETS

Nutrient budgets for dissolved ions and dissolved silica for the Hubbard Brook watershed-ecosystems were determined from the difference between the meteorologic input per hectare and the geologic output per hectare (Bormann and Likens, 1967). Input was calculated from the product of the ionic concentration (mg/liter) and the volume (liters) of water as precipitation (Likens, *et al.*, 1967 and Fisher, *et al.*, 1968). Additional input from applications of herbicides was added to the precipitation input. Output was calculated as the product of the volume (liters) of water draining from the watershed-ecosystems and its ionic concentration (mg/liter). Budgets for all of the ions and substances measured are given in Table 5.

Net losses were greatly increased after deforestation and herbicide treatment for all ions except ammonium, sulfate, and bicarbonate. Two factors are involved in the removal of nutrients from the deforested watershed: 1) increased runoff and 2) increased ionic concentrations in stream water. If the concentrations had not increased from the undisturbed condition, increased runoff would have accounted for a 39% increase in gross export the first year and a 28% increase the second year after deforestation. However, the gross outputs for 1967–68 were greater than the undisturbed watershed, W6 (Table 5), by 7.6-fold for Ca^{++}, 5.5-fold for Mg^{++}, 15.2-fold for K^+, 2.2-fold for Na^+, 46-fold for NO_3-N, 1.8-fold for Cl^-, 7.9-fold for Al^{+++} and 1.9-fold for SiO_2-Si, clearly indicating that increased stream water concentrations are primarily responsible for the increased nutrient loss from the ecosystem.

Nitrogen losses from W2 after deforestation, although very large already, do not take into account volatilization. Allison (1955) reported volatilization losses averaging 12 percent of the total nitrogen losses from 106 fallow soils. However, denitrification is an anaerobic process and requires a nitrate substrate generated aerobically (Jansson, 1958); consequently, for substantial denitrification to occur in fields, aerobic and anaerobic conditions must exist in close proximity. The large increases in subsurface flow of water from the deforested watershed suggests that such conditions may have been more common than in the undisturbed ecosystem. Moreover, Alexander (1967) points out, "When ammonium oxidation takes place at a pH lower than 5.0 to 5.5, or where the acidity produced in nitrification increases the hydrogen ions to an equivalent ex-

tent, the formation of nitrite can lead to a significant chemical volatilization of nitrogen."

Net losses of SO_4-S from the deforested ecosystem were about 40% lower in 1966–67, and 100% lower in 1967–68 than from undisturbed watersheds. In fact, the 1967–68 budget for SO_4-S in W2 was balanced in contrast to the undisturbed situation (Table 5). Precipitation is by far the major source of sulfate for the undisturbed watersheds (Fisher, *et al.*, 1968). Although the amount of sulfate added by precipitation in 1967–68 was increased slightly relative to previous years, the net export of sulfate was zero, with sulfate input in precipitation exactly balancing streamwater export (Table 5). The decreases in streamwater sulfate concentration and gross export from the ecosystem occurred concurrently with the increases in streamwater nitrate concentration and gross export after forest cutting (Figs. 7 and 8).

Average sulfate concentrations in stream water were 3.8 and 3.7 mg/liter during 1966–67 and 1967–68 (Table 4), far below the 6.4 and 6.8 mg/liter values recorded in 1964–65 and 1965–66 before cutting (Fisher, *et al.*, 1968). Much of this change can be explained by two facts, 1) the 39 to 28% increase in streamwater discharge from 1966 to 1968, which resulted from the elimination of transpiration by deforestation, and 2) the elimination of sulfate generation by sources internal to the ecosystem. If the decreases in sulfate concentrations were wholly due to increased runoff after deforestation, concentrations calculated on the basis of expected runoff (*i.e.*, normal for the undisturbed system, Table 2) and measured gross sulfate lost from W2 during 1966–67 and 1967–68 (Table 5), should approximate the weighted streamwater concentrations for the undisturbed period, 1964–66. However, these calculated concentrations (5.3 and 4.7 mg/liter) equal only 79 and 70% respectively of the average weighted concentrations for 1964–66. These differences in concentration may be due to some year-to-year variation, but are largely explained by a sharp reduction in the internal release of sulfate from the ecosystem, which we earlier attributed to chemical weathering and biological activity (Fisher, *et al.*, 1968, Likens, *et al.*, 1969). The average annual internal release of sulfate (i.e., an amount equivalent to net loss) supplies about 10 kg/ha in the undisturbed watersheds (Table 5). Removal of this source of sulfate would account for the lower than expected adjusted sulfate concentrations mentioned above.

Thus, apparently, the normal, relatively small release of $SO_4^=$ from the ecosystem by chemical weathering and microbial activity (Table 5) probably became negligible following forest cutting. There are at least two possible mechanisms, operating simultaneously or separately, which may account for this:

i) There may be decreased oxidation of various sulfur compounds to $SO_4^=$. Waksman (1932) has suggested that a high concentration of nitrate is very toxic to sulfur oxidizing bacteria, such as *Thiobacillus thiooxidans*. This species may be important in sulfate oxidation in the deforested watershed since *T. thiooxidans* is capable of active growth at low pH (Alexander 1967). In the undisturbed watersheds we have observed a highly significant inverse linear relationship between the concentration of nitrate and sulfate in drainage water. This relationship is particularly clear in plots of sulfate concentrations against nitrate concentration using data from November through April, when the vegetation is dormant (Fig. 13). The inverse relationship between NO_3^- and

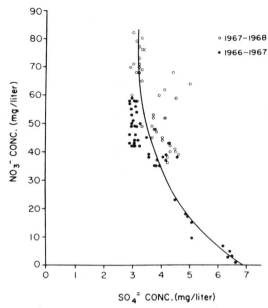

FIG. 14. Relationship between nitrate and sulfate concentrations in stream water from Watershed 2 during 1966–67 and 1967–68. Nitrate values less than 25 mg/liter indicate the chemical transition period (1 June 1966 through 31 July 1966, Fig. 7) between undisturbed and deforested conditions.

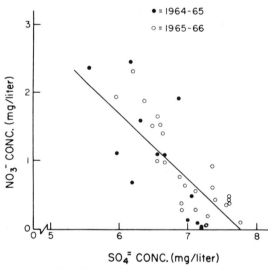

FIG. 13. Relationship between nitrate and sulfate concentrations in stream water from Watershed 2. Data were obtained during November through April of 1964–65 and 1965–66, which was prior to the increase in nitrate concentration resulting from clearing of the forest vegetation (Fig. 7). The F-ratio for this regression line is very highly significant ($p < 0.001$) and the correlation coefficient is 0.79.

$SO_4^=$ concentrations is very obvious in the first water-year after deforestation, 1966–67, when nitrate concentrations in stream water increased from normal (undisturbed) values to very high concentrations (Fig. 14). During the second wa-

ter-year after cutting, 1967–68, the nitrate concentration in stream water from W2 increased even more, whereas the sulfate concentration decreased very little and coincided with the concentration of sulfate in precipitation after adjustment for water loss by evaporation. Perhaps there is an intricate feedback mechanism between the toxicity of the nitrate concentrations and microbial oxidation of sulfur compounds within the soil. Another possibility is that the number of sulfur oxidizing bacteria have been selectively reduced by the herbicides in the deforested watershed.

ii) Although somewhat unlikely, there may be increased sulfate reduction brought about by more anaerobic conditions, particularly in the lower more inorganic horizons of the soil (*e.g.*, Waksman, 1932). That is, an increased zone of water saturation in the deeper layers and in topographic lows on the cutover watershed probably has less free oxygen than in the undisturbed situation, promoting sulfur reduction. One difficulty, however, is that the growth of the most important sulfur reducing bacteria (*Desulfovibrio* spp.) is greatly retarded by acid conditions (Alexander, 1967). Also, molecular hydrogen released by anaerobic bacterial decomposition of organic matter may be used for the reduction of sulfate (Postgate, 1949; Rankama and Sahama, 1950).

The chloride budget for the undisturbed watershed during 1966–67 showed that input in precipitation exceeded the gross output, whereas the budget was essentially balanced during the 1967–68 water-year (Table 5). However after deforestation, significant net losses of chloride were observed (Table 5). The application of Bromacil in 1966–67 potentially added the equivalent of 3.0 kg Cl/ha or about 50% of the chloride input as precipitation. From the pattern of chloride changes in stream water following the addition of this herbicide (Fig. 7), it would appear that the herbicide and/or its degradation products were lost from the watershed quite gradually throughout the year. Measurements of Bromacil in stream water seemed to confirm this (Pierce, 1969). Since the Bromacil (1966–67) and possibly 2, 4, 5–T (1967–68) were not all flushed from the ecosystem within a year, then the internal release of chloride from the ecosystem probably represented an even greater percentage of the gross annual output (Table 5). Based upon streamwater concentrations in W2 and W6 (Fig. 7), it would appear that the internal reservoir (plus external inputs from herbicides) of chloride within the ecosystem has been essentially exhausted in two years following deforestation.

GENERAL DISCUSSION AND SIGNIFICANCE

The intrasystem cycle of a terrestrial ecosystem links the organic, available nutrient, and soil and rock mineral compartments through rate processes including decomposition of organic matter, leaching and exudate from the biota, nutrient uptake by the biota, weathering of primary minerals, and formation of new secondary minerals (Fig. 15). The deforestation experiment was designed to test the effects of blockage on a major ecosystem pathway, i.e., nutrient and water uptake by vegetation, on other components of the intrasystem cycle and on the export behavior of the system as a whole. The block was imposed by cutting all of the forest vegetation and subsequently preventing regrowth with herbicides. We hoped that this experimental procedure would provide information about the nature of the homeostatic capacity of the ecosystem. The deforested condition has been maintained since 1 January 1966.

Forest clearing and herbicide treatment had a profound effect on the hydrologic and nutrient relationships of our northern hardwood ecosystem. Annual runoff (water export) increased by some 33 cm or 39% in the first year and 27 cm or 28% in the second year over that expected. Moreover, the discharge pattern was altered so

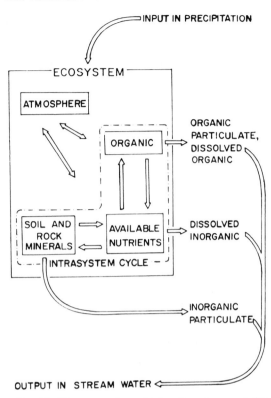

FIG. 15. Diagrammatic model for sites of accumulation and pathways of nutrients in the Hubbard Brook ecosystem (after Bormann, et al., 1969).

that sustained, higher flows occurred in the summer months and the snow pack melted earlier in the spring. This overall increase in stream runoff is large compared to the average increase (about 20 cm) found for other such experiments throughout the world (Hibbert, 1967), but is less than the maximum increase of 41 cm found for clearcut watersheds in North Carolina (Hoover, 1944).

No previous comprehensive measurements have been made of the homeostatic ability of a watershed-ecosystem to retain nutrients despite major shifts in the hydrologic cycle, including increased discharge, following deforestation (Odum, 1969).

Our results showed that cation and anion export did not change for the first 5 months (winter and spring) after deforestation, but then the ionic concentrations increased spectacularly, and remained at high levels for the 2 years of observation (Tables 4 and 5, Figs. 7, 8, 9 and 10). Annual net losses in kg/ha amounted to about 142 for nitrate-nitrogen, 90 for calcium, 36 for potassium, 32 for dissolved silica, 24 for aluminum, 18 for magnesium, 17 for sodium, and 4 for chlorine during 1967–68. These losses are much

TABLE 8. Comparative net gains or losses of dissolved solids in runoff following clear-cutting of Watershed 2 in the Hubbard Brook Experimental Forest for the period 1 June to 31 May. In metric tons/km²-yr

	1966-67		1967-68	
	W2	*W6*	*W2*	*W6*
Ca	−7.5	−0.8	−9.0	−0.9
K	−2.3	−0.1	−3.6	−0.2
Al	−1.7	−0.1	−2.4	−0.3
Mg	−1.6	−0.3	−1.8	−0.3
Na	−1.7	−0.6	−1.7	−0.7
NH₄	+0.1	+0.2	+0.2	+0.3
NO₃	−43.0	+1.5	−62.8	+1.1
SO₄	−0.5	−0.8	0	−1.0
HCO₃	−0.1	−0.2	0	−0.3
Cl	−0.1	+0.2	−0.4	0
SiO₂—aq	−6.6	−3.6	−6.9	−3.6
Total	−65.0	−4.6	−88.4	−5.9

greater than for adjacent undisturbed ecosystems (Table 5). Ammonium-nitrogen was essentially unchanged relative to the undisturbed condition during this period and showed an annual net gain of about 1 to 2 kg/ha. In comparison with the undisturbed watershed-ecosystems the greatest changes occurred in nitrate-nitrogen and potassium export. Nitrate-nitrogen is normally accumulated in the undisturbed ecosystem in contrast to this very large export, and the net potassium output increased about 18-fold. The total net export of dissolved inorganic substances from the deforested ecosystem is 14–15 times greater than from the undisturbed ecosystem (Table 8).

The terrestrial ecosystem is one of the ultimate sources of dissolved substances in surface water. The contribution of dissolved solids (gross export) by our undisturbed forest ecosystems, 13.9 metric tons/km² (Bormann, et al., 1969) is only about 25% of the dissolved load predicted by Langbein and Dawdy (1964) for regions with 75 cm of annual runoff. Their estimates were based on data from watersheds of the north Atlantic slope, which probably include areas disturbed by agriculture or logging. The difference between our undisturbed forest ecosystems and the regional prediction is credited in part to the operation of various regulating biotic factors associated with mature undisturbed forest (Bormann, et al., 1969).

Deforestation markedly altered the ecosystem's contribution of dissolved solids to the drainage waters. Total gross export, exclusive of dissolved organic matter, was about 75 metric tons/km² in 1966–67 and 97 metric tons/km² in 1967–68. These figures exceed the regional prediction of Langbein and Dawdy (1964). However, it should be noted that the accelerated export of

dissolved substances results primarily from mining the nutrient capital of the ecosystem and cannot be sustained indefinitely.

Surprisingly, the net export of dissolved inorganic substances from the cutover watershed is about double the annual value estimated for particulate matter removed by debris avalanches in the White Mountains (Bormann, et al., 1969). Thus, the effects of deforestation may have almost twice the importance of avalanches in short-term catastrophic transport of inorganic materials downslope in the White Mountains.

Coupled with this increase in gross and net export of dissolved substances, there has been at least a 4-fold increase in the export of inorganic and organic particulate matter from the deforested ecosystem. This increase indicates that the biotic mechanisms that normally minimize erosion and transport (Bormann, et al., 1969) are also becoming less effective.

The greatly increased export of nutrients from the deforested ecosystem resulted primarily from an alteration of the nitrogen cycle within the ecosystem. Whereas in the undisturbed system, nitrogen is cycled conservatively between the living and decaying organic components, in the deforested watershed, nitrate produced by microbial nitrification from decaying organic matter, is rapidly flushed from the system in drainage waters. In fact, the increased nitrate output accounts for the net increase in total cation and anion export from the ecosystem (Likens, et al., 1969). With the increased availability of nitrate ions and hydrogen ions from nitrification, cations are readily leached from the system. Cations are mobilized as hydrogen ions replace them on the various exchange complexes of the soil and as organic and inorganic materials decompose. Based upon the increased output for sodium, chemical decomposition of inorganic materials in the deforested ecosystem is also accelerated by about 3-fold.

If the streamwater concentrations had remained constant after deforestation, increased water output alone would have accounted for 39% of the increased nutrient export the first year and 28% of the increased nutrient export the second year. However, the very large increase in annual export of dissolved solids from the deforested ecosystem occurred primarily because the streamwater concentrations were vastly increased, mostly as a direct result of the increased nitrification. The increased output of nutrients originated predominantly from the organic compartment of the watershed-ecosystem (Fig. 15).

Our study shows that the retention of nutrients within the ecosystem is dependent on constant and efficient cycling between the various components of the intrasystem cycle, i.e., organic, available nutrients, and soil and rock mineral compartments (Fig. 15). Blocking of the pathway of nutrient uptake by destruction of one sub-component of the organic compartment, i.e., vegetation, leads to greatly accelerated export of the nutrient capital of the ecosystem. From this we may conclude that one aspect of homeostasis of the ecosystem, i.e., maintenance of nutrient capital, is dependent upon the undisturbed functioning of the intrasystem nutrient cycle, and that in this ecosystem no mechanism acts to greatly delay loss of nutrients following sustained destruction of the vegetation.

The increased output of water from the deforested watershed was readily visible during the summer months, however the increased ion and particulate matter concentrations were not. The stream water from the deforested watershed appeared to be just as clear and potable as that from adjacent, undisturbed watersheds. However this was not the case. By August, 1966, the nitrate concentration in stream water exceeded (at times almost doubled) the concentration recommended for drinking water (Public Health Service, 1962).

The high nutrient concentrations, plus the increased amount of solar radiation (absence of forest canopy) and higher temperature in the stream, resulted in significant eutrophication. A dense bloom of *Ulothrix zonata* (Weber and Mohr) Kütz, has been observed during the summers of 1966 and 1967 in the stream of W2. In contrast the undisturbed watershed streams are essentially devoid of algae of any kind. This represents a good example of how an overt change in one component of an ecosystem, alters the structure and function, often unexpectedly, in another part of the same ecosystem or in another interrelated ecosystem. Unless these ecological interrelationships are understood, naive management practices can produce unexpected and possibly widespread deleterious results.

Conclusions

1. The quantity and quality of drainage waters were significantly altered subsequent to deforestation of a northern hardwoods watershed-ecosystem. All vegetation on Watershed 2 of the Hubbard Brook Experimental Forest was cut, but not removed, during November and December of 1965; and vegetation regrowth was inhibited by periodic application of herbicides.

2. Annual runoff of water exceeded the expected value, if the watershed were undisturbed, by 33 cm or 39% during the first water-year after deforestation and 27 cm or 28% during the second water-year. The greatest increase in water discharge, relative to an undisturbed situation, occurred during June through September, when runoff was 414% (1966–67) and 380% (1967–68) greater than the estimate for the untreated condition.

3. Deforestation resulted in large increases in streamwater concentrations of all major ions except NH_4^+, $SO_4^=$ and HCO_3^-. The increases did not occur until 5 months after the deforestation. The greatest increase in streamwater ionic concentration after deforestation was observed for nitrate, which increased by 41-fold the first year and 56-fold the second year above the undisturbed condition.

4. Sulfate was the only major ion in stream water from Watershed 2 that decreased in concentration after deforestation. The 45% decrease the first year (1966–67) resulted mostly from increased runoff of water and by eliminating the generation of sulfate within the ecosystem. The concentration of sulfate in stream water during 1967–68 equalled the concentration in precipitation after adjustment for water loss by evaporation. Sulfate concentrations were inversely related to nitrate concentrations in stream water in both undisturbed and deforested watersheds.

5. In the undisturbed watersheds the stream water can be characterized as a very dilute solution of sulfuric acid (pH about 5.1 for W2); whereas after deforestation the stream water from Watershed 2 became a relatively stronger nitric acid solution (pH 4.3), considerably enriched in metallic ions and dissolved silica.

6. The increase in average nitrate concentration in precipitation for the Hubbard Brook area compared to data from 1955–56, as well as the consistent annual increase observed from 1964–1968, may be some measure of a general increase in air pollution.

7. The greatly increased export of dissolved nutrients from the deforested ecosystem was due to an alteration of the nitrogen cycle within the ecosystem. Whereas nitrogen is normally conserved in the undisturbed ecosystem, in the deforested ecosystem nitrate is rapidly flushed from the system in drainage water. The mobilization of nitrate from decaying organic matter, presumably by increased microbial nitrification, quantitatively accounted for the net increase in

total cation and anion export from the deforested ecosystem.

8. Increased availability of nitrate and hydrogen ions resulted from nitrification. Cations were mobilized as hydrogen ions replaced them on various exchange complexes of the soil and as organic and inorganic materials were decomposed. Chemical decomposition of inorganic materials in the deforested ecosystem was accelerated about 3–fold. However, the bulk of the nutrient export from the deforested watershed originated from the organic compartment of the ecosystem.

9. The total net export of dissolved inorganic substances from the deforested ecosystem was 14–15 times greater than from undisturbed ecosystems. The increased export occurred because the streamwater concentrations were vastly increased, primarily as a direct result of the increased nitrification, and to a much lesser extent because the amount of stream water was increased.

10. The deforestation experiment resulted in significant pollution of the drainage stream from the ecosystem. Since August, 1966, the nitrate concentration in stream water has exceeded, almost continuously, the maximum concentration recommended for drinking water. As a result of the increased temperature, light and nutrient concentrations, and in sharp contrast to the undisturbed watersheds, a dense bloom of algae has appeared each year during the summer in the stream from Watershed 2.

11. Nutrient cycling is closely geared to all components of the ecosystem; decomposition is adjusted to nutrient uptake, uptake is adjusted to decomposition, and both influence chemical weathering. Conservation of nutrients within the ecosystem depends upon a functional balance within the intrasystem cycle of the ecosystem. The uptake of water and nutrients by vegetation is critical to this balance.

LITERATURE CITED

Alexander, M. 1967. Introduction to Soil Microbiology. John Wiley and Sons, Inc., New York, 472 pp.

Allison, F. E. 1955. The enigma of soil nitrogen balance sheets. Advan. Agron. 7: 213–250.

Anderson, D. H., and H. E. Hawkes. 1958. Relative mobility of the common elements in weathering of some schist and granite areas. Geochim. Cosmochim. Acta 14(3): 204–210.

Bormann, F. H., and G. E. Likens. 1967. Nutrient cycling. Science 155(3761): 424–429.

Bormann, F. H., G. E. Likens, D. W. Fisher, and R. S. Pierce. 1968. Nutrient loss accelerated by clearcutting of a forest ecosystem. Science 159: 882–884.

Bormann, F. H., G. E. Likens, and J. S. Eaton. 1969. Biotic regulation of particulate and solution losses from a forest ecosystem. BioScience 19(7): 600–610.

Boswell, J. G. 1955. The microbiology of acid soils. IV. Selected sites in Northern England and Southern Scotland. New Phytol. 54(2): 311–319.

Federer, C. A. 1969. Radiation and snowmelt on a clear-cut watershed. E. Snow Conf. Proc., Boston, Mass. (1968) pp. 28–41.

Fisher, D. W., A. W. Gambell, G. E. Likens, and F. H. Bormann. 1968. Atmospheric contributions to water quality of streams in the Hubbard Brook Experimental Forest, New Hampshire. Water Resources Res. 4(5): 1115–1126.

Gambell, A. W., and D. W. Fisher. 1966. Chemical composition of rainfall, eastern North Carolina and southeastern Virginia, U.S. Geol. Survey Water-Supply Paper 1535K: 1–41.

Harvey, H. H., and A. C. Cooper. 1962. Origin and treatment of a supersaturated river water. Internat. Pacific Salmon Fish. Comm., Prog. Rept. No. 9: 1–19.

Hart, G., R. E. Leonard, and R. S. Pierce. 1962. Leaf fall, humus depth, and soil frost in a northern hardwood forest. Forest Res. Note 131, Northeastern For. Exp. Sta., Durham, N. H.

Hibbert, A. R. 1967. Forest treatment effects on water yield. pp. 527–543. In: Proc. Internat. Symposium on Forest Hydrology, ed. by W. E. Sopper and H. W. Lull, Pergamon Press, N. Y.

Hoover, M. D. 1944. Effect of removal of forest vegetation upon water yields. Trans. Amer. Geophys. Union, Part 6: 969–975.

Hornbeck, J. W., and R. S. Pierce. 1969. Changes in snowmelt run-off after forest clearing on a New England watershed. E. Snow Conf. Proc., Portland, Maine (1969). (In press).

Hornbeck, J. W., R. S. Pierce, and C. A. Federer. Streamflow changes after forest clearing in New England. (In preparation).

Iwasaki, I., S. Utsumi, and T. Ozawa. 1952. Determination of chloride with mercuric thiocyanate and ferric ions. Chem. Soc. Japan Bull. 25: 226.

Hewlett, J. D., and A. R. Hibbert. 1961. Increases in water yield after several types of forest cutting. Quart. Bull. Internatl. Assoc. Sci. Hydrol. Louvain, Belgium, pp. 5–17.

Jansson, S. L. 1958. Tracer studies on nitrogen transformations in soil with special attention to mineralisation-immobilization relationships. Kungl. Lantbrukshögskolans Annaler. 24: 105–361.

Järnefelt, H. 1949. Der Einfluss der Stromschnellen auf den Sauerstoff-und Kohlensäuregehalt und das pH des Wassers im Flusse Vuoksi. Verh. Internat. Ver. Limnol. 10: 210–215.

Johnson, N. M., G. E. Likens, F. H. Bormann, and R. S. Pierce. 1968. Rate of chemical weathering of silicate minerals in New Hampshire. Geochim. Cosmochim. Acta. 32: 531–545.

Johnson, N. M., G. E. Likens, F. H. Bormann, D. W. Fisher, and R. S. Pierce. 1969. A working model for the variation in streamwater chemistry at the Hubbard Brook Experimental Forest, New Hampshire. Water Resources Res. 5(6): 1353–1363.

Juang, F. H. F., and N. M. Johnson. 1967. Cycling of chlorine through a forested watershed in New England. J. Geophys. Research 72(22): 5641–5647.

Junge, C. E. 1958. The distribution of ammonia and nitrate in rain water over the United States. Trans.

Amer. Geophys. Union **39**: 241–248.

——. 1963. Air Chemistry and Radioactivity. Academic Press, N. Y. 382 pp.

——, and R. T. Werby. 1958. The concentration of chloride, sodium, potassium, calcium and sulphate in rain water over the United States. J. Meteorol. **15**: 417–425.

Langbein, W. B., and D. R. Dawdy. 1964. Occurrence of dissolved solids in surface waters in the United States. U. S. Geol. Survey Prof. Paper **501–D**: D115–D117.

Lieberman, J. A., and M. D. Hoover. 1948a. The effect of uncontrolled logging on stream turbidity. Water and Sewage Works **95**(7): 255–258.

——. 1948b. Protecting quality of stream flow by better logging. Southern Lumberman: 236–240.

Likens, G. E., F. H. Bormann, N. M. Johnson, and R. S. Pierce. 1967. The calcium, magnesium, potassium, and sodium budgets for a small forested ecosystem. Ecology **48**(5): 772–785.

Likens, G. E., F. H. Bormann and N. M. Johnson. 1969. Nitrification: Importance to nutrient losses from a cutover forested ecosystem. Science **163**(3872): 1205–1206.

Lindroth, A. 1957. Abiogenic gas supersaturation of river water. Arch. für Hydrobiol. **53**: 589–597.

Macan, T. T. 1958. The temperature of a small stony stream. Hydrobiologia **12**: 89–106.

Marshall, C. E. 1964. The physical chemistry and minerology of soils. Vol. I. Soil materials. Wiley and Sons, N. Y. 388 pp.

Mason, B. 1966. Principles of geochemistry. 3rd ed. Wiley and Sons, N. Y. 329 pp.

McConnochie, K., and G. E. Likens. 1969. Some Trichoptera of the Hubbard Brook Experimental Forest in central New Hampshire. Canadian Field Naturalist. **83**(2): 147–154.

Minckley, W. L. 1963. The ecology of a spring stream, Doe Run, Meade County, Kentucky. Wildlife Monogr. **11**: 1–124.

Nye, R. H., and D. J. Greenland. 1960. The soil under shifting cultivation. Commonwealth Bureau of Soils, Harpenden, England, Tech. Bull. No. **51**, 156 pp.

Odum, E. P. 1969. The strategy of ecosystem development. Science **164**(3877): 262–270.

Pierce, R. S. 1969. Forest transpiration reduction by clearcutting and chemical treatment. Proc. Northeastern Weed Control Conference. **23**: 344–349.

Postgate, J. R. 1949. Competitive inhibition of sulphate reduction by selenate. Nature (**4172**): 670–671.

Rankama, K., and T. G. Sahama. 1950. Geochemistry. Chicago Univ. Press, 912 pp.

Ruttner, F. 1953. Fundamentals of Limnology. Univ. of Toronto Press. (Transl. by D. G. Frey and F. E. J. Fry). 242 pp.

Scott, D. R. M. 1955. Amount and chemical composition of the organic matter contributed by overstory and understory vegetation to forest soil. Yale Univ. School of Forestry. Bull. No. **62**, 73 pp.

Smith, W., F. H. Bormann, and G. E. Likens. 1968. Response of chemoautotrophic nitrifiers to forest cutting. Soil Science **106**(6): 471–473.

Tebo, L. D. 1955. Effects of siltation, resulting from improper logging, on the bottom fauna of a small trout stream in the southern Appalachians. Prog. Fish Culturist **12**(2): 64–70.

Thiessen, A. H. 1923. Precipitation for large areas. Monthly Weather Review **51**: 348–353.

Trimble, G. R., and R. S. Sartz. 1957. How far from a stream should a logging road be located? J. Forestry **55**(5): 339–341.

U. S. Forest Service. Northeastern Forest Experiment Station. 1964. Hubbard Brook Experimental Forest. Northeast. For. Exp. Sta., Upper Darby, Penn. 13 pp.

U. S. Public Health Service. 1962. Drinking water standard. U.S. Public Health Service Publ. **956**. Washington, D. C.

Waksman, S. A. 1932. Principles of soil microbiology. 2nd ed. Williams and Wilkins Co., Baltimore. 894 pp.

Weber, D. F., and P. L. Gainey. 1962. Relative sensitivity of nitrifying organisms to hydrogen ions in soils and in solutions. Soil Sci. **94**: 138–145.

Whitehead, H. C., and J. G. Feth. 1964. Chemical composition of rain, dry fallout, and bulk precipitation at Menlo Park, California, 1957–1959. J. Geophys. Res. **69**(16): 3319–3333.

Woods, W. J. 1960. An ecological study of Stony Brook, New Jersey. Ph.D. Thesis, Rutgers Univ. New Brunswick. 307 pp.

Part V

RATES OF EROSION

Editors' Comments
on Papers 23 Through 27

23 EVEREST
*Some Observations on the Quantity of Earthy Matter Brought
Down by the Ganges River*

24 DOLE and STABLER
Denudation

25 HOLEMAN
The Sediment Yield of Major Rivers of the World

26 LI
Denudation of Taiwan Island Since the Pliocene Epoch

27 MEYBECK
*Excerpt from Concentrations des eaux fluviales en éléments
majeurs et apports en solution aux océans*

Parts I–IV have presented much information on erosion rates, particularly in the papers by Fournier, Langbein and Schumm, Gibbs, and Dunne. This final part brings together a number of papers that provide a contemporary summary of erosion rates on the subcontinental-world scale.

Much of what is known currently of the spatial and temporal variations in erosion rates is based on measurements and estimates of sediment yield. A wide range of techniques has been used to determine and predict sediment yields, (see, for example, the multitude of techniques currently in use by different U.S. government agencies; ARS, 1975). Reservoir sediment deposition surveys, such as those reported by Dendy and Champion (1973), are the most reliable method of determining sediment yields because the deposits include all of the traction load and most of the suspended load. The high cost of such surveys and the absence of reservoirs in many basins preclude using this method frequently.

Erosion rates have been estimated by a variety of other methods. For example, Clark and Jager (1969) used geochronological and heat flow data to estimate a denudation rate of 0.4–1 mm yr^{-1} for the Alps; Marchand (1972) estimated a rate of 1–3 cm per 1,000 years from depths of ground lowering from beneath the projected base of a radiometrically dated basalt flow in the White Mountains, California; Gordon (1979) used sediment volumes in Long Island Sound to estimate mean denudation rates in central New England; and Carrara and Carroll (1979) used tree root exposure to conclude that erosion rates in the Piceance Basin, Colorado, have been increasing during the last 400 years.

Data on river discharge and sediment concentration are used to estimate sediment yield and erosion rates more frequently than the other methods because in many countries they are routinely collected for the purpose of water resource investigation and management and therefore are readily available at little or no additional cost. For example, Harris (1962) compiled sediment load data for the United States and Puerto Rico for the period 1950–1960 using 118 different published and unpublished reports. Although apparently straightforward, this method of evaluating mean rates of material export from drainage basins has a number of sources of error, reviewed in particular by Meade (1969), Janda (1971), and Walling (1978). A major difficulty is that only data for suspended sediment concentrations are frequently available, although bed load may exceed 20 percent of the total load in some streams (Gregory and Walling, 1973, table 4.4c), and dissolved load may be completely dominant in many others (Corbel, 1964). Errors may arise from failure to take into account the sediment and solute loads introduced by human activity; these have been briefly considered in Part IV. Atmospheric inputs of solutes and dust may similarly account for a significant proportion of the material exported from a basin. At the world scale, Meybeck estimated that 15 percent of fluvial solute loads are derived from the atmosphere (Paper 27). Furthermore, Walling (1978) demonstrated that depending on the precise procedures used to analyze a given set of data, severe errors may result.

Paper 23 is a pioneering attempt at quantification in the earth sciences, although it presents data uncorrected for the factors mentioned above. The Reverend Everest estimated an annual sediment discharge by the Ganges River of 355 million tons, which is substantially less than the estimates of 1450 million t yr^{-1} (1600 million short t yr^{-1}) listed by Holeman (Paper 25) or the more recent figure of 703 million t yr^{-1} given by Subramanian (1979). He may perhaps be forgiven his inaccuracy in view of the magnitude of the task of measuring sediment transport in a major river like the Ganges. The data Subramian used are little more sophisticated than those collected by Everest 150 years before.

There were other infrequent attempts to quantify rates of fluvial sediment transport during the decades following Everest's report, such as that by Humphreys and Abbott (1861) for the Mississippi River. The first thorough study was reported by Dole and Stabler in 1909 (Paper 24); this was a forerunner of a series of reports of river water chemistry by staff of the U.S. Geological Survey (Clarke, 1910, and 1924; Livingstone, 1963). Using data collected by a number of agencies of the U.S. government, Dole and Stabler estimated a total erosion rate for rivers of the United States of 720 million t yr^{-1} (783 million short t yr^{-1}), which was uncorrected for precipitation inputs. Roughly one-third of this was in solution and the remainder in suspension; mean specific erosion rates are 31 t km^{-2} yr^{-1} and 59 t km^{-2} yr^{-1}, respectively, for the whole country. By comparison, Meybeck estimated a specific erosion rate of 38 t km^{-2} yr^{-1} for dissolved load (Paper 27), and Holeman a rate of 86 t km^{-2} yr^{-1} for suspended load (Paper 25), both for the whole of North America. Hence Dole and Stabler obtained a figure for dissolved load in agreement with that of Meybeck (note the different areas considered) working seventy years later, and a figure for suspended load differing by only 30 percent from that of Holeman. Neither figure is necessarily correct; Holeman himself asserts that there is a possible influence of soil conservation, stream-bank stabilization, and reservoir construction upon fluvial sediment yields. That such figures may not be representative of long-term conditions is exemplified by the suspended sediment yield data for the Yuma station of the Colorado River, which shows a reduction from 966 t mi^{-2} during the 1911–1916 period to 0.63 t mi^{-2} during the period 1965–1967 as a result of dam construction (Curtis, Culbertson, and Chase, 1973).

The data Holeman presented have been refined in recent years, but we still have little information about erosion and sediment yields from large areas of the earth's surface, particularly in the developing countries. For example, estimates for two major rivers, the Amazon and the Ganges, are based on a very small sample of data (Gibbs, Paper 17; Meade et al., 1979; Subramanian, 1979), and no data at all are available for many others. Nevertheless, major advances are being made, particularly since the international hydrological decade. Thus, Holeman presented sediment yield data for but one New Zealand river, the Waipaoa, which has been severely affected by agricultural development. During the last decade, however, a large quantity of sediment concentration data has become available, permitting Adams (1979), Griffiths (1979), and Thompson and Adams (1979) to present figures that give a much more reliable overall picture of erosion in New Zealand.

Paper 26 is a benchmark in that it presents perhaps the highest value of specific suspended sediment yield yet recorded in medium-sized drainage basins. Li shows average suspended sediment yields (his PDR in units of mg cm^{-2} yr^{-1} equivalent to 10 t km^{-2} yr^{-1}) to be in excess of 13,000 t km^{-2} yr^{-1} in the tectonically active Taiwan central range. In the New Zealand southern Alps, another tectonically active zone at the margin of the Pacific plate, Griffiths (1979) has estimated a mean annual specific suspended sediment yield of 17,000 t km^{-2} yr^{-1} in a 352 km^2 drainage basin. Both the Taiwan and the New Zealand data are for forested watersheds only slightly modified by human activity; the geologic, topographic, and climatic environment is responsible for these extremes. Although many of the studies included in this book have indicated the importance of vegetation cover, and particularly of a forest cover, in minimizing erosion rates, in these and other similar extreme cases it proves to be negligible, as contended by Fournier (see Paper 14).

To accompany the paper by Holeman, which presents data for suspended sediment yields, we include Meybeck's recent review of dissolved matter transport (Paper 27). Meybeck produced an estimate for total dissolved matter discharge, including a very considerable anthropogenic input, to the oceans of 3483 million t yr^{-1} (Paper 27, table 3). In comparison, Holeman's worldwide estimate for suspended sediment discharge was 18,160 million t yr^{-1}, roughly five times as great.

REFERENCES

Adams, J., 1979, Sediment Loads of North Island Rivers, New Zealand—A Reconnaissance, *Jour. Hydrology (New Zealand)* **18**:36–48.

Agricultural Research Service, 1975, Present and Prospective Technology for Predicting Sediment Yields and Sources, *Sediment Yield Workshop Proc. ARS-S-40*, U.S. Dept. of Agriculture Sedimentation Laboratory, Mississippi, 285p.

Carrara, P. E., and T. R. Carroll, 1979, The Determination of Erosion Rates from Exposed Tree Roots in the Piceance Basin, Colorado, *Earth Surf. Proc.* **4**:307–317.

Clark, S. P., and E. Jager, 1969, Denudation Rate in the Alps from Geochronological and Heat Flow Data, *Am. Jour. Sci.* **267**:1143–1160.

Clarke, F. W., 1910, A Preliminary Study of Chemical Denudation, *Smithsonian Misc. Colln.* **56**:1–19.

Clarke, F. W., 1924, Data of Geochemistry, *U.S. Geol. Survey Bull. 770*, pp. 114–121.

Corbel, J., 1964, L'érosion terrestre. Etude quantitative, *Annales Géographie* **73**:385–412.

Curtis, F. W., J. K. Culbertson, and E. B. Chase, 1973, Fluvial-Sediment Discharge to the Oceans from the Conterminous United States, *U.S. Geol. Survey Circ. 670*, 17p.

Dendy, F. E., and W. A. Champion, 1973, Summary of Reservoir Sediment Deposition Surveys Made in the United States Through 1970, *U.S. Dept. Agriculture Misc. Pub. 1266*, 82p.

Gordon, R. B., 1979, Denudation Rate of Central New England Determined from Estuarine Sedimentation, *Am. Jour. Sci.* **279**:632–642.

Gregory, K. J., and D. E. Walling, 1973, *Drainage Basin Form and Process*, Wiley, New York.

Griffiths, G. A., 1979, High Sediment Yields from Major Rivers of the Western Southern Alps, New Zealand, *Nature* **282**:61–63.

Harris, K. F., 1962, Inventory of Published and Unpublished Sediment-Load Data, United States and Puerto Rico, 1950–1960, *U.S. Geol. Survey Water-Supply Paper 1547*, 117p.

Humphreys, A. A., and H. L. Abbott, 1861, Report on the Physics and Hydraulics of the Mississippi River, *U.S. Army Corps Engineers Prof. Paper 13*, 146p.

Janda, R. J., 1971, An Evaluation of Procedures Used in Computing Chemical Denudation Rates, *Geol. Soc. America Bull.* **82**:67–80.

Livingstone, D. A., 1963, Chemical Composition of Rivers and Lakes, *U.S. Geol. Survey Prof. Paper 440-G*, 64p.

Marchand, D. E., 1972, Rates and Modes of Denudation, White Mountains, East California, *Am. Jour. Sci.* **270**:109–135.

Meade, R. H., 1969, Errors in Using Modern Stream-Load Data to Estimate Natural Rates of Denudation, *Geol. Soc. America Bull.* **80**:1265–1274.

Meade, R. H., C. F. Nordin, W. F. Curtis, F. M. Costa Rodrigues, D. M. Do Vale, and J. M. Edmond, 1979, Sediment Loads in the Amazon River, *Nature* **278**:161–163.

Subramanian, V., 1979, Chemical and Suspended-Sediment Characteristics of Rivers in India, *Jour. Hydrology* **44**:37–55.

Thompson, S. M., and J. Adams, 1979, Suspended Sediment Load in Some Major Rivers of New Zealand, in *Physical Hydrology—New Zealand Experience*, ed. D. L. Murray and P. Ackroyd, New Zealand Hydrological Society, Wellington, pp. 213–229.

Walling, D. E., 1978, Reliability Considerations in the Evaluation and Analysis of River Loads, *Zeitschr. Geomorphologie* **29**:29–42.

23

Reprinted from *Asiatic Soc. Bengal Jour.* 1:238–242 (1832)

SOME OBSERVATIONS ON THE QUANTITY OF EARTHY MATTER BROUGHT DOWN BY THE GANGES RIVER

R. Everest

[Read before the Physical Class, 13th June, 1832.]

In the course of last summer, I made some attempts to ascertain the weight of solid matter contained in a given quantity of Ganges water, both in the dry and rainy season, but I found the weight so variable on different days (when little difference might have been expected) that I can hardly consider the observations numerous enough to give a correct average. Such as they are, however, they may not be without interest in the absence of other information on the subject. I therefore take the liberty of laying them before the Society, and shall, if opportunity offers, endeavour to add some further data, both for the weight of solid matter, and also for the rise and velocity of the river.

1. A quantity of Ganges' water taken 27th May, 1831, gave when evaporated, a solid residuum of 1.084 grains per wine quart.

2. July 21st. There had been little rain for some days, and the river was low for the season: a wine quart contained of soluble matter 2.0 grains; of insoluble, 16.2;—total 18.2.

3. August 2nd. The river being much higher, the same quantity of water gave insoluble matter 28.7, soluble 3.0;—total 31.7.

4. August. 13th. The river had reached its maximum, and gave from the same quantity of water, insoluble matter 73.5 grains; soluble 2.7;—total 76.2.

5. August 20th. The water had hitherto been taken from the side, but as it was evident that the quantity of matter held in suspension in the middle of the current was much greater than towards the bank, where the water was nearly still, I took two separate portions as before, and obtained, from the middle, 40 grs. of insoluble residuum; from the side 20 grains ditto: add for soluble matter, suppose two grains to each, the middle gives 42, the side 22 grs. The river to-day was at the same height as on the 13th.

6. Sept. 7th. Two portions taken this day, as before, gave for the middle 19, the side 15.

7. Sept. 21st. The middle gave 22, the side 20.

The different proportion in this case was occasioned by a strong east wind agitating the water near the shore.

8. Oct. 4th. The middle gave 9.6, the side 5.1.

The numbers therefore stand thus :

303

	Middle.		Side.
20th August,....	42	....	22
7th Sept........	19	....	15
21st, ...	22	....	20
4th Oct........	9.6	....	5.1
Average	23.1		15.5

These numbers are to each other nearly as 3 to 2, so that we may correct the previous observations as follows :

21st July......	27.3
2nd August....	47.5
13th———.....	114.3

The first rain fell on the afternoon of the 14th June, and we have unfortunately no observation for the first month of their duration, but I think 27.3, the quantity on the 21st July, would be rather a low than a high average for it. For the second month, viz. from the 15th July to the 15th August, we have three observations, which give an average of 63 grs. For the third month, we have two, with an average of 30.5 grs. For the fourth month we have two, giving an average of 15.8. So that the average for the whole period, from the 15th June to the 15th October, would be 34 grs. On the 15th March, 1832, a quantity of water taken in the middle of the stream gave 6 grains per quart: a mean between this and 9.6 (the weight obtained on the 4th October), or 7.8 grs., may be taken as the average from the 15th October to the 15th March, or for five months.

1.084 per quart was the weight obtained on the 27th May, from water taken at the side. If we correct this as before, it gives 1.63, and the mean between this and 6, or 3.8 will be the average for the remaining three months, from the 15th March to the 15th June.

I will here add such data as I possess respecting the breadth, depth, and velocity of the stream.

I found the breadth on the 15th March 1832, to be 660 yards. The distance up the bank to the maximum height of the rains was 158 yards more on the southern shore, and 38 yards on the northern ; which reduced to a level, would give a total breadth of 2563 feet, for the maximum breadth of the rains ; 1980 feet, being the least breadth in the dry weather.

On the same day, the total perpendicular fall from the maximum height of the rains was 28 feet. The whole breadth of 1980 feet, sounded at intervals of 300 feet, gave an average of 17 feet 8 inches, which added to 28, gives 45 feet 8 inches for the maximum depth in the rains. The river had reached this maximum on the 13th August, and continued at the same until the 29th. After that it subsided as follows :

	ft.	in.
On the 26th August the fall was	7	6
7th Sept.	5	2
21st Sept.	5	6
27th Sept.	13	0
4th Oct.	15	4
25th Oct.	22	10

So that we may call the average depth for the third month of the rains 4 feet 3 inches below maximum, or in round numbers 41 feet. For the fourth month, viz. from the 15th Sept. to 15th Oct. the average depth would be 11 feet 3 inches below maximum, or 34 feet. For the first two months of the rains, the rise of the river was not measured, but from the quantity of rain that fell and the apparent increase, I cannot believe that the average depth would be less than the average for the fourth month, viz. 34 feet. If then we assume this number for the first two months, we have an average depth of 36 feet for the four months of the rains. A mean between 22 feet 10 inches (the fall measured on the 25th Oct.), and 28 feet (the fall measured on the 15th March), gives 24 feet 11 inches for the fall below the maximum in the intervening period, or in round numbers 20 feet for the actual depth during that period. For the three months of dry weather ensuing, 17 feet 8 inches may be taken as the depth.—We have seen that the least breadth in the dry weather was 1980 feet, and the greatest in the rains 2563 feet ; leaving a difference of 583 feet. So that while the depth diminished 28 feet, the breadth diminished 583 feet, or 21 feet of breadth nearly for one of level. This proportion gives us 2383 feet for the average breadth during the four rainy months. Owing to the diminished velocity near the bank, and the diminished quantity of matter held in suspension there, this excess must be again reduced. Probably 2080 feet may be reckoned as the fair average breadth for the rainy months, 1780 for the winter, and 1730 for the hot months. The velocity, by the mean of two measurements on the 2nd and 14th April, 1832, at the surface, was 4425 feet per hour. I have no similar measurement for the maximum velocity in the rains, but while the river was at its greatest height I came a computed distance of 10 miles in an hour and a half, and from other observations of the same kind, both by myself and others, I am induced to estimate the maximum velocity at $6\frac{1}{2}$ miles, or 34320 feet per hour. Assuming that the velocity varies as the depth, we have 23,800 feet for the average velocity during the rains, 7435 feet for the winter, and 4425 feet for the hot months. I make no allowance for the decreased velocity of the stream near the bottom, because it is in all probability compensated by the increased weight of matter held in suspension there ; for the decreased velocity at the sides I have allowed by reducing the breadth. Our whole data therefore stands thus :

Season.	Depth. ft. in.	Average breadth as reduced.	Velocity. ft. per hour	Cubic feet discharged per second.
Rains, 4 months,..	36	2080	23800	494,208
Winter, 5 months, ...	20	1780	7435	71,200
Hot weather, 3 months,..	176	1730	4445	36,330

34 grains per wine quart was found to be the average for the rains. Now as a wine quart of water weighs 14544 grains, we have about $\frac{1}{428}$th part of solid matter by weight. But as the specific gravity of this cannot be stated at less than 2, we have $\frac{1}{856}$th part in bulk for the solid matter discharged, or 577 cubic feet per second. This gives a total of 6,082,041,600 cubic feet for the discharge in the 122 days of the rains :—7.8 grains per wine quart was the weight determined for the five winter months or $\frac{1}{1855}$th part in weight, and $\frac{1}{3718}$th part in bulk, which gives 19 cubic feet per second, or a total of 247,881,600 cubic feet for the whole 151 days of that period :—3.8 grains per wine quart was the weight allowed for the three hot months, which gives a $\frac{1}{3827}$th part by weight, and a $\frac{1}{7654}$th part by bulk, or about 4.8 cubic feet per second for the discharge of solid matter, and a total of 38,154,240 cubic feet for the discharge during the 92 days. The total annual discharge then would be 6,368,077,440 cubic feet.

In comparing these observations with some previous ones, I was glad to find that my average for the rains of $\frac{1}{856}$th part in bulk is nearly the same as that obtained by Captain Forbes, viz. 2 cubic inches in 1728, or 1 cubic foot.

I have stated the discharge for the hot months at 36,330 c. f. It is stated in the GLEANINGS at 20,000 at Benares. On looking over the data from which the estimates are drawn, I see that the product of the breadth and depth there given is greater than my own, the two products being to each other nearly as 5 to 8, but that the velocity I have found is to what is there given, nearly as 15 to 5. My measurements were made with care, and as I have been unable to detect any mistake in them, I have given the result of them in the hope that some one may be induced either to verify or contradict them. Again, there is a great difference between the discharge there estimated for the rains and my own results.

The former amounts to 1,372,500, or nearly three times mine, which is only 494,208. Had I made the estimate for three months of rain instead of four, my average of course would have been larger.   There is too another reason why my estimate may be considered as lower than the truth. There is in the rains a small back stream, which forms an island of the opposite shore here. I examined this in the rains, but found the velocity of it so trifling, that I was induced to pass it over. Neither of these causes could raise mine to within one half of the Benares estimate.

24

Reprinted from *U.S. Geol. Survey Water-Supply Paper 234*, 1909, pp. 78–93

DENUDATION.

By R. B. DOLE and H. STABLER.

INTRODUCTION.

The accompanying tables present estimates of the rate of denudation in the United States. The figures show the rate at which the earth's crust is being moved as solid particles carried in suspension by streams and as matter carried in aqueous solution. The first table is a summary of the estimated denudation for the whole United States and for the primary drainage basins; the other tables contain detailed estimates for smaller areas. The map indicates graphically the rates of denudation in different parts of the country.

SOURCES OF DATA.

The computations of denudation factors are based on figures representing the amount of mineral matter carried by streams, the size of the areas tributary to the streams, and the quantity of water discharged by the streams. The run-off data are derived principally from measurements made by the water-resources branch of the United States Geological Survey; some of the measurements, especially in lower Mississippi Valley and on the Great Lakes, were made by the Engineer Corps, U. S. Army; the Weather Bureau, Department of Agriculture, has contributed series of gage heights in several streams; and estimates of run-off based on the best available information have been made for areas regarding which no measurements are at hand. The estimates of the size of the drainage basins are either copied from printed reports or measured from the best available maps. The greater part of the chemical data are derived from complete mineral analyses of river waters, about 5,000 in all, performed for the water-resources branch under the direction of R. B. Dole by W. M. Barr, F. W. Bushong, C. K. Calvert, W. D. Collins, F. M. Eaton, J. R. Evans, P. L. McCreary, Chase Palmer, J. L. Porter, M. G. Roberts, F. C. Robinson, Walton Van Winkle, and A. J. Weith. Daily samples of water were collected for one year from about 150 rivers and lakes in California and in the States east of the one hundredth meridian. Ten consecutive samples were then united and the composites thus obtained were subjected to complete mineral analysis. The information regarding dissolved and suspended matter in the streams of the arid States, except California, has been furnished by the United States Reclamation Service from work done under the direction of W. H. Heileman, but unfortunately the detailed analyses could not be procured and therefore the comparative accuracy of the estimates is unknown. In the California and the Reclamation Service work 50 cubic centimeters of filtered and 50 cubic centimeters of unfiltered sample were evaporated, dried at 110° C., and weighed to determine suspended and dissolved solids. In the other analyses the suspended matter in 500 cubic centimeters of the sample was removed by filtration through a Gooch crucible, dried at 180° C., and weighed for suspended solids; the filtrate was then evaporated, dried at 180° C., and weighed for dissolved solids. Some error is introduced by comparing solids determined at 180° with solids determined at 110°. The size of the error varies with the amount of organic matter and water of crystallization present and with the character of the solids themselves, and is probably variable for different rivers and for the same river from time to time; consequently no correction for it has been made. In most cases, however, this circumstance introduces less than 10 per cent error.

The figures for Kennebec River are computed from analyses by G. C. Whipple and E. C. Levy. The dissolved solids for Connecticut River and for Housatonic River are from reports of the Connecticut state board of health; for Merrimac River, from reports of the Massachusetts state board of health; for Blackstone River, from reports of the Rhode Island state board of health. The data for Hudson River at Albany, N. Y.; Potomac River at Great Falls, Md.; and Allegheny River at Pittsburg, Pa., are from reports of water-supply investigations at those places. The figures for Colorado River at Yuma, Ariz., are computed from analyses by R. H. Forbes and from others made in continuation of his work by the Reclamation Service. The data for Ohio River at Louisville, Ky., and at Cincinnati, Ohio, are quoted from reports by G. W. Fuller. The data at West Alton and at Jefferson Barracks, Mo., are quoted from analyses by A. W. Palmer in connection with the case of the State of Missouri against the State of Illinois and the sanitary district of Chicago. The estimates of suspended and dissolved solids at New Orleans, La., are computed from fifteen years' sediment determinations by the Engineer Corps, U. S. Army, and from five years' determinations by the New Orleans water and sewerage board. In all the quoted analyses the estimates are given by weight and not by volume, and in nearly all cases they show the average condition of the water for one or more years.

METHODS OF COMPUTATION.

The estimates of denudation, computed from the above-mentioned data by H. Stabler, indicate the removal in solution and suspension of solids from the primary and secondary drainage basins of the United States as classified by the Geological Survey. The determinations on which the computations are based are given in detail. The eleven primary basins are designated by Roman numerals. Under the secondary basins, which are designated by Arabic figures, important tributaries and the sampling stations are indicated.

The second, third, fourth, and fifth columns of the tables give, respectively, the areas of the basins, the dissolved solids, the suspended solids, and the annual run-off per square mile. In columns 6 and 7 the annual denudation in tons per square mile for the areas above the points at which samples were collected is computed by multiplying together solids in parts per 1,000,000, run-off in second-feet per square mile, and 0.985; in all other regions the denudation is estimated from the data for known areas. Columns 8 and 9 show the total denudation in thousands of tons per year, computed for the secondary areas by multiplying denudation in tons per square mile per year by the drainage area in thousands of square miles.

The depth in millionths of an inch per year covered by the material removed is found by dividing the tons per square mile per year by 0.1917 and the last three columns bear reciprocal relations to columns 10, 11, and their sum. Any attempt to estimate erosion in volumetric terms from determinations of dry suspended matter and dissolved solids involves the use of factors which are by no means absolute. The actual specific gravity of the mineral substance carried in streams in the United States is not greatly different from 2.6. This figure is practically identical with that commonly assumed for the specific gravity of the earth's crust and corresponds to a weight of 165 pounds per cubic foot. Each 165 pounds of substance found in water, therefore, represents the erosion of approximately 1 cubic foot of the crust of the earth, and estimates of ultimate rock losses based upon these figures are probably not in error more than 8 or 10 per cent. Common earth or loam, however, contains a large amount of air space, or voids, and dry earth is estimated as weighing 80 to 110 pounds per cubic foot.. If an estimate of erosion be made upon this basis the error for a large area will probably not be great, but may amount to 20 per cent or more when calculations are made for small areas. Finally a third factor for calculation is based upon an attempt to determine the volume of river sediment or mud banks that a given weight of suspended matter may form. Investigators working upon different streams in the United States have obtained results indicating that a cubic foot of sediment may be produced by 50 to 125 pounds of dry material. The compactness of the mud is so variable that an estimate of this nature based upon an average of 90 pounds per cubic foot is likely to be in error by about 45 per cent. In view of the widely divergent values given for river sediment and for surface loam, the estimates for denudation expressed in millionths of an inch in depth from the entire drainage area and in years required for the erosion of 1 inch from the drainage area are based upon the assumption that 165 pounds of suspended or dissolved solids represent the removal of 1 cubic foot of the earth's crust. The factor 0.1917 has been computed on that basis. For the convenience of those who prefer a different basis of calculation, the following table of ratios is presented. The figures given in this report for denudation in inches can be recomputed by dividing them by the ratio corresponding to the weight given in column 1.

Ratios for recomputing denudation in inches.

Weight of material per cubic foot.	Corresponding specific gravity.	Ratio.
Pounds.		
165	2.64	1.00
150	2.40	.91
140	2.24	.85
130	2.08	.79
120	1.92	.73
110	1.76	.67
100	1.60	.61
90	1.44	.55
80	1.28	.49
70	1.12	.42
60	.96	.36
50	.80	.30

The estimates for the primary basins and for the whole United States are based upon the estimates for the secondary areas.

PROBABLE ACCURACY.

The majority of the figures for run-off per square mile per year are based upon discharge measurements extending over periods of seven years or more and are, with a few exceptions, within 10 per cent of the true average values. The drainage areas in square miles are so nearly correct that their errors are negligible in these calculations. The estimates of dissolved solids represent the average condition of the water for a period of one year or more. By comparison of annual averages for several streams in different years it has been estimated that the average for one year is generally not over 10 per cent from the correct average value, and it is undoubtedly true that the average total solids does not vary from year to year nearly so much as the average run-off. On the other hand, suspended solids are subject to greater variation than run-off. Floods are always attended by extreme rises in suspended solids, and the amount of suspended matter is subject to enormous variation from place to place on account of changes in stream velocity, the character of the river bed, and other features. In consequence of these facts, it is possible that denudation estimates, based on average suspended matter for one year, may be in error as much as 50 per cent. When all these facts are considered in conjunction with the distance of the sampling stations from the mouths of the rivers, it may be estimated that the calculated figures for denudation are generally within 20 per cent of the true values. Only two significant figures, therefore, have been retained in the last five columns. In many streams of the arid and semiarid regions the yearly fluctuations in run-off are very great and the mineralization of the waters is very high, so that estimates of denudation in that part of the country are not so reliable as similar estimates for eastern rivers. As the analytical data for the northern Pacific, Hudson Bay, and western Gulf of Mexico basins are extremely meager, the estimated denudations for those areas may be more than 20 per cent in error, and further investigations in those regions must be made before reliable values can be calculated.

DISCUSSION OF SUMMARY.

The summary presents in tabular form denudation estimates for the primary drainage basins and for the whole United States. The figures for dissolved solids practically represent material carried into the ocean; the figures for suspended solids, on the other hand, represent more properly material carried to tide water, because the decrease in stream velocity at that point occasions a gradual deposition of the matter transported in solid form. The tons per square mile per year removed from different basins show interesting comparisons. In respect to dissolved matter, the southern Pacific basin heads the list with 177 tons, the northern Atlantic basin being next with 130 tons. The rate for Hudson Bay basin, 28 tons, is lowest; that for the Colorado and western Gulf of Mexico basins is somewhat higher. The denudation estimates for the southern Atlantic basin correspond very closely to those for the entire United States. The amounts are generally lowest for streams in the arid and semiarid regions, because large areas there contribute little or nothing to the run-off. The southern Pacific basin is an important exception to this general rule, presumably because of the extensive practice of irrigation in that area. The amounts are highest in regions of high rainfall, though usually the waters in those sections are not so highly mineralized as the waters of streams in arid regions.

Colorado River brings down the most suspended matter, delivering 387 tons per year for each square mile of its drainage basin. Though many small streams bring silt into the Great Lakes, sedimentation clears the water, and practically no suspended matter is transported by St. Lawrence River. In general much less suspended matter is carried by northern than by southern rivers, a phenomenon influenced probably more by the texture of the soil and the subsoil and the geologic character of the rocks than by stream velocity.

The detailed estimates throw considerable light upon the progress of erosion in different sections of the river valleys. The Mississippi, for instance, apparently discharges more material than is brought in by its tributaries, thus indicating that its lower valley is still being eroded. The lower Colorado, however, appears to be receiving deposits from both dissolved and suspended matter taken from its upland drainage area. The Rio Grande is similar to the Colorado in this respect.

The estimates reveal that the surface of the United States is being removed at the rate of thirteen ten-thousandths of an inch per year, or 1 inch in 760 years. Though this amount seems trivial when spread over the surface of the country, it becomes stupendous when considered as a total, for over 270,000,000 tons of dissolved matter and 513,000,000 tons of suspended matter are transported to tide water every year by the streams of the United States. This total of 783,000,000 tons represents more than 350,000,000 cubic yards of rock substance, or 610,000,000 cubic yards of surface soil. If this erosive action had been concentrated upon the Isthmus of Panama at the time of American occupation, it would have excavated the prism for an 85-foot level canal in about seventy-three days.

Summary.

Drainage basin.	Physical and chemical data.				Estimated denudation.								
	Area (square miles).	Solids (parts per million).		Run-off (second-feet per square mile).	Tons of solids per square mile per year.		Thousand tons of solids per year.		Millionths of an inch per year.		Years required for 1 inch.		
		Dissolved.	Suspended.		Dissolved.	Suspended.	Dissolved.	Suspended.	Dissolved.	Suspended.	Dissolved.	Suspended.	Total.
I. Northern Atlantic	159,400				130	39	20,800	6,100	680	200	1,500	5,000	1,100
II. Southern Atlantic	123,900				94	176	11,700	21,800	490	920	2,000	1,100	710
III. Eastern Gulf of Mexico	142,100				117	144	16,600	20,400	580	750	1,700	1,300	750
IV. Western Gulf of Mexico	315,700				36	72	11,400	22,700	190	380	5,300	2,700	1,800
V. Mississippi River	1,265,000				108	269	136,400	340,500	560	1,400	1,800	710	500
VI. Laurentian Basin	175,000				116	1	20,400	180	600	5	1,700	200,000	1,610
VII. Colorado River	220,000				51	387	11,700	89,000	270	2,000	3,800	500	440
VIII. Southern Pacific	72,700				177	75	12,800	5,500	920	390	1,100	2,600	760
IX. Northern Pacific	270,000				100	20	27,000	5,400	520	100	1,900	9,600	1,600
X. Great Basin	223,000				90	50	20,100	11,200	470	290	2,100	3,800	1,400
XI. Hudson Bay	62,000				28	21	1,700	1,300	150	110	6,800	9,100	3,900
Total	3,048,900						290,600	524,080					
The United States	3,088,500				94	170	290,600	524,080	490	890	2,040	1,130	730
The United States (Great Basin denudation being taken as zero)	3,088,500				87	166	270,500	512,880	460	870	2,190	1,150	760

Detailed estimates.

Drainage basin.	Physical and chemical data.			Estimated denudation.									
	Area (square miles).	Solids (parts per million).		Run-off (second-feet per square mile).	Tons of solids per square mile per year.		Thousand tons of solids per year.		Millionths of an inch per year.		Years required for 1 inch.		
		Dis-solved.	Sus-pended.		Dis-solved.	Sus-pended.	Dis-solved.	Sus-pended.	Dis-solved.	Sus-pended.	Dis-solved.	Sus-pended.	Total.
I. Northern Atlantic	159,400				130	39	20,795	6,147	680	200	1,500	5,000	1,100
1. St. Johns (in United States)	7,500			1.76	85	7	638	53	440	37	2,300	27,000	2,100
2. St. Croix (in United States)	1,100			2.61	100	8	110	9	520	42	1,900	24,000	1,800
3. Penobscot	8,500			1.96	90	8	765	68	470	42	2,100	24,000	2,000
4. Kennebec	5,970			1.84	85	7.3	508	42	440	37	2,300	27,000	2,100
Waterville, Me.	4,270	48	4.0	1.76	83	8							
5. Androscoggin	3,500			2.14	100	7	350	28	520	42	1,900	24,000	1,800
Brunswick, Me.	3,500	48		2.14	101	8							
6. Presumpscot	600			1.68	85	7	51	4	440	37	2,300	27,000	2,100
7. Saco	1,720			2.36	100	8	172	14	520	42	1,900	24,000	2,100
8. Merrimac	5,020			1.62	65		326	35	340	37	3,000	27,000	2,700
Lawrence, Mass.	4,550	41		1.58	64								
9. Connecticut	11,100			1.99	90	8	999	89	470	42	2,100	24,000	2,000
Goodspeeds, Conn.	10,900	45		1.99	88	.7							
10. Blackstone	460				125	7	58	3	650	37	1,500	27,000	1,500
Valley Falls, R. I.	440	74		1.71	125	8							
11. Thames	1,450				125	8.5	181	10	650	37	1,500	27,000	1,500
12. Housatonic	1,980		4.0	2.17	195		386	16	1,000	42	980	24,000	940
Gaylordsville, Conn.	1,020			2.15									
Derby, Conn.	1,920	92		2.15	195								
13. Hudson	13,400			1.90	190		2,544	402	990	160	1,000	6,400	870
Albany, N. Y.	8,800	107	17.0	1.80	189	30							
Hudson, N. Y.	9,530	108	16.0	1.82	193	30							
Mechanicsville, N. Y.	4,500			1.81		29							
Dunsback Ferry (Mohawk River)	3,440												
14. Passaic	950			1.73	180	75	171	71	940	390	1,100	2,600	750
15. Raritan	1,110			2.17	180	75	200	83	940	390	1,100	2,600	750
Bound Brook, N. J.	800	85	36.0	2.15	179	76							
16. Delaware	13,100			2.14	150	55	1,964	721	780	290	1,300	3,500	940
Lambertville, N. J.	6,860	70	26.0	1.97	151	56							
17. Susquehanna	27,400			2.19	165	35	4,520	960	860	180	1,200	5,500	900
Danville, Pa.	11,070	112	21.0	1.56	188	35							
Williamsport, Pa.	5,640	74	18.0	1.71	123	30							
18. Potomac	14,300			1.68	125	95	1,787	1,358	650	500	1,500	2,000	870
Piedmont, W. Va.	410			1.21									

Detailed estimates—Continued.

Drainage basin.	Physical and chemical data.				Estimated denudation.								
	Area (square miles).	Solids (parts per million). Dis-solved.	Solids (parts per million). Sus-pended.	Run-off (second-feet per square mile).	Tons of solids per square mile per year. Dis-solved.	Tons of solids per square mile per year. Sus-pended.	Thousand tons of solids per year. Dis-solved.	Thousand tons of solids per year. Sus-pended.	Millionths of an inch per year. Dis-solved.	Millionths of an inch per year. Sus-pended.	Years required for 1 inch. Dis-solved.	Years required for 1 inch. Sus-pended.	Years required for 1 inch. Total.
I. Northern Atlantic—Continued.													
18. Potomac—Continued.													
Cumberland, Md.	620	130	29.0	1.70	217	49							
Millville, W. Va. (Shenandoah River).	3,000	140	39.0	1.07	147	41							
Point of Rocks, Md.	9,650	115	85.0	1.07	127	94							
Great Falls, Md.	11,400			1.12	115	95	1,195	988	600	500	1,700	2,000	900
19. James.	10,400			1.35	117	93	3,870	1,193	680	210	1,500	4,800	1,100
Cartersville, Va.	6,230	89	71.0	1.33	130	40							
Richmond, Va.	6,820			1.33									
20. Minor Chesapeake Bay.	12,800			1.59									
21. Minor North Atlantic.	17,000			2.18									
II. Southern Atlantic.	123,900				94	176	11,688	21,775	490	920	2,000	1,100	710
1. Chowan.	5,000			1.40	85	80	425	400	440	420	2,300	2,400	1,200
2. Roanoke.	9,750			1.25	101	256	984	2,490	530	1,300	1,900	750	540
South Boston, Va. (Dan River).	2,750	71	264.0	1.34	94	348							
Randolph, Va.	3,080	79	127.0	1.39	108	174							
3. Tar.	4,360			1.26	85	80	371	349	440	420	2,300	2,400	1,200
4. Neuse.	5,550			1.32	86	80	477	444	450	420	2,200	2,400	1,200
Raleigh, N. C.	1,000	73	68.0	1.19	86	80							
Selma, N. C.	1,170			1.23									
5. Cape Fear.	9,030			1.23	69	25	623	226	300	130	2,800	7,700	2,000
Wilmington, N. C.	9,030	57	21.0	1.60	69	25							
6. Pedee (Yadkin).	10,600			1.69	113	154	1,198	1,630	590	800	1,700	1,200	720
Salisbury, N. C.	3,400	69	94.0	1.66	113	154							
Pee Dee, N. C.	6,100			1.66									
7. Santee.	14,800			1.44	107	233	1,583	3,445	560	1,200	1,800	820	560
Camden, S. C. (Wateree River).	4,500	73	214.0	1.85	104	303							
Waterloo, S. C. (Saluda River).	1,060			1.85	113	98							
Columbia, S. C. (Saluda River).	2,350	62	54.0	1.38	98	232							
8. Savannah.	11,100			1.66	98	232	1,088	2,575	510	1,200	2,000	830	580
Augusta, Ga.	7,300	60	142.0		98	232							

Locality													
9. Ogeechee	5,140			1.27	90	225	463	1,156	470	1,200	2,100	850	610
10. Altamaha	14,100			1.26	85	215	1,198	3,030	440	1,100	2,300	880	640
Macon, Ga. (Oconee River)	2,420	69	174.0	1.31	86	224							
Dublin, Ga. (Oconee River)	4,180	68	171.0	1.24	83	209							
11. Minor South Atlantic	34,500				95	175							
III. Eastern Gulf of Mexico	142,100				117	144							
1. Suwanee	9,970			1.50	100	150	3,278	6,030	500	910	2,000	1,100	710
2. Apalachicola	18,800			1.61	103	139	16,610	20,430	580	750	1,700	1,300	750
Albany, Ga. (Flint River)	5,000	67	58.0	1.56	103	89	997	1,500	520	780	1,900	1,300	770
West Point, Ga. (Chattahoochee River)	3,300	52	136.0		102	265	1,936	2,988	540	830	1,900	1,200	730
3. Mobile	42,100			1.98	129	151	5,190	6,350	670	790	1,500	1,300	680
Selma, Ala. (Alabama River)	15,400	82	100.0	1.57	146	178							
Epes, Ala. (Tombigbee River)	8,830	94	100.0	1.81	98	104							
4. Pearl	8,650			1.06	75	58	649	502	390	300	2,600	3,300	1,400
Jackson, Miss.	3,120	59	46.0	1.42	75	58							
5. Minor Eastern Gulf	62,700			1.29	125	145	7,838	9,090	650	760	1,500	1,300	710
IV. Western Gulf of Mexico	315,700			1.50	36	72	11,370	22,740	190	380	5,300	2,700	1,800
1. Brazos	48,800			.200	92	96	2,970	3,100					
Waco, Tex.	32,300	1,136	1,188.0	.082	24	26							
2. Colorado (of Texas)	39,000			.097	24	26	878	952					
Austin, Tex.	35,000	321	351.0	.075	25	78							
3. Rio Grande	248,000			.032	21	418	3,250	10,140					
United States area	130,000	700	14,140.0	.030	25	78							
El Paso, Tex.	38,000			.032	90	60							
Laredo, Tex. (United States area)	123,000	791	2,475.0		94	68							
Pecos River	56,000			.032	92	11							
Dayton, N. Mex.	16,300	2,988	2,150.0	.032									
Carlsbad, N. Mex.	19,800	2,680	331.0	.035									
V. Mississippi River													
1. Main stream—													
Minneapolis, Minn	19,600	200	7.9	.810	159	6.3							
Quincy, Ill.	135,500	203	119.0	.538	108	63							
Jefferson Barracks, Mo.	700,700	206	964.0	.263	53	250							
Menard, Ill.	711,900	209	634.0	.263	70	164							
Memphis, Tenn.	941,000	202	519.0	.546	109	279							
New Orleans, La.	1,261,000	190	600.0	.560									
At mouth	1,265,000				105	331	3,120	124	830	33	1,200	30,000	1,200
2. Minor lower eastern— Tributaries	29,000				80	60							
3. Ohio River	214,000			1.71	190	200	2,320	1,740	420	310	2,400	3,200	1,400
Tennessee River	39,000			1.60	159	200	40,000	42,750	990	1,000	1,000	900	490
Chattanooga, Tenn.	21,400	101	127.0		159	200							
Gilbertsville, Ky	38,400				171	227							
Cumberland River	18,000			1.40	171	227							
Kutawa, Ky	18,200	124	165.0										

Detailed estimates—Continued.

Drainage basin.	Physical and chemical data.				Tons of solids per square mile per year.		Estimated denudation.						
	Area (square miles).	Solids (parts per million).		Run-off (second-feet per square mile).			Thousand tons of solids per year.		Millionths of an inch per year.		Years required for 1 inch.		
		Dis-solved.	Sus-pended.		Dis-solved.	Sus-pended.	Dis-solved.	Sus-pended.	Dis-solved.	Sus-pended.	Dis-solved.	Sus-pended.	Total.
V. Mississippi River—Continued.													
3. Ohio River—Continued.													
Green River	8,580				165	225							
Kentucky River	7,870	104	142.0	1.60	164	220							
Frankfort, Ky.	5,140				164	223							
Licking River	3,300				160	220							
Big Sandy River	4,000				160	200							
Kanawha River	11,000				160	180							
Monongahela River	7,620	81	84.0	1.80	217	203							
Elizabeth, Pa.	5,580				143	149							
McKeesport, Pa. (Youghiogheny River)	1,770	197	163.0	2.30	446	369							
Allegheny River	11,100				160	60							
Kittanning, Pa.	8,690	87	30.0	1.62	139	48							
Brilliant, Pa.	11,000	116	48.0		185	77							
Beaver River	3,100				180	65							
Muskingum River	7,740	244	70.0	.966	232	67							
Zanesville, Ohio	5,830				232	67							
Scioto River	6,400				220	68							
Great Miami River	5,400	289	94.0		210	68							
Dayton, Ohio	2,450				210	68							
Wabash River	33,010			.737	249	86							
Terre Haute, Ind.	12,200	336	193.0		244	140							
Vincennes, Ind.	12,720			.737	290	38							
White River, Indianapolis, Ind. (West Fork)	10,800	450	38.0	.792	351	30							
Azalia, Ind. (East Fork)	1,520	279	48.0	.900	247	43							
Shoals, Ind. (East Fork)	2,100			.924									
Embarass River	4,900	283	66.0	.76	212	49							
Lawrenceville, Ill.	2,450				212	49							
Little Wabash River	2,390	176	68.0	.77	133	52							
Carmi, Ill.	2,600				133	52							
Minor tributaries	2,500												
Main stream:	56,800												
Wheeling, W. Va.	23,800			1.72									

Locality													
Cincinnati, Ohio	72,400	120	230.0	1.00	189	363	400	230	890	510	1,100	2,000	720
Louisville Ky.	85,000	150	350.0	1.50	221	517						2,100	700
4. Big Muddy River	2,370												
Murphysboro, Ill.	2,250	225	129.0	.77	170	98							
5. Kaskaskia River	5,880						1,060	540	940	480	1,100	1,600	570
Carlyle, Ill.	2,720	248	126.0	.74	170	98							
6. Illinois River	27,900						6,040	3,290	1,100	620	880	1,100	500
Kampsville, Ill.	26,730	267	145.0	.83	180	92							
7. Rock River	10,970						2,190	1,970	1,000	940	960		
Rockton, Ill.	6,150	267	236.0	.773	180	92							
Rock Falls, Ill.	8,640				218	118							
8. Wisconsin River	12,280				218	118	1,720	86	730	37	1,400	27,000	1,300
Necedah, Wis.	5,800				200	180							
Portage, Wis.	8,600	98	4.6	1.45	203	179							
9. Chippewa River	9,570				140	7	1,110	48	600	26	1,700	38,000	1,600
Eau Claire, Wis.	6,740	90	3.7	1.31	140	6.6							
10. St. Croix River	7,660				116	5	920	38	630	26	1,600	38,000	1,500
11. Minor upper.					116	4.7							
12. Minnesota River	10,000				120	5	1,800	900	940	470	1,100	2,100	710
Eastern tributaries	16,100				180	90	1,420	418	460	140	2,200	7,400	1,700
Mankato, Minn.	13,400	480	142.0	0.186	89	26							
Shakopee, Minn.	15,100	247	120.0	.573	88	26							
13. Wapsipinicon River	2,570				145	50	370	128	760	260	1,300	3,800	980
14. Iowa River	12,400				145	50	1,800	620	760	260	1,300	3,800	980
Iowa City, Iowa	3,320	228	61.0	.656	139	68							
Cedar Rapids, Iowa (Cedar River)	6,320				147	39							
15. Des Moines River	14,700				169	349	2,480	5,130	880	1,800	1,100	550	370
Keosuqua, Iowa	14,300	312	642.0	.551	169	349							
16. Missouri River	528,000				50	290	26,400	153,100	260	1,500	3,800	660	560
Florence, Nebr.	323,000	454	2,059.0	.154	69	312							
Omaha, Nebr.	323,000	426	2,032.0	.154	60	286							
Kansas City, Mo.	430,000	346	1,890.0	.143	53	292							
Ruegg, Mo.	528,000	294	1,138.0	.157	45	176							
West Alton, Mo.	528,000	712	1,331.0	.157	25	46							
Milk River	26,700												
Havre, Mont.	7,300	375	853.0	.035	70	158							
Yellowstone River	69,700												
Glendive, Mont.	66,090	302	392.0	.188	39	51							
Platte River	83,800												
Fremont, Nebr.	71,400	383	612.0	.132	72	115							
Columbus, Nebr.	70,400				120	140							
Kansas River	61,400												
Lecompton, Kans.	58,600			.191	57	77							
Holliday, Kans.	61,100				122	145							
17. Arkansas River	189,000				160	240	22,700	26,650	630	730	1,600	1,400	740
Arkansas City, Kans.	44,500	1,284	1,736.0	.045	157	244							
Little Rock, Ark.	158,000	630	748.0	.197									
18. Red River	90,000						14,000	21,000	840	1,300	1,200	800	480
Shreveport, La.	56,900	561	870.0	.285	145	50							
19. Minor western tributaries	63,000						9,140	3,150	700	260	1,300	3,800	980

Detailed estimates—Continued.

Drainage basin.	Physical and chemical data.						Estimated denudation						
	Area (square miles).	Solids (parts per million).		Run-off (second-feet per square mile).	Tons of solids per square mile per year.		Thousand tons of solids per year.		Millionths of an inch per year.		Years required for 1 inch.		
		Dissolved.	Suspended.		Dissolved.	Suspended.	Dissolved.	Suspended.	Dissolved.	Suspended.	Dissolved.	Suspended.	Total.
V. Mississippi River—Continued.													
Cumulative summary:													
1. Above Quincy, Ill.:													
Mississippi at Minneapolis, Minn.	19,600				159	6.3	3,120	124	830	33	1,200	30,000	1,200
Minnesota River	16,100				88	26	1,420	418	460	140	2,200	7,400	1,700
Wapsipinicon River	2,570				145	50	370	128	760	260	1,300	3,800	980
Iowa River	12,400				145	50	1,800	620	760	260	1,300	3,800	980
Des Moines River	14,700				169	349	2,480	5,130	880	1,800	1,100	550	370
St. Croix River	7,660				120	5	920	38	630	26	1,600	38,000	1,500
Chippewa River	9,570				116	5	1,110	48	600	26	1,700	38,000	1,600
Wisconsin River	12,280				140	7	1,720	86	730	37	1,400	27,000	1,300
Rock River	10,970				200	180	2,190	1,970	1,000	940	960	1,100	500
Minor streams	29,650				157	63	4,660	1,880	820	330	1,200	3,000	870
Mississippi at Quincy, Ill	135,500												
Estimated from tributaries					146	77	19,790	10,442	760	400	1,300	2,500	860
Actually determined					108	63	14,640	8,540	560	330	1,800	3,000	1,100
Average					127	65	12,215	9,491	660	340	1,500	3,000	1,000
2. Above Menard, Ill.:													
Mississippi at Quincy, Ill	135,500				146	77	19,790	10,442	760	400	1,300	2,500	860
Illinois River	27,900				218	118	6,080	3,290	1,100	620	880	1,600	570
Missouri River	528,000				50	290	26,400	153,100	260	1,500	3,800	680	560
Kaskaskia River	5,880				180	92	1,060	540	940	480	1,100	2,100	700
Minor streams	14,620				145	50	2,120	731	760	260	1,300	3,800	980
Mississippi at Menard, Ill	711,900						55,450	168,103					
Estimated from tributaries					78	236	55,450	168,103	410	1,200	2,500	810	610
Actually determined					70	164	49,850	116,700	340	860	2,700	1,200	820
Average					74	200	52,650	142,402	390	1,000	2,600	990	700

	Area (sq. miles)											
3. Above Memphis, Tenn.:												
Mississippi at Menard, Ill.	711,900			78	236	55,450	168,103	410	1,200	2,500	810	610
Big Muddy River	2,370			170	98	400	250	800	510	1,100	2,000	720
Ohio River	214,000			190	200	40,600	42,750	900	1,000	1,000	900	490
Minor streams	12,730			145	50	1,840	636	760	200	1,300	3,800	980
Mississippi at Memphis, Tenn.	941,000											
Estimated from tributaries				104	225	98,290	211,719	540	1,200	1,800	850	580
Actually determined				109	279	102,500	262,400	570	1,500	1,800	690	490
Average	941,000			106	252	100,395	237,060	550	1,300	1,800	760	540
4. Mouth of river:												
Mississippi at Memphis, Tenn.	941,000			104	225	98,290	211,719	540	1,200	1,800	850	580
Arkansas River	189,000			120	140	22,700	26,450	630	730	1,600	1,400	740
Red River	90,000			160	240	14,400	21,600	840	1,300	1,200	800	480
Minor streams	45,000			103	56	4,640	2,540	540	290	1,900	3,400	1,200
At mouth	1,265,000											
Estimated from tributaries				111	208	140,030	262,309	580	1,100	1,700	920	600
Determined from analyses at New Orleans				105	331	132,820	418,720	550	1,700	1,800	580	440
Average				108	269	136,425	340,514	560	1,400	1,800	710	510
VI. Laurentian Basin: [a]												
United States area	175,000											
1. Lake Superior.												
United States area	76,100											
Saulte Ste. Marie, Mich.	48,600	.60	1.02	116	1	20,360	181	600	5	1,700	200,000	1,600
2. Lakes Michigan-Huron.	137,800		Trace.	60	1	2,920	49	310	5	3,200	200,000	3,100
United States area	71,300		.874	126	.8	8,950	59	660	4	1,500	250,000	1,500
3. Lakes Superior-Michigan-Huron			.924	99	.9	11,870	108	520	5	1,900	200,000	1,900
United States area	213,900	130	.022		.0							
Chicago, Ill. (drainage canal)	119,900	108	Trace.	96	.9							
Port Huron, Mich. (St. Clair River)	213,900	130	.902									
4. Lake Erie.	213,900		.539		.5							
United States area	40,800		.863	184		4,310	11	960	3	1,000	330,000	1,000
5. Lakes Superior-Michigan-Huron-Erie.	23,400		.018	113	.8	16,180	119	590	4	1,700	250,000	1,700
United States area	254,700			113	.8							
Chicago, Ill. (drainage canal)	143,300	130	Trace.	2.3	.0							
Buffalo, N. Y. (Niagara River and New York canals)	254,700	133	.845	111	.8							

a Figures are based upon assumption that unit denudation above Ogdensburg is the same for the United States and Canada.

Detailed estimates—Continued.

Drainage basin.	Area (square miles).	Solids (parts per million). Dissolved.	Solids (parts per million). Suspended.	Run-off (second-feet per square mile).	Tons of solids per square mile per year. Dissolved.	Tons of solids per square mile per year. Suspended.	Thousand tons of solids per year. Dissolved.	Thousand tons of solids per year. Suspended.	Millionths of an inch per year. Dissolved.	Millionths of an inch per year. Suspended.	Years required for 1 inch. Dissolved.	Years required for 1 inch. Suspended.	Years required for 1 inch. Total.
VI. Laurentian Basin—Continued.													
6. Lake Ontario......	33,000												
United States area......	18,000			1.04	144	1	2,600	18	750	5	1,300	200,000	1,300
7. Lakes Superior-Michigan-Huron-Erie-Ontario......	287,700				116	.9							
United States area......	161,300			.884	116	.9	18,780	137	600	5	1,700	200,000	1,000
Chicago, Ill. (drainage canal)	287,700	130	Trace.	.016	2.1	.0							
Buffalo, N. Y. (New York canals)	287,700	133	Trace.	.008	1.1	.0							
Ogdensburg, N. Y. (St. Lawrence River)	287,700	134	Trace.	.859	113	.9							
8. St. Lawrence River (below Ogdensburg, N. Y.)—													
United States area......	13,660			1.72	116	3.2	1,580	44	600	17	1,700	59,000	1,000
Oswegatchie River—Ogdensburg, N. Y......	1,580	77	10.0		130	17							
Lake Champlain......	7,900	66	Trace.		113	1.7							
Rouses Point, N. Y......	7,750			1.74									
VII. Colorado River:													
United States area......	230,000			.0733	51	387	11,730	89,010	270	2,000	3,800	500	440
Yuma, Ariz......	225,000	710	5,354.0		51	387							
1. Gila......	71,100				80	a 225	5,690	16,000	420	1,200	2,400	800	630
Salt River......	13,000			.238	150	235							
Roosevelt, Ariz......	5,700	646	1,300.0	.141	151	304							
McDowell, Ariz. (Verde River)......	6,000	363	1,137.0	.189	50	158							
Mesa, Ariz......	12,260	1,090	1,274.0		203	237							
Gila River......		811	4,682.0		13	210							
San Carlos, Ariz......	13,500	811	4,682.0	.015	12	69							
Florence, Ariz......	17,850	944	14,130.0	.015	14	209							
2. Little Colorado......	31,000				11	134	340	4,150	57	700	18,000	1,400	1,300
Woodruff, Ariz......	6,000	602	7,149.0	.019	11	134							
3. Grand......	26,200				198	470	5,190	12,310	1,000	2,400	970	410	a 250
Palisades, Colo......	8,550	a 429	b 1,358.0	.546	230	730							
Whitewater, Colo. (Gunnison River)......	7,870	484	553.0	.350	167	190							

Locality													
4. Green	41,000				90	165	3,690	6,760	470	860	2,100	1,200	750
Jensen, Utah	26,600	369	679.0	.247	90	165							
VIII. Southern Pacific	72,700				177	75	12,850	5,460	920	390	1,100	2,600	760
1. Minor Southern Pacific	22,800				196	71	4,470	1,620	1,000	370	980	2,700	720
Soledad, Cal. (Arroyo Seco)	215	284	33.0	.325	222	26							
Santa Maria, Cal. (Santa Maria River)	1,000	2,412	1,302.0	.793	204	110							
Santa Barbara, Cal. (Santa Ynez River)	207	714	64.0	.495	348	31							
Calabasas, Cal. (Malibu Creek)	97	717	61.0	.375	265	23							
Mentone, Cal. (Santa Ana River)	182	139	53.0	.409	56	21							
Pala, Cal. (San Luis Rey)	318	321	142.0	.352	111	49							
2. San Francisco Bay	49,900				168	77	8,380	3,840	880	400	1,100	2,500	780
Sacramento River	31,400			1.45	172	86							
Sacramento, Cal	29,500	121	60.0	1.45	172	86							
San Joaquin River [c]	16,500			1.02	102	60							
Lathrop, Cal	12,500	161	60.0	1.02	162	60							
IX. Northern Pacific	[d]270,000				100	20	27,000	5,400	520	100	1,900	9,600	1,600
X. Great Basin	[d]223,000				90	50	20,070	11,150	470	260	2,100	3,800	1,400
XI. Hudson Bay (in United States)	62,000				28	21	1,740	1,300	150	110	6,800	9,100	3,900
Red River of the North—													
Total area in United States	50,000				28	21	1,400	1,050	150	110	6,800	9,100	3,900
Grand Forks, N. Dak.	25,000	[e]200	[e]150.0	.141	28	21							

[a] Suspended matter probably far too low.
[b] Analyses cover only portion of year. Results appear too high for an average.
[c] Excluding Tulare Lake.

[d] Analytical data extremely meager.
[e] Estimated from four analyses.

25

Reprinted from Water Resources Research 4:737–747 (1968)

The Sediment Yield of Major Rivers of the World

JOHN N. HOLEMAN

Engineering Division, Soil Conservation Service
U. S. Department of Agriculture, Hyattsville, Maryland 20782

Abstract. The amount of suspended sediment transported by rivers to the seas each year is tabulated. The major rivers are ranked in order of tons of sediment transported per year and drainage area and water discharge data are included. The rivers are listed by continents in subsequent tables with data on drainage area, annual sediment yields in tons, sediment production rates in tons per square mile per year, the years of sediment measurement, and the sources of data. This sample represents more than one-third of the land contributing water-borne sediment to the seas and, if representative, indicates an annual world sediment yield of 20 billion (20 × 10⁹) tons. The data suggest that Africa, Europe, and Australia have very low sediment yields (<120 tons per square mile per year), South America's rate is low, North America's is moderate, and Asia's is high to the degree of yielding up to 80% of the sediment reaching the oceans annually.. (Key words: Data; erosion; runoff; sediment; world)

INTRODUCTION

Knowledge of the amount of sediment transported by the various rivers of the world is important for many reasons. For example, reservoirs should be designed with enough space to store the sediment expected to accumulate in them and yet retain full effectiveness during their design life, and sediment yield is an indication of the rate of erosion in the drainage basin. A summation of sediment yields by basins may indicate a regional, continental, and even an approximate world rate.

The amount of sediment transported by water to ocean level has been estimated by a number of people. These estimates have covered a rather wide range. For example, *Fournier* [1960] calculated more than 64 × 10⁹ (64 billion) tons a year. The estimate of *Lopatin* [1952] was about 14 × 10⁹ short tons annually. The data compiled here represent more than a third of the land contributing to the sea. Direct extrapolation of these data by continents gives a world total sediment yield to the marine environment of 20.2 × 10⁹ tons per year.

The validity of some of the data is questionable because of short records and variability of sediment sampling methods. In addition, some of the data may include estimates of sediment transported as bedload, whereas most only include suspended loads. This compilation is not a complete listing of all of the major rivers discharging to the sea whose sediment load has been measured; however, it does include the major rivers (excepting some in the United States) with published data that the writer could locate at this time.

The data were reported in many different forms, such as cubic meters per year, metric tons per square kilometer per year, and kilograms per second, and some large area denudation rates were given in feet per thousand years and millimeters per year. The various weights were converted into short tons (2000 lbs) and areas into square miles. For the conversion of volume of rock eroded into weight, the average specific gravity of the rock was assumed to be 2.65.

GENERAL

The Earth's surface is nearly 200 million square miles (actually 196.9). Of this, 71% is water. Land makes up the remaining area of 57.5 million million square miles. Of this, about 5 million are in Antarctica, and about 13.2 million square miles are closed basins and deserts contributing no runoff. According to *Livingstone* [1963], this leaves 39.2 million square miles contributing runoff and water-borne sediment to the seas.

Strahler [1963, p. 264] presents a world map

of average annual precipitation, and *Putnam* [1964] estimated the average annual rainfall on land to be 24,000 cubic miles a year (between 26 and 27 inches). *Langbein and Schumm* [1958] found that the maximum sediment yield in the United States occurs when annual effective precipitation is between 10 and 14 inches. The sediment yield dropped sharply as annual rainfall decreased from 10 inches because of the deficiency of runoff. On the other hand, sediment yield decreased generally with rainfall above 14 inches, because the increased rainfall produced a greater density of vegetative cover and therefore less erosion.

This generalization may or may not apply to other environments of the world, but the relation between rainfall and vegetative cover seems universal. A world map showing the distribution of types of vegetation is in the 1941 Yearbook of Agriculture, *Climate and Man* [*U. S. Dept. Agr.*, 1941]. The vegetation reflects rather closely the annual rainfall patterns. The rates of runoff are also closely tied in with precipitation. In 1960, Durum and others estimated the annual worldwide runoff to be 40 million cfs, or approximately 1 cubic foot per second per square mile (csm). Recently, with additional data, the figure has been refined downward to 0.82 csm (W. H. Durum, personal communication, 1968). This is 11.1 inches of worldwide runoff.

A summary of sediment yields on a worldwide basis was not available, and this, in large part, is what prompted preparing this compilation. Of the 16 rivers having drainage areas greater than 400,000 square miles, no sediment data were located for the Lena and Amur in northeastern Asia, the Mackenzie and Nelson in Canada, and the Zambezi in southeast Africa. The Nelson River flows from Lake Winnipeg to the Hudson Bay. Sediment data are included for the two principal tributaries of Lake Winnipeg, the Red and Saskatchewan rivers. The Murray River in Australia has about the same drainage area as the Nelson, and sediment yield is estimated for it (see page 741). Table 1 lists 16 rivers that transport more than 100 million tons of sediment to the sea annually. To facilitate comparisons of the sediment production rates of the world's river basins, the following arbitrary but convenient limits were used: those exceeding 500 tons/sq mi/yr as high; those between 200 and 500 as moderate; and those under 200 as low.

TABLE 1. Selected Rivers of the World Ranked by Sediment Yield ($>10^8$ tons/yr)

Name	Location	Total Drainage Area,* 10^3 sq mi	Average Annual Suspended Load (thousand tons)	(tons/sq mi)	Average Discharge at Mouth,* 10^3 cfs
1. Yellow	China	260	2,080,000	7,540	53
2. Ganges	India	369‡	1,600,000	4,000	415
3. Bramaputra	East Pakistan	257‡	800,000	3,700	430
4. Yangtze	China	750	550,000	1,400	770
5. Indus	West Pakistan	374‡	480,000	1,300	196
6. Ching (Yellow trib.)	China	22†	450,000	20,500	2
7. Amazon	Brazil	2,230	400,000	170	6,400
8. Mississippi	USA	1,244	344,000	280	630
9. Irrawaddy	Burma	166	330,000	2,340	479
10. Missouri (Miss. trib.)	USA, Missouri	529	240,000	450	69
11. Lo (Yellow trib.)	China	10†	210,000	20,200	...
12. Kosi (Ganges trib.)	India	24‡	190,000	7,980	64‡
13. Mekong	Thailand	307‡	187,000	1,240	390
14. Colorado	USA	246§	149,000	1,080	5.5§
15. Red	North Viet Nam	46‡	143,000	3,090	138‡
16. Nile	Egypt	1,150	122,000	100	100

Incomplete, as data for many rivers are lacking. For sources consult Tables 2, 3, 4, and 5.
* Data in these two columns are from AGI Data Sheet No. 32 (rev. 1964), unless otherwise noted.
† U. N. 1953.
‡ U. N. 1965.
§ *Durum et al.*, 1960.

TABLE 2. Suspended Sediment Yield of Selected Rivers, North America

River	Location	Drainage Area, sq. mi.	Average Annual Suspended Load (10³ tons)	Average Annual Suspended Load (tons/sq mi)	Average Water Discharge at Mouth, 10³ cfs	Period of Sediment Record	Source
Mississippi	Baton Rouge, Louisiana†		262,487			10/49-9/66	Hobbs, B. L. (personal communication, 1968)
Atchafalaya	Simmesport, Louisiana		121,400			10/51-9/65	Hobbs, B. L. (personal communication, 1968)
Mississippi	(sum of two above for 14 yrs.)	1,244,000	344,000	277	630‡	10/51-9/65	Hobbs, B. L. (personal communication, 1968)
Missouri*	Hermann, Missouri	528,200	239,562	454	69.3‡	1949-54	Hobbs, B. L. (personal communication, 1968)
St. Lawrence	mouth, Canada	498,000	4,000	8	500‡	...	Corbel [1959]
Colorado	Grand Canyon, Arizona	137,800	149,000	1,082	5.5	10/25-9/57	Judson and Ritter [1964]
Saskatchewan	The Pas, Canada	125,064	4,600	37	80§	1954-60; 62-64	IASH [1967]
Red	Lockport, Canada	110,782	1,770	16	80§	1956-58; 61-64	IASH [1967]
Snake	Central Ferry, Washington	103,500	13,100	127	48.6‡	4/50-7/52	Judson and Ritter [1964]
Columbia	Pasco, Washington	102,600	10,300	100	256‡	6/50-7/52	Judson and Ritter [1964]
Ohio*	Cincinnati, Ohio	76,580	15,000	196	258.1¶	10/41-9/42	Hobbs, B. L. (personal communication, 1968)
Yellowstone*	Glendive, Montana	66,880	30,300	455	...	1929-31	Fournier [1960]
Brazos	Richmond, Texas	34,800	34,800	1,000	5.2‖	1924-1950	Fournier [1960]
Rio Grande	San Acacia, New Mexico	26,770	9,420	352	2.7‖	10/47-9/56	Judson and Ritter [1964]
Alabama	Claiborne, Alabama	22,000	2,130	97	31.9‡	10/52-9/60	Judson and Ritter [1964]
Potomac	mouth, Maryland & Virginia	14,600	2,500	170	11.1**	...	Sedimentation [1967]
San Joaquin	Vernalis, California	14,010	347	25	4.7††	10/56-9/60	Judson and Ritter [1964]
Rio San Juan	Santa Rosalia, Mexico	12,000	5,340	445	0.3‖	1934-1941	Fournier [1960]
Delaware	Trenton, New Jersey	6,780	998	147	20.1‡	10/49-9/57	Judson and Ritter [1964]
Sabine	Logansport, Louisiana	4,860	730	150	8.8‖	1932-1950	Fournier [1960]
Pecos	Puerto de Luna, New Mexico	3,970	2,720	685	0.6‖	10/48-9/57	Judson and Ritter [1964]
Eel	Scotia, California	3,113	18,200	5,846	7.0††	10/57-9/60	Judson and Ritter [1964]
Total (without tributaries)		2,464,649	603,955	Average 245			

* Tributary to a listed river.
† Sampling at Red River Landing after Jan. 1, 1958.
‡ AGI Data Sheet #32 (rev. 1964).
§ AGI #32, mouth of Nelson River.
‖ U. S. Geol. Surv. Water-Supply Paper 1732.
¶ U. S. Geol. Surv. Water-Supply Paper 1725.
** U. S. Geol. Surv. Water-Supply Paper 1722.
†† U. S. Geol. Surv. Water-Supply Paper 1735.

NORTH AMERICA

North America is the continent with the most data collected on sediment transported by water. Most of the data are from the United States and are derived from secondary sources as indicated (Table 2). The original data used by these sources were collected mostly by the U. S. Geological Survey and the U. S. Corps of Engineers.

The rivers are listed in descending order of drainage area. The Mississippi since 1949 has transported an average of 262 million tons a year past Red River Landing and Baton Rouge, Louisiana. This average is based on the U. S. Corps of Engineers measurements for 1949–1966. Before 1952, an annual load of 500 million tons at this location was not unusual; since then, no more than 325 million tons have been measured in any one water year. Soil conservation practices, streambank stabilization measures, and new reservoirs have been the major factors in this reduced sediment transport.

Brown [1950] reported that the Missouri River at Yankton, South Dakota, carried an average of 134 million tons a year. This figure was based on nine years of record. In 1953 storage began at Ft. Randall and Garrison Dams, 82 and 615 river miles upstream from Yankton, respectively. In June 1955, Gavins Point Dam, only a few miles upstream from Yankton, was closed. The effect of these three dams upstream can be noted in the sharp decline of the suspended load to an annual average (1955–1963) of 2,363,000 tons per year (D. C. Bondurant, Corps of Engineers, personal communication, 1968).

The St. Lawrence has a large drainage area, but much of it consists of the Great Lakes and, as one might expect, the St. Lawrence carries a very small amount of sediment. The Colorado discharges a large load, formerly much more than it does now. The 149 million tons (Table 2) is a 32-year average.

Additional rivers with decreasing drainage areas and sediment yields are listed. There is only one river in Mexico listed, and for Canada, besides the St. Lawrence, only the Saskatchewan and the Red are listed. Sediment in these two rivers in south central Canada was measured upstream from their debouchment at Lake Winnipeg. The average annual sediment load was measured at 37 and 16 tons per square

mile, respectively. The average sediment load for the rivers listed in Table 2 is 283 tons per square mile per year; after deleting the rivers that are tributaries, the average for North America becomes 245 tons per square mile per year.

SOUTH AMERICA

The continental divide in South America, or crest of the Andes mountains, is near the Pacific Ocean. The rain falling on the eastern slopes of the Andes has to travel down the Amazon about 3000 miles to the Atlantic Ocean. The Amazon has the largest drainage area of any river in the world and by far the largest water discharge. Because of the heavy rainfall on the large basin, the Amazon is believed to discharge anywhere from 11% [*Oltman*, 1964] to 18% [*Davis*, 1964] of the world's total annual volume of water discharged into the oceans. On the other hand, it transports only about 2% of the sediment reaching the oceans annually. According to *Gibbs* [1967], 82% of the suspended sediment reaching the mouth of the Amazon comes from the Andes mountains and highlands, only 12% of the drainage area; and about 6% of the total sediment discharge is bedload.

There are other very large rivers in South America. The Orinoco discharges the third largest volume of water in the world, exceeded only by the Amazon and Congo, but its sediment yield is moderate (Table 3). The Parana' to the south ranks eighth in the world in both drainage area and water discharge, but sediment yield is low: only 100 tons per square mile. An estimate of the sediment yield for South America is 160 tons per square mile (Table 3).

AFRICA

On the basis of this sample of five rivers, Africa has less than one-half the sediment production rate of South America: about 70 tons per square mile. The rate of the Medjerdah in Tunisia is high, the Cheliff in Algeria is moderate, the Nile is low, and the Niger and Congo are remarkably low (see Table 3).

Africa has four main rivers, each discharging to one of the four points of the compass: the Nile to the north, the Niger to the south, the Zambezi to the east, and the Congo to the west. The table indicates low sediment yields in central Africa. Although evidence of excessive ero-

TABLE 3. Suspended Sediment Yields of Selected Rivers
South America—Africa—Australia

River	Location	Drainage Area, sq. mi.	Average Annual Suspended Load (10³ tons)	(tons/sq mi)	Average Water Discharge at Mouth,† 10³ cfs	Period of Sediment Record	Source
South America							
1 Amazon	mouth, Brazil	2,368,000	400,000	170	6,400		Ames, F. C. (personal communication, 1967)
2 Parana	mouth, Argentina	±890,000	90,000	100	526		Corbel [1959]
3 Orinoco	mouth, Venezuela	366,700	95,340	260	800		Van Andel [1967]
4 Uruguay	Concordia, Argentina	150,000	15,000	100	140		Babb [1893]
5 Negro*	Primera Angustura, Argentina	36,670	1,487	405	36	1935–1965	IASH [1967]
6 Caroni*	at Orinoco, Venezuela	±35,000	52,500	1,500			Koelzer, V. A. (personal communication, 1967)
7 Colorado	Pichi Mahuida, Argentina	9,000	7,600	880		1938–1964	IASH [1967]
	Total (less tributary)	3,820,370	609,427	Average 160			
Africa							
1 Congo	mouth, Congo	1,550,000	71,300	46	1,400		Corbel [1959]
2 Nile	delta, Egypt	1,150,000	121,854	100	100	1958–1964	Simons, D. B. (personal communication, 1967)
3 Niger	Baro, Nigeria	430,000	4,960	12	215		Dekker [1961]
4 Cheliff	mouth, Algeria	8,600	3,400	395		16 years	Tixeront [1960]
5 Medjerdah	mouth, Tunisia	8,080	14,750	1,825		5 years	Tixeront [1960]
	Total	3,146,680	216,264	Average 70			
Australia, et al.							
30 streams	Eastern Australia			85		2 years	Douglas [1967]
Murray-Darling	mouth, Australia	414,000	35,190				
Waipaoa	Kanakanaia, New Zealand	610	12,160	19,930	13	1960–1964	IASH [1967]
	Total	414,610	47,350	Average 115			

* Tributary to listed river.
† AGI Data Sheet No. 32 (rev. 1964).
‡ Estimate based on the 85 tons/sq. mi mentioned in text.

sion in South Africa exists, no quantitative data were located.

AUSTRALIA

Sediment yield data for Australia are very limited. According to Fournier [1960], eastern Australia appears to average about 850 tons per square mile. *Douglas* [1967] disputed this rate. He has observed the sediment load in 30 rivers in eastern Australia (New South Wales and Queensland) and believes that Fournier's figure is too high by a whole order of magnitude. Therefore, a figure of 85 tons per square mile may be more in order, although it is only an estimate. The vast central part of this island continent is a desert.

EUROPE

Europe has none of the 16 major rivers of the world no matter whether they are ranked by drainage, water discharge, or sediment yield (Table 1). In terms of drainage area, the Volga, which empties into the Caspian Sea, ranks seventeenth. It carries a low suspended sediment load. The second largest is the Danube, flowing from southwestern Germany eastward into the Black Sea. Its sediment load is also low (Table 4).

The Rhine carries a heavy load of sediment as it enters Lake Constance in Switzerland, but by the time it reaches the flatlands of Holland it carries little sediment. The Po, the Tiber, and the Arno in Italy carry relatively high amounts of sediment, but generally erosion is not a big problem in Europe. This is because rainfall is distributed well throughout the year, and low-intensity, gentle rains are typical. The average sediment production rate is about 90 tons per square mile based on these data.

ASIA

The largest continent, Asia, also produces by far the most sediment reaching the sea (Table 5). According to a United Nations publication of 1953, *The Sediment Problem* [*United Nations*, 1953], the Yellow River of China carries more than 2×10^9 short tons of sediment to the Yellow Sea each year on the average. The principal source is the yellow loess of north central China, giving the river and sea its color and name. The Yangtze River to the south of the Yellow River also flows from west to east. Although the Yangtze drains a much larger area than the Yellow River, it transports approximately a fourth of the amount of sediment. The Lo and Ching rivers are tributaries to the Yellow and carry nearly three times as much sediment per square mile as any of the other rivers of the world with equal or larger drainage areas listed here (20 thousand tons per square mile). At times, flows in the Yellow River have been reported to be 40% sediment by weight; that is, 400,000 ppm [*Todd and Eliassen*, 1938]. The Yellow River carries the largest amount of sediment of any river in the world.

The Ganges River in India is the second largest carrier of sediment, discharging more than $1\frac{1}{2} \times 10^9$ tons of sediment into the Bay of Bengal each year. The Ganges arises along the southwest side of the Himalaya Mountains and flows south of them roughly parallel to their crest. The runoff from most of the south side of this highest mountain range in the world flows into the Ganges. This area is subject to intense erosion because of the high relief and the monsoon climate. Some of the maximum annual rainfalls and most intense storms of the world have occurred in the foothills of the Himalayas. As an example, at Cherrapunji, India, in the Khasi Hills, 241 inches of rain fell in the month of August 1841, 40.8 inches of which fell in one day. The average annual rainfall at Cherrapunji is 426 inches [*U. S. Dept. Agr.*, 1941, p. 664]. The Kosi, tributary to the Ganges, alone transports on the average 190 million tons of sediment a year.

The Bramaputra River originates on the north side of the Himalayas in Tibet. It flows eastward and also roughly parallels the trend of the mountains. On the eastern flank of the Himalayas the Bramaputra breaks through to the south, then reverses directions and flows west along the southeastern side of the mountains. It empties into the Bay of Bengal, forming a joint delta with the Ganges. These two rivers are thought to rank as number 2 and 3 among the world's rivers in the amount of sediment discharged. The Indus River in West Pakistan transports almost as much sediment as the Bramaputra.

Eleven of the 16 rivers in Table 1 with sediment yields $>10^3$ tons are in southern Asia. Undoubtedly, as the store of data increases,

TABLE 4. Suspended Sediment Yields of Selected Rivers
Europe

River	Location	Drainage Area, sq. mi.	Average Annual Suspended Load (10³ tons)	Average Annual Suspended Load (tons/sq mi)	Average Water Discharge at Mouth, 10³ cfs	Period of Sediment Record	Source
Volga	Dubovka, USSR	521,490	20,770	40	283‡	1934, 35, 38–40	Lopatin, 1952
Danube	mouth, USSR	315,000	21,420	68	218‡	1938–39	Babb, 1883; Corbel, 1959
Duepr	Verkhnedneprovsk, USSR	167,520	1,210	7	194‡	1938–39	Lopatin, 1952
Don* (Volga trib.)	Razdorskaya, USSR	146,020	5,360	37		1932–40, 46, 47	Lopatin, 1952
Ural	Topolinski, USSR	74,790	1,760	24		1936–41, 47	Lopatin, 1952
Vistula	Tczew, Poland	74,560	1,690	23	38‡	1946–53	Jarocki, 1963 (p. 80)
Tisza*	at Danube River, Hungary	60,310	11,040	180		not cited	Fournier, 1960
Rhine	mouth, Holland	56,000	504	9	78‡	not cited	Corbel, 1959
Rhine*	Lake Constance, Switzerland	4,600	9,632†	2,094†		not cited	Corbel, 1959
Loire	Nantes, France	46,740	467	10	30§	not cited	Corbel, 1959
Oder	Gozdowice, Poland	42,230	147	4		1961–64	IASH, 1967
Po	Pontelagoscuro, Italy	20,960	16,770	800		1956–62	IASH, 1967
Seine	Paris, France	17,140	1,220†	71†	51‡	pre 1865	Corbel, 1959
Tiber	Rome, Italy	6,390	6,420	1,005		1933–46, 49–63	IASH, 1967
Drin	Can Deje, Albania	4,770	16,220	3,400		1960–63	IASH, 1967
Garonne	Toulouse, France	3,860	2,760	715	24§	1839–46	Fournier, 1960
Inn*	Reisach, West Germany	3,767	3,515	933		1953–1960	IASH, 1967
Arno	San Giovanni alla Vena, Italy	3,160	2,430	770		1936–42; 54–64	IASH, 1967
Semani	Urae Kucit, Albania	2,040	24,190	11,844		1961–63	IASH, 1967
Simento	Giarretta, Sicily	707	3,960	5,605		1936–42; 57–63	IASH, 1967; Gazzolo and Bassi, 1960
Total (less tributaries)		1,357,357	121,938	Average 90			

* Tributary to a listed river.
† Includes bed load.
‡ AGI Data Sheet No. 32 (rev. 1964).
§ Durum, W. H. (personal communication, 1968).

TABLE 5. Suspended Sediment Yields of Selected Rivers
Asia

River	Location	Drainage Area, sq. mi.	Average Annual Suspended Load (10³ tons)	(tons/sq mi)	Average Water Discharge at Mouth, 10³ cfs	Period of Sediment Record	Source
1. Yenisei	Igarka, USSR	954,200	11,600	12	614†	1942, 43	Lopatin [1952]
2. Ob	Salekhard, USSR	945,300	15,700	17	441†	1938–44	Lopatin [1952]
3. Ganges	delta, East Pakistan	409,200	1,600,000	4,000	498‡	1874–1879	Fournier [1960]
4. Yangtze	Chikiang, China	395,650	552,000	1,400	770†		U. N. [1953]
5. Indus	Kotri, West Pakistan	370,000	481,000	1,300	239‡	1902–25	Fournier [1960]
6. Indus*	Kalabagh, West Pakistan	117,700	751,000	6,370			U. N. [1965]
7. Yellow	Shenhsien, China	276,000	2,083,000	7,545	53†	1934–42	U. N. [1953]
8. Bramaputra	delta, East Pakistan	216,000	800,000	3,700	706		U. N. [1965]
9. Mekong	Mukdaham, Thailand	151,000	187,000	1,240	530		U. N. [1965]
10. Irrawaddy	Prome, Burma	141,700	331,000	2,340	479†		U. N. [1953]
11. Pearl-West	Wuchow, China	120,800	30,800	250	277		U. N. [1953]
12. Mahanadi	Naraj, India	51,000	67,800	1,330	101		U. N. [1953]
13. Euphrates	Tabqa, Syria	46,570	4,750	100	51§	1962–64	IASH [1967]
14. Red	Hanoi, North Viet Nam	46,300	143,000	3,080	138		U. N. [1965]
15. Chao Phya	Nakornsawan, Thailand	41,100	12,490	300		1956–63	IASH [1967]
16. Kabul*	Nowshera, West Pakistan	34,880	26,000	745	24‡	1961, 1962	Koelzer [1967]
17. Tigris	Bagdad, Iraq	30,800	57,600	1,870	51§	1918, 1919	Fournier [1960]
18. Kosi*	Chatra, India	23,800	190,000	7,980	64		U. N. [1965]
19. Ching*	Changchiashan, China	22,000	450,000	20,500	2	1932–45	U. N. [1953]
20. Chenab*	Alexandria Bridge, W. Pakistan	12,560	55,000	4,380		1961, 1962	Koelzer [1967]
21. Lo*	Chuantou, China	10,400	210,000	20,200		1934–45	U. N. [1953]
22. Damodar	Rhondia, India	7,680	31,300	4,050	11		U. N. [1953]
23. Ishikari	Ebestu, Japan	4,900	1,926	393		1960	IASH [1967]
24. Tone	Matsudo, Japan	4,630	3,160	778		1951	IASH [1967]
Total (less tributaries)		4,212,830	6,414,576	Average 1,530			

* Tributary to a listed river.
† AGI Data Sheet No. 32 (rev. 1964).
‡ U. N. 1965.
§ Tigris, Euphrates, and Karun Rivers, Geotimes, March 1962.

there will be changes in this list. Southern Asia; however, will remain the principal contributor of sediment to the oceans.

Northern Asia, consisting of Siberia and Mongolia, has little sediment data available. Yet four of the world's largest rivers, in terms of drainage area, are in northern Asia. From west to east they are the Ob, the Yenisei, and the Lena, which flow north into the Arctic Ocean, and the Amur, which flows east to the Pacific. The best indication of sediment yield in this area is the report of *Lopatin* [1952] that the average annual denudation for the USSR is 0.027 mm. This is equivalent to 1.06 inches per 1000 years, or about 200 tons per square mile per year. The estimated annual rate of sediment reaching the oceans from Asia is 1530 tons per square mile (closed basins and deserts excluded).

SUMMARY

The measured and estimated sediment yield are summarized by continents in Table 6: Africa, Europe, and Australia appear very low, averaging 70, 90, and 115 tons per square mile each year, respectively; South America is low with 160 tons per square mile; North America is a moderate sediment producer with 245; and the high producer of sediment is Asia, with 1530 tons per square mile.

Table 6b shows the total sediment yield to the oceans as extrapolated directly from the above data. Using the total areas draining to the oceans and multiplying by the average rates ascertained, the total is 20.2×10^9 tons. Of this, 80% comes from Asia, which makes up one-quarter of the land area draining to the oceans. Table 7 compares this figure with some other world estimates.

To visualize better the 20.2 billion tons of sediment, consider it by volume rather than by weight. From reservoir sedimentation surveys of the United States, the volume-weight of submerged sediment deposits may be assumed

TABLE 6a.   Summary of Measured Annual Sediment Yields of Rivers to Oceans
(tributaries deleted)

Continent	Measured Drainage Area, Mi^2	Annual Suspended Sediment Discharge	
		(Tons) (1000 tons)	(Tons/Mi^2)
North America	2,464,649	603,955	245
South America	3,820,370	609,427	160
Africa	3,146,680	216,264	70
Australia	414,610	47,350	115
Europe	1,357,357	121,938	90
Asia	4,212,830	6,414,576	1,530
Total	15,416,496	8,013,510	Average 520 tons/mi^2

TABLE 6b.   Total Sediment Yield to Oceans Extrapolated from Above Data

Continent	Total Area Draining to Oceans,* Mi^2	Annual Suspended Sediment Discharge	
		(Tons/Mi^2)	(10^9 Tons)
North America	8,000,000	245	1.96
South America	7,500,000	160	1.20
Africa	7,700,000	70	0.54
Australia	2,000,000	115	0.23
Europe	3,600,000	90	0.32
Asia	10,400,000	1,530	15.91
Total	39,200,000		20.16

Total suspended sediment discharge to oceans is about 20 billion tons a year.

* *Livingstone*, 1963.

TABLE 7. Comparison of Estimates of Sediment Reaching the Oceans Annually

Reference		Denudation or Sediment Yield		Equivalent*, tons/sq mi
		various units	10⁹ short tons	
Fournier	1960	51.1 × 10⁹ metric tons	= 64*	1630
Kuenen	1950	32.5 × 10¹² kilograms	= 35.8	915
Gilluly	1955	12.0 km³ of rock†	= 35.0	895
Pechinov	1959	0.09 mm/yr	= 26.7	680
Schumm	1963	0.25 ft/1000 yr‡	= 22.6	575
Holeman	1968	20.16 × 10⁹ tons	= 20.2	520
Lopatin¶	1952	12.695 × 10⁹ tons (metric)	= 14.0	355

* Adjusted to 39.2 milllion square miles contributing sediment to the oceans.
† A refinement of Kuenen's estimate.
‡ Based on drainage areas <1500 sq mi and effective precipitation of less than 40 inches per year.
¶ From *Jarocki* [1963, p. 8].

to average about 60 pounds per cubic foot, or about 1300 tons per acre-foot. On this basis the sediment reaching the sea each year is enough to cover France, Belgium, the Netherlands, Luxembourg, Switzerland, and Portugal with an inch of mud.

As a last observation about these sediment yield data, it is important to remember that these figures are a small per cent of the soil material that is moved each year. The Potomac River will serve as an illustration of this. The Potomac River Basin has been studied in greater detail than most of the areas included here. The Sedimentation and Erosion Sub-Task Force of the Federal Interdepartmental Task Force [*Sedimentation*, 1967] has estimated that about 2.5 million tons of sediment are discharged into the Potomac estuary annually. This group points out that gross erosion within the basin is over 50 million tons a year. In other words, only about 5% of the products of erosion in this watershed reach tidewater.

REFERENCES

Babb, C. C., The sediment of the Potomac River, *Science, 21,* 342–343, 1893.
Brown, C. B., Sediment transportation, in *Engineering Hydraulics,* edited by H. Rouse, 769–857, John Wiley & Sons, Inc., 1950.
Corbel, J., Vitesse de l'erosion, *Z. Geomorphologie, 3,* 1–28, 1959.
Davis, L. C., The Amazon's rate of flow, *Nat. History, 73*(6), 5–19, 1964.
Dekker, G., Sediment transport measurements and computations on the Niger, *Inter-African Conf. Hydrol., Publ. 66,* London, Comm. Tech. Cooperation in Africa So. of Sahara, 256–265, 1961.

Douglas, I., Man, vegetation, and the sediment yields of rivers, *Nature, 215,* 925–928, August 26, 1967.
Durum, W. H, S. G. Heidel, and L. J. Tison, World-wide runoff of dissolved solids, *Intern. Assn. Sci. Hydrol., Publ. 51,* 618–628, 1960.
Fournier, F., *Climat et erosion,* Presses Universitaires de France, Paris, 1960.
Gazzolo, T., and G. Bassi, Contribution a l'etude des degre d'erosion des sals constituent les bassins versants des cours d'eau italiens, *Intern. Assn. Sci. Hydrol., Gen. Assembly, Helsinki,* 112–134, 1960.
Gibbs, R. J., The geochemistry of the Amazon River system. Part 1, *Geol. Soc. Am. Bull. 78,* 1203–1232, 1967.
Gilluly, J., Geologic contrasts between continents and ocean basins, *Geol. Soc. Am. Spec. Paper 62,* 7–18, 1955.
International Association of Scientific Hydrology, Transport total des sediments aux oceans, *IASH Bull. 12(3),* 100–110, Sept. 1967.
Jarocki, W., A study of sediment (Badanie rumowiska) OTS 60-21273, English Transl., Warsaw, Poland, 254, 1963.
Judson, S., and D. F. Ritter, Rates of regional denudation in the U. S., *J. Geophys. Res., 69* (16), 3395–3401, 1964.
Kuenen, P. H., *Marine Geology,* John Wiley & Sons, New York, 568, 1950.
Langbein, W. B., and S. A. Schumm, Yield of sediment in relation to mean annual precipitation, *Trans. Am. Geophys. Union, 39*(6), 1076–1084, 1958.
Livingstone, D. A., Chemical composition of rivers and lakes, in Data of Geochemistry, sixth ed., *U. S. Geol. Surv. Profess. Paper 440-G,* 64, 1963.
Oltman, R. E., H. O. Sternberg, F. C. Ames, and L. C. Davis, Amazon River Investigations, reconnaissance measurements of July 1963, *U. S. Geol. Surv. Circ. 486, 15,* 1964.
Lopatin, G. V., Sediment deposits in the rivers of the USSR, *Izd. Geograficheskoi Literatury,* Moscow, 1952.

Pechinov, D., Water erosion and solids discharge (Vodna eroziya i to"rd ottok), *Priroda* (Sofia, Bulgaria), *8*(1), 49–52, 1959.

Putnam, W. C., *Geology,* Oxford University Press, New York, 480, 1964.

Schumm, S. A., The disparity between present rates of denudation and orogeny, *U. S. Geol. Surv. Profess. Paper 454-H,* Washington, D. C., 1963.

Sedimentation and Erosion Sub-Task Force of the Federal Interdepartmental Task Force, *Report on the Potomac,* Soil Conservation Service, U. S. Department of Agriculture, Hyattsville, Maryland, 31, 1967.

Strahler, A. N., *The Earth Sciences,* Harper & Row, New York, 681, 1963.

Tixeront, J., Debit solide des cours d'eau en Algerie et en Tunisie, *Intern. Assn. Sci. Hydrol., Gen. Assembly Helsinki,* 26–41, 1960.

Todd, O. J., and S. Eliassen, The Yellow River problem, *Proc. Am. Soc. Civil Engrs., 64,* 1921–1991, 1938.

United Nations, compendium of international rivers in the ECAFE region, *Proc. Sixth Reg Conf. Water Res. Devel. in Asia, ECAFE,* New York, 1965.

United Nations, Soil erosion in various countries of the region, *The Sediment Problem,* ECAFE, Bangkok, 11–16, 1953.

United States Department of Agriculture, *Climate and Man,* 1941 Yearbook of Agriculture, U. S. Government Printing Office, Washington, D. C., 1248, 1941.

Van Andel, TJ. H., The Orinoco delta, *J. Sedimen. Petrol., 37*(2), June 1967.

(Manuscript received April 19, 1968; revised May 8, 1968.)

26

Denudation of Taiwan Island since the Pliocene Epoch

Yuan-Hui Li
Swiss Federal Institute of Technology (EAWAG)
CH-8600 Dübendorf, Switzerland

ABSTRACT

The average denudation rate of the Central Range of Taiwan Island has been increasing since the Pliocene Epoch. Today it is at least 1,365 mg/cm^{2}yr, which is probably the highest known value in the world.

INTRODUCTION

The reversal of the downthrust of the colliding Eurasia and Philippine Sea plates occurs near the area east of Taiwan Island (Wu, 1970; Lee and Hilde, 1975, unpub. data); therefore, Taiwan Island has been very active in a tectonic sense since early in its formation (mainly the present Central Range) during the Paleocene Epoch. As will be shown by the present study, the acceleration of the denudation rate since Pliocene time attests to the rapid upward movement of Taiwan Island.

PRESENT DENUDATION RATE OF TAIWAN ISLAND

The present-day physical denudation rate (hereafter, PDR) at each observational station located in the upstream drainage areas of Taiwan was calculated from extensive data of suspended-particle loadings on rivers throughout the island (Comm. Water Resource Control, 1973). Similarly, the chemical denudation rate (CDR) of upstream drainage areas at more than 30 stations was also estimated by using the bimonthly chemical data (Hsü, 1964) and the water-flow data (Comm. Water Resource Control, 1970) for the rivers. Figures 1A and 1B show the values of the PDR and the CDR, respectively.

$$\text{PDR} = \sum_{i=1}^{365} a \cdot Q_i^{b+1}/A$$

and

$$\text{CDR} = (Q_T \cdot \sum^{n} Q_i \cdot C_i)/(A \cdot \sum^{n} Q_i),$$

where Q_i = water-flow rate in cubic metres per day; a and b = empirical constants from the relation $C_i = a \cdot Q_i^b$ in which C_i = concentration of suspended particles for the case of PDR and total dissolved salts for the case of CDR; Q_T = annual water-flow rate; and n = number of observations. The calculated PDR values should be considered as minima, because the coarser sand grains and pebbles transported near and (or) on the river beds were not included in the measurements. The calculated CDR values, on the other hand, represent the maxima, because in the calculation, one-half of bicarbonate ions in river waters was assumed to have been entirely derived from the dissolution of carbonate rocks and the remainder from the air. (In the case of chemical weathering of silicate minerals, all bicarbonate ions in river waters were directly or indirectly—for example, oxidation of organic materials—supplied by the air and should not be considered as components derived from weathered rocks.)

The dashed line in Figure 1A represents the 600-mg/cm^{2}yr contour of the PDR. The area to the right of the dashed line has an average PDR of about 1,300 mg/cm^{2}yr, and the area to the left, about 325 mg/cm^{2}yr. The great change in the PDR can be explained by the change in relief and (or) in the bedrock geology. For example, in the Central Range of Taiwan (that is, the area of the pre-Miocene formations, enclosed by the dotted line in Fig. 1A), the relief and the relief ratio are extremely high (the highest peak is 3,997 m, and the relief ratio is no less than one-tenth) and so are the PDR values. One exception is the upstream area of river 1 and the area between rivers 2 and 3 (locations in Fig. 1A) where the bedrock consists mainly of compact formations of Eocene and Oligocene age. Outside the Central Range, the drainage area between rivers 4 and 5 is the only location with a comparably high PDR, although the

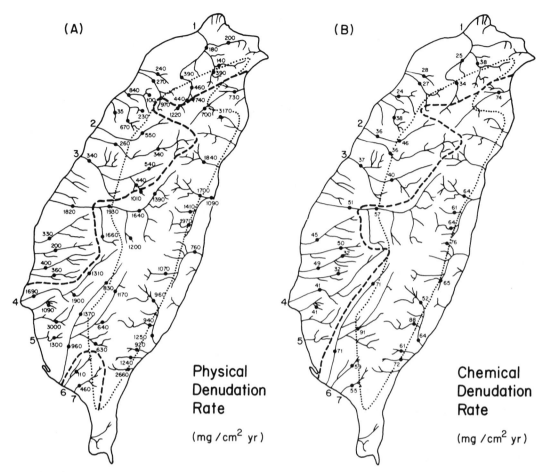

Figure 1. Physical denudation rates (A) and chemical denudation rates (B) of upstream drainage areas at each station. Values in milligrams per square centimetre per year. Dashed lines represent boundary between low and high denudation rates; area within dotted line represents Central Range.

relief ratio is relatively lower. Most of the bedrock in this area consists of loosely bound siltstone and mudstone of late Miocene and Pliocene ages.

In comparison with the average PDR for the Asian continent (30 mg/cm²yr; Garrels and Mackenzie, 1971) for the Yellow River drainage of China (290 mg/cm²yr; Comm. Water Resource Control, 1973) and for the Alpine Rhine (115 mg/cm²yr; Li and Erni, 1974), the average PDR for Taiwan Island is extremely high. The average PDR for the Central Range (1,300 mg/cm²yr) is probably the highest known physical denudation rate in the world. At this high rate of denudation, the Central Range could be leveled within half a million years, if the orogeny ceases.

The dashed line in Figure 1B is the 50-mg/cm²yr contour of the CDR. The area to the right of this contour has an average CDR of about 65 mg/cm²yr, and the area to the left, about 38 mg/cm²yr. The contrast is not as pronounced as in the case of the PDR. For comparison, the average CDR is only about 3 mg/cm²yr for the Asian continent (Garrels and Mackenzie, 1971) and 18 mg/cm²yr for the Alpine Rhine (Li and Erni, 1974).

The areas of high and low chemical denudation rates in general correspond well to the areas of high and low physical denudation rates (the high and low values are relative to the dashed line in Figs. 1A and 1B). However, the CDR is low in the drainage area between rivers 4 and 5, whereas the PDR is high. Another minor exception is the drainage area of rivers 6 and 7 where the PDR is low but the CDR is high; this situation possibly indicates some large carbonate outcrops in the area. In the regions with low values of both PDR and CDR, the ratio of

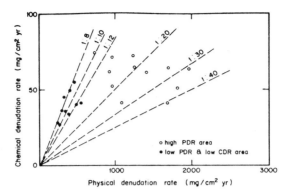

Figure 2. Ratio of chemical denudation rate to physical denudation rate at various stations.

PDR to CDR clusters around 10 ± 2 (Fig. 2), which is similar to the average for the Asian continent (Garrels and Mackenzie, 1971). The PDR/CDR ratio ranges from 10 to 40 (Fig. 2) in regions with high PDR values.

DISCUSSION

Integrating the thickness contours of the Pleistocene Toukoshan formation and the Pliocene formation deposited in the basin west of the Central Range down to a 500-m contour line as given by Chou (1973) and taking the mean density of sedimentary rocks to be 2.5 g/cm³ allow the mass of the deposited Pleistocene and Pliocene sediment to be estimated at about 7.0×10^{19} g and 8.5×10^{19} g, respectively. According to Chou (1973), the main source area for the sediment was the Central Range.

If it is assumed that (1) an equal amount of sediment was also discharged to the Pacific Ocean on the east side of the Central Range; (2) only a relatively small amount of the Pleistocene and Pliocene formations has been removed by erosion; (3) 1 to 2 m.y. were required for the deposition of the Pleistocene Toukoshan formation, and 10 m.y. were required for deposition of the Pliocene formation; and (4) the area of the Central Range since Pliocene time was about the same as it is today (that is, 18×10^{13} cm²), then the average physical denudation rate of the Central Range during the Pliocene and the Pleistocene would be 96 mg/cm²yr and 385 to 770 mg/cm²yr, respectively. In comparison with the present-day minimum value of 1,300 mg/cm²yr (some human effects cannot be excluded altogether, although most of the population and the agricultural activities are concentrated in the coastal plains around the island), it is apparent that the denudation rate of the Central Range has been accelerating since the Pliocene. This implies that the Central Range has been rising rapidly; such movement is also indicated by the uplifted coastal terraces of several stages and by the raised beach deposits and coral reefs found at the northern and southern ends of Taiwan today (Chou, 1973).

According to the work of Lin (1969) and the data compiled by Bonilla (1975), the minimum uplift rate of the recent coastal terraces (less than 9,000 yr old) around the island ranges from 2 mm/yr to 6.4 mm/yr and averages about 4 mm/yr. On the other hand, the average denudation rate (physical plus chemical) outside the area with high physical denudation rates (Fig. 1) is about 1.5 mm/yr (if it is assumed that the density of the formations is about 2.5 g/cm³). Therefore, the actual average uplift rate of the coastal terraces (minimum uplift rate plus denudation rate) may be about 5.5 mm/yr. If the uplift rate of 5.5 mm/yr also applies to the Central Range, the height of the Central Range would not change much today, since the average denudation rate of the Central Range is also about 5.5 mm/yr (if it is assumed that the density of rock is about 2.5 g/cm³).

One can conclude that the high denudation rate of the Central Range today is a direct consequence of the rapid upward movement of Taiwan Island, but the rapid denudation of the Central Range, in turn, aids the further uplift of the island by an isostatic adjustment.

REFERENCES CITED

Bonilla, M. G., 1975, A review of recently active faults in Taiwan: U.S. Geol. Survey Open-File Rept. 75–41.
Chou, J. T., 1973, Sedimentology and paleogeography of the upper Cenozoic System of western Taiwan: Geol. Soc. China Proc., no. 16, p. 111–143.
Committee for Water Resource Control, 1970, Taiwan hydrological year book: Taiwan, Bur. Economy.
——— 1973, The preliminary estimation of the suspended particle loading of rivers in Taiwan: Taiwan, Bur. Economy, 58 p. (in Chinese).
Garrels, R. M., and Mackenzie, F. T., 1971, Evolution of sedimentary rocks: New York, Norton Co., Inc., 397 p.
Hsü, Y. P., 1964, The water quality of irrigation waters in Taiwan: Taipei, Natl. Taiwan Univ., 142 p. (in Chinese with English summary).
Li, Y. H., and Erni, P. E., 1974, Erosionsgeschwindigkeit im Einzugsgebiet des Rheins: Vom Wasser, v. 43, p. 15–42.
Lin, C. C., 1969, Holocene geology of Taiwan: Acta Geol. Taiwanica, no. 13, p. 83–126.
Wu, F. T., 1970, Focal mechanisms and tectonics in the vicinity of Taiwan: Seismol. Soc. America Bull., v. 60, p. 2045–2056.

ACKNOWLEDGMENTS

Reviewed by S. Emerson, F. T. Wu, and M. Gosh.

MANUSCRIPT RECEIVED OCTOBER 9, 1975
MANUSCRIPT ACCEPTED DECEMBER 5, 1975

27

Copyright © 1979 by Masson Editeur S. A.

Reprinted from pages 215–237 and 243–246 of *Rev. Geol. Dyn. Geog. Phys.* **21**:
215–246 (1979)

Concentrations des eaux fluviales en éléments majeurs et apports en solution aux océans

par Michel MEYBECK *

RÉSUMÉ. — La moyenne mondiale de la composition naturelle des eaux superficielles (en mg.l^{-1} : SiO_2 10,4 ; Ca^{++} 13,4 ; Mg^{++} 3,35 ; Na^+ 5,15 ; K^+ 1,3 ; Cl^- 5,75 ; SO_4^{--} 8,25 ; HCO_3^- 52, pour le drainage exoréique) a été établie sur 60 grands fleuves, ou groupes de fleuves, représentant 63 % du débit des rivières aux océans, et, pour la partie restante, sur une typologie générale des transports en solution, établie sur l'écoulement et la température moyenne du bassin, avec une correction d'intensité de relief. Les facteurs lithologiques, non pris en compte, influencent toutefois les valeurs typiques des transports bien que définies sur des superficies étendues (10^6 km^2). Le bilan résultant s'écarte souvent des précédents (D.A. Livingstone ; O.A. Alekin et L.V. Brazhnikova), notamment pour certains éléments (le potassium) ou pour des régions (l'Afrique), en raison d'une meilleure information sur la zone tropicale et d'une considération particulière des apports anthropiques. Ces derniers ont été estimés d'après l'évolution de certains grands fleuves en régions industrielles et d'après des mesures directes : ils représentent déjà 12 % de l'apport minéral naturel (jusqu'à 30 % pour Na^+, Cl^- et SO_4^{--}). Les autres formes d'apports aux océans, liées aux glaciers, au volcanisme actif et aux eaux souterraines, sont négligeables par rapport aux fleuves. Environ 15 % des éléments dissous des eaux superficielles proviennent de l'atmosphère. Les régions montagneuses humides (12 % de la superficie exoréique) qui subissent une dénudation chimique élevée (30 mm/1 000 ans) sont la source de 15 % des apports en solution aux océans. La majeure partie de la silice (74 %) provient de la zone tropicale. Les apports spécifiques — par km^2 ou par km^3 — varient pour chaque océan, et sont maximum pour l'Arctique et l'Atlantique. La décomposition géochimique des eaux superficielles révèle que 57 % de HCO_3^- proviennent du CO_2 atmosphérique et que l'érosion chimique actuelle des continents (9 mm/1 000 ans pour leur partie exoréique) s'effectue surtout aux dépens des roches sédimentaires — elles libèrent 89 % des produits de l'érosion, dont les 2/3 pour les roches carbonatées — en raison de leurs vitesses relatives de dissolution, très élevées par rapport aux roches cristallines (évaporites $\simeq$ 80, roches carbonatées $\simeq$ 20, roches cristallines $\simeq$ 1,2, roches détritiques = 1). Il en résulte que 95 % des eaux fluviales sont bicarbonatées calciques, avec des variétés majeures : sulfatée-magnésienne (13 % des eaux) typique des régions sédimentaires, et chlorurée-sodique (34 %) typique des régions cristallines. La teneur en silice, qui varie avec la température moyenne du bassin, permet d'introduire une différenciation supplémentaire. L'extrapolation aux temps géologiques des valeurs d'érosion actuelles, qui ne devraient être déduites des transports en solution qu'après une triple correction (pollution, apports océaniques, CO_2 atmosphérique), s'avère délicate.

Mots-clés : Composition chimique, Rivières, Eléments majeurs, Bilan mondial, Océans, Erosion.

ABSTRACT. — *Major elements contents of river waters and dissolved inputs to the oceans :* The natural average composition of surface waters (in mg.l^{-1} : SiO_2 10,4 ; Ca^{++} 13,4 ; Mg^{++} 3,35 ; Na^+ 5,15 ; K^+ 1,3 ; Cl^- 5,75 ; SO_4^{--} 8,25 ; HCO_3^- 52, for the exoreic drainage) is based on 60 big rivers, or groups of rivers, which represent 63 % of the total river discharge to the ocean, and, for the remaining 37 %, on a general typology of river dissolved transports established on runoff and average watershed temperature, with a relief correction. Lithological factors, not taken into account, may still influence the typical transport although these are defined on extended area (10^6 km^2). Some dissimilarities between this balance and the previous ones (D.A. Livingstone, O.A. Alekin and L.V. Brazhnikova), can be found (K^+, Africa) due to a better information on tropical rivers and a special attention paid to anthropogenic inputs. These are estimated from the evolution of some big rivers in industrialized regions and from direct measurements : the natural dissolved mineral input is already exceeded by 12 % (up to 30 % for Na^+, Cl^- and SO_4^{--}). Other sources of dissolved material to the ocean (glaciers, leaching of fresh volcanic rocks, underground waters) are negligible with regards to the rivers. About 15 % of dissolved elements in surface waters take their origin in atmosphere. About 15 % of the dissolved inputs to the ocean originate in the humid mountainous regions (12 % of the exoreic drainage), which are exposed to intensive chemical denudation (30 mm/1 000 years), and 74 % of the dissolved silica comes from the tropical zone. Specific inputs to oceans — per km^2 or km^3 — are variable and maximum values are observed for the Arctic and Atlantic oceans. According to the geochemical decomposition of surface waters, 57 % of HCO_3^- derives from atmospheric CO_2, and present-day chemical erosion of continents (9 mm/1 000

* Laboratoire de Géologie, E.N.S., 16 rue d'Ulm, (F) 75230 Paris Cédex 05.

years for their exoreic part) is mainly accomplished to the detriment of sedimentary rocks (they release 89 % of chemical erosion products, of which 2/3 derive from carbonate rocks) due to their relative rates of dissolution with regards to the crystalline rocks (evaporites $\simeq$ 80, carbonate rocks $\simeq$ 20, crystalline rocks $\simeq$ 1.2, shales and sandstones = 1). Therefore 95 % of surficial waters are of the calcium carbonate type, with two major varieties : magnesium-sulphate (43 % of waters) found in sedimentary regions and sodium chloride (34 %) found in crystalline regions. Silica content, which is temperature dependent, is another means of differentiation. The extrapolation to past geologic times of present-day erosion rates − which should be obtained through three corrections (pollution, oceanic cyclic salts, atmospheric CO_2) − will certainly be a difficult task.

Key-words : Chemical Composition, Rivers, Major Element, World Balance, Ocean, Erosion.

Introduction.

La première étude détaillée des apports en solution des rivières aux océans fut réalisée par F.W. Clarke (1924). Depuis, de nombreux auteurs ont amélioré cette estimation (E.J. Conway, 1943 ; O.A. Alekin et L.V. Brazhnikova, 1960 ;D.A. Livingstone, 1963). Les transports de matériaux dissous effectués par les rivières sont en effet largement utilisés pour quantifier l'érosion chimique (J. Corbel, 1959a, 1964 ; I. Douglas, 1964,) pour étudier les types d'altération actuels (J.Y. Gac et M. Pinta, 1973 ; Y. Tardy, 1971), enfin pour évaluer les apports à l'océan et, d'une façon générale, pour étudier la partie externe du cycle géologique (N.M. Strakhov, 1967 ; K.K. Turekian, 1969, 1971 ; A.P. Lisitzin, 1972 ; R.M. Garrels et F.T. Mackenzie, 1971 ; H.D. Holland, 1978). D'autres publications font maintenant état de modifications profondes, dues aux activités humaines, des processus d'altération, des transports en solution et des taux d'érosion des continents (R.H. Meade, 1969 ; E.D. Goldberg, 1972 ; S. Judson, 1968 ; N.M. Johnson et al., 1972 ; W.C. Ackerman et al., 1970).

On distingue généralement la matière particulaire de la matière dissoute. La quantité totale de matière particulaire transportée à l'océan a fait l'objet de nombreuses estimations (F. Fournier, 1960 ; J.N. Holeman, 1968 ; J.M.L. Jansen et R.B. Painter, 1974 ; U.S.S.R. Committee for I.H.D., 1978), mais sa composition chimique n'est encore estimée que sur une vingtaine de fleuves (V.V. Gordeev et A.P. Lisitzin, 1978 ; J.M. Martin et M. Meybeck, 1979). La composition chimique de la phase dissoute a été beaucoup plus étudiée : outre les études déjà citées, mentionnons celles de W.H. Durum et al. (1960), de R.J. Gibbs (1967, 1972), de J. Kobayashi (1960) pour les éléments majeurs, et de G.S. Konovalov et A.A. Ivanova (1970), K.K. Turekian et M. Scott (1967) pour les éléments traces. En ce qui concerne les éléments majeurs : SiO_2, Ca^{++}, Mg^{++}, Na^+, K^+, Cl^-, SO_4^{--}, HCO_3^-, qui représentent environ 99 % de la matière minérale en solution dans les eaux superficielles, deux références sont essentiellement citées : D.A. Livingstone (1963), et O.A. Alekin et L.V. Brazhnikova (1960).

Depuis la publication de ces deux études, de nombreuses données nouvelles sont disponibles qui justifient l'essai d'un nouveau bilan. A la suite des travaux de la Décennie Hydrologique Internationale (Unesco, 1969 et 1971), l'estimation du bilan hydrique mondial s'est améliorée (M.I. Lvovich, 1972 ; A. Baumgartner et E. Reichel, 1975 ; U.S.S.R. Committee for the I.H.D., 1978). Par exemple, les nouvelles mesures de débit sur l'Amazone ont porté celui-ci de 100 000 $m^3.s^{-1}$ à 175 000 $m^3.s^{-1}$, entraînant une rectification de tout le bilan de l'Amérique du Sud. De plus, des analyses d'eaux de rivières sont maintenant disponibles dans presque toutes les régions du globe. La fiabilité d'un nouveau bilan est donc augmentée puisqu'on dispose d'informations sur environ 63 % de la superficie des continents drainée par les fleuves vers les océans (écoulement exoréique). Par rapport aux bilans précédents, quatre éléments nouveaux seront présentés ici :

1) une typologie de la variation géographique des transports en solution,

2) un bilan des apports à l'océan pour les quatre zones climatiques suivantes : froide, tempérée, aride, et tropicale,

3) un bilan séparé des apports à chaque océan,

4) une estimation des différentes origines des apports actuels (érosion des continents, CO_2 atmosphérique, précipitations d'origine océanique, apports anthropiques).

Vu la pauvreté de l'information sur les éléments nutritifs et sur la matière organique dissoute, leur bilan n'a pas été tenté. Quant aux éléments-traces, on trouvera dans une publication récente (J.M. Martin et M. Meybeck, 1979) des estimations pour une quarantaine d'éléments.

I. − MÉTHODES DE CALCUL DES APPORTS DISSOUS AUX OCÉANS.

A. Principaux problèmes posés.

Il convient tout d'abord de distinguer l'*érosion des continents*, le *transport par les rivières* et l'*apport aux océans* (I. Douglas, 1964 ; R. Meade, 1969 ; R.J. Janda, 1971). En effet, une partie des matériaux libérés par l'altération et par l'érosion mécanique n'est pas transportée par les rivières jusqu'à l'océan (dépôt des matériaux détritiques plus grossiers sur les versants et dans les plaines d'inondation ; précipitation de certains éléments dissous, notamment dans les lacs). De plus, une partie des éléments dissous transportés par les rivières ne provient pas de l'altération (apports atmosphériques d'origine marine, volcanique ou anthropique ; CO_2 atmosphérique utilisé dans les réactions d'altération ; rejets directs d'effluents urbains, industriels ou miniers, etc.). Enfin tous les éléments transportés par les rivières n'atteignent pas les océans : la superficie non glacée des continents est de $133,1.10^6 km^2$, mais $33,2.10^6 km^2$ sont drainés vers l'intérieur des continents (*écoulement endoréique ou interne*) (A. Baumgartner et E. Reichel, 1975). Certains éléments dissous peuvent également être précipités ou adsorbés sur les particules en suspension dans le milieu estuarien, et réduire ainsi les apports en solution à l'océan. Un bilan complet doit aussi prendre en compte les autres formes d'apports en solution (glaciers, lessivage de roches volcaniques récentes, eaux souterraines, aérosols d'origine continentale).

La distinction entre matière dissoute et matière particulaire est parfois difficile. Pour certains éléments comme Al et Fe, elle dépend de la porosité du filtre utilisé (V.C. Kennedy *et al.*, 1974). Ainsi les études plus anciennes de F.W. Clarke (1924) ou de D.A. Livingstone (1963) considéraient Al et Fe comme des éléments dissous majeurs aux teneurs proches de 1 mg.l^{-1}. En fait, une grande partie de ces éléments est sous une forme colloïdale, retenue sur les filtres plus fins utilisés aujourd'hui, et qui précipite dans le milieu estuarien (G. Figuères, J.M. Martin et M. Meybeck, 1979). Toutefois, pour les éléments étudiés ici, la distinction état dissous - état particulaire ne pose pas de problème.

Le bilan des apports en solution à l'océan implique une extrapolation pour les régions où l'analyse chimique des eaux n'est pas connue. Deux types d'extrapolation peuvent être faits.

● extrapolation sur la base d'un bilan hydrique connu et d'une concentration en solution estimée ;

● extrapolation sur la base d'une superficie drainée connue et d'un transport dissous spécifique estimé.

Ce qui s'écrit :

$$Md = \Delta\tau \sum_j A_j Td_j$$

ou/et :

$$Md = \Delta\tau \sum_j Cd_j Q_j$$

où Md = masse transportée en solution pendant le laps de temps $\Delta\tau$ (généralement en t.an^{-1}) ;

A_j = superficie du bassin de la rivière j ;

Td_j = transport dissous spécifique (masse par unité de temps et de surface, généralement en t.km^{-2}.an^{-1} ; 1 t = 10^6 g) ;

Cd_j = concentration moyenne en solution (en mg.l^{-1} ou g.m^{-3}) ;

Q_j = débit moyen (m^3.s^{-1} ; km^3.an^{-1}).

L'extrapolation est nécessaire puisque, si les cinquante plus grands fleuves apportent environ 49 % de l'eau aux océans, l'addition de 150 fleuves n'amène cette proportion qu'à 58 %. Pour couvrir 70 % des apports fluviatiles, il faudrait considérer un millier de rivières (Unesco, 1979).

Les bilans peuvent aussi différer par la simple expression des résultats, en particulier pour les bicarbonates. Les auteurs soviétiques ont généralement exprimé les bicarbonates en CO_3^{--}, ce qui diminue la *concentration ionique globale* (Σ_i) obtenue par addition des teneurs en ions majeurs. Dans les travaux plus récents (U.S.S.R. Committee for the I.H.D., 1978), les deux expressions sont utilisées.

Le tableau I résume les résultats et les caractéristiques des principaux bilans précédents. On remarquera la grande diversité de l'information utilisée et du mode de calcul. On note que F.W. Clarke, ne disposant pas d'un bilan hydrique mondial, a établi des transports spécifiques moyens, mais n'a pas pu proposer des concentrations moyennes dans les rivières, la proportion relative des éléments étant seule donnée. O.A. Alekin et L.V. Brazhnikova ont basé leur bilan sur une excellente connaissance des transports ioniques en U.R.S.S., mais n'ont pu donner qu'une concentration ionique totale pour l'ensemble des fleuves, le détail pour chaque ion n'est pas présenté. Le dernier bilan de O.A. Alekin (U.S.S.R. Committee for the I.H.D., 1978, tableau 163) ne porte aussi que sur la concentration ionique globale. Il diffère assez nettement du précédent : Q = 40 300 km^3.an^{-1}, A = 100,3.10^{6}km^2, Σ_i = 78,4 mg.l^{-1}. Par contre le transport ionique reste le même : Td_i = 31,5 t.km^{-2}.an^{-1}, mais le détail du bilan fait toujours défaut.

Seul D.A. Livingstone a calculé à la fois les concentrations et les transports spécifiques sur l'ensemble du globe. Mais son information de base est souvent peu fiable ou insuffisante (Amérique du Sud, Afrique, Sud-Est asiatique). Mon précédent bilan (M. Meybeck, 1977), dont seul le résultat global était présenté, disposait d'une information proche de celui présenté ici, mais l'extrapolation aux régions inconnues était faite par analogie — de la même façon que F.W. Clarke et D.A. Livingstone — et non pas sur une typologie ; de plus, les apports d'origine anthropique ont été revus.

340

B. Les bases du bilan proposé.

1. *Le principe du bilan.*

La détermination du bilan a été faite en six étapes.

a) calcul des concentrations moyennes, des transports spécifiques et des masses transportées, soit pour des grands fleuves, ou pour des régions bien homogènes (annexe 1). Pour ces dernières les moyennes régionales sont basées sur un nombre de rivières variant de 2 à 200.

b) calcul des apports à l'océan des rivières étudiées (annexe 3) : ils correspondent à 63 % du débit et à 63 % de la superficie du drainage exoréique.

c) détermination suivant des critères morphoclimatiques de 15 catégories de transports en solution effectués par les eaux de surface. Cette typologie a utilisé une grande partie des rivières étudiées (tabl. II). Les transports spécifiques moyens sont donnés au tableau III. Les importantes variations de la composition des eaux en fonction du substrat géologique sont alors minimisées en utilisant des secteurs représentatifs de l'ordre de 10^6 km².

d) cartographie des différents types de transports en solution (fig. 1) d'après les critères retenus (tabl. II) et estimations des superficies correspondant à chaque type (annexe 2).

e) extrapolation aux régions inconnues des transports en solution types. Les superficies correspondantes ont été déterminées par continent et pour chaque type. Cette opération a été effectuée de telle façon que les débits totaux (régions connues + inconnues) ainsi obtenus se rapprochent à 5 % près de l'écoulement de chaque continent. Ce type d'extrapolation a été réalisé d'une part pour le calcul des apports en provenance de chaque continent (tabl. V), d'autre part pour celui des apports arrivant à chaque océan (tabl. VII).

f) les rivières utilisées étant déjà contaminées par des rejets anthropiques, une correction basée sur la population et son degré de développement industriel a été appliquée pour obtenir une estimation des apports naturels et des apports d'origine anthropique (tabl. VII).

L'extrapolation basée sur les transports en solution est plus fiable que celle basée sur les concentrations (O.A. Alekin et L.V. Brazhnikova, 1960), puisque les superficies des bassins sont des données plus faciles à déterminer que les débits. De plus, la variabilité géographique des transports en solution des grands fleuves mondiaux est plus faible (1 à 30) que celle des concentrations (1 à 100) (M. Meybeck, 1976). Néanmoins, quand on veut repasser des apports en solution aux concentrations moyennes, on est obligé de considérer les bilans hydriques. Ceux, très détaillés, de A. Baumgartner et E. Reichel (1975) nous ont constamment servi de base. Nous en avons déduit les apports des glaciers arctiques (312 km³.an⁻¹ pour 1,73.10⁶ km² pour le Groënland) et antarc-

tiques (1980 km³.an⁻¹ pour 14,1.10⁶ km²). Les apports d'eau liquide (écoulement de surface + écoulement souterrain direct) sont alors de 37 400 km³.an⁻¹ pour une superficie exoréique de 99,9.10⁶ km². On note que l'estimation des apports fluviatiles est en constante augmentation (tabl. I).

2. *Données utilisées.*

Les caractéristiques hydrologiques des fleuves et régions étudiés sont fournies à l'annexe 1. La plupart des références correspondantes figurent dans une étude antérieure (M. Meybeck, 1976) ou proviennent des inventaires récents (Unesco, 1969, 1971, 1979 ; U.S.S.R. Committee for the I.H.D., 1978).

Les analyses chimiques moyennes utilisées sont pour la plupart récentes, et plus des deux tiers n'étaient pas à la disposition de D.A. Livingstone (1963), en particulier celles des grands fleuves suivants : Mackenzie, Orénoque, Parana, Amazone, Uruguay, Chari, Congo, Niger, Zambèze, Danube, Gange, Indus, Brahmapoutre, Mékong. La représentativité des analyses reportées est très variable, allant d'un échantillon annuel jusqu'à une moyenne pondérée par les débits basée sur 50 échantillons par an (ann. 1). J'ai recalculé les moyennes pondérées de certains fleuves (Chao Phrya, Gange, Brahmapoutre, Nil, Indus), la différence entre les moyennes arithmétiques et pondérées étant quelquefois de 30 % pour certains éléments. Le détail complet des valeurs régionales est fourni à l'annexe 1 ; les moyennes reportées sont pondérées par les débits des différentes rivières utilisées pour certaines régions (Deccan ; Alpes, Ibérie, Islande, Pologne ; Golfe de Carpentaria ; Canada Central, Colombie Britannique, Québec, Territoires du N.W., Texas ; Guyanes). La balance ionique a été vérifiée pour chaque analyse et quelquefois corrigée, ou utilisée pour estimer des éléments non analysés. Pour la silice des rivières soviétiques, j'ai pris une estimation unique de 6 mg $SiO_2.l^{-1}$ (V.P. Zverev, communication orale). Pour ces mêmes rivières, O.A. Alekin et L.V. Brazhnikova (1960) ne différencient pas Na et K, aussi ai-je exprimé le potassium d'après les rapports K/Na de N.P. Morozov (1969) ; toutefois, ces valeurs de K semblent trop faibles. Enfin, les carbonates des rivières soviétiques ont été transcrits en HCO_3^-, par le facteur 1,36 préconisé par O.A. Alekin et L.V. Brazhnikova (1960) ; les valeurs obtenues sont semblables à celles recalculées par V.P. Zverev et V.Z. Rubeikin (1973).

Comme D.A. Livingstone (1963), j'ai éliminé au maximum les rivières soumises à une forte pollution minérale (Côte est des Etats-Unis, Europe de l'Ouest) de façon à baser le bilan et la typologie sur des données plus proches des valeurs naturelles. Ainsi les analyses du Mississippi, du Saint-Laurent, du Rhin et des rivières polonaises datent-elles d'avant 1900. Malgré cette précaution, on verra que les analyses utilisées sont déjà légèrement affectées par la pollution.

II. — ESSAI DE TYPOLOGIE DES TRANSPORTS EN SOLUTION À LA SURFACE DU GLOBE.

La composition chimique des eaux superficielles et les transports en solution effectués par les rivières sont

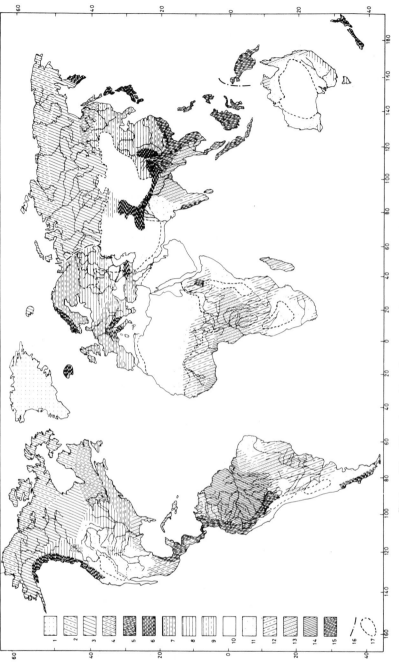

Fig. 1. – **Répartition géographique des types morphoclimatiques de transports en solution.**

Les types sont basés (voir tableau II) sur les températures des bassins (moyennes de janvier et de juillet d'après l'Atlas de l'U.S.S.R. Acad. of Science, 1964) et sur les cartes de débits spécifiques d'A. Baumgartner et E. Reichel (1975). Dans les zones de forts gradients de température (Himalaya) ou d'écoulement (Mexique, Chili), les types intermédiaires n'ont pas été considérés. Cette carte, réalisée pour extrapoler les taux de transports en solution dans les régions inconnues, donne aussi une idée générale de la répartition de l'érosion chimique, les figurés les plus foncés correspondant généralement aux érosions les plus fortes déterminées au tableau IX.

1 : glaciers ; 2 : toundra ; 3 : taïga ; 1 : taïga humide ; 5 : taïga très humide ; 6 : tempéré très humide ; 7 : tempéré humide ; 8 : tempéré ; 9 : tempéré semi-aride ; 10 : désert ; 11 : aride ; 12 : tropical contrasté ; 13 : tropical humide ; 11 : tropical très humide (plaine) ; 15 : tropical très humide (montagne) ; 16 : limites choisies de l'Asie ; 17 : régions endoréiques.

Fig. 1. – **Geographic distribution of morphoclimatic types of dissolved matter transport.**

These types are based (see table II) on average air temperature (mean of January and July temperatures according to the Atlas of U.S.S.R. Acad. of Science, 1964) and on runoff maps of A. Baumgartner and E. Reichel (1975). In regions of high temperature (Himalaya...) or runoff (Mexico, Chili) gradient, intermediate types have not been considered. This map, realized to extrapolate the dissolved transport rates in region where not data are available, gives a pattern of the chemical erosion distribution : darker areas usually correspond to higher rates (see table IX).

étroitement liés à trois groupes de facteurs très difficiles à séparer : la lithologie, le relief, le climat. L'effet de la *lithologie* est prépondérant sur des petits bassins où les concentrations en solution de rivières voisines peuvent varier d'un facteur 10 suivant la nature des roches superficielles (J. Kobayashi, 1960 ; J.P. Miller, 1961). La minéralisation des eaux drainant des roches sédimentaires est environ 4 fois plus forte que celle des eaux sur roches effusives, et 8 fois plus forte que celle sur roches plutoniques. Si l'on considère le détail des éléments, les teneurs en Ca^{++}, Mg^{++}, HCO_3^-, SO_4^{--} sont bien plus élevées dans les bassins sédimentaires, alors que les régions volcaniques libèrent le plus de silice dissoute (J.D. Hem, 1970 ; M. Meybeck, 1979).

Sur les bassins des grands fleuves, cet effet s'estompe le plus souvent du fait de la plus grande variabilité des roches rencontrées, et j'ai cherché à le minimiser en me basant pour chaque type de transport sur des superficies les plus étendues possible.

Dans les régions de fort *relief*, les transports en solution sont en général plus élevés pour un même débit spécifique (O.A. Alekin et L.V. Brazhnikova, 1962 ; M. Meybeck, 1976). Il y a sans doute ici convergence de plusieurs causes : la circulation des eaux est souvent plus profonde qu'en plaine, ce qui favorise la dissolution des roches ; l'érosion mécanique, toujours très élevée, permet le décapage constant des

		SiO_2	Ca^{++}	Mg^{++}	Na^+	K^+	Cl^-	SO_4^{--}	HCO_3^-	Σi	Σm	Bilan hydrique Q	Aire drainée A	Échant. connu Échant. total	
														Q %	A %
F.W. Clarke (1)	Cd	–	–	–	–	–	–	–	–	–	35.0	ND	102.4	15	20
1924	Td	3.1	5.4	0.9	1.53	0.56	1.5	3.2	18.9	31.9					
O.A. Alekin et (2)	Cd	–	–	–	–	–	–	–	–	89		32400	101.6	44	48
L.V. Brazhnikova 1960, 1966, 1968	Td	–	–	–	–	–	–	–	–	31.2					
D.A. Livingstone (3)	Cd	13.1	15	4.1	6.3	2.3	7.8	11.2	58.4	105.1	118.2	36450	101	13	22
1963	Td	4.2	4.8	1.3	2.0	0.73	2.5	3.6	18.6	33.5	37.7				
J. Corbel (4)	Cd										31.6	ND	ND	30	32
1964	Td														
M. Meybeck (5)	Cd	11.6	14.6	3.8	5.1	1.35	5.3	8.5	57.7	96.4	108	37400	99.9	57	52
1977	Td	4.35	5.5	1.4	1.9	0.5	2.0	3.15	21.6	36.0	40.4				
Cette étude (6)	Cd	10.4	13.4	3.35	5.15	1.3	5.75	8.25	52	89.2	99.6	37400	99.9	63	63
«Valeurs naturelles»	Td	3.9	5.0	1.25	1.9	0.48	2.15	3.1	19.4	33.3	37.2				
«Valeurs actuelles»	Cd	10.4	14.7	3.65	7.2	1.4	8.25	11.5	53	99.7	110.1				

TABLEAU I. — **Principaux bilans originaux des apports en solution aux océans.**

La similitude de la minéralisation globale des eaux (Σi et Σm) masque d'importantes variations sur certains ions (K^+) et sur les moyennes de chaque continent (tabl. V) dues aux bilans hydriques et aux modes de calcul différents mais surtout au manque de données dans la zone tropicale humide et à la prise en compte variable de l'influence anthropique. Cd concentrations en solution ($mg.l^{-1}$) ; Td transports spécifiques en solution ($t.km^{-2}.an^{-1}$) ; Σi concentration ionique globale ($mg.l^{-1}$) ; Σm concentration minérale globale ($\Sigma i + SiO_2$) ($mg.l^{-1}$) ; Q écoulement des rivières vers l'océan ($km^3.an^{-1}$) ; A superficie du drainage exoréique, glaciers exclus (10^6 km^2) ; N.D. non déterminé. (1) transport en solution moyen extrapolé d'après les valeurs connues en Europe et aux USA ; (2) bilan complet pour l'URSS + typologie pour 14 régions morphoclimatiques puis extrapolation du transport ionique global d'après les types définis ; (3) bilan des régions connues, transports dissous dans les régions inconnues extrapolés par analogies ; concentrations moyennes pondérées par les débits et extrapolées régions par régions ; analyses anciennes prises en compte pour les régions industrielles ; (4) transport en solution moyen défini sur 12 grands fleuves, puis extrapolation ; (5) bilan des régions connues, transports dissous extrapolés par analogie, déduction globale des apports anthropiques.

TABLE I. — *Major determinations of river dissolved inputs to the ocean.*

If total dissolved contents of each balance are similar, important variations are observed for some ions (K^+) and for the individual values of each continent (table V). These discrepancies are due to different water budgets and computation methods but also to a general lack of data in the humid tropics and to various considerations of Man impact. Cd dissolved contents ($mg.l^{-1}$) ; Td specific dissolved transport ($t.km^{-2}.year^{-1}$) ; Σi total ionic content ($mg.l^{-1}$) ; Σm total mineral content ($\Sigma i + SiO_2$) ($mg.l^{-1}$) ; Q total river runoff ($km^3.year^{-1}$) ; A exoreic area, glaciers excluded (10^6 km^2).

couches superficielles altérées, devenues plus résistantes à l'érosion chimique; enfin, les massifs montagneux présentent une proportion élevée de roches sédimentaires plus vulnérables à l'érosion chimique.

Le climat est la principale cause de variation des transports en solution, son influence portant à la fois sur la nature de l'altération et sur l'intensité d'évacuation des éléments libérés. Le transport en solution T_d croit en effet avec l'écoulement (mesuré par le débit spécifique q), mais moins vite que ce dernier, et selon une relation du type $T_d = aq^b$ avec $0 < b < 1$ (M. Meybeck, 1976). La variabilité des transports en solution est de deux ordres de grandeur (5 à 500 t.km^{-2}.an^{-1}), tandis que l'écoulement varie de près de trois ordres de grandeur. La relation entre le type d'altération et le climat est bien connue (G. Pedro, 1968; R.M. Garrels et F.T. Mackenzie, 1971; Y. Tardy, 1971). Elle se traduit notamment par une augmentation des teneurs en silice rencontrées dans les eaux superficielles en fonction de la température moyenne de la région (M. Meybeck, 1979).

La typologie présentée ici est basée sur le débit spécifique et la température. Quinze types de transports en solution ont ainsi été définis d'après 4 classes de température et 5 classes de débit spécifique (tabl. II). L'influence du relief n'a été prise en compte que pour le type « tropical très humide » pour lequel j'ai distingué les régions de plaine de celles de montagne : il y a en effet concordance entre les régions de fort relief et celles d'écoulement élevé dans les autres zones

q mm.an^{-1} / t (°C) l.s^{-1}. km^{-2}	< 30 / < 1	30-120 / 1-4	120-280 / 4-9	280-630 / 9-20	> 630 / > 20
< 4			**Toundra et taïga** Bering, Kara, Laptev, Yukon, Nelson, Canada central Mackenzie, Territoire du Nord-Ouest	**Taïga humide** Fraser, Barentz, Finlande, Québec.	**Taïga très humide** *(fort relief)* Colombie Britan. Norvège, Islande
4-15		**Tempéré semi-aride** Aral, Caspienne, Mer Noire, Colorado.	**Tempéré** Pologne, Iberia Mississipi, Danube, Baltique, Shatt el Arab.	**Tempéré humide** Alpes, Columbia, Suède, Saint-Laurent.	**Tempéré très humide** *(fort relief)* Brahmapoutre, Japon, Nouvelle-Zélande.
15-25	**Aride** Orange, Texas, Nil Blanc, Murray.				
> 20		**Tropical contrasté** Parana, Zambèze, Niger, Golfe Carpentaria, Chari, Sénégal, Volta.		**Tropical humide** Chao Phrya, Sri Lanka, Birmanie, Congo.	**Tropical très humide** *(plaine)* Guyanes, Orénoque. **Tropical très humide** *(montagne)* Mékong, Magdalena, Papouasie N.G.

TABLEAU II. — **Types de transports en solution basés sur des critères morphoclimatiques : débits spécifiques, températures et relief.**
Chaque type est caractérisé par des rivières pour lesquelles ont été déterminées : les températures moyennes (t en °C) des bassins — d'après les cartes des isothermes de janvier et de juillet (U.S.S.R. Acad. Sci., 1964), et les écoulements q = Q/A, d'après les valeurs fournies à l'annexe 1 (avec les analyses chimiques). Une distinction particulière de relief a été faite pour le type tropical très humide. La lithologie n'a pas été prise en compte ici.

TABLE II. — *Types of dissolved transport based on morphiclimatic criteria : specific runoff, air temperature and relief.*
Each type is characterized by rivers for which average air temperature (t in °C) have been determined from isotherms maps in January and July (U.S.S.R. Acad. Sci., 1964) and runoff q from the Q and A data given in annex 1 together with the chemical analyses. The tropical rainy type is splitted into two classes according to relief intensity. Lithological factors are not considered here.

tempérée et froide. Pour chaque type j'ai identifié des fleuves représentatifs (tabl. II) en laissant de côté certains grands bassins trop hétérogènes (Nil, Indus, Gange, Amazone) ou situés à la limite de deux types (Uruguay, Deccan). Par contre les bassins endoréiques ont été utilisés.

La lithologie n'a pas été considérée ici. Pour minimiser son influence, les types ont été définis sur des régions très étendues et aussi diversifiées que possible. Mais il est certain que la nature des roches soumises à l'érosion joue encore un grand rôle dans certains types. Ainsi les eaux des types tempérés présentent des te-

		SiO_2	Ca^{++}	Mg^{++}	Na^+	K^+	Cl^-	SO_4^{--}	HCO_3^-	Σi	q	A
Toundra	Cd	(3.7)	17.7	4.7	7.0	0.8	8.9	13.6	62.3	115		
et taïga	Td	(0.75)	3.5	0.95	1.4	0.15	1.75	2.7	12.4	22.9	6.3	18.5
Taïga	Cd	4.0	9.65	2.1	2.8	0.6	2.7	9.3	31.9	59.3		
humide	Td	1.6	3.85	0.82	1.1	0.24	1.1	3.7	12.7	23.6	12.6	2.64
Taïga très	Cd	5.3	5.15	1.35	3.8	0.6	3.6	4.7	22.2	41.4		
humide	Td	5.75	5.55	1.45	4.1	0.63	3.9	5.1	23.9	44.6	34	0.54
Tempéré	Cd	(8)	14.0	3.45	5.0	2.0	5.85	6.9	56.6	93.8		
très humide	Td	(10)	17.9	4.45	6.35	2.6	7.5	8.8	72.3	120	40	1.22
Tempéré	Cd	6.0	24.8	4.4	5.15	1.5	6.4	18.2	79	140		
humide	Td	2.35	9.7	1.7	2.0	0.58	2.5	7.1	31	54.7	12.3	2.47
Tempéré	Cd	7.1	37.6	9.9	12.2	2.2	14.3	30.2	126	234		
	Td	1.4	7.3	1.9	2.35	0.42	2.75	5.8	24.4	45.0	6.1	5.76
Tempéré	Cd	(7)	54	10.8	24.8	1.85	26.2	74.5	140	332		
semi-aride	Td	(0.8)	6.3	1.25	2.9	0.2	3.05	8.7	16.3	38.7	3.7	4.63
Aride	Cd	15.4	38.3	13.5	64.5	5.5	63.4	54.9	153	393		
	Td	0.25	0.63	0.22	1.05	0.09	1.05	0.9	2.5	6.4	0.5	5.19
Tropical	Cd	14.4	6.3	2.6	5.0	1.8	4.2	3.65	30.0	55.5		
contrasté	Td	2.35	1.0	0.42	0.81	0.3	0.68	0.43	5.35	9.0	5.1	6.9
Tropical	Cd	11.4	8.25	3.5	4.9	1.65	4.2	4.0	45.4	71.8		
humide	Td	4.15	3.0	1.3	1.8	0.6	1.5	1.4	16.5	26.0	11.4	4.9
Tropical très	Cd	11.4	3.15	1.0	1.8	0.6	3.1	3.1	11.2	23.9		
hum. (plaine)	Td	11.4	3.15	1.0	1.8	0.6	3.1	3.1	11.2	23.9	31	1.19
Tropical très	Cd	14.5	15.4	3.15	5.4	1.4	7.7	6.5	61.8	102		
hum. (montagne)	Td	16.4	17.4	3.55	6.1	1.6	8.7	7.3	69.9	115	36	1.44

TABLEAU III. — **Valeurs moyennes des transports en solution (Td en t.km^{-2}.an^{-1}) et des concentrations (Cd en mg.l^{-1}) pour 12 types morphoclimatiques.**

Les transports en solution sont directement liés au débit spécifique (q en l.s^{-1}.km^{-2}). Les deux pôles maximum « tempéré très humide) et « tropical très humide de montagne », sont séparés par des transports de plus en plus faibles lorsqu'on se rapproche de la zone aride où la minéralisation des eaux est par contre la plus forte. Les eaux les plus diluées se rencontrent dans les plaines de la zone tropicale très humide. Malgré les superficies (A en 10^6 km^2) très grandes sur lesquelles les types sont définis, une influence de la lithologie demeure : les fortes concentrations des types tempérés reflètent la plus grande abondance des roches sédimentaires, l'inverse s'observe pour les types tropicaux. Les données utilisées ici n'ont pas été corrigées d'une possible pollution. Les taux d'érosions chimiques réels sont présentés au tableau IX.

TABLE III. — *Average dissolved transports (Td in t.km^{-2}. year^{-1}) and concentrations (Cd in mg.l^{-1}) for 12 morphoclimatic types.*

Dissolved transports are directly linked to specific runoff (q in l.s^{-1}.km^{-2}). The two observed maximums (temperate very humid, and tropical very humid in mountains) are separated by lower transports observed in the more arid zones, where water mineralization is at its maximum. Most dilute waters are found in humid tropical plains. Despite the very large area (A in 10^6 km^2) on which the types were defined, there are still some influences of lithology : higher concentrations found for the temperate types are due to greater abundance of sedimentary rocks while the reverse is observed for the tropical types. Data used here were not corrected for any possible pollution. True chemical erosion rates are given in table IX.

345

neurs en Ca^{++}, HCO_3^- et SO_4^{--} nettement plus élevées que la moyenne, en raison de l'abondance plus grande des roches sédimentaires en Amérique du Nord et en Europe où ils sont définis. Inversement, les eaux du type tropical contrasté sont particulièrement peu minéralisées, vu la prépondérance des bassins cristallins utilisés. Cependant, si ces types sont biaisés, cela affecte probablement peu le bilan, car l'extrapolation est en général réalisée sur des bassins de lithologie analogue à celle des types.

Les types ont été nommés d'après des critères climatiques, sauf pour les régions froides pour lesquelles les termes toundra et taiga sont préférés. Une cartographie générale des types a été effectuée (fig. 1). Les types toundra et taiga différenciés sur la figure ont finalement été regroupés dans une catégorie unique, les compositions des eaux étant très semblables (tabl. II et III). De même, le type tropical contrasté englobe deux classes de débits spécifiques. Le type taiga très humide, rencontré uniquement au Kamchatka, a été approché en utilisant les bassins les plus froids du type voisin (tempéré très humide). La répartition des superficies de chaque type a été obtenue pour chaque continent par planimétrage (annexe 2).

Les transports spécifiques et les concentrations moyens caractéristiques de chaque type ont été calculés en utilisant les fleuves représentatifs (tabl. III). Il est difficile de comparer ces résultats avec les essais de typologie précédents. O. Alekin et L.V. Brazhnikova (1960) ont utilisé leur étude des rivières soviétiques comme base en prenant le rapport pluviosité/évaporation comme indice caractéristique. Si l'on constate une bonne concordance pour les régions tempérées, mes estimations pour les régions froides sont plus fortes que les leurs, tandis que mes valeurs pour les forts reliefs sont un peu plus faibles. La comparaison avec les travaux de J. Corbel (1964) est encore plus délicate du fait de l'influence importante de la lithologie sur les petits bassins choisis et de l'amalgame des transports en solution et en suspension.

		SiO_2	Ca^{++}	Mg^{++}	Na^+	K^+	Cl^-	SO_4^{--}	HCO_3^-	Réf.
Évolution des grands fleuves	mp	ND	51	17	85	6	100	136	ND	(1)
Effluents de mixte	mp	ND	17.2	6	36	3.5	49	78	64	(2)
Montréal résidentiel	mp	ND	3.2	0.65	6.6	1.0	8.2	13.5	24	(2)
Effluents urbains (USA)	mp	2.4	3	1.5	14	2	15	6	20	(3)
Bassin du Merrimack	mp	ND	18	3.9	65	4.7	ND	ND	ND	(4)
Moyenne choisie	mp		*50*	*10*	*85*	*5*	*100*	*135*	*(80)*	
Apport anthropique à l'océan (5)	Mp		53	10.6	91	6	108	148	124	
Charge par unité démophorique	mud		*0.7*	*0.15*	*1.2*	*0.08*	*1.4*	*1.85*	*(1.5)*	
Apport anthropique à l'océan (6)	Mp		47	10.5	78	5	93	124	100	

TABLEAU IV. – **Bilan des apports en solution d'origine anthropique à l'océan.**

Deux bilans (Mp en 10^6 t.an^{-1}) sont présentés. Le premier (5) est basé sur les rejets par habitant (mp en kg.habitant^{-1}.an^{-1}) d'après (1) l'étude des grands fleuves (voir annexe 4), et des études ponctuelles ; (2) l'étude d'effluents urbains résidentiels, ou résidentiels et industriels, à Montréal (A. Caillé et al., 1973) ; (3) L.W. Weinberger et al. (1966) ; (4) J. Caesar et al. (1976). Dans le deuxième bilan (6), plus juste, on fait intervenir le degré de développement des pays, mesuré par la consommation d'énergie par habitant (indice démophorique de J.R. Vallentine, 1978) (m.u.d. = rejet par unité démophorique en kg.an^{-1}). Environ 500.10^6 t de sels dissous parviennent chaque année aux océans en raison des diverses activités humaines. Pour Na^+, Cl^-, SO_4^{--} ces apports correspondent à une augmentation de plus de 30 % des valeurs naturelles.

TABLE IV. – *Balance of anthropogenic dissolved inputs to the ocean.*

Two balances (Mp in 10^6 t.year^{-1}) are proposed. The first (5) is based on per capita loadings (mp in kg.year^{-1} per capita) determined mainly from the evolution of six major rivers in Europe and North America (1) (annex 4) and from direct determinations on sewers (2) (3) (4). The second (6) takes into account the stage of development of countries, as measured by energy consumption (demophoric index of J.R. Vallentine, 1978) (m.u.d. = annual loading per demophoric unit, in kg). About 500.10^6 t of dissolved salts resulting from various Man activities reach each year the oceans. These inputs have increased by more than 30 % the natural values for Na^+, Cl^- and SO_4^{--}.

III. – BILAN DES TRANSPORTS EN SOLUTION DES RIVIÈRES.

Le bilan proprement dit a été effectué en trois parties : bilan partiel des bassins connus, extrapolation à l'ensemble du drainage exoréique en utilisant la typologie précédente, enfin déduction des concentrations d'origine anthropique présentes dans les analyses utilisées. Un bilan particulier pour chaque océan est présenté à part.

A. Drainage exoréique (externe).

Le bilan partiel des transports en solution des rivières étudiées est donné à l'annexe 3 ; le bilan global (annexe 3, également) comprend ce bilan partiel plus l'estimation pour chaque continent des apports restant calculés sur la base de la répartition des superficies de l'annexe 2 et des transports caractéristiques de chaque type définis au tableau III. Les concentrations et les transports moyens sont obtenus en divisant la masse globale transportée par les débits et les superficies totaux de chaque continent. Ces valeurs sont encore faussées par la présence d'apports anthropiques, surtout en Europe ; le bilan final présentant les valeurs corrigées est au tableau VI.

Remarquons que l'échantillon des régions connues (64.10^6 km^2) correspond à un débit spécifique très voisin de celui de l'ensemble des régions exoréiques ($11,7$ contre $11,8$ l.s^{-1}.km^{-2}). Cet échantillon partiel est donc bien représentatif et peut alors servir pour étudier la répartition relative des types d'eaux superficielles (voir V.A). D'autre part, je pense qu'*il ne faut pas accorder trop d'intérêt à des valeurs moyennes de concentrations ou de transports en solution à l'échelle des continents.* Ceux-ci sont souvent très hétérogènes, notamment en ce qui concerne le climat (Asie, Amérique du Nord, Océanie) ; il n'est donc pas fondé de discuter ces valeurs. Tout au plus peut-on noter que l'abondance des roches sédimentaires en Europe (87 % contre 66 % pour l'ensemble des continents, d'après H. Blatt et R.L. Jones, 1975) est sans doute responsable de la valeur plus élevée des concentrations des eaux de ce continent ; l'inverse, observé en Afrique, est dû à la plus grande abondance de roches cristallines (42 % contre 34 % en moyenne).

B. Drainage endoréique (interne).

Les analyses chimiques des eaux superficielles dans les régions endoréiques sont rares, et le bilan hydrique y est mal établi. Si ces bassins n'ont pas d'incidence sur les apports à l'océan, ils contribuent cependant à la dénudation globale des continents et doivent donc être pris en compte. L'annexe 3 fournit les valeurs des transports en solution pour les trois plus grands bassins endoréiques non désertiques (Caspienne, Mer d'Aral et Chari). La concentration moyenne résultante est trois fois plus élevée que celle du drainage externe. Elle est très proche de celle du type tempéré semi-aride (tabl. III) défini en partie par les bassins de la Caspienne et de la Mer d'Aral. Pour estimer le transport en solution global de l'ensemble des régions endoréiques, on peut utiliser la répartition des types de transport et leur valeur typique (annexe 2, tabl. III) ; on trouve ainsi pour l'ensemble de ces régions un transport ionique global de l'ordre de 200.10^6 t.an^{-1}, et un transport en silice de l'ordre de 10.10^6 t.an^{-1}, soit un transport total moyen pour toutes les régions endoréiques de $6,5$ t.km^{-2}.an^{-1}. L'érosion chimique y serait de l'ordre de 5 t.km^{-2}.an^{-1}. Cette moyenne très faible est essentiellement due à la présence des régions aréiques ou arides ($28,6.10^6$ km^2). Dans les rares régions endoréiques montagneuses et très humides, les transports en solution et l'érosion chimique peuvent être très élevés (Pamir et Caucase), comme dans les autres régions du type tempéré très humide.

C. Les apports en solution d'origine anthropique.

Les apports dissous d'origine anthropique ont parfois modifié complètement la composition chimique des eaux superficielles, et cette influence doit être prise en compte pour rectifier le bilan basé sur les analyses actuelles des rivières. Ces apports parviennent aux rivières soit par les précipitations (particulièrement les sulfates ; C.E. Junge, 1963 ; L. Granat et al., 1976), soit par le ruissellement (notamment les éléments nutritifs), soit par les rejets directs (urbains, industriels, miniers). De plus, l'action de l'homme sur le climat (l'acidité des précipitations augmente dans les régions industrielles ; N.M. Johnson et al. 1972), et sur les sols (pratiques agricoles, déboisement) modifie également ce bilan (R. Meade, 1969 ; G.E. Likens et al., 1970). Ainsi, E.P. Odum (1971) a pu comparer l'homme aux volcans puisque de son action résulte une production de CO_2, de SO_2 et de poussières minérales. *La plupart des actions humaines conduisent donc à augmenter la minéralisation des rivières et les apports en solution à l'océan.*

Deux cas, qui sont sans doute encore de peu d'importance à l'échelle globale, échappent à cette tendance : construction de réservoirs et irrigation. En effet, il peut se produire dans les lacs artificiels une rétention de matière dissoute (précipitation de carbonates, formation de matière organique). L'irrigation intensive peut conduire à une rétention de sels dans les sols. Les apports à l'océan de sels dissous du Nil et de l'Indus ont été ainsi considérablement diminués.

Quand on dispose d'analyses anciennes fiables, *on note dans les rivières des pays industrialisés une augmentation nette des concentrations des éléments majeurs depuis 1900. Les éléments les plus affectés sont Na$^+$, Cl$^-$, SO$_4^{--}$ et Ca^{++}.* L'évolution de six grands fleuves est reportée à l'annexe 4. La variation la plus forte est observée dans le Rhin où les chlorures ont décuplé, et où le sodium est passé de 5 à 120 mg.l^{-1} en un siècle, en raison des rejets de saumures des mines de potasse en Alsace, et de sel en Lorraine, auxquels s'additionnent tous les effluents urbains et industriels déversés depuis Bâle jusqu'à la frontière hollandaise. Nous nous sommes limités ici aux éléments majeurs mais l'évolution des éléments nutritifs ou de certains métaux (Pb, Zn, Cu) dans les rivières des pays industrialisés montre des tendances comparables sinon plus marquées. Ainsi les bilans d'azote et de phosphore dissous à l'océan seraient actuellement de l'ordre de 3 et 5 fois les bilans naturels (J. Van Bennekom et W. Salomons, 1979).

Il est très difficile de prendre en compte séparément toutes ces influences à l'échelle de la planète ; aussi ai-je opéré d'une manière globale en cinq étapes :

● J'ai essayé de reconstituer une composition moyenne naturelle des précipitations (annexe 5). La contribution de la pollution atmosphérique aux eaux de surface n'est donc pas calculée directement (voir IV.B.1).

● Les modifications de la qualité des eaux de surface sont jugées sur l'évolution depuis un siècle de grands fleuves en Amérique du Nord (Mississippi, Saint-Laurent), et en Europe (Rhin, Oder et Vistule, Seine), et sur des études directes de pollution.

● On définit ainsi des charges annuelles apportées aux eaux de surface par habitant *(per capita)* caractéristiques des pays à fort développement industriel.

● Pour tenir compte du stade de développement des pays, on introduit un indice proportionnel à la consommation d'énergie (J.R. Vallentine, 1978). On obtient ainsi des charges *per capita* corrigées.

● On peut alors obtenir d'une part une estimation de la quantité globale de sels dissous arrivant par les rivières à l'océan due aux influences anthropiques, d'autre part une correction du bilan effectué à l'annexe 3 compte tenu de la population et du développement des pays concernés.

Pour chaque rivière, j'ai calculé la masse additionnelle transportée et je l'ai rapportée à la population vivant actuellement (1968) sur le bassin (annexe 4).

Les rejets *per capita* sont tous du même ordre de grandeur, les seules différences importantes sont observées pour Na$^+$ et Cl$^-$, et sont à mettre au compte des activités minières intensives sur les bassins du Rhin, de l'Oder et de la Vistule. On peut ainsi définir des rejets moyens *per capita* correspondant à 180.10^6 habitants, et les comparer à des mesures directes (tabl. IV). L'étude des effluents de Montréal (A. Caillé *et al.*, 1973) a porté sur une trentaine d'effluents urbains, soit purement résidentiels, soit mixtes (résidentiels + industriels). J'ai obtenu les rejets *per capita* en soustrayant l'analyse moyenne de l'eau de distribution à celle des effluents. Ces valeurs sont très semblables à celles obtenues par L.W. Weinberger *et al.* (1966) pour les Etats-Unis (localisation non précisée) et par J. Caesar *et al.* (1976) sur la rivière Merrimack. La différence entre les rejets calculés d'après les grands fleuves et obtenus directement sur des effluents est sans doute à mettre au compte des activités minières. On peut ainsi proposer des rejets *per capita* typiques de l'ensemble des activités humaines (tabl. IV).

La silice pose un problème particulier, car, jusqu'ici on ne pensait pas que sa concentration dans les eaux pouvait être influencée par les activités humaines. Or l'évolution du Rhin de 1875 à 1971 montre une forte augmentation en silice dissoute (annexe 4). L'analyse de 1875 peut être sujette à caution, mais l'évolution de 1964 à 1971 confirme cette tendance (Commission du Rhin). Une variation identique a été observée dans la Charente (F. Salvadori, 1976). Elle est due aux rejets d'eaux usées d'une usine de traitement d'engrais. De même, certains effluents de l'agglomération new-yorkaise présentent des teneurs en silice nettement plus élevées que celles de l'Hudson (H.J. Simpson *et al.*, 1973). Toutefois, on ne dispose pas d'assez d'information pour estimer cet apport supplémentaire à l'échelle de la planète.

Un premier bilan d'apports anthropiques (tabl. IV) a été effectué en considérant que la population des pays industrialisés était de 700 millions d'habitants, correspondant à toutes les formes de rejets, à laquelle j'ai ajouté une population de 400 millions d'habitants de pays développés moins industrialisés, à laquelle j'ai affecté un rejet de type résidentiel (effluent résidentiel de Montréal). Ce type de bilan ne tient que très grossièrement compte du stade d'industrialisation des pays et ne permet pas la correction des analyses déjà utilisées.

Le deuxième bilan (tabl. IV) est basé sur les rejets par habitant corrigés de leur consommation d'énergie. J.R. Vallentine (1978) a défini un « indice démophorique » D, comme étant le rapport de l'énergie consommée par un pays à l'énergie nécessaire à des fins strictement physiologiques, et il en donne la valeur pour tous les pays. D varie de moins de l'unité (pays où même les besoins élémentaires en énergie ne sont pas satisfaits) à 97 (Etats-Unis). La population de chaque pays peut alors s'exprimer en « unités démophoriques » = population x indice D. L'ensemble de la planète correspond actuellement à 66,8.10^9 unités démophoriques, dont 20,4.10^9 pour les Etats-Unis et 24,5.10^9 pour l'Europe. On peut calculer les rejets par unité démophorique (m.u.d.) basés sur les rejets *per capita* des grandes rivières, et un indice démophorique correspondant de 70 (tabl. IV), en faisant deux hypothèses : 1) les rejets sont liés aux populations et à leur consommation d'énergie ; 2) cette relation est linéaire. Cette deuxième hypothèse n'est valable qu'en première approximation, puisqu'il est peu probable qu'à 1 million de personnes à l'indice D < 1 correspondent des rejets identiques à ceux de 10 000 Nord-Américains (D ≃ 90).

L'apport actuel aux océans d'éléments minéraux en solution d'origine anthropique est de l'ordre de 450 10^6 t.an^{-1}. Cela représente une augmentation de la minéralisation globale des rivières de 12 % et, pour certains éléments (Na$^+$, Cl$^-$, SO$_4^{--}$), de 30 %. Remarquons que, du fait du mode de calcul, ces apports englobent au moins en partie les rejets directement déversés dans les océans, notamment dans la zone estuarienne.

D. Apports naturels en solution.

Beaucoup de rivières utilisées dans le bilan de l'annexe 3 subissent déjà une influence anthropique (Pô, Danube, Columbia, Suède, Japon, etc.), aussi une correction est-elle nécessaire. J'ai effectué, pour chaque continent, une estimation des « populations démophoriques » prises en compte précédemment, soit en tout 22 . 10^9 unités démophoriques provenant surtout d'Europe, d'Amérique du Nord et d'Asie (Japon, U.R.S.S.). Les concentrations corrigées sont données aux tableaux I et V. On note que les différences avec les valeurs de l'annexe 3 sont très réduites, et sans doute inférieures à l'erreur commise sur ces bilans, sauf pour l'Europe (Na$^+$, Cl$^-$ et SO$_4^{--}$).

On peut donc comparer ce bilan avec celui de D.A. Livingstone (1963 : tableau 81) : les moyennes mondiales ne diffèrent pas de plus de 25 %, sauf pour le potassium, mais les résultats divergent beaucoup si l'on considère les continents un à un. Ainsi les concentrations moyennes de Livingstone pour l'Afrique sont beaucoup plus fortes (98 mg.l^{-1} pour la concentration ionique globale). Il semble que cela soit dû surtout à la mauvaise représentativité des rivières disponibles, ce qui conduit à une surestimation des apports du type tropical contrasté. D'une façon générale, les valeurs de D.A. Livingstone sont plus élevées, peut-être est-ce dû à un effet des apports anthropiques, son bilan étant compris entre nos estimations des concentrations naturelles et actuelles (tabl. I).

Le dernier bilan soviétique (U.S.S.R. Committee for the I.D.H., 1978, tableau 163) ne porte que sur la concentration ionique globale de chaque continent : Asie 72,9 mg.l^{-1} ; Afrique 151 mg.l^{-1} ; Amérique du Nord 90 mg.l^{-1} ; Amérique du Sud 45 mg.l^{-1} ; Europe 116,7 mg.l^{-1} ; Océanie 70 mg.l^{-1} ; moyenne mondiale 78,4 mg.l^{-1}. Les auteurs ne donnant pas de détail sur leur mode de calcul et leurs sources, il est difficile de voir l'origine des divergences avec mon bilan, en particulier pour l'Afrique. Par rapport à mon bilan précédent (1977), moins documenté et moins rigoureux, les valeurs présentées ici sont plus faibles ; cela est dû à une meilleure estimation des apports des zones tropicales humides, et des influences anthropiques.

E. Apports en solution des continents aux océans.

Le bilan des apports en solution aux quatre océans, Atlantique, Pacifique, Indien, et Glacial Arctique est un paramètre important de leur équilibre géochimique qui dépend à la fois de leur bilan hydrique, mais aussi de la composition chimique des eaux continentales qu'ils reçoivent. Dans un tel bilan, il est nécessaire d'examiner toutes les formes d'apports (glaciers, eaux souterraines, atmosphère, lessivage des produits du volcanisme actif), ainsi que les modifications possibles des apports des rivières lors de leur passage dans la zone estuarienne.

1. Apports des glaciers.

Les apports d'eau provenant des glaciers sont estimés à 2 300 km^3.an^{-1}, dont 1 980 km^3.an^{-1} pour l'Antarctique seul (A. Baumgartner et E. Reichel, 1975). Certains océanographes (J.D. Burton et P. Liss, 1973) ont vu dans l'Antarctique une source majeure d'éléments dissous à l'océan. Environ 700.10^6 t de silice proviendraient chaque année d'une mise en solution partielle de la farine glaciaire. Cette estimation théorique repose sur les calculs de flux dans les divers compartiments océaniques où les masses engagées sont nettement plus grandes que les apports terrigènes. J.M. Edmond (1973) et D.C. Hurd (1977) ont démontré qu'il n'était pas nécessaire de chercher un apport massif de silice dissoute venant de l'Antarctique, et que celui-ci ne pouvait en aucun cas en être à l'origine, D.C. Hurd l'estimant entre 0,02 et 1,8.10^6 t.an^{-1}.

Les quelques analyses chimiques de glace montrent des concentrations en solution très faibles (annexe 6), mais ne peuvent pas rendre compte d'une éventuelle mise en solution dans l'océan de la farine glaciaire. Dans les eaux douces comme dans l'océan, cette dissolution est très restreinte, puisque les teneurs en éléments majeurs de l'eau de fonte des glaciers du type alpin sont généralement très faibles, quoique nettement supérieures à celle des glaces (annexe 6). En prenant

349

les valeurs des eaux de fonte, on aboutit à une estimation maximum : *les apports en solution dus aux calottes glaciaires représentent moins de 1 % des apports dissous fluviaux.*

2. Apports des eaux souterraines.

Le bilan hydrique de A. Baumgartner et E. Reichel (1975) utilisé ici inclut les eaux souterraines se déversant directement dans les océans, les apports de celles-ci sont donc contenus dans ce bilan. Toutefois, j'ai assimilé les concentrations des eaux souterraines à celles des eaux superficielles. Comme ces dernières sont toujours moins élevées (J.D. Hem, 1970), le bilan global est donc sous-estimé. Le dernier bilan soviétique (U.S.S.R. Committee for the I.H.D., 1978) estime à 5 % les apports d'eaux souterraines directs à l'océan. Le chiffre semble élevé puisque R. G. Dzhamalov, I.S. Zektzer et A.V. Meskheteli (1977) estiment cette proportion à 1,1 % pour le bassin de la Caspienne. En prenant cette valeur pour l'ensemble des continents et en attribuant aux eaux souterraines une minéralisation triple de celles des eaux superficielles, l'erreur commise serait de 2 %. Vu les autres incertitudes sur le bilan, on peut ne pas les prendre en considération.

3. Apports atmosphériques continentaux à l'océan.

Jusqu'ici, la plupart des géochimistes ont considéré les océans comme une source d'apports atmosphériques aux continents, et non l'inverse. En effet, une grande partie des sels dissous trouvés dans les précipitations est de cette provenance (C.E. Junge, 1963). Les apports éoliens de poussières continentales à l'océan sont par contre connus depuis longtemps et estimés de 100 à 500.10^6 t.an^{-1} (E.D. Goldberg, 1972). Les aérosols d'origine continentale pourraient également constituer un apport en solution à l'océan. Toutefois, ce phénomène a été surtout étudié jusqu'ici dans des régions déjà modifiées par l'homme (H.L. Windom, 1979) et son flux global n'est pas quantifié. G. Crozat (1978) a pu mettre en évidence dans l'atmosphère un rejet de potassium provenant de certains végétaux. Toutefois il est peu probable que ces apports puissent constituer une source importante d'éléments majeurs à l'océan.

	SiO_2	Ca^{++}	Mg^{++}	Na^+	K^+	Cl^-	SO_4^{--}	HCO_3^-	Σi
Afrique	12.0	5.25	2.15	3.8	1.4	3.35	3.15	26.7	45.8
Amérique du Nord	7.2	20.1	4.9	6.45	1.5	7.0	14.9	71.4	126.3
Amérique du Sud	10.3	6.3	1.4	3.3	1.0	4.1	3.5	24.4	44.0
Asie	11.0	16.6	4.3	6.6	1.55	7.6	9.7	66.2	112.5
Europe	6.8	24.2	5.2	3.15	1.05	4.65	15.1	80.1	133.5
Océanie	16.3	15.0	3.8	7.0	1.05	5.9	6.5	65.1	104.3
Monde	10.4	13.4	3.35	5.15	1.3	5.75	8.25	52	89.2

TABLEAU V. — **Compositions moyennes des eaux de chaque continent.**

Ces concentrations (mg.l^{-1}) ne correspondent qu'au drainage exoréique et sont obtenues après déduction des apports anthropiques dans le bilan général de l'annexe 3. Seule l'Europe présente des corrections significatives (pour Na$^+$, Cl$^-$ et SO$_i^-$). Les variations observées d'un continent à l'autre résultent des climats et des reliefs différents mais reflètent aussi les lithologies dominantes : roches sédimentaires en Europe et Amérique du Nord (malgré le Bouclier Canadien) et roches cristallines en Afrique et en Amérique du Sud. Toutefois l'hétérogénéité de la plupart des continents limite beaucoup l'utilisation de ces moyennes. Ces valeurs peuvent différer des bilans précédents, en particulier pour l'Afrique dont les eaux sont ici deux et trois fois moins minéralisées que d'après D.A. Livingstone (1963) (98 mg.l^{-1}) et d'après les auteurs soviétiques (U.S.S.R. Comm. for I.H.D., 1978) (151 mg.l^{-1}).

TABLE V. — *Average composition of river waters for each continent.*

These concentrations (mg.l^{-1}) only correspond to the exoreic runoff and are obtained after deduction of anthropogenic inputs from the general balance (annex 3). Europe is the only one for which significant corrections were found (for Na$^+$, Cl$^-$ and SO$_i^-$). The differences between contents from one continent to another originate from the variability of climates and of relief intensities, but also reflects the influence of lithology : sedimentary rocks dominate in Europe and North America while crystalline rocks greatly influence water quality in Africa and South America. However the heterogeneity of most continents greatly limits the use of these averages. These values may differ from previous estimates, up to 200 % or 300 % for african waters : 98 mg.l^{-1} according to D.A. Livingstone (1963) and 151 mg.l^{-1} according the U.S.S.R. Committee for I.H.D. (1978).

4. Lessivage des produits du volcanisme actif.

Le volcanisme produit de l'ordre de 10^9 t de cendres par an. Le lessivage de la part retombant sur le continent constitue une source d'éléments en solution qui est peu prise en compte dans ce bilan, la plupart des fleuves étudiés ici ne comportant pas de zones volcaniques actives. P.S. Taylor et R.E. Stoyber (1973), estiment ce type d'apport, silice exclue, à 30.10^6 t.an^{-1}, soit moins de 1 % des apports des rivières (tabl. VI). La production de SO_2 et H_2S rejetés dans l'atmosphère par les volcans serait de l'ordre de 30.10^6 t.an^{-1} SO_4^{--}, dont une grande part retombe sur les océans (L. Granat et al., 1976); l'autre part est une des sources naturelles des sulfates atmosphériques (voir IV.B.1).

5. Passage des apports fluviaux dans la zone estuarienne.

Dans la zone estuarienne, les eaux fluviales rencontrent des conditions physico-chimiques et biologiques nouvelles : force ionique élevée, milieu souvent plus réducteur, turbidités très fortes, cycle et production biologiques différents, etc. Il peut en résulter des modifications des formes sous lesquelles les matériaux d'origine continentale sont apportés aux océans. De nombreuses études ont porté depuis vingt ans sur l'évolution des éléments dissous dans les estuaires, en particulier celle de la silice et des éléments nutritifs.

La silice est le seul élément majeur pour lequel les teneurs fluviales sont plus élevées que la valeur océanique, et son comportement présente une grande diversité : comportement conservatif (J. Kobayashi, 1967), défaut de silice dissoute jusqu'à 30 % (F.S. Li et al., 1964), ou même excès de silice (F. Salvadori, 1977). Si l'excès de silice, rarement observé, est sans doute d'origine anthropique (H.J. Simpson et al., 1975), le défaut de silice a tantôt été attribué à la production de diatomées dans l'estuaire, tantôt à la floculation suivie de précipitation de silice polymère, ou à l'adsorption de silice sur la matière particulaire de l'estuaire. Une revue récente de cette question (P.S. Liss, 1976) aboutit aux éléments suivants : la précipitation de silice est observée surtout dans les estuaires où les eaux fluviales sont riches en silice dissoute, mais existe aussi dans les autres; elle est peu liée à l'activité biologique. Un mécanisme de régulation est avancé par P.S. Liss qui constate que les teneurs typiques atteintes dans les estuaires sont proches de 14 mg $SiO_2.l^{-1}$, soit une valeur semblable à celle des rivières d'après D.A. Livingstone (1963). Les taux de précipitation les plus couramment observés sont de l'ordre de 10 à 20 %. Si l'on affectait ces taux aux rivières riches en silice (régions volcaniques et zone tropicale), le bilan global des apports passerait de 390 à 360 ou même 320.10^6 t.an^{-1}.

Les autres éléments majeurs (Na^+, Mg^{++}, SO_4^{--}) étudiés ont généralement un comportement conservatif dans les estuaires. C'est-à-dire que les concentrations en solution suivent une loi de dilution linéaire, le mélange relatif des eaux étant apprécié par un élément assumé être non-réactif, en général la chlorinité. Toutefois, une précipitation de $CaCO_3$ dans les estuaires est possible (R. Wollast, comm. pers.), elle serait due à la baisse de pH observée dans ce milieu (W.G. Mook et B.K.S. Koene, 1975) qui peut modifier la distribution des carbonates et occasionner un transfert de CO_2 dans l'atmosphère. Comme les eaux superficielles sont en moyenne loin de la saturation en $CaCO_3$, aux teneurs en CO_2 rencontrées, ce phénomène ne doit pas influencer le bilan à l'océan d'une manière significative.

6. Bilan des apports des rivières à chaque océan.

Les transports des rivières peuvent donc être utilisés pour estimer en première approximation les flux de matériaux continentaux en solution à chaque océan. Seules les valeurs obtenues pour la silice devront être regardées avec prudence, puisqu'un apport réel de 80 à 90 % du transport des rivières est possible.

Cette estimation a été conduite avec une approche analogue à celle du bilan général présenté à l'annexe 3 : calcul du total des apports fluviaux connus à chaque océan, puis extrapolation basée sur les superficies restantes et les types de transports. Le découpage des terres émergées a été réalisé par bassin versant océanique, et non plus par continent, et ajusté de façon à obtenir le même bilan hydrique que celui proposé par A. Baumgartner et E. Reichel (1975) pour chaque océan. Seules sont calculées la concentration ionique globale et la silice. Les rapports ioniques sont en effet très peu variables à cette échelle; par contre, le caractère zonal des teneurs en silice et son importance sur le bilan géochimique des océans sont les raisons d'un bilan particulier.

Les apports totaux (M_d) et les apports spécifiques par unités de superficie (m_s) et de volume (m_v) océaniques varient largement d'un océan à l'autre (tabl. VII). Les océans Atlantique et Arctique reçoivent relativement plus de silice et d'ions majeurs que les océans Indien et Pacifique : l'océan Atlantique reçoit en effet, à lui seul, 49 % des apports d'eau continentale, contre 31 % seulement à l'océan Pacifique et 13 % à l'océan Indien. Par contre, c'est la petite taille de l'océan Arctique par rapport à son bassin versant qui est responsable de ses apports spécifiques élevés malgré les concentrations en silice de ses tributaires deux fois plus faibles que la moyenne mondiale. Les temps de séjour des éléments majeurs dans les divers océans pourraient donc être très différents si on considérait uniquement les apports continentaux, à l'exclusion des échanges entre océans et des apports internes (volcanisme sous-marin, diffusion des eaux interstitielles etc.).

IV. – IMPLICATIONS GÉODYNAMIQUES DU BILAN DES ÉLÉMENTS DISSOUS.

Les transports en solution et les apports des rivières aux océans peuvent être utilisés pour mieux connaitre les phénomènes d'altération, d'érosion, de sédimentation océanique actuels, et devraient également fournir des indications précieuses sur l'évolution qualitative et quantitative de ces phénomènes au cours des temps géologiques. J'essaierai de dégager ici quelques conclusions de cette étude concernant l'origine de la composition chimique des eaux de surface, et l'érosion des continents.

A. Influence de la lithologie sur la composition des eaux superficielles.

L'échantillon global des rivières utilisées ici, vu sa taille (63 % des apports en eau), son débit spécifique, identique à celui de l'échantillon total, et la présence de rivières de tous les types, peut être considéré comme bien représentatif de l'ensemble des eaux s'écoulant à la surface des continents. On peut ainsi étudier la répartition, basée sur les débits des fleuves, des différents types d'eaux définis par leurs ions principaux (analyses exprimées en meq.l^{-1}).

L'essentiel des eaux de surface (98 %) est bicarbonaté calcique. Le sodium lié à Cl^-, SO_4^{--} ou à HCO_3^- n'est le cation principal que de 1,7 % des eaux seulement. Cette répartition minimise donc les types chlorurés sodiques soit dilués (« rain dominated »), soit très concentrés (« evaporation-crystallization »), définis par R.J. Gibbs (1970) ; l'ensemble des eaux correspond à son type « rock dominated », c'est-à-dire bicarbonaté calcique. On peut distinguer au sein de ce dernier type trois grands sous-groupes :

● *sous-groupe sulfaté-magnésien (47 % des eaux superficielles) avec :*
$$Ca^{++} > Mg^{++} > Na^+ > K^+ ;$$
$$HCO_3^- > SO_4^{--} > Cl^-$$
caractéristique des terrains sédimentaires, particulièrement rencontré en Europe, en Amérique du Nord et en Asie.

● *sous-groupe chloruré-sodique (33 % des eaux) avec :*
$$Ca^{++} > Na^+ > Mg^{++} > K^+ ;$$
$$HCO_3^- > Cl^- > SO_4^{--}$$
caractéristique des régions cristallines et métamorphiques. Une partie des eaux africaines, et la plupart des eaux sud-américaines sont de ce type.

	SiO$_2$	Ca^{++}	Mg^{++}	Na$^+$	K$^+$	Cl$^-$	SO$_4^{--}$	HCO$_3^-$	Σm	Réf.
Érosion chimique	388	500	119	137	46	115	292	834	2431	(1)
CO$_2$ atmosphérique								1106	1106	(2)
Précipitations océaniques		2	7	55	2	100	15		181	(3)
Total des apports naturels	388	502	126	192	48	215	307	1940	3718	(4)
Apports actuels	*388 (8)*	*549*	*136*	*270*	*53*	*308*	*431*	*2040*	*4187*	(5)
Apports des glaciers (maxi)	1.8	9.2	1.6	1.15	0.9	0.7	7	23	45	(6)
Volcans actifs	?	2.1	3.1	0.93	0.05	8.4	15.9	?	(30)	(7)

TABLEAU VI. – **Flux de matière dissoute des rivières aux océans comparés aux autres apports (10^6 t.an^{-1}).**

Les produits de la dénudation chimique des continents (1) ne correspondent qu'à 65 % des matériaux transportés à l'océan (4), le reste provient des aérosols océaniques (5 %) (3) et du CO$_2$ atmosphérique (30 %) (2). La pollution (5) a augmenté de 12 % ces apports naturels. Les apports des glaciers (6) et des volcans (7) sont mineurs par rapport à ceux des rivières. Σm : apports minéraux totaux. (1) Les produits dissous de l'érosion éolienne sont inclus ; (2) totalité de HCO$_3^-$ transporté en pays cristallin + moitié de HCO$_3^-$ transporté en pays calcaire ; (3) calculés en normalisant à Na$^+$ supposé d'origine océanique à 100 % ; (4) concentration moyenne du tableau V × 37 400 km^3.an^{-1} ; (5) apports naturels + rejets calculés sur la base des unités démophoriques (tabl. IV) ; (6) cf. annexe 6 ; (7) P.S. Taylor et R.E. Stoyber (1973) ; (8) Les apports en silice dissoute pourraient être diminués de 20 % après passage dans la zone estuarienne.

TABLE VI. – *Dissolved material fluxes from rivers to oceans as compared to other sources (10^6 t.year^{-1}).*

The products of continental chemical denudation (1) represent only 65 % of the material naturally carried by rivers to ocean (4). The remaining part originates for 5 % from oceanic aerosols (3), and for 30 % from atmospheric CO$_2$ (2). Pollution (5) minus (4), has increased by 12 % these natural loads. Glaciers (6) (from annex 6) and active volcanoes (7) are minor inputs. (8) Dissolved silica flux could be lowered by 20 % during the transit through the estuarine zone.

● *sous-groupe chloruré-magnésien (15 % des eaux)* avec :
$$Ca^{++} > Mg^{++} > Na^+ > K^+ ;$$
$$HCO_3^- > Cl^- > SO_4^{--}$$
très abondant en Afrique et dans les rivières provenant de l'Himalaya.

L'abondance du calcium s'explique facilement pour les eaux drainant les roches sédimentaires, mais étonne un peu pour les eaux drainant des roches cristallines. L'alteration des roches cristallines libère également du Ca^{++} (Y. Tardy, 1971) ; mais surtout il suffit que 10 % des roches affleurant dans un bassin soient d'origine sédimentaire pour que le calcium demeure l'élément dominant. En effet les teneurs en Ca^{++} sont 10 fois plus fortes dans ces eaux de drainage que dans celles provenant de roches cristallines (M. Meybeck, 1979).

L'origine des bicarbonates s'explique plus facilement puisque même l'altération des roches cristallines, dans les conditions physico-chimiques de la surface actuelle des continents, conduit à des eaux bicarbonatées, le CO_2 atmosphérique ou présent dans les sols fournissant la totalité de cet anion. Dans la dissolution des roches carbonatées, c'est la moitié des bicarbonates qui provient de l'atmosphère (R.M. Garrels et F.T. Mackenzie, 1971).

Il est remarquable de noter que le potassium n'est jamais abondant. Il provient surtout de l'altération des minéraux silicatés potassiques rencontrés dans les roches cristallines et dans les roches sédimentaires détritiques. L'origine du sodium est bien plus variée : minéraux silicatés, dépôts d'évaporites, halite contenue dans les roches sédimentaires d'origine marine, aérosols marins et même pollution. Ces deux éléments ont donc des abondances et des variabilités très différentes dans les eaux de surface.

Environ 15 % des ions transportés par les eaux superficielles proviennent des précipitations atmosphériques (voir IV.B.1 et V.P. Zverev et V.Z. Rubeikin, 1973). Cet apport est responsable de variations secondaires de la composition ionique, particulièrement en Na^+ et Cl^-. Dans quelques régions cristallines proches des cotes les apports atmosphériques peuvent dominer les produits de l'altération, comme en Finlande, dans les Territoires du Nord-Ouest du Canada (voir annexe 1) et en Amazonie (R.J. Gibbs, 1970).

Les eaux de caractères évaporitiques sont rarement observées dans les rivières pérennes. L'influence de l'évaporation s'observe d'abord par l'augmentation relative des sulfates (eaux du type tempéré semi-aride ; tabl. III), puis des chlorures (eaux du type aride ; tabl. III). Les eaux franchement sulfatées ou chlorurées à forte minéralisation ($\Sigma i > 500$ mg.l^{-1}) n'existent

			Q	SiO$_2$	Σi
Atlantique					
	rivières étudiées	Md	12950	125	900
	extrapolation	Md	5560	63	570
	bassin total	Md	18510	188	1470
		Cd		10	80
		ms		2.2	17
		mv		0.6	4.5
Glacial Arctique					
	rivières étudiées	Md	2350	13.2	302
	extrapolation	Md	260	0.15	4.6
	bassin total	Md	2610	13.4	306
		Cd		5.1	117
		ms		1.4	32
		mv		1.0	24
Indien					
	rivières étudiées	Md	2780	35	429
	extrapolation	Md	1930	27	179
	bassin total	Md	4710	62	608
		Cd		13.1	129
		ms		0.85	8.3
		mv		0.22	2.1
Pacifique					
	rivières étudiées	Md	5310	66	458
	extrapolation	Md	6260	71	710
	bassin total	Md	11570	137	1170
		Cd		11.9	101
		ms		0.83	7.0
		mv		0.20	1.7

TABLEAU VII. — **Apports en solutions à chaque océan.**

Les apports naturels totaux (Md en 10^6 t.an^{-1}) et les concentrations moyennes (Cd en mg.l^{-1}) de silice et d'ions majeurs (Σi) ont été déterminés pour chaque océan. En raison des tailles relatives des océans (valeurs de H.W. Menard et S.M. Smith, 1966) et de leurs bassins versants, l'Atlantique et l'Arctique reçoivent nettement plus d'apports dissous spécifiques par unité de surface (ms en t.km^{-2}.an^{-1}) ou par unité de volume (mv en t.km^{-3}.an^{-1}). Les apports en silice aux océans Atlantique et surtout Indien et Pacifique peuvent être diminués de 20 % si on considère la précipitation de silice très souvent observée dans les estuaires.

TABLE VII. — *Dissolved loads to each ocean.*

Natural loads (Md in 10^6 t.year^{-1}) and average concentrations (Cd in mg.l^{-1}) from land to each ocean have been computed for silica and total ions (Σi). Due to the relative sizes of oceans (according to H.W. Menard and S.M. Smith, 1966) and of their drainage basins, Atlantic and Arctic oceans receive much higher specific loads per unit area (ms in t.km^{-2}.year^{-1}) or per unit volume (mv in t.km^{-3}.year^{-1}). Silica loads to Atlantic, Indian and Pacific oceans could be lowered by 20 % if the silica precipitation, commonly observed in estuaries, were considered.

guère que dans les rivières intermittentes rencontrées dans les zones désertiques (P. Blanc et G. Conrad, 1968) ou dans des régions où affleurent des dépôts évaporitiques comme le Sud-Ouest des U.S.A. (J.H. Feth, 1971), et le Rio Huallaga dans les Andes, responsable d'une grande partie des transports en NaCl de l'Amazone (J.M. Edmond, communication personnelle).

Il existe deux grandes catégories d'eaux superficielles, en effet *la répartition des fréquences des concentrations ioniques dans les eaux superficielles est bimodale*. Le premier mode se situe vers une concentration ionique globale de 35 mg.l^{-1} et correspond aux *eaux de la zone tropicale humide* (Afrique occidentale, Amazonie, Indonésie, etc.) du type bicarbonaté calcique, souvent du sous-groupe chloruré-sodique. Le deuxième mode est à 120 mg.l^{-1} environ et correspond d'une part aux *eaux des bassins sédimentaires en zones tempérée et froide* (Mississippi, Danube, Ob,...) d'autre part aux *eaux des régions montagneuses* quelle que soit leur situation (Rocheuses, Alpes, Andes, Himalaya...); ces eaux sont bicarbonatées calciques, généralement du sous-groupe sulfaté-magnésien.

La silice permet d'introduire un critère supplémentaire dans la caractérisation des eaux. En effet, son abondance est liée à la lithologie, mais aussi au type d'altération largement gouverné par des facteurs climatiques. Cette relation apparaît nettement dans les teneurs moyennes en silice des différents types d'eau (tabl. III). Les valeurs les plus élevées ($SiO_2 > 20$ mg.l^{-1}) correspondent généralement au drainage de roches volcaniques dans la zone tropicale humide. Bien que moins bien étudiées dans la zone froide, *les teneurs en silice présentent une répartition inversée par rapport aux concentrations ioniques*, puisque les eaux tropicales sont plus riches (mode vers 12 mg SiO_2.l^{-1}) que les eaux des zones tempérée et froide (mode vers 5 mg SiO_2.l^{-1}). Seules les eaux provenant de régions montagneuses de la zone tropicale humide présentent à la fois des concentrations ioniques et des teneurs en silice élevées.

B. Origines des éléments transportés par les rivières.

L'origine des éléments dissous donne des renseignements sur les cycles des éléments, sur l'érosion des roches superficielles, le recyclage des roches sédimentaires, etc. R.M. Garrels et F.T. Mackenzie (1971), et H.D. Holland (1978) ont utilisé les données de D.A. Livingstone (1963) pour décomposer l'analyse des eaux suivant leurs origines. Ces essais ont été quelquefois poussés plus loin que ne le permettaient les marges

d'erreurs du bilan de Livingstone. Avant d'étudier à mon tour les produits dissous de l'altération, il convient tout d'abord de soustraire aux rivières la part des éléments dissous ne provenant pas de l'érosion des continents.

1. *Les apports atmosphériques.*

Les précipitations ont été depuis longtemps (F.W. Clarke, 1924) reconnues comme une source majeure des éléments dissous des eaux superficielles, mais leur étude s'est surtout développée récemment (P.J. Viro, 1953; E. Gorham, 1961; E. Eriksson, 1959; C.E. Junge et R.T. Werby, 1958; V.P. Zverev, V.Z. Rubeikin, 1973 et J. Morelli, 1978). L'origine des apports atmosphériques est multiple : aérosols marins, poussières continentales, gaz et poussières volcaniques, pollution, etc. J'ai tout d'abord cherché la composition moyenne des précipitations naturelles en éliminant les valeurs provenant des régions industrielles. L'annexe 5 présente les apports atmosphériques (en t.km^{-2}.an^{-1}) de certaines grandes régions du globe (U.R.S.S., Etats-Unis) et de quelques stations isolées. J'ai calculé la moyenne pondérée des apports spécifiques et la composition moyenne des précipitations en tenant compte de la répartition des types climatiques définis pour les rivières. Pour l'ensemble du globe les apports atmosphériques représentent 15 % des transports ioniques naturels des rivières. Une proportion semblable a été trouvée pour l'ensemble de l'U.R.S.S. par V.P. Zverev et K.Z. Rubeikin (1973).

Il est difficile d'étudier directement les origines des apports atmosphériques. Aussi est-on forcé de formuler des hypothèses simplificatrices : il n'y a pas d'influence anthropique dans les valeurs moyennes données à l'annexe 5, elles ont été prises en compte précédemment avec l'ensemble des apports anthropiques (voir III. C). Les gaz et poussières volcaniques sont considérés comme provenant de l'érosion éolienne des continents. Le sodium est considéré comme entièrement d'origine océanique et les aérosols océaniques se forment en conservant les rapports ioniques de l'eau de mer. Cette dernière hypothèse permet, en normalisant les précipitations au sodium, d'en obtenir la partie d'origine marine. Elle est peut-être inexacte pour certains ions comme le potassium, dont J. Morelli (1978) a mis en évidence le fractionnement différentiel lors de la formation des aérosols.

Avec ces hypothèses, *les apports d'origine océanique ne correspondraient qu'à 35 % des précipitations totales*. Sur les 515.10^6 t.an^{-1} d'apports atmosphériques aux régions exoréiques, 180.10^6 t sont d'origine océanique, le reste provient de l'érosion éolienne des continents et du volcanisme. Cette proportion peut sembler faible, mais il est certain qu'à l'intérieur des continents la majeure partie des éléments dissous des précipitations est d'origine continentale (C.E. Junge et R.T. Werby, 1958; V.P. Zverev et V.Z. Rubeikin, 1973). En soustrayant les masses des apports océaniques de celles transportées par les rivières, on obtient les produits de l'érosion chimique des continents en y incluant le CO_2 atmosphérique utilisé dans les réactions d'altération (tabl. VI). En divisant ces dernières données par le débit total des fleuves, on obtient les concentrations moyennes des produits provenant de l'érosion (tabl. VIII) déduction faite des apports anthropiques et océaniques.

2. Les produits de l'altération.

La méthode de décomposition des concentrations proposées ici (tabl. VIII) est proche de celles de R.M. Garrels et F.T. Mackenzie (1971) et de H.D. Holland (1978), mais utilise également les rapports chimiques caractéristiques des eaux de drainage des divers types de roches. Le principe général est le suivant : l'ensemble des chlorures − avec une quantité égale de sodium − est attribué à la dissolution de la halite contenue dans les roches sédimentaires ou affleurant en gisement. Les sulfates proviennent à la fois de la dissolution des sulfates calciques et magnésiens des roches sédimentaires et de l'oxydation des sulfures présents dans les roches sédimentaires ou cristallines,

l'acide sulfurique ainsi formé réagit sur les roches en libérant des cations (Ca^{++}, Mg^{++}, Na^+, K^+). Les bicarbonates se répartissent entre la dissolution des carbonates calciques et magnésiens et l'utilisation du CO_2 lors de l'altération. L'altération des silicates présents dans les roches cristallines et sédimentaires libère la totalité de la silice et du potassium, le reste du sodium, du calcium et du magnésium. La décomposition présentée ici ne correspond qu'aux eaux du drainage exoréique. Les analyses sont d'abord exprimées en $\mu eq.l^{-1}$. Tous les chlorures ($87 \mu eq.l^{-1}$) sont retirés avec Na^+ correspondant. Il faut noter ici que l'abondance de NaCl provenant de l'érosion des roches sédimentaires dépend de l'estimation précédente de la part du NaCl d'origine océanique trouvé dans les eaux.

La teneur en SO_4^{--} des eaux de drainage des terrains sédimentaires est environ 10 fois plus forte que pour les terrains cristallins (M. Meybeck, 1979). Vu l'abondance des roches sédimentaires à l'affleurement (66 % d'après H. Blatt et R.L. Jones, 1975), elles sont donc à l'origine de 95 % des sulfates, le restant provenant des sulfures des roches cristallines. Ceux-ci réagissent avec les cations dans des proportions supposées être celles des eaux drainant les roches cristallines ($5 Ca^{++}$, $3 Mg^{++}$, $1 Na^+$, $1 K^+$ en eq.l^{-1}). Dans les

	SiO$_2$	Ca^{++}	Mg^{++}	Na$^+$	K$^+$	Cl$^-$	SO$_4^{--}$	HCO$_3^-$	
Concentration naturelle	*173*	*670*	*259*	*155*	*32*	*87*	*163*	*846*	(1)
NaCl				87		87			(2)
S^{--}		24	14	5	3		46		(3)
(Ca, Mg) SO$_4$		88	29				117		(4)
Altération des silicates									(5)
en montmorillonite	45	4	3	26	12			45	
en kaolinite	128	6	4	37	17			64	
(Ca, Mg) CO$_3$		548	209					737	(6)

TABLEAU VIII. − **Origines des produits de l'érosion chimique.**

La dénudation chimique des continents concerne essentiellement les roches sédimentaires (89 % des produits de l'érosion dont 55 % pour les roches carbonatées). Les vitesses d'altération des principaux types de roche estimées en comparant les produits libérés et les proportions d'affleurement sont très variables : roches détritiques 1, roches cristallines ≃ 1,5, roches carbonatées 20, évaporites 80. (1) Concentrations moyennes (en $\mu eq.l^{-1}$ pour les ions et $\mu mole.l^{-1}$ pour SiO$_2$) correspond aux flux (1) moins (3) du tableau VI ;(2) présent dans les roches sédimentaires; (3) 80 % des sulfures altérés proviennent des roches sédimentaires et 20 % des roches cristallines; (4) le rapport Ca/Mg est celui des eaux riches en SO$_4^{--}$ drainant des roches sédimentaires; (5) la proportion des deux formes d'altération est celle des apports en silice des zones tropicale (74 %) et tempérée (26 %); (6) le rapport Ca^{++}/Mg^{++} est très proche de celui des eaux drainant des roches sédimentaires. La balance ionique du départ est légèrement déficitaire en anions.

TABLE VIII. − *Origins of chemical erosion products.*

Chemical denudation products mainly originate (89 %) from sedimentary rocks, specially from carbonate rocks (55 %). Weathering rates, as estimated by the comparison of weathering products with the distribution of surficial rocks, are highly variable : detrital sedimentary rocks 1, crystalline rocks ≃ 1.5, carbonate rocks 20, evaporites 80. (1) the average contents (in $\mu eq.l^{-1}$, and $\mu mole.l^{-1}$ for silica) correspond to fluxes, (1) minus (3), from table VI; (2) in sedimentary rocks; (3) 80 % of weathered sulfides from sedimentary rocks, 20 % from crystalline rocks; (4) Ca^{++}/Mg^{++} as in sulfate − rich waters draining sedimentary rocks; (5) the ratio between the two major weathering processes is similar to the proportion of silica loads from the tropical (74 %) and temperate (26 %) zones; (6) Ca^{++}/Mg^{++} as observed in waters draining sedimentary rocks.

roches sédimentaires, 23 % du soufre se trouve sous forme de sulfures (H.D. Holland, 1978) qui réagissent comme ceux des roches cristallines. Les sulfates des roches sédimentaires se dissolvent en donnant 3 Ca^{++} pour 1 Mg^{++} (en $eq.l^{-1}$), proportion trouvée dans les eaux très sulfatées. La décomposition finale donne (en $\mu eq.l^{-1}$) :

	SO_4^{--}	Ca^{++}	Mg^{++}	Na^+	K^+	
sulfures.. {	8	1	2,5	1	0,5	R. cristallines
	38	20	11,5	4	2,5	}
sulfates ..	117	88	29	.	.	} R. sédimentaires
Total....	163	112	13	5	3	

La décomposition des silicates fait intervenir le type d'altération. Schématiquement, selon que l'on prend comme terme de l'altération la formation de montmorillonite (bisiallitisation de G. Pedro, 1968) ou celle de kaolinite (monosiallitisation), la quantité de silice libérée pour la même quantité d'ions varie du simple au double (R.M. Garrels et F.T. Mackenzie, 1971). La décomposition individuelle des fleuves étudiés ici montre que, dans leur grande majorité, les eaux de la zone tropicale correspondent à la monosiallitisation, tandis que celles des zones tempérée et froide correspondent à la bisiallitisation, ce qui confirme les travaux de G. Pedro. Comme le bilan géographique montre que

Types de transports dissous	% superf. (1)	% écoulem. (1)	Transport de silice		Transport ionique		Érosion chimique Ed (3)
			Md (2)	%	Md	%	
Toundra et taïga	20.0	10.7	15.0	3.9	466	13.1	14
Taïga humide	3.15	3.4	5.0	1.3	74	2.1	15.5
Taïga très humide	0.2	0.6	1.1	0.2	9	0.25	32
Tempéré pluvieux	4.5	15.3	45	11.8	540	15.4	80
Tempéré humide	7.45	7.75	17.5	4.6	407	12.0	35
Tempéré	6.7	3.35	9.4	2.5	301	8.8	28
Tempéré semi-aride	3.35	1.05	2.7	1.0	130	3.7	24
Tropical contrasté	13.25	5.85	31.1	8.2	119	3.4	6.4
Tropical humide	9.2	8.85	38.2	10.2	239	8.0	15.5
Tropical très humide (plaine)	6.9	18.45	78.6	20.8	165	4.8	22
Tropical très humide (montagne)	7.95	24.05	130	34.4	908	25.6	67
Aride + désert	17.2	0.65	3.9	1.0	3457	2.8	3 (4)
Ensemble de la zone tropicale	*37.3*	*57.2*	*278*	*73.6*	*1431*	*41*	*(5)*
Régions pluvieuses à fort relief	**12.65**	**40**	**176**	**47**	**1457**	**42**	$\simeq$ **74 (6)**

TABLEAU IX. — **Origines géographiques des apports en solution à l'océan, et variations de l'érosion chimique suivant les différents types morphoclimatiques.**

L'érosion chimique est maximum dans les régions de montagne tempérées humides et tropicales humides. Il en résulte que les apports en ions et en silice à l'océan proviennent surtout des régions de montagne (6) et de l'ensemble de la zone tropicale en ce qui concerne la silice seule (5). L'érosion moyenne est de 19 $t.km^{-2}.an^{-1}$ (7 mm/1 000 ans), pour l'ensemble des continents et de 24 $t.km^{-2}.an^{-1}$, pour la partie exoréique seule. Pour chaque type morphoclimatique (tabl. II) ont été calculés : (1) la part de la superficie exoréique totale ($99,9.10^6 km^2$) et de l'écoulement correspondant ($37~100~km^3.an^{-1}$) ; (2) les apports en silice et en ions majeurs aux océans (Md en $10^6 t.an^{-1}$) ainsi que leur importance relative dans l'apport total ; enfin (3) l'érosion chimique ($t.km^{-2}.an^{-1}$) (transports dissous naturels moins les apports atmosphériques et le CO_2 atmosphérique). Ce bilan, calculé d'une façon différente que le précédent (annexe 3, tabl. V et VI) lui est presque identique ; (4) Ed correspond au type aride seul.

TABLE IX. — *Geographic origins of dissolved loads to the ocean and variations of chemical erosion according to various morphoclimatic types.*

For each morphoclimatic types the following were determined : (1) % of exoreic area and of exoreic runoff; (2) dissolved loads to the oceans (Md in $10^6 t.year^{-1}$) both for silica and for total ionic content; (3) chemical erosion (Ed in $t.km^{-2}.year^{-1}$) computed as natural transport rates minus oceanic cyclic salts and atmospheric CO_2. Ed is maximum in mountainous regions of the humid temperate and tropical zones. As a result the dissolved ionic loads to the ocean mainly originate from mountainous area (6), while 74 % of silica comes from the tropical zone alone (5).

74 % de la silice dissoute transportée provient des rivières de la zone tropicale (tabl. IX), j'ai décomposé la concentration en silice dans cette proportion. En supposant que Na^+ restant et tout K^+ proviennent de l'altération des silicates, et que Ca^{++} et Mg^{++} sont présents dans la proportion $5/3$ en eq.l^{-1}, on obtient la décomposition détaillée au tableau VIII.

Le restant de Ca^{++}, Mg^{++} et HCO_3^- est attribué à la dissolution des carbonates. Remarquons que le rapport $Ca^{++}/Mg^{++} = 2,6$ en eq.l^{-1} trouvé ici est très proche de celui des eaux drainant les roches sédimentaires (2,5 d'après M. Meybeck, 1979).

Le CO_2 atmosphérique est à l'origine d'environ 57 % des bicarbonates des eaux superficielles, ce qui correspond à 33 % des éléments libérés lors de l'érosion chimique. *On ne peut donc estimer la dénudation chimique des continents par les transports en solution des rivières qu'après trois corrections essentielles : la pollution, les apports atmosphériques d'origine océanique, et le CO_2 atmosphérique.*

Suivant les hypothèses choisies, plusieurs types de décomposition pourraient être faits, mais l'imprécision des moyennes prises au départ (sans doute supérieure à 20 %) ne justifie pas d'entrer dans le détail. Ici, par exemple, il y a un défaut d'anion de 20 μeq.l^{-1} que j'ai attribué aux bicarbonates : il est dû au déséquilibre de la balance ionique (cations 1 116 μeq.l^{-1}, anions 1 096 μeq.l^{-1}). De même il est vraisemblable que la part de Ca^{++} et de Mg^{++} attribuée à l'altération des silicates est supérieure à celle qui est proposée ici, puisque les eaux de drainage des roches cristallines sont quelquefois plus riches en Ca^{++} et Mg^{++} qu'en Na^+ et K^+. Pour rétablir les proportions, il faudrait une teneur moyenne en silice plus élevée ou une proportion monosiallitisation/bisiallitisation différente.

C. L'érosion des continents.

L'étude de l'érosion des continents basée sur les flux de matière mesurables actuellement repose sur l'hypothèse de base que le système érosion-transport est à l'équilibre. Il y a plusieurs objections majeures à cette hypothèse : une partie des matériaux solides transportés actuellement dans les zones froide et tempérée proviennent encore de l'érosion glaciaire; si les matériaux dissous provenant des régions de fort relief représentent bien un phénomène d'érosion actuel, ceux provenant des régions de plaine soumises à de très longues périodes d'altération ne correspondent plus à l'érosion d'une roche-mère, mais à celle de sols quelquefois très développés, comme ceux de la zone tropicale. Enfin, l'action de l'homme a modifié largement le bilan des transports solides des rivières, ceux-ci n'étant souvent plus du tout en équilibre avec les produits de l'érosion. Malgré ces restrictions, les flux de matériaux transitant à la surface des continents représentent un des moyens les plus sûrs d'étudier l'érosion.

1. *Vitesses relatives d'érosion chimique des roches superficielles.*

Les roches sédimentaires fournissent la plus grande part des produits de l'érosion chimique. Il résulte en effet du tableau VIII que, pour les régions exoréiques sur lesquelles porte cette décomposition, 89 % des éléments proviennent des roches sédimentaires (halite 7,8 %, sulfures sédimentaires 4,0 %, sulfates 11,9 %, silicates des roches détritiques 10,2 %, carbonates 55 %). Les roches cristallines correspondent à seulement 11 % des produits de l'érosion (sulfures 0,8 %, silicates 10,2 %). Ces proportions sont calculées après soustraction du CO_2 atmosphérique. La comparaison de cette répartition avec celle des roches affleurant à la suface des continents va fournir les vitesses relatives d'érosion chimique de chaque type de roches. En combinant les proportions données par H. Blatt et R.L. Jones (1975), et par R.M. Garrels et F.T. Mackenzie (1971, p: 207), on trouve les pourcentages d'affleurement suivants : roches cristallines : 34 %, roches détritiques : 55 %, carbonates : 10 % et évaporites < 1 %. Les produits de l'érosion chimique correspondants sont : 11 %, 14,2 %, 55,1 % et 19,7 %. Les rapports % produits d'érosion/% affleurement sont les suivants : évaporites 80, carbonates 20, roches cristallines 1,2, roches détritiques 1. Ces valeurs sont bien sûr approximatives, vu les incertitudes cumulées dans les calculs, mais fournissent une idée de l'érosion chimique différentielle de ces terrains. Ces proportions sont tout à fait comparables à celles des minéralisations des eaux drainant ces types de roches. Il est vraisemblable que la vitesse d'érosion des roches cristallines par rapport aux roches détritiques est légèrement plus élevée, dans le rapport 1,5 à 2 qui est celui des concentrations en silice des eaux de drainage pour une même zone climatique.

2. *Géographie de l'érosion chimique.*

Pour chaque type morphoclimatique précédemment défini on peut calculer la valeur moyenne de l'érosion chimique (tabl. IX) en déduisant des transports types du tableau III, les apports océaniques (en moyenne 5,5 % des transports ioniques) et la part du CO_2 atmosphérique (57 % des bicarbonates en moyenne). Sur des superficies très étendues (10^6 km^2) l'érosion chimique moyenne varie de 3 à 80 t.km^{-2}.an^{-1} si on exclut les régions désertiques. Mais des valeurs maximales bien plus fortes peuvent être observées sur des petits bassins. Ainsi, dans les terrains calcaires et gypseux, l'érosion chimique peut aller au moins jusqu'à 340 t.km^{-2}.an^{-1} (Dranse du Chablais, recalculé d'après M. Meybeck, 1972). Cette valeur est sans doute dépassée en cas d'affleurement de dômes de sel en région humide (cas de l'Huallaga en Amazonie; J.M. Edmond, communication personnelle).

La répartition géographique de l'érosion chimique est très inégale et peut être approchée par celle des transports dissous (fig. 1 et tabl. IX). Il apparaît ainsi trois maximums : dans les régions tempérées au relief élevé de l'hémisphère nord (du 35° N au 55° N) et de l'hémisphère sud (Chili, Nouvelle-Zélande), et enfin dans les régions équatoriales montagneuses (Andes, Insulinde, Nouvelle-Guinée). Ces zones sont généralement séparées par des régions sèches ou désertiques à très faibles transports (Sud-Ouest des Etats-Unis, Sahara, bassins de la Caspienne et de la mer d'Aral, Nord du Chili et de l'Argentine). Une exception majeure se situe en Chine où il y a passage graduel du type tempéré humide au type tropical humide. La géographie des transports dissous est ainsi très proche de celle de la végétation présentée par M.D.F. Udwardy (1975).

L'ensemble des régions humides montagneuses, qui ne représente que 12,5 % du drainage exoréique, fournit 45 % de la silice et 41 % des ions majeurs aux océans. Les apports des différents types morphoclimatiques définis au tableau II sont en effet calculables en considérant l'abondance et l'intensité de ceux-ci (tabl. IX). Cette influence des régions montagneuses s'explique facilement par la concordance d'un écoulement élevé et de fortes concentrations ioniques, et en plus, dans la zone tropicale, de fortes teneurs en silice. *La zone tropicale, qui ne représente que 37 % des superficies drainées vers l'océan, assure 74 % des apports en silice.*

Sur l'ensemble des continents *l'érosion chimique est en moyenne de 19 t.km^{-2}.an^{-1}* et pour les régions exoréiques seules elle est de 24 t.km^{-2}.an^{-1}. Ces valeurs correspondent, si on prend une densité moyenne de 2,6 pour les roches superficielles, à des dénudations chimiques de 7 mm/1 000 ans et de 9 mm/1 000 ans. Les produits correspondants de l'altération sont de 2 420.10^6 t.an^{-1} pour le drainage exoréique (99,9.10^6 km^2) et de 120.10^6 t.an^{-1} pour le drainage endoréique (33,2.10^6 km^2).

3. Importance relative de l'érosion chimique et de l'érosion mécanique.

Si l'on compare les masses totales des matériaux libérés par les érosions mécanique et chimique et transportés par les rivières, les premières sont 6 à 8 fois plus importantes que les secondes dans les régions exoréiques. Il n'est cependant pas possible d'extrapoler ce rapport aux érosions elles-mêmes pour les raisons exposées précédemment. Les transports des matériaux solides des rivières vers les océans sont généralement esti-

més entre 15 et 20.10^9 t.an^{-1} (J.M. Jansen et R.B. Painter, 1971 ; U.S.S.R. Committee for the I.H.D., 1978 ; J.N. Holeman, 1968), soit un transport solide spécifique de 150 à 200 t.km^{-2}.an^{-1}. Il est possible que le double de matériaux soit effectivement libéré par l'érosion des massifs montagneux.

La plus grande partie des matériaux détritiques transportés à l'océan provient des régions de montagne où l'érosion mécanique peut être 10 à 100 fois supérieure à celle des régions de plaine. Si on considère les grands fleuves (M. Meybeck, 1976), la répartition plurimodale des fréquences des transports solides T$_s$ reflète cette origine : il y a un premier mode pour les régions de plaine, type Congo ou Ob, vers 5 à 10 t.km^{-2}.an^{-1}, un autre vers 50 à 100 t.km^{-2}.an^{-1} pour les grands bassins dont la partie supérieure est montagneuse, comme l'Amazone, et un dernier vers 300 à 500 t.km^{-2}.an^{-1} pour les bassins situés essentiellement en montagne, comme le Brahmapoutre ou le Mékong et pour certains bassins de régions arides (Colorado, Huang Ho). Géographiquement, le transport solide domine beaucoup moins. Ainsi pour tout le territoire de l'U.R.S.S., le rapport T$_s$/T$_d$ est inférieur à 1 pour 64 % du territoire (F.A. Makarenko *et al.*, 1973). Pour l'ensemble du drainage exoréique, on a les proportions suivantes : T$_s$/T$_d$ < 1 pour 35 %, 1 < T$_s$/T$_d$ < 2 pour 25 % et 2 < T$_s$/T$_d$ pour 40 % des superficies (M. Meybeck, 1976).

Sous l'effet de l'érosion chimique, la matière particulaire transportée par les rivières s'enrichit en minéraux résiduels, aux fortes teneurs en Al, Fe et Ti, et s'appauvrit en minéraux les plus solubles dans lesquels Na, Ca, Mg sont préférentiellement lessivés. Cet effet sera le mieux observé dans les régions de plaine où les matériaux résultant de l'altération ne sont pas masqués par l'abondance des débris de roche provenant de l'érosion mécanique seule (J.M. Martin et M. Meybeck, 1978 et 1979). La composition chimique des suspensions des rivières est comparable à celles des différents types de sédiments détritiques donnés par R.M. Garrels et F.T. Mackenzie (1971) : les suspensions du Colorado sont très proches des grès, la moyenne des suspensions provenant des montagnes et de la zone tempérée (Mississippi, Danube, Mackenzie, Gange, etc...) est très proche de celle des roches détritiques fines récentes, des grauwackes et des sédiments océaniques récents, celle des suspensions de la zone tropicale (Amazone, Congo, Orénoque, etc...) est semblable aux roches détritiques fines du Précambrien et aux argiles rouges océaniques. La moyenne générale des suspensions actuelles a une composition analogue à celle des argiles marines côtières.

Les divers éléments majeurs transitent de façon différente entre les continents et les océans. Pour chacun, on peut calculer le rapport transport en solution/transport total (particules + solution) : S 80 %, Cl 85 %, Ca 62 %, Na 63 %, Mg 40 %, K 14 %, Si 4,4 %, Mn 2 %, Fe 0,2 %, Al 0,13 %. Des rapports élevés ont également été notés pour des éléments-traces As, B, Br, Li, Sb, Sr (J.M. Martin et M. Meybeck, 1979). On voit ainsi l'importance du transport en solution dans le cycle géochimique de ces éléments.

Si l'on combine les deux formes d'érosion, les vitesses d'érosion totale des roches superficielles sont beaucoup plus homogènes que celles résultant de la seule érosion chimique. En effet la plus grande part des matériaux particulaires érodés provient des roches sédimentaires détritiques, et la vitesse globale d'érosion de celles-ci se rapproche de celle des roches sédimentaires solubles. Le recyclage global des roches carbonatées et des évaporites ne s'effectue donc pas beaucoup plus vite que celui des autres roches sédimentaires. On retrouve la succession de R.M. Garrels et F.T. Mackenzie (1971, p. 272) concernant les âges moyens des roches sédimentaires (en millions d'années : évaporites 200, roches carbonatées 300, ensembles des roches sédimentaires 600). Par contre, dans les régions de plaine où l'érosion chimique égale souvent l'érosion mécanique, le recyclage préférentiel des roches les plus solubles reste valable.

Conclusions.

Ce bilan semble proche des précédents (D.A. Livingstone, 1963 ; O.A. Alekin et L.V. Brazhnikova, 1960) en ce qui concerne les masses totales transportées, mais il s'en éloigne beaucoup pour certains éléments (K, SiO_2) ou pour certains continents, comme l'Afrique dont les chiffres avaient été surestimés d'un facteur deux ou trois.

Les données utilisées ici sont en effet très différentes : 1) le bilan hydrique utilisé est plus fiable, il inclut les nouvelles mesures de débit sur l'Amazone, et fournit des cartes d'écoulement détaillées servant de base à l'extrapolation des transports en solution ; 2) les analyses chimiques récentes et plus fiables existent sur 63 % des rivières se jetant aux océans. A part le Yang Tsé Kiang et le Huang Ho, tous les grands fleuves sont considérés ici ; 3) l'extrapolation des transports en solution aux régions inconnues est fondée sur une typologie déterminée sur plus de $50.10^6 km^2$; 4) l'influence anthropique a été spécialement prise en compte par plusieurs méthodes.

L'amélioration du bilan est possible par une meilleure connaissance de certaines grandes rivières et de la zone tropicale humide, qui contribue le plus aux apports à l'océan, et par une typologie plus fine prenant aussi en compte les principaux caractères lithologiques.

Il est également nécessaire d'avoir des données plus précises en Na^+, K^+ et SiO_2 dans les grands fleuves sibériens de même que sur le comportement de SiO_2 dans les estuaires.

Les transports en solution de douze types morphoclimatiques, définis sur la base des régimes thermiques, de l'écoulement spécifique et avec une correction de relief ont pu être déterminés. Le quatrième facteur du transport, la lithologie, a été volontairement minimisé en intégrant les valeurs moyennes sur des grandes superficies ($10^6 km^2$), toutefois son influence est encore visible sur certains types définis en zone tempérée et tropicale.

Le bilan par zones climatiques montre d'une part que l'essentiel (74 %) de la silice apportée aux océans provient de la zone tropicale (37 % de la superficie exoréique), d'autre part que les régions humides de fort relief (12,5 % de la superficie exoréique) sont responsables de 43 % des apports en silice et de 41 % des apports ioniques. Les bilans aux quatre océans (Arctique, Atlantique, Indien, Pacifique) sont très différents si on normalise les apports aux superficies ou aux volumes océaniques : l'Arctique et l'Atlantique bénéficient des apports en solution les plus élevés, le premier en raison de ses dimensions modestes par rapport à un bassin versant étendu, le deuxième à cause de son bilan hydrique (50 % des eaux fluviales lui arrivent).

Les autres apports en solution dûs aux eaux souterraines se déversant directement aux océans, aux calottes glaciaires, au lessivage des cendres volcaniques fraîches, sont négligeables vis-à-vis des apports fluviatiles. Environ 15 % des transports naturels des rivières proviennent des retombées atmosphériques dont un tiers est d'origine océanique. Comme une grande partie (57 %) des bicarbonates dissous provient du CO_2 atmosphérique, et que les rivières des régions développées reçoivent maintenant des sels provenant de multiples pollutions, il est impossible d'assimiler les transports en solution des rivières à des mesures de dénudation chimique sans effectuer de multiples corrections. D'autres précautions doivent être prises en aval dans l'estimation des apports à l'océan en raison de la précipitation de la silice dans les estuaires.

La décomposition de l'analyse moyenne des eaux de surface montre que l'essentiel (89 %) des sels dissous provenant de la dénudation chimique des continents dérive des roches sédimentaires qui se dégradent beaucoup plus vite que les roches cristallines dans l'ordre suivant : évaporites >> roches carbonatées >> roches détritiques ⩾ roches cristallines. L'abondance des roches à l'affleurement et leurs vitesses relatives de dissolutions sont à l'origine du caractère bicarbonaté calcique observé dans la presque totalité (98 %) des eaux fluviatiles. Les eaux chlorurées sont très rares et les eaux sulfatées sont observées dans les régions semi-arides. Deux grandes variétés d'eau apparaissent : le sous-type chloruré sodique (33 % des eaux), de faible minéralisation (concentration ionique totale de 35 mg.l^{-1}) est rencontré dans les terrains cristallins et abonde dans la zone tropicale où les teneurs en silice sont alors élevées (10 à 15 mg.l^{-1}); le sous-type sulfaté-magnésien (47 % des eaux), de forte minéralisation (120 mg.l^{-1}) résulte du drainage des terrains sédimentaires, il est très fréquent dans la zone tempérée, où la silice présente des teneurs plus faibles (4 à 8 mg.l^{-1}).

L'érosion chimique actuelle moyenne est de 24 t.km^{-2}.an^{-1} pour les régions drainées vers les océans — soit une dénudation de 9 mm/1 000 ans pour une densité moyenne de 2,6 — et de 19 t.km^{-2}.an^{-1} pour l'ensemble des terres émergées non glacées. Pour l'ensemble du globe, ce taux est de l'ordre de 10 fois inférieur à la dénudation mécanique. Mais, dans de vastes régions de plaines, les deux formes d'érosion ont des vitesses comparables : les roches évaporitiques et carbonatées, les plus solubles, sont alors érodées préférentiellement.

La pollution générale de l'hydrosphère s'observe également pour les eaux de surface où les éléments majeurs ont globalement augmenté de 12 % par rapport aux valeurs naturelles. Certains éléments (Na$^+$, Cl$^-$, SO$_1^{--}$) ont même évolué de plus de 30 %.

L'extrapolation des phénomènes actuels d'érosion chimique et de transport à l'océan dans les temps géologiques est très délicate car les inconnues se multiplient : superficie totale des continents ; répartition paléogéographique des écoulements (drainage endoréique/exoréique), des températures et du relief. Nous avons vu en particulier l'importance des zones chaudes et humides dans le transport de la silice et des régions d'orogénèses récentes dans le bilan des apports à l'océan. Enfin, la nature des roches superficielles, qui a considérablement changé depuis le Cambrien (R.M. Garrels et F.T. Mackenzie, 1971), joue également un grand rôle. Le moyen le plus simple reste l'utilisation d'une typologie (celle-ci pouvant être améliorée) correspondant à des entités géographiques facilement identifiables, et pouvant être projetée dans le passé.

Remerciements.

Je suis particulièrement heureux de remercier toutes les personnes et administrations qui m'ont communiqué de précieuses données de qualité des eaux, et en particulier : R. Laaksonen (Finlande), B.K. Handa (Inde), S. Solomon (Canada), R.J. Davies-Colley et G.L. Leary (Nouvelle-Zélande), J. Michalczewski (Pologne), l'Hydrometeorological Service, Min. of Works (Guyana), l'Internationalen Kommission zum Schutze des Rheins (Koblenz, R.F.A.), J.L. Kennedy et P. Dunstan (Australie) et l'Australian Water Resources Council, R. Tirtotjonro (Indonésie), S.H. Whitlow (Environment, Canada), G.K. Seth et S. Banerji (Inde), T. Petr (Australie), le Bureau of Water Resources (Papua, New-Guinea), et la Division des Sciences de l'Eau de l'Unesco. Je tiens également à remercier MM. G. Bocquier, G. Pedro et R. Wollast pour leur critique d'une première version de ce travail. Les analyses inédites ont été effectuées par le Centre de Recherches Géodynamiques de Thonon (France). Certains échantillons ont été aimablement prélevés par J.-M. Martin, E. Benedetti et J.F. Jarrige.

[*Editors' Note:* Appendixes 1 through 6 have been omitted.]

RÉFÉRENCES

ACKERMAN W.C., HARMESON R.H. et SINCLAIR R.A. (1970). – Some long-term trends in water quality of rivers and lakes. *Eos, Trans. Amer. Geoph. Union*, vol. 51, n° 6, p. 516-522.

AL DROUBI A., FRITZ B., GAC J.V. et TARDY Y. (1977). – Prediction of the chemical evolution of natural waters during evaporation. *Proc. 2nd Int. Symp. Water-Rock, Interactions*, Strasbourg 17-25 août 1977, vol. 2, p. 13-22.

ALEKIN O.A. et BRAZHNIKOVA L.V. (1960). – Contribution à la connaissance de l'écoulement des substances dissoutes de la surface terrestre du globe. *Gidrochim. Mat.*, vol. 32, p. 12-34 (in Russian).

ALEKIN O.A. et BRAZHNIKOVA L.V. (1962). – Sur la corrélation entre le transport ionique et celui des substances en suspension. *Dokl. Acad. Nauk. SSSR*, vol. 146, n° 1, p. 203-206 (in Russian).

ALEKIN O.A. et BRAZHNIKOVA L.V. (1968). – Dissolved matter discharge and mechanical and chemical erosion. *Ass. Int. Hydrol. Sci.*, Publ. 78, p. 35-41.

ALVAREZ HERRERO C., BUSTEO ARAGON A. et CATALAN LAFUENTE J.G. (1969). 8 Distribucion de elementos traza en la cuenca del Duero. *Agua*, vol. 5, p. 47-57.

ANDRULEWICZ E., DUBROWSKI R. et ZMUDZINSKI L. (1972). – Preliminary investigations to determine the influence of Polish rivers on the Baltic Sea pollution. 8th Conf. Baltic Oceanographers, Copenhagen, paper 6.

ARMANNSSON H. *et al.* (1963). – Efnarann sokn vatns. Vatnasvid Hvitar-Olfusar. Orkustofnun (National Energy Authority), Reykjavik, Islande.

AYERS R.S. et ASCOT D.W. (1976). – Water Quality for Agriculture. Irrigation and Drainage Paper no. 29, FAO, Rome.

BAUMGARTNER A. et REICHEL E. (1975). – The world water balance. Elsevier, 179 p.

BENNEKOM A.J. van, BERGER G.W., HELDER W. et DE VRIES R.T.P. (1978). – Nutrient distribution in the Zaire estuary and river plume. *Neth. J. Sea Res.*, vol. 12, n° 3-4, p. 296-323.

BENNEKOM A.J. van et SALOMONS W. (1979). – Pathways of nutrients and organic matter fron. land to ocean through rivers. SCOR Worshop on RIOS, Rome, 26-30 March 1979, extended abstract.

BLANC P. et CONRAD G. (1968). – Evolution géochimique des eaux de l'Oued Saoura (Sahara nord-occidental). *Rev. Géogr. phys. Géol. dynam.*, vol. 10, fasc. 5, p. 415-428.

BLATT H. et JONES R.L. (1975). – Proportion of exposed igneous, metamorphic ans sedimentary rocks. *Geol. Soc. Amer. Bull.*, vol. 86, p. 1085-1088.

BRIAT M. (1974). – Dosage du chlore, du sodium, et du manganèse par activation neutronique dans le névé antarctique : origine et retombée de ces éléments. Thèse de 3e cycle, Université de Grenoble, 111 p.

BRUNSKILL G.J. *et al.* (1975). – The chemistry, mineralogy and rates of transport of sediments in the Mackenzie and Porcupine rivers watersheds, NWT and Yukon 1971-73. Technical report no. 566, Fisheries and Marine Service, Environment Canada.

BURTON J.D. et LISS P.S. (1973). – Process of supply and removal of dissolved silicon in the oceans. *Geochim. Cosmochim. Acta*, vol. 37, p. 1761-1773.

CAILLE A., CAMPBELL P., MEYBECK M. et SASSEVILLE J.L. (1973). – Etude du fleuve Saint-Laurent : Effluents urbains de l'agglomération de Montréal. Rapport Interne, INRS-Eau, Québec, 203 p.

CANALI L. (1964). – Transport de matériel en solution et en suspension du Pô. *Ass. Int. Hydrol. Sci. Bull.*, vol. 9, n° 1, p. 17-26.

CAESAR J., COLLIER R., EDMOND J.M., FREY F., MATISOFF G., NG A.C. et STALLARD R.F. (1976). – Chemical dynamics of a polluted watershed : the Merrimack river in Northern New England. *Env. Sci. Technol.*, vol. 10, fasc. 7, p. 697-704.

CARBONNEL J.P. et MEYBECK M. (1975). – Quality variation of the Mekong river at Phnom Penh, Cambodia, and chemical transport in the Mekong basin. *J. Hydrol.*, vol. XXVII, p. 249-265.

CARROLL D. (1970). – Rock weathering. Plenum Press New York, 203 p.

CLARKE F.W. (1924). – Data of geochemistry, Vth edition. *U.S.G.S. Survey Bull.*, n° 770, 841 p.

COCHE A.G. et BALON E.K. (1974). – Lake Kariba, a man-made tropical ecosystem in Central Africa. *Ecolog. Monographs. Junk*, 767 p.

CONWAY E.J. (1943). – The chemical evolution of the ocean. *Irish Acad. Proc.*, vol. 48 B, p. 161-212.

CORBEL J. (1959a). – Vitesse de l'érosion. *Z. Geomorphol.*, vol. 3, p. 1-28.

CORBEL J. (1959b). – Erosion en terrain calcaire. *Ann. Géogr.*, vol. 366, p. 97-120.

CORBEL J. (1964). – L'érosion terrestre. Etude quantitative. *Ann. Géogr.*, vol. 398, p. 385-412.

CROZAT G. (1978). – L'aérosol atmosphérique en milieu naturel. Etude des différentes sources de potassium en Afrique de l'Ouest (Côte d'Ivoire). Thèse état, Univ. P. Sabatier-Toulouse, 70 p.

DEPETRIS P.J. (1976). – Hydrochemistry of the Panama river. *Limnol. Oceanogr.*, vol. 21, n° 5, p. 736-739.

DESSEVRE-DELEPOULLE A. (1978). – Les eaux de la région rouennaise. Thèse de Doctorat 3e cycle, Université de Paris VI, 152 p.

DE VILLIERS P.R. (1962). – The chemical composition of the Orange river at Violsdrif, Cape Province. *Ann. Geol. Survey Pretoria*, vol. V, n° 1, p. 198-206.

DOUGLAS I. (1964). – Intensity and periodicity in denudation processes with special reference to the removal of material in solution by rivers. *Z. Geomorphol.*, vol. 8, n° 4, p. 453-472.

DURUM W.H., HEIDEL G. et TISON L.J. (1960). – Worldwide run-off of dissolved solids. *Ass. Int. Hydrol. Sci.*, Publ. 51, p. 618-628.

DZHAMALOV R.G., ZEKTZER I.S. et MESKHETELI A.V. (1977). – The role of sub-surface ionic run-off in the salt balance of inland seas. Proc. 2nd Water-Rock Interaction Symp., 17-25 August 1977, Strasbourg. I-54 - I-63.

EDMOND J.M. (1973). – The silica budget of the Antarctic circum-polar current. *Nature*, vol. 241, n° 5389, p. 391-393.

EISMA D., BENNEKOM A.J. Van (1979). − Verlag van chemisch onderzoele in de estuaria van Suriname river en Marowijne. *Neth. Inst. Sea Res.,* Texel, report 1969-7.

ERIKSSON E. (1969). − The yearly circulation of chloride and sulphur in nature : meteorological, geochemical, and geological implications. Part I. *Tellus,* n° 11, p. 375-403.

FETH J.H. (1971). − Mechanisms controlling world water chemistry : evaporation-crystallization process. *Science,* vol. 172, p. 871-872.

FIGUERES G., MARTIN J.M. et MEYBECK M. (1978). − Iron behaviour in Zaire estuary. *Neth J. Sea Res.,* vol. 12, fasc. 3/4, p. 329-337.

FOURNIER F. (1960). − Débit solide des cours d'eau. Essai d'estimation de la perte en terre subie par l'ensemble du globe terrestre. *Ass. Int. Hydrol. Sci.,* Publ. 53, p. 19-25.

GAC D.Y. et PINTA M. (1973). − Bilan de l'érosion et de l'altération en climat tropical humide. *Cahiers ORSTOM,* série Géologie, vol. 5, n° 1,p. 83-96.

GARRELS R.M. et MACKENZIE F.T. (1971). − Evolution of sedimentary rocks. W.W. Norton New-York, 397 p.

GARRELS R.M., MACKENZIE F.T. et HUNT C. (1973). − Chemical Cycles and the Global Environment. William Kaufmann inc., Los Altos, California, 206 p.

GIBBS R.J. (1967). − The geochemistry of the Amazon River System. Part I : the factors that control the salinity and the composition and concentration of the suspended solids. *Geol. Soc. Amer. Bull.,* vol. 78, p. 1203-1232.

GIBBS R.J. (1970). − Mechanism controlling world water chemistry. *Science,* vol. 170, p. 1088-1090.

GIBBS R.J. (1972). − Water Chemistry of the Amazon River. *Geochim. Cosmochim. Acta,* vol. 36, p. 1061-1066.

GOLDBERG E.D. (1972). − Man's role in the major sediment cycle. *In :* The changing chemistry of the oceans, D. Dryssen &D. Jagner Ed., J. Wiley-Interscience, p. 267-288.

GOMIS C., ALVAREZ HERRERO C. et BARIEL MARTI F. (1969). − Analisis quimico de uno de los principales rios espagnoles. *Agua,* vol. 5, p. 137-159.

GORDEEV V.V. et LISITZIN A.P. (1978). − Composition chimique moyenne des substances en suspension des rivières dans le monde et alimentation des océans par des matériaux sédimentaires fluviatiles (en Russe). *Dokl. Akad. SSSR,* vol. 238, n° 1, p. 225-228.

GORHAM E. (1961). − Factors influencing the supply of major ions to inland waters with special reference to the atmosphere. *Amer. Geol. Soc. Bull.,* vol. 72, p. 795-840.

GRANAT L., RODHE H. et HALBERG R.O. (1976). − The global sulphur cycle. *In :* Nitrogen, phosphorus and sulphur - global cycles, B.H. Svensson & R. Söderlund Ed., *Ecol. Bull.,* vol. 22, fasc. 89, p. 89-134.

GROVE A.T. (1972). − The dissolved and solid load carried by some West African rivers : Senegal, Niger, Benue, and Chari. *J. Hydrology,* vol. XVI, n° 4, p. 277-300.

HANDA B.K. (1971). − Chemical composition of monsoon rain water over Bankipur. *Indian J. Meteorol. Geoph.,* vol. 22, p. 603.

HANDA B.K. (1972). − Geochemistry of the Ganga river water. *Indian Geohydrology,* vol. 8, n° 2, p. 71-78.

HANDA B.K. (1973). − Geochemistry of Indian river waters. Int. Symp. on Recent Researches in Geochemistry, Patna, India, in press.

HEM J.D. (1970). − Study and interpretation of the chemical characteristics of natural water. U.S. Geol. Surv. Wat. Supply Paper 1473, 363 p.

HOLEMAN J.N. (1968). − The sediment yield of major rivers of the world. *Water Resources Res.,* vol. 4, n° 4, p. 737-747.

HOLLAND H.D. (1978). − The Chemistry of the atmosphere and oceans. J. Wiley, 351 p.

HOLTAN H. (1976). − Chemical conditions and variations in river water. *In :* Fresh water and on the sea, p. 27-32, Assoc. Norw. Oceanographers, Oslo, Norway.

HURD D.C. (1977). − The effect of glacial weathering on the silica budget of Antarctic waters. *Geochim. Cosmochim. Acta,* vol. 41, fasc. 9, p. 1213-1222.

HURST H.E. (1957). − The Nile. Constable, London, 331 p.

IMEVBORE A.M.A. (1970). − The chemistry of the river Niger in the Kainji reservoir area. *Arch. Hydrobiol.,* vol. 67, n° 3, p. 412-431.

JANDA R.J. (1971). − An evaluation of procedures used in computing chemical denudation rates. *Geol. Soc. Amer. Bull.,* vol. 82, n° 1, p. 67-80.

JANSEN J.M.L. et PAINTER R.B. (1971). − Predicting sediment yield from climate and topography. *J. Hydrology,* vol. 21, p. 371-380.

JOHNSON N.M., REYNOLDS R.C. et LIKENS G.F. (1972). − Atmospheric sulfur : its effect on the chemical weathering of New Hampshire. *Science,* vol. 177, p. 514-516.

JUDSON S. (1968). − Erosion of the land (what's happening to our continents ?). *Amer. Scientist,* vol. 56, p. 356-374.

JUNGE C.E. (1963). − Air chemistry and radioactivity. Academic Press, New-York, 382 p.

JUNGE C.E. et WERBY R.T. (1958). − The concentration of chloride, sodium, potassium, calcium and sulphate in rainwater over the United States. *J. Meteorology,* vol. 15, p. 417-425.

KELLER W.D. (1972). − Glacial milk. *In :* Encyclopedia of Geochemistry and Environmental Sciences, R.W. Fairbridge Ed., Van Nostrand, New-York, p. 463-466.

KENNEDY V.C., ZELLWEGER G.W. et JONES B.F. (1974). − Filter pore-size effect on the analysis of Al, Fe, Mn and Ti in water. *Water Resources Res.,* vol. 10, fasc. 4, p. 785-790.

KOBAYASHI J. (1960). − A chemical study of the average quality and characteristics of river waters of Japan. *Ber. Ohara Inst. Landwirt. Biol.,* vol. 11, n° 3, p. 313-357.

KOBAYASHI J. (1967). − Silica in fresh waters and estuaries. Proc. IBP Symp. Amsterdam, 1966, p. 41-55.

KONOVALOV G.S. et IVANOVA A.A. (1970). − River discharge of microelements from the territory of the U.S.S.R. to sea basins. *Oceanol.,* vol. 10, fasc. 4. p. 482-488.

LAAKSONEN R. (1971). − Quality of inland waters in Finland. *Aqua Fennica,* p. 47-57.

LEOPOLD J.B., WOLMAN M.G. et MILLER J.P. (1964). − Fluvial processes in geomorphology. Freeman, San Francisco, London, 522 p.

LI F.S., WU Y.D., WANG L.F. et CHEN Z.H. (1964). − Physicochemical processes of silicates in the estuarial region. *Oceanol. Limnol. Sinica,* vol. 6, p. 311-322.

LIEPOLT R. (1967). − Limnologie der Donau. Schweizerbart'che Verlag-buchhandlung, Stuttgart.

LIKENS G.E., BORMANN F.H., JOHNSON N.M., FISHER D.W. et PIERCE R.S. (1970). – Effect of forest cutting and herbicide treatment on nutrient budget in the Hubbard Brook Watershed ecosystem. *Ecolog. Monogr.*, vol. 40, p. 23-47.

LISITZIN A.P. (1972). – Sedimentation in the world ocean. *Soc. Econ. Paleo. Min.*, special publ. 17, Tulsa, Oklahoma.

LISS P.S. (1976). – Conservative and non-conservative behaviour of dissolved constituents during estuarine mixing. *In :* Estuarine Chemistry, J.D. Burton, P.S. Liss Ed., Academic Press, London, 229 p., p. 93-130.

LIVINGSTONE D.A. (1963). – Chemical composition of rivers and lakes. Data of geochemistry chapter G, U.S. Geol. Survey Prof. Paper 440G, G1-G64.

LVOVICH M.I. (1972). – World water balance (General report). *In :* World Water Balance, Proc. of the Reading Symp. July 1970. *Studies and reports in Hydrology* 11, Unesco Press, p. 401-415.

MAKARENKO F.A., ZVEREV V.P. et KONONOV V.I. (1973). – Subsurface chemical runoff in the USSR area. Proc. Ist Symp. Hydrogeochem. Biogeochem., Nov. 1970, Clarke Cy Washington D.C., 567-573.

MARTIN J.M. et MEYBECK M. (1978). – Major element content in river dissolved and particulate loads. *In :* Biogeochemistry of estuarine sediments, Proc. of Melreux Symp. Nov. 1976, E.D. Goldberg Ed., Unesco, p. 111-126.

MARTIN J.M. et MEYBECK M. (1979). – Elemental mass-balance of material carried by world major rivers. *Marine Chemistry*, vol. 7, p. 173-206.

MEADE R.H. (1969). – Errors in using modern strean load data to estimate natural rates of denudation. *Geol. Soc. Amer. Bull.*, vol. 80, n° 7, p. 1265-1274.

MENARD H.W. et SMITH S.M. (1966). – Hypsometry of ocean basin provinces. *J. Geoph. Res.*, vol. 71, fasc. 18, p. 4305-4325.

MEYBECK M. (1972). – Bilan hydrochimique et géochimique du Lac Léman. *Verh. Int. Ver. Limnol.*, vol. 18, p. 442-453.

MEYBECK M. (1976). – Total dissolved transport by world major rivers. *Hydrol. Sci. Bull.*, vol. 21, n° 2, p. 265-284. Erratum *21*, 4, 651.

MEYBECK M. (1977). – Dissolved and suspended matter carried by rivers : composition, time and space variations, and world balance. *In :* Interaction between sediments and fresh waters, H.L. Golterman Ed., Amsterdam, Sept. 6-10, 1976, 473 p., Junk and Pudoc, Amsterdam, p. 25-32.

MEYBECK M. (1978). – Note on dissolved elemental contents of the Zaire river. *Neth. J. Sea Res.*, vol. 12, fasc. 3/4, p. 293-295.

MEYBECK M. (1979). – Pathways of major elements from land to ocean through rivers. SCOR Workshop on RIOS, Rome, 26-30 March 1979, extended abstract.

MILLER J.P. (1961). – Solutes in small streams draining single rock types, Sangre de Cristo Range, New Mexico. U.S. Geol. Surv. Wat. Supply Paper 1535 F, 23 p.

MOOK W.G. et KOENE B.K.S. (1975). – Chemistry of dissolved inorganic carbon in estuarine and coastal brackish waters. *Est. Coast. Mar. Sci.*, vol. 3, p. 325-336.

MORELLI J. (1978). – Données sur le cycle atmosphérique des sels marins. *J. Rech. Océanogr.*, vol. 3, fasc. 4, p. 27-49.

MOROZOV N.P. (1969). – Geochemistry of the alkali metals in rivers. *Geokhimiya*, vol. 6, p. 729-739.

ODUM E.P. (1971). – Fundamentals of Ecology. Saunders, Philadelphia, Pa, 3rd ed., 574 p.

PEDRO G. (1968). – Distribution des principaux types d'altération chimique à la surface du globe. *Rev. Géogr. phys. Géol. dynam.*, vol. X, fasc. 5, p. 457-470.

PETR T. (1976). – Some chemical features of two Papuan fresh waters. (PNG). *Austr. J. Mar. Freshwater Res.*, vol. 27, p. 467-474.

REYNOLDS R.C. et JOHNSON N.M. (1972). – Chemical weathering in the temperate glacial environment of the Northern Cascade Mountains. *Geochim. Cosmochim. Acta*, vol. 36, p. 537-554.

ROCHE M.A. (1975). – Geochemistry and natural and isotopic tracing : two complementary ways to study the natural salinity regime of Lake Tchad. *J. Hydrol.*, vol. 26, p. 153-171.

ROCHE M.A., DUBREUIL P. et HOEPFFNER M. (1974). – Dynamique des eaux, des sels et des sédiments en suspension dans les estuaires du Mahury et de l'Approuague. Rapport Int. Orstom, section hydrologie, Paris, 80 p.

SALVADORI F. (1976). – Etude de quelques paramètres chimiques de la qualité des eaux dans l'estuaire de la Charente. Thèse de Doctorat de 3ᵉ cycle, Univ. Paris 6, 95 p.

SIMPSON H.J., HAMMOND D.E., DECK B.L. et WILLIAMS S.C. (1975). – Nutrients budgets in the Hudson river estuary. *In :* Marine Chemistry in the Coastal Environment, T.M. Church Ed., Am. Chem. Soc. Symp. Series 18, Washington D.C., p. 618-635.

SLATT R.M. (1972). – Geochemistry of melt water streams from nine alaskan glaciers. *Geol. Soc. Amer. Bull.*, vol. 83, p. 1125-1132.

SOUCHEZ R.A., LARRAIN R.D. et LEMMENS M.N. (1973). – Refreezing of interstitial water in a sub-glacial cavity of an alpine glacier as indicated by the chemical composition of ice. *J. Glaciol.*, vol. 12, p. 453-459.

STRAKHOV N.M. (1967). – Principles of lithogenesis, part. I, chapter I. Consultants Bureau and Olivier and Boyd, 245 p.

TARDY Y. (1971). – Characterization of the principal weathering types by the geochemistry of waters from some European and African crystalline massifs. *Chemical Geology*, vol. 7, p. 253-271.

TAYLOR P.S. et STOYBER R.E. (1973). – Soluble material on ash from active Central American volcanoes. *Geol. Soc. Amer. Bull.*, vol. 84, p. 1031-1042.

TOLLAN A. (1976). – River runoff in Norway. *In :* Fresh Water on the Sea, p. 11-13, Assoc. Norw. Oceanographers, Oslo, Norway.

TUREKIAN K.K. (1969). – Oceans, streams and atmosphere. *In :* Handbook of Geochemistry, K.H. Wedepohl Ed., Springer Verlag, chap. 10.

TUREKIAN K.K. (1971). – Rivers, tributaries and estuaries. *In :* Impingement of man on the oceans, D.W. Hood Ed., John Wiley and Sons, chap. 2.

TUREKIAN K.K. et SCOTT M. (1967). – Concentrations of Cr, Ag, Mo, Ni, Co and Mn in suspended material in streams. *Env. Sci. Technol.*, vol. 1, p. 940-942.

UDWARDY M.D.F. (1975). – A classification of the biogeographical provinces of the world. IUCN occasional paper no. 18. Int. Union Conserv. Nature, Morges, Suisse, 48 p.

UNESCO (1969 and 1971). – Discharge of selected rivers of the world. *Studies and reports in Hydrology,* vol. I, II, III.

UNESCO (1979). – World register of rivers discharging to the oceans. Division of Water Sciences, Technical Paper.

U.S.S.R. Academy of Sciences (1964). – Physical-geographic atlas of the world. Moscow. Translated in *Soviet Geography* (1965), vol. 6, fasc. 5-6, 401 p.

U.S.S.R. Committee fot the International Hydrological Decade (1978). – World water balance and water resources of the earth. *Studies and reports in Hydrology,* 25, Unesco Press, 663 p.

VALLENTINE J.R. (1978). – Today is yesterday's tomorrow. *Verh. Internat. Ver. Limnol.,* vol. 20, p. 1-12.

VIRO P.J. (1953). – Loss of nutrients and the natural balance of the soil of Finland. *Comm. Inst. Forest. Fenniae,* vol. 42, n° 1, p. 1-45.

WEILER R.R., CHAWLA V.K. (1969). – Dissolved mineral quality of the Great Lakes waters. Proc. 12th Conf. Great Lakes Res., p. 801-818, Intern. Ass. Great Lakes Res.

WEINBERGER L.W., STEPHAN D.G. et MIDDLETON F.M. (1966). – Solving our water problems - water renovation and reuse. *Ann. New-York Acad. Sci.,* vol. 136, fasc. 5, p. 131-154.

WINDOM H.L. (1979). – Comparison of RIOS and atmospheric transport of inorganic materials to the coastal zone. SCOR Workshop on RIOS, Rome, 26-30 March 1979, extended abstract.

ZVEREV V.P. et RUBEIKIN V.Z. (1973). – The role of atmospheric precipitation in circulation of chemical elements between atmosphere, hydrosphere and lithosphere. Proc. Ist Symp. Hydrogeochem. Biogeochem., Tokyo, Sept. 1970, Clarke Cy Washington D.C., p. 613-620.

RAPPORTS ET ANNUAIRES HYDROLOGIQUES

I. PAR PAYS

CANADA (Québec) Annuaires hydrologiques de la qualité des eaux. Ministère des richesses naturelles, Québec.

CANADA Water Quality Branch, Inland Water Directorate, Environment Canada. Comm. Personnelle et « Données sur la qualité des eaux des territoires du Nord-Ouest, 1960-1973 ». Ottawa 1976.

ESPAGNE Analisis de Calidad de Aguas. Direccion General De Obras Hidraulicas, Madrid.

ÉTATS UNIS US Geological Survey. Quality of surface waters of the United States. In US Geol. Surv. Water Supply Papers.

GUYANA Records of surface water quality 1966-1972. Hydrometeorological Service, Georgetown.

INDES International Hydrological Decade Newsletters no. 13 et 14, 1971.

PAKISTAN Annual report of river and climatological data of West Pakistan 1970 Vol. I River discharge, sediment and quality data. Water and Power Development Authority, Lahore.

PORTUGAL Campanha de Caracterização da Poluição do Rio Tejo (1973-74). Direcção General des Serviços Hidraulicos, Lisboa 1974.

SUÈDE Goffeng G. 1971. Hydrological Data Norden. National Committees for the IHD in Denmark, Finland, Iceland, Norway and Sweden.

URUGUAY Comision tecnica mixta de Salto Grande. Segunda reunión sobre aspectos de desarollo ambiental en el proyecto Salto Grande, 4-8 mayo 1976, Salto, Uruguay.

VENEZUELA Instituto Nacional de Obras Sanitarias. Dep. Est. Proyectos. Laboratorio de Aguas. Caracas.

II. PAR RIVIÈRES

DANUBE Limnologia sectorului Romanesc al Dunarii, 1967. Academic Rep. Soc. Romania, 651 p. Bucarest.

RHIN Rijnnota 1973. Uitgave Vereniging Milieudefensie, Amsterdam.

RHIN Commission internationale pour la protection du Rhin contre la pollution. Rapport sur les analyses physico-chimiques de l'eau du Rhin, 1966-1971.

*Manuscrit déposé le 5 février 1979,
accepté le 2 juillet 1979.*

AUTHOR CITATION INDEX

Abbott, H. L., 302
Ackerman, W. C., 361
Adams, J., 10, 301, 302
Agricultural Research Service, 301
Ahnert, F., 284
Al Droubi, A., 361
Alekin, O. A., 361
Alexander, M., 294
Allen, J. R. L., 240
Allison, F. E., 294
Alvarez, H. C., 361, 362
Ames, F. C., 208, 333
Ampferer, O., 72
Anderson, D. H., 204, 294
Anderson, H. W., 174, 240
Andrews, E. D., 10
Andrews, H. N., Jr., 240
Andrulewicz, E., 361
Armannsson, H., 361
Ascot, D. W., 361
Axelsson, V., 285
Ayers, R. S., 361

Babb, C. C., 330, 333
Bagnold, R. A., 10, 32, 82, 123
Baker, B. H., 284
Baker, V. R., 10, 11
Balon, E. K., 361
Bariel, M. F., 362
Barrell, J., 72
Barton, D. C., 82
Bascom, W. N., 32
Bassi, G., 333
Bauer, F., 50
Baumgartner, A., 361
Baver, L. D., 105
Beasley, R. P., 144
Beaty, C. B., 11
Beckinsale, R. P., 11
Beerbower, J. R., 240
Belt, G. H., 253
Bennekom, A. J. van, 361, 362

Berger, G. W., 361
Berkaloff, E., 43
Berry, L., 253, 285
Bisal, F., 82
Bjerknes, C. A., 207
Blackwelder, E., 82
Blaisdell, J. P., 189
Blanc, P., 361
Blatt, H., 361
Blench, T., 32
Bolton, G. C., 284
Bonilla, M. G., 337
Bormann, F. H., 153, 154, 253, 294, 295, 363
Borst, H. L., 105
Boswell, J. G., 294
Boughton, W. C., 252
Bout, P., 50
Bouyoucos, C. J., 174
Boyle, J. R., 83
Brannock, W. W., 99
Brater, E. F., 240
Braudeau, G., 43
Brazhnikova, L. V., 361
Briat, M., 361
Bricker, O. P., 153, 225
Brindley, G. W., 225
Brosky, D. L., 265
Brown, C. B., 32, 189, 333
Brown, G., 225
Brown, G. W., 252
Browne, W. R., 240
Brune, G. M., 62, 153, 174, 189, 252, 284
Brunengo, M. J., 253, 284
Brunsden, D., 11
Brunskill, G. J., 361
Burton, J. D., 361
Busch, D. A., 240
Busteo, A. A., 361
Butler, B. E., 240

Cady, W. M., 11
Caesar, J., 361

Caille, A., 361
Cailleux, A., 62, 241
Caine, N., 11, 50
Campbell, D. A., 240
Campbell, P., 361
Canada Department of Agriculture, 123
Canali, L., 361
Carbonnel, J. P., 361
Carrara, P. E., 301
Carroll, D., 361
Carroll, T. R., 301
Carson, M. A., 2, 11, 84
Carter, R. W., 265
Catalan Lafuente, J. G., 361
Cavaille, A., 50
Cayeux, L., 240
Chamberlin, R. T., 72
Champion, W. A., 302
Chan, C. K., 144
Charlesworth, J. K., 62
Chase, E. B., 302
Chawla, V. K., 364
Chen, Z. H., 362
Cheng, E. D. H., 83
Chepil, W. S., 83, 123
Cheronis, N. D., 84
Chorley, R. J., 11
Chou, J. T., 337
Chow, V. T., 32
Christiansson, C., 285
Claridge, G. G. C., 252
Clark, S. P., 301
Clarke, F. W., 302, 361
Cleaves, E. T., 153, 225
Clements, F. E., 189
Clyde, C. G., 83
Coche, A. G., 361
Colby, B. R., 32, 83, 265
Coldwell, A. E., 153
Collier, C. R., 252
Collier, R., 361
Committee for Water Resource Control, 337
Conrad, G., 361
Conrad, V. A., 189
Conway, E. J., 361
Cook, H. L., 105
Cooke, R. U., 83
Cooper, A. C., 294
Cooper, W. S., 32
Corbel, J., 50, 62, 72, 153, 240, 252, 302, 326, 330, 333, 361
Corps of Engineers, U.S. Army, 174
Correns, C. W., 83, 99
Costa, J. E., 252
Costa Rodrigues, F. M., 302

Craig, L. C., 240
Crozat, G., 361
Culbertson, J. K., 32, 302
Cummins, W. A., 50
Curtis, C. D., 153
Curtis, W. F., 302

David, L. C., 208
David, T. W. E., 240
Davis, G. H., 153
Davis, L. C., 333
Davis, W. M., 62
Dawdy, D. R., 295
Deck, B. L., 363
Dekker, G., 333
Dendy, F. E., 284, 302
Depetris, P. J., 361
Derruau, M., 50
Dessevre-Delepoulle, A., 361
Devderiani, A. S., 42
De Villiers, P. R., 361
De Vries, R. T. P., 361
Dietrich, W. E., 253, 284
Diseker, E. G., 160, 284
Dobson, G. C., 284
Doeglas, D. J., 240
Dole, R. B., 50, 62, 72
Douglas, I., 50, 153, 240, 252, 284, 333, 361
Do Vale, D. M., 302
Drake, C. L., 72
Dubreuil, P., 363
Dubrowski, R., 361
Duley, F. L., 105, 160
Dunbar, C. O., 240
Dunford, E. G., 189
Dunn, A. J., 11
Dunn, I. S., 83, 144
Dunne, T., 153, 225, 253, 284
Durum, W. H., 32, 240, 333, 361
Dury, G. H., 11
Dutt, G. N., 240
Dyrness, C. T., 285
Dzhamalov, R. G., 361

Eardley, A. J., 50
Eaton, J. S., 153, 154, 253, 294
Edgar, D. E., 11
Edmond, J. M., 302, 361
Eiler, J. P., 33
Einstein, H. A., 83, 125, 130
Eisma, D., 362
Eliassen, S., 334
Ellison, W. D., 105
Elwell, H. M., 160
Emmett, W. W., 50
Engelhardt, W. von, 99

Eriksson, E., 362
Erni, P. E., 337
Eskola, P., 240
Ewing, M., 72
Eyles, R. J., 284

Favre, H., 130
Federal Inter-Agency River Basin Com-
    mission, 62, 189
Federer, C. A., 294
Fell, A., 50
Feltz, H. R., 225
Ferrell, W. R., 32
Feth, J. H., 153, 295, 362
Figueres, G., 362
Fisher, D. W., 153, 225, 294, 363
Flaxman, E. M., 62, 83
Foster, G. R., 83, 84
Fournier, F., 42, 154, 189, 240, 284, 326, 331,
    333, 362
Franjii, K. K., 32
Frank, D. M., Jr., 144
Frederickson, A. F., 154
Fredriksen, R. L., 253
Free, E. E., 123
Free, G. R., 84
Fritz, B., 361
Frey, F., 361

Gabert, P., 42
Gac, J. V., 361, 362
Gainey, P. L., 295
Gambell, A. W., 294
Garner, H. F., 240
Garrels, R. M., 337, 362
Gazzolo, T., 333
Gerlach, T., 50
Gerson, R., 11, 12
Gibbs, R. J., 154, 207, 333, 362
Gilbert, G. K., 84, 130, 253
Gill, D., 253
Gilluly, J., 11, 50, 62, 72, 240, 333
Glock, W. S., 62
Glymph, L. M., 11
Godfrey, A. E., 225
Godillot, T., 43
Goldberg, E. D., 362
Goldich, S., 154
Goldsmith, J. W., 72
Gomis, C., 362
Gordeev, V. V., 362
Gordon, R. B., 302
Gorham, E., 362
Gottschalk, L. C., 174, 189, 265
Gould, H. R., 62
Government of Kenya, 284

Graf, W. H., 84
Graf, W. L., 253
Granat, L., 362
Graustein, W. C., 11
Gravlee, G. C., 12
Greenland, D. J., 295
Greenway, P. J., 285
Gregory, K. J., 284, 302
Griffiths, G. A., 302
Groot, J. J., 72
Grove, A. T., 362
Gunn, R., 62
Gutenberg, B., 62
Guy, H. P., 265
Gwynne, M. D., 285

Hack, J. T., 11, 225
Hadding, A., 240
Hadley, R. F., 2, 63, 241, 265
Haffty, J., 204
Haggett, P., 284
Haites, T. B., 241
Halberg, R. O., 362
Hamblin, W. K., 240
Hammer, T. R., 253
Hammond, D. E., 363
Handa, B. K., 362
Happ, S. C., 253, 284
Harbeck, G. E., 32
Harmeson, R. H., 361
Harris, K. F., 302
Hart, G., 294
Harvey, H. H., 294
Haupt, H. F., 253
Hawkes, H. E., 204, 294
Hays, O. E., 160
Heidel, S. G., 240, 333, 361
Helder, W., 361
Helgeson, H. C., 84
Hem, J. D., 204, 362
Hembree, C. H., 32, 83, 154
Hendrickson, B. H., 105
Henin, S., 43
Hess, H. H., 72
Hewlett, J. D., 225, 294
Hibbert, A. R., 225, 294
High, R. D., 62
Hill, H. O., 160
Hills, E. S., 240
Hjulström, F., 84
Ho, R., 284
Hobba, R. L., 241
Hoepffner, M., 363
Holeman, J. N., 50, 284, 362
Holland, H. D., 362
Hollingworth, S. E., 240

367

Holmes, A., 62, 72, 240
Holtan, H., 362
Hoover, M. D., 294, 295
Hopson, C. A., 225
Hornbeck, J. W., 284, 294
Horton, R. E., 105, 174
Hsu, K. J., 62
Hsü, Y. P., 337
Hubbell, D. W., 83
Hubbert, M. K., 72
Humphreys, A. A., 302
Hung, C. S., 84
Hunt, C., 362
Hurd, D. C., 362
Hurst, H. E., 362
Hutner, S. H., 207

Imevbore, A. M. A., 362
Inglis, C. C., 32
International Association of Hydrological
    Sciences, 284
International Association of Scientific
    Hydrology, 253, 326, 330, 333
Ivanova, A. A., 362
Iwasaki, I., 294

Jackli, H., 42
Jacobs, J. A., 62
Jager, E., 301
Jahn, A., 50
Jahns, R. H., 32
Janda, R. J., 253, 302, 362
Janda, R. S., 284
Jansen, J. M. L., 154, 253, 284, 362
Jansson, S. L., 294
Järnefelt, H., 294
Jarocki, W., 330, 333
Jennings, J. N., 50
Jenny, H., 174
Johnson, D. W., 72
Johnson, J. W., 32
Johnson, N. M., 154, 225, 253, 294, 295, 362,
    363
Jones, B. F., 362
Jones, R. L., 361
Journaux, A., 50
Juang, F. H. T., 225, 294
Judson, S., 50, 253, 285, 326, 333, 362
Junge, C. E., 204, 294, 295, 362

Kaiser, E., 240
Kaitera, P., 72
Keller, F. J., 265
Keller, W. D., 84, 154, 225, 362
Kellerhals, R., 253
Kelly, L. L., 105

Kennedy, V. C., 362
Ketner, K. B., 72
Khosla, A. N., 50, 62
King, L. C., 50, 62
Kirkby, M. J., 11, 285
Kittredge, J., 240
Kobayashi, J., 362
Kochel, R. C., 11
Koene, B. K. S., 363
Kononov, V. I., 363
Konovalov, G. S., 362
Kramer, J., 189
Kraus, E., 72
Kraus, R. K., 84
Krumbein, W. C., 32
Krynine, P. D., 241
Kuenen, P. H., 72, 333
Kulp, J. L., 62
Kunholtz-Lordat, G., 42

Laaksonen, R., 362
Lagache, M., 84, 99
Lamarch, V. C., 50
Lambe, T. W., 144
Lane, E. W., 144
Langbein, W. B., 32, 63, 189, 241, 285, 295,
    333
Laronne, J. B., 84
Larrain, R. D., 363
Lawes, E. F., 285
Laws, J. O., 105
Lawson, A. C., 72
Lees, G. M., 63
LeGrande, H. E., 225
Lemmens, M. N., 363
Leonard, R. E., 294
Leopold, L. B., 11, 32, 33, 50, 189, 362
Lewis, L. A., 285
Li, F. S., 362
Li, Y. H., 285, 337
Lichty, R. W., 241
Lieberman, J. A., 295
Liepolt, R., 362
Likens, G. E., 153, 154, 253, 294, 295, 362, 363
Lin, C. C., 337
Lindroth, A., 295
Linton, D. L., 50
Lisitzin, A. P., 362, 363
Liss, P. A., 361, 363
Lisytsina, K. N., 11
Livingstone, D. A., 208, 302, 333, 363
Longwell, C. R., 63
Lopatin, G. V., 330, 331, 333
Lougee, R. J., 63
Love, S. K., 32
Lovering, T. S., 225

Lowdermilk, W. C., 105
Lvovich, M. I., 363
Lyles, L., 84

Macan, T. T., 295
McConnochie, K., 295
McDougal, D. T., 189
McKee, E. D., 72, 241
Mackenzie, F. T., 337, 362
Mackin, J. H., 32, 189, 265
MacLachlan, J. C., 72
MacLachlan, M. E., 72
McPherson, H. J., 253
Maddock, T., Jr., 32, 189
Maillet, R., 72
Makarenko, F. A., 363
Malina, F. J., 33
Maner, S. B., 11, 285
Marchand, D. E., 154, 302
Marshall, C. E., 84, 295
Marston, A. F., 33
Martin, C. W., 253
Martin, J. M., 362, 363
Mason, B., 295
Matisoff, G., 361
Meade, R. H., 50, 144, 253, 302, 363
Mech, S. J., 160
Meckel, L. D., 241
Megahan, W. F., 253
Melhorn, W. N., 11
Melton, F. A., 33
Menard, H. W., 63, 363
Meshcheryakov, Y. A., 63
Meskheteli, A. V., 361
Meybeck, M., 361, 362, 363
Meyer, L. D., 83, 84
Meyer-Peter, E., 130
Michon, X., 43
Middleton, F. M., 364
Middleton, H. E., 174
Miller, J. P., 11, 33, 204, 285, 362, 363
Miller, R. B., 225
Miller, R. L., 285
Milne, R. A., 123
Minckley, W. L., 295
Monke, E. J., 84
Montgomery, A., 204
Moody-Stuart, M., 241
Mook, W. G., 363
Moore, D. G., 253
Moore, W. L., 144
Morelli, J., 363
Morozov, N. P., 363
Mosley, M. P., 11
Moss, J. H., 11
Mudge, M. R., 72

Murray, G. E., 72
Murray-Rust, D. H., 285
Musgrave, G. W., 160, 189
Musser, J. J., 252
Myrick, R. M., 50

Neal, J. H., 105, 160
Ng, A. C., 361
Nichols, R. L., 33
Nielsen, K. F., 82
Nikolov, S., 84
Niyogi, D., 241
Nordin, C. F., 302
Norris, L. A., 253
Norton, R. A., 160
Nye, R. H., 295

O'Brien, M. P., 33
Odum, E. P., 295, 363
Ollier, C. D., 84
Oltman, R. E., 207, 333
Ongweny, G. S. O., 284, 285
Onstad, C. A., 83
Oriel, S. S., 72
Owens, L. B., 11
Ozawa, T., 294

Paces, T., 84
Painter, R. B., 154, 253, 284, 362
Palmer, V. J., 160
Pannekoek, A. J., 63
Parans de Ceccaty, R., 72
Partheniades, E., 84
Patnaik, N., 84
Pechinov, D., 334
Pedro, G., 363
Pels, S., 241
Penck, W., 63
Petr, T., 363
Pettijohn, F. J., 241
Pickering, R. J., 252
Pickup, G., 12
Pierce, R. S., 154, 253, 294, 295, 363
Pinta, M., 362
Pitty, A. F., 50, 154
Polynov, B. B., 204
Polzer, W., 99, 153
Pope, J. B., 160
Postgate, J. R., 295
Potter, H. R., 50
Potter, P. E., 241
Pratt, D. J., 285
Pugh, J. C., 63
Putnam, W. C., 334

Rainwater, F. H., 32, 154, 204

Rankama, K., 295
Rao, P., 285
Rapp, A., 42, 50, 253, 285
Raudkivi, A. J., 84
Rayner, J. N., 50
Rector, R. L., 33
Reed, J. C., 11
Reesman, A. L., 225
Reeves, C. C., 253
Reichel, E., 361
Reinhart, K. G., 284
Remenieras, G., 43
Renfro, G. W., 12
Reynolds, R. C., 154, 362, 363
Richardson, E. C., 284
Rittenhouse, G., 284
Ritter, D. F., 50, 285, 323, 326
Roberson, C. E., 153
Robertson, T., 241
Roche, M. A., 363
Rodhe, H., 362
Roehl, J. W., 12, 285
Roose, E. J., 285
Rosa, J. M., 241
Rosenan, E., 33
Rouse, H., 124
Rubeikin, V. Z., 364
Rubey, W. W., 84, 125, 128
Runcorn, S. K., 72
Russell, I. C., 63
Russell, R. D., 62
Russell, R. J., 241
Ruttner, F., 295

Sabatier, G., 84, 99
Sahama, T. G., 295
Salomons, W., 361
Salvadori, F., 363
Sanders, J. E., 72
Santos, H. S., 225
Santos, P. S., 225
Sartz, R. S., 295
Sasseville, J. L., 361
Sawhney, B. L., 83
Schatz, A., 84
Schatz, V., 84
Schick, A. P., 12, 226, 254, 265, 285
Schumm, S. A., 12, 32, 50, 63, 241, 265, 285, 333, 334
Schwarzbach, M., 241
Scott, D. R. M., 295
Scott, K. M., 12
Scott, M., 363
Sedimentation and Erosion Sub-Task Force of the Federal Interdepartmental Task Force, 334

Seed, H. B., 144
Selby, M. J., 12
Shapiro, L., 99
Shen, H. W., 84
Shields, A., 124
Simpson, H. J., 363
Sinclair, R. A., 361
Sitter, L. U. de, 63
Slack, K. V., 225
Slatt, R. M., 363
Slaymaker, H. O., 253
Slosser, J. W., 160
Smerdon, E. T., 144
Smith, D. D., 33, 85, 154, 160, 285
Smith, S. M., 363
Smith, T. G., 189
Smith, W., 295
Smyth, C. H., 204
Snedecor, G. W., 174
Snyder, G. G., 253
Soons, J. M., 50
Souchez, R. A., 363
Southwick, D. L., 225
Stabler, H., 50, 62, 72
Stallard, R. F., 361
Stallings, J. H., 84
Starkel, L., 12, 50
Steiner, A., 50
Stephan, D. G., 364
Sternberg, H. O., 333
Sternberg, O'R., 208
Stoddart, D. R., 50
Stokes, W. L., 241
Stone, R., 63
Storey, H. C., 241
Stoyber, R. E., 363
Strahler, A. N., 63, 334
Strakhov, N. M., 363
Subramanian, V., 302
Sundborg, A., 285
Sutherland, A. J., 84
Sutton, G. H., 72
Swann, D. H., 241
Swanson, F. J., 285

Taillefer, F., 43
Tamm, C., 84
Tanaka, M., 12
Tardy, Y., 361, 363
Taylor, C. M., 285
Taylor, P. S., 363
Tebo, L. D., 295
Temple, P., 253, 285
Thatcher, L. L., 204
Thiadens, A. A., 241
Thiessen, A. H., 295

Thom, H. C. S., 33
Thompson, S. M., 302
Thornbury, W. D., 50, 63
Thornes, J. B., 11
Tilley, C. E., 73
Tison, L. J., 240, 333, 361
Tixeront, J., 43, 334
Todd, O. J., 334
Tollan, A., 363
Trelawny, G. S., 84
Tricart, J., 43, 241
Trimble, G. R., 295
Trimble, S. W., 254, 285
Trobitz, H. K., 240
Tsuboi, C., 63
Turekian, K. K., 363
Twenhofel, W. H., 241

Udden, J. A., 123
Udwardy, M. D. F., 363
U.S.S.R. Academy of Sciences, 364
U.S.S.R. Committee for the International
    Hydrological Decade, 364
United Nations, 329, 331, 334
UNESCO, 364
U.S. Army, 33
U.S. Department of Agriculture, 265, 325,
    334
U.S. Forest Service, 189, 295
U.S. Navy Hydrographic Office, 33
U.S. Public Health Service, 295
U.S. Soil Conservation Service, 285
U.S. Waterways Experiment Station,
    134
U.S. Weather Bureau Monthly Climatolog-
    ical Summaries, 225
Ursic, S. J., 226
Utsumi, S., 294

Vallentine, J. R., 364
Van Andel, TJ. H., 334
Vanoni, V. A., 2, 85
Vening-Meinesz, F. A., 73
Viro, P. J., 364
Voigt, G. K., 83
Voronov, P. S., 50

Wahlstrom, E. E., 63
Waksman, S. A., 295
Walker, T. R., 241
Walling, D. E., 2, 284, 302
Wang, L. F., 362

Warner, R. F., 12
Waters, A. C., 62
Watson, J. P., 11
Weaver, J. E., 189
Weaver, P., 73
Weber, D. F., 295
Wegman, E., 50, 63
Weiler, R. R., 364
Weinberger, L. W., 364
Weir, G. W., 241
Werby, R. T., 204, 295, 362
Whitehead, H. C., 295
Whitehouse, F. W., 242
Whitman, I. L., 265
Whitt, D. M., 160
Williams, M. A. J., 50
Williams, P. W., 50
Williams, S. C., 363
Wilm, H. G., 189
Wilson, J. T., 62, 73
Wilson, L., 154, 254, 285
Windom, H. L., 364
Wischmeier, W. H., 33, 85, 154, 285
Wollast, R., 99
Wollny, E., 105
Wolman, M. G., 11, 12, 32, 33, 85, 226, 254,
    265, 285, 362
Woodburn, R., 105
Woodford, A. O., 62
Woodruff, C. M., 160
Woods, W. J., 295
Woolley, R. R., 33
Wu, F. T., 337
Wu, Y. D., 362
Wundt, W., 50
Wyart, J., 84, 99

Yalin, M. S., 85
Yang, C. T., 85
Yarnell, D. L., 160, 189
Yoder, H. S., Jr., 73
Yoder, R. E., 160
Young, A., 50

Zektzer, I. S., 361
Zellweger, G. W., 362
Zeuner, F. E., 63
Zingg, A. W., 33
Zmudzinski, L., 361
Zonneveld, J. I. S., 242
Zverev, V. P., 363, 364

SUBJECT INDEX

Abalation, 34. *See also* Denudation
Aeolian erosion control, 120–122
Aeolian processes, 5, 31, 79–80. *See also*
    Dust storms
Aeolian sediment transport, 13, 22–23, 26–
    27, 47, 79–80, 108–123
  surface-supported
    rolling, 112
    saltation. *See* Saltation
    surface creep, 109, 112, 116, 118
Aggradation, 189
Angle of descent, 115
Antecedent moisture, 82, 100, 137. *See also*
    Cohesion
Arroyo, 30
Atmospheric solute input, 45, 149, 202, 217,
    221–222, 259, 299, 354–355
Atterberg Limits, 82, 137

Badlands, 58
Bankfull stage, 13, 25
Bernoulli effect, 112–113

Capacity, 7, 282
Catastrophic events, 5, 13–16, 26, 29–31,
    37–38 152, 225
Catastrophism, 4–6
Channel fill, 236–237
Channel shape, 24, 30, 231–239, 261–264
Chelation, 77
Clastic load. *See* Sediment, transport, bed-
    load; Sediment, transport, suspend-
    ed load
Climate, 19, 21–22, 49, 52, 148, 150, 175–189
  arid, 5, 79, 183, 230
  cold, 45–48, 52, 150–151, 205–207
  semiarid, 21–22, 52, 54, 230, 239
  temperate, 22, 35, 152
  tropical, 35, 150–151, 205–207, 229, 267–
    285
Climatic erosion factor, 177–178
Climatic shift, 2, 8–9, 10, 50, 188–189

Coastal processes, 5, 27–31, 69–70
Cohesion, 24, 31, 82, 137–145
Colloids, 36
Colluviation, 39–40
Colluvium, 35, 40–41
Competence, 13–15, 23, 25–27, 31, 36, 40,
    76, 80–81, 113, 116, 124–125, 137
Critical tractive force. *See* Competence

Davisian cycle, 9, 57–59
Degradation, 189, 255
Denudation, 4–5, 9–10, 13, 22, 34–41, 45–73,
    200–203, 224, 299, 308–323, 333, 335–
    337
  chronology, 49–50, 251, 281, 298–299
Depression storage, 155
Downcutting. *See* Entrenchment
Drainage area, 7, 13, 17–18, 51–53, 57, 169–
    170, 180, 183–184, 201, 276–277, 282
Dunes, 13, 23, 26–27, 31, 109
Dust fall, 9, 109–111
Dust storms, 13, 23, 111, 116–119

Effective diameter, 134–135
Effective discharge, 24
Effective force, 13–14, 20, 29
Effective precipitation, 52–53, 181
Entrenchment, 5, 30, 38–41, 58, 238
Equilibrium, 8–9, 13, 26–29, 255–256
  profile, 27
Erodibility, 22, 79, 147, 170. *See also*
    Cohesion
Erosibility. *See* Erodibility
Erosion
  accelerated, 2. *See also* Land use
  aeolian. *See* Aeolian sediment transport
  anthropogenic, 2. *See also* Land use
  bank, 24, 30, 40–41, 148, 166, 172, 262–264
  channel bed, 30, 41, 58
  coastal. *See* Coastal processes
  cycle. *See* Davisian cycle
  lateral, 5

in mountains, 9, 35, 37, 45–50, 53, 59, 205–207. *See also* Sediment, yield, maximum
potential, 162, 167, 172
raindrop, 78–79, 100–107
slopewash, 47, 49, 79
soil, 78. *See also* Aeolian sediment transport; Erosion, raindrop; Erosion, slopewash; Rills
spatial variability, 6, 34–35, 38–42
temporal variability, 35–38. *See also* Frequency
Erosional activity, 100
Erosional damage, 42, 100
Erosivity, 79, 147, 281

Fallout, 259
Fans, 30–31, 35, 40–41, 46
Flocculation, 82
Floodplain, 24–26, 30, 40, 152
Floods, 14–16, 26, 30, 36, 192. *See also* Catastrophic events
peak, 14–16
surge, 6
Frequency, 5–6, 8, 13–33, 36, 152, 281–283

Glaciation, 6, 26, 45, 49, 50, 60, 193
Glacier, 45
Gradient. *See* Slope
Grain path, 102, 114–117
Groundwater chemistry, 5, 20, 35–40, 217–221, 350
Gully, 30, 39, 50, 53, 283

Hillslope form, 58, 157
Hurricane, 26
Hydrolysis, 77. *See also* Solution

Incision. *See* Entrenchment
Incongruent dissolution. *See* Solution
Infiltration, 35, 39, 100, 104–107, 155
Interflow, 219
Isostasy, 59–62. *See also* Tectonism

Karst, 5

Laminar boundary layer, 129, 133
Land use, 8, 50, 149, 165–168, 172, 182–183, 186, 213, 230–231, 248–252, 256–295, 346–349
Langbein-Schumm rule, 148–149, 182, 229, 250

Mass movements, 5–7, 45–48, 229
avalanches, 40, 46–48
creep, 7, 39, 46–48, 278
landslides, 30, 38, 40, 45–48
mudflows, 30–31, 46
rockfalls, 46–48
solifluction, 37–39, 46–48
Multiple terraces, 60

Nutrients. *See* Solute load

Oceans, chemical composition of, 10
Orogeny. *See* Tectonism
Overland flow, 7, 35, 100

Paleohydrology, 230–232. *See also* Climatic shift
Peneplain, 9, 59, 202
Pool, 25–26, 41

Rainfall
intensity, 19, 100–107, 185
seasonality factor, 148, 177–180, 184
simulator, 100, 155, 159
Raindrop erosion. *See* Erosion
Reaction rate. *See* Solution, kinetics
Recovery rate, 6, 60
Recurrence interval, 5, 6, 13, 15, 16, 18–19, 37, 159. *See also* Frequency
Relative mass transfer, 45. *See also* Sediment, yield
Relative mobility, 149, 190–192, 198, 200–201
Relief. *See* Slope
Relief ratio, 274–277
Reservoir sedimentation, 34–35, 49, 52–54, 183–184, 298, 300, 332
Resistance, 8, 59, 81. *See also* Erodibility
Riffle, 25–26, 38, 41
Rills, 6, 39, 79
Roughness, 120, 129–130, 135, 137

Salinity. *See* Solute load
Saltation, 36, 79–80, 109–116
Scour, 30–31, 135, 139. *See also* Erosion
Scour criteria. *See* Competence
Sediment
delivery ratio, 6–7, 276
recycling, 10, 68–69, 238, 249, 313
size, 7, 25–27, 30, 40
source, 7, 66, 69, 173, 224–225, 238, 278–281
storage, 6–7, 36, 41, 153, 249, 276, 282, 333. *See also* Erosion, spatial variability
supply-limited, 7
transport
bedload, 18–19, 34, 37–38, 60, 80–81, 124–136, 207, 272, 287, 299

dissolved load. *See* Solute load
suspended load, 6, 15–20, 22, 24, 36–38, 40–41, 49–52, 116–119, 149, 161–174, 205–207, 213, 271–292, 299–307, 312–337, 358–359. *See also* Dust storms
transport-limited, 7
waves, 30
yield, 51
 errors, 299, 312
 geologic, 6, 10, 38–39, 41, 49–50, 51–73, 153, 224, 231–232, 251, 277–278
 maximum, 53–54, 148, 183, 283, 301, 335–337
 units, 7–8, 45, 53, 301, 324
Sedimentation, 10, 63–66
Settling velocity, 128
Shear stress, 14
Sheet flow. *See* Overland flow
Slope, 7, 14, 19, 25–40 passim, 49–50, 53–58, 79, 148, 155–160, 164–165, 169, 274–277
Soil
 aggregates, 100–107
 erosion. *See* Erosion
 moisture, 31, 160
Solute load, 7, 9, 20–22, 34–38, 45–48, 52, 67–70, 77–78, 149–151, 190–207, 289–294, 301, 311–323, 335–367
 storage, 35–36, 150, 224
Solution, 5, 7, 149–152
 kinetics, 77–78, 86–99

Step length, 81, 126–127
Stream order, 169
Surface aggregation ratio, 171

Talus, 45. *See also* Colluvium
Tectonism, 9, 51–64, 71, 237
Threshold, 6, 13–14
 of sediment transport. *See* Competence
Time independence, 8–9, 58
Tsunami, 13

Underfit streams, 9
Uniformitarianism, 4–6, 51, 228
Universal Soil Loss Equation, 79, 146–147

Vegetal density, 186–188, 230, 275
Vegetation, 22–30, 151–153, 185–189, 209–210, 213, 218–219, 224, 228–232, 237–239, 251–252, 287, 293–294, 301. *See also* Land use

Watershed area. *See* Drainage area
Wave steepness, 27–29
Weathering, 9, 45, 76–78, 86, 98–99, 149–152, 190, 198, 209–217, 222–223, 355–357
 frost, 47–48, 77, 193
 insolation, 47
 solution. *See* Solution; Solute load

About the Editors

JONATHAN B. LARONNE received the B.Sc. in geography, geology, and climatology in 1970 from the Hebrew University of Jerusalem, the M.Sc. in geography in 1973 from McGill University, and the Ph.D. in earth resources in 1976 from Colorado State University. After a one-year period of postdoctoral research into water quality of surface runoff, he returned to Israel to lecture at the Ben-Gurion University of the Negev. His present research interests are salt loading from diffuse sources, sediment transport in gravel-bedded channels, and depression storage and infiltration losses from reservoirs in the Negev Desert. Dr. Laronne has published a number of papers in these and other fields.

M. PAUL MOSLEY received the B.A. in geography from Cambridge University in 1970 and the M.S. and Ph.D. in earth resources from Colorado State University in 1972 and 1975. Since then he has worked for a number of land management agencies in New Zealand, conducting research into a wide range of problems relating to erosion, sedimentation, water quality, and water yield. He is currently leader of the Environmental Hydrology Group, Water and Soil Science Centre of the Ministry of Works and Development, with responsibility for studying the impact of water resource developments upon river channel behavior and morphology and upon related in-stream resources. Dr. Mosley has published almost thirty papers on geomorphic, erosional, and hydrologic processes.